Study Guide and Solutions Manual

Peter M. Mirabito

University of Kentucky

For

Genetic Analysis: An Integrated Approach

First Edition

Mark F. Sanders

University of California at Davis

John L. Bowman

Monash University, Melbourne, Australia

University of California at Davis

PEARSON

Boston   Columbus   Indianapolis   New York   San Francisco   Upper Saddle River
Amsterdam   Cape Town   Dubai   London   Madrid   Milan   Munich   Paris   Montréal   Toronto
Delhi   Mexico City   São Paulo   Sydney   Hong Kong   Seoul   Singapore   Taipei   Tokyo

Editor-in-Chief: Beth Wilbur
Executive Director of Development: Deborah Gale
Senior Acquisitions Editor: Michael Gillespie
Project Editor: Anna Amato
Managing Editor, Production: Mike Early
Production Project Manager: Jane Brundage
Production Management and Composition: Progressive Publishing Alternatives
Design Manager: Mark Ong
Image Lead: Donna Kalal
Cover Production: Seventeenth Street Studios
Manufacturing Buyer: Michael Penne
Executive Marketing Manager: Lauren Harp
Text and Cover Printer: Edwards Brothers Malloy

Cover Photo Credit: EM Unit, UCLA Medical School, Royal Free Campus/Wellcome Images

5 6 7 8 9 10—EBM—16 15 14 13

ISBN 10: 0-13-174167-5; ISBN 13: 978-0-13-174167-6

Contents

1

The Molecular Basis of Heredity, Variation, and Evolution

1.01 Genetics Problem-Solving Toolkit

Key Terms and Concepts

- **Genome:** All the genetic information in a cell or organism.

- **Chromosome:** One double-stranded DNA molecule and its associated proteins.

- **Genetic locus:** A location on a chromosome.

- **Gene:** A segment of DNA that contains the information for the synthesis of a specific RNA, which in many cases is an mRNA that specifies the amino acid sequence of a protein.

- **Allele:** One specific sequence of a gene or genetic locus.

- **DNA (deoxyribonucleic acid):** A polymer of deoxynucleotides attached by phosphodiester bonds. DNA is typically double stranded, composed of two antiparallel deoxynucleotide chains that are held together by base-pairing and base-stacking interactions.

- **RNA (ribonucleic acid):** A polymer of ribonucleotides, attached to each other by phosphodiester bonds. RNA is typically single stranded.

- **Deoxynucleotides:** A deoxyribose sugar attached to a nitrogenous base (thymine, cytosine, adenine, or guanine) and one, two, or three phosphates.

- **Ribonucleotides:** A ribose sugar attached to a nitrogenous base (uracil, cytosine, adenine, or guanine) and one, two, or three phosphates.

- **Replication:** Synthesis of a copy of a DNA molecule by DNA polymerase, which uses the strands of the original DNA molecule as templates to determine the order of deoxynucleotides in the new DNA strands.

- **Transcription:** Synthesis of an RNA copy of a DNA sequence by RNA polymerase, which uses one strand of DNA as a template to determine the order of ribonucleotides in the RNA.

- **Translation:** Synthesis of a polypeptide by a ribosome, tRNAs, and many accessory factors, which use the sequence of codons in an mRNA to specify the order of amino acids in the polypeptide.

- **Mutation:** A change in DNA sequence.

- **Natural selection:** Differential reproduction among individuals in a population due to interactions of their phenotype with the environment.

- **Random genetic drift:** Differential reproduction among individuals in a population due to random effects.

- **Migration:** Movement of members of a species from one population into another.

- **Phylogenetic tree:** A display of the evolutionary relationships among species determined by morphological or molecular information.

- **Parsimony:** States that the simplest explanation for differences among species is the most likely.

Key Analytical Tools

Phylogenetic Tree Using Parsimony

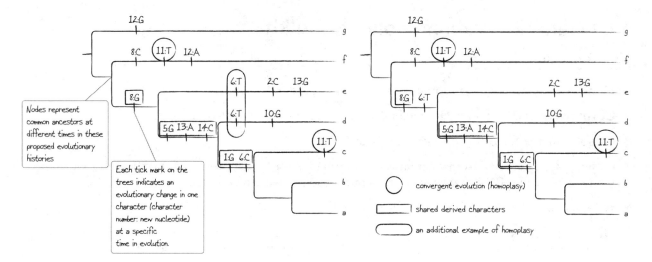

Key Genetic Relationships

Genetic Code

Second Position

First Position (5′ end)	U	C	A	G	Third Position (3′ end)
U	UUU ⎤ UUC ⎦ Phe (F) UUA ⎤ UUG ⎦ Leu (L)	UCU ⎤ UCC UCA UCG ⎦ Ser (S)	UAU ⎤ UAC ⎦ Tyr (Y) UAA – stop UAG – stop	UGU ⎤ UGC ⎦ Cys (C) UGA – stop UGG – Trp (W)	U C A G
C	CUU ⎤ CUC CUA CUG ⎦ Leu (L)	CCU ⎤ CCC CCA CCG ⎦ Pro (P)	CAU ⎤ CAC ⎦ His (H) CAA ⎤ CAG ⎦ Gln (Q)	CGU ⎤ CGC CGA CGG ⎦ Arg (R)	U C A G
A	AUU ⎤ AUC AUA ⎦ Ile (I) AUG – Met (M)	ACU ⎤ ACC ACA ACG ⎦ Thr (T)	AAU ⎤ AAC ⎦ Asn (N) AAA ⎤ AAG ⎦ Lys (K)	AGU ⎤ AGC ⎦ Ser (S) AGA ⎤ AGG ⎦ Arg (R)	U C A G
G	GUU ⎤ GUC GUA GUG ⎦ Val (V)	GCU ⎤ GCC GCA GCG ⎦ Ala (A)	GAU ⎤ GAC ⎦ Asp (D) GAA ⎤ GAG ⎦ Glu (E)	GGU ⎤ GGC GGA GGG ⎦ Gly (G)	U C A G

1.02 Types of Genetics Problems

1. Describe the levels of organization of genetic information.
2. Analyze the structure of nucleic acids.
3. Describe the processes of gene expression.
4. Describe the forces driving evolutionary change.
5. Construct and analyze phylogenetic trees.

1. Describe the levels of organization of genetic information.

Problems of this type ask you to organize your thinking about genetic material into a hierarchy that reflects the biological organization of genes. They require you to hone your definitions of the terms *genome, chromosome, locus, gene, DNA,* and *allele.* You may be asked to relate the terms to each other. You may also be asked to develop an analogy explaining the relationships of these terms using materials or concepts from everyday life.

Example Problem: Use the terms *genomes, chromosomes, genes, genetic loci,* and *alleles* to create a concept map. Place *genes* in the center and the other four terms at four corners around *genes,* with chromosomes and alleles on opposite corners of the box (not near each other). Draw arrows going to and from each pair of terms except for between genomes and genetic loci and between chromosomes and alleles. Use the following phrases to describe the relationships between these terms: *contain, have different, are contained in,* and *are forms of.*

Solution Strategies	Solution Steps
Evaluate	
1. Identify the topic this problem addresses, and explain the nature of the requested answer.	1. This problem tests your understanding of the relationships among different levels of genetic organization. It asks you to develop a concept map using the terms *genomes, chromosomes, genes, genetic loci,* and *alleles.*
2. Identify the critical information given in the problem.	2. The map should contain the word *genes* in the center and the other terms at the corners of a box surrounding *genes.* The phrases *contain, have different, are contained in,* and *are forms of* should be used to describe the relationships between neighboring pairs of terms.
Deduce	
3. Determine which phrase accurately describes the relationship between each pair of terms.	3. Chromosomes *contain* genes and genetic loci. Genomes *contain* chromosomes, genes, and alleles. Genes *are contained in* genomes, chromosomes, and genetic loci. Genes *have different* alleles. Genetic loci *are contained in* chromosomes and *contain* genes and alleles. Alleles *are forms of* genes and genetic loci and *are contained in* genomes.
Solve	
4. Draw the concept map and label the arrows with the relationships identified in step 3.	4.

For more practice, see Problems 5 and 6.

2. Analyze the structure of nucleic acids.

Problems of this type ask you to recall the structure of DNA and RNA and analyze samples of nucleic acids to make predictions about their structure, replication, or chemical composition. You may be given nucleic acid sequences and asked to analyze them, or you may be given data on the composition of unknown nucleic acids and asked to infer their structure from the data.

Example Problem: Use the results shown in the table to determine whether the nucleic acid is double-stranded DNA, single-stranded DNA, double-stranded RNA, or single-stranded RNA. For samples with inconclusive data, indicate which, if any, of the possibilities are ruled out.

Sample Number	Data
1	25% of the bases are thymine.
2	65% of the bases are uracil.
3	35% of the bases are adenine.
4	60% of the bases are thymine.
5	35% of the bases are thymine, 15% are guanine, and 35% are adenine.
6	100% of the sugars are ribose.

Solution Strategies	Solution Steps
Evaluate	
1. Identify the topic this problem addresses, and explain the nature of the requested answer.	1. This problem tests your understanding of the structure of nucleic acids. The possible answers are double-stranded DNA, single-stranded DNA, double-stranded RNA, and single-stranded RNA.
2. Identify the critical information given in the problem.	2. You are given information on the base composition of samples 1 through 5 (1 = 25% T; 2 = 65% U; 3 = 35% A; 4 = 60% T, 5 = 35% T, 15% G, 35% A) and the sugar composition of sample 6, which is 100% ribose. For sample 5, the results imply that C = 15%.
Deduce	
3. Examine the data for each sample and determine which of the possible answers are inconsistent with the result. **TIP:** DNA contains deoxyribose and thymine but not uracil, whereas RNA contains ribose and uracil but not thymine. In double-stranded nucleic acids, the % G = % C and % T (or U) = % A.	3. The data rules out some possible answers for samples 1, 2, 4, 5, and 6. Sample 1 is not RNA, because it contains thymine. Sample 2 is not DNA, because it contains uracil; it is not double stranded, because the % U cannot equal the % A. Sample 4 is not RNA, because it contains thymine; it is not double stranded, because the % T cannot equal the % A. Sample 5 is not RNA, because it contains thymine. Sample 6 is not DNA, because it contains ribose.
4. Examine the data for each sample to determine which of the remaining possible answers is most consistent with the results.	4. The data for sample 5 is most consistent with double-stranded DNA because the % T = % A and the % G = % C.

(continued)

Solution Strategies	Solution Steps
Solve	
5. Formulate the answer by indicating which type of nucleic acid is most consistent with the results. List all answers that are equally likely.	5. **Answer:** Sample 1 could be single-stranded or double-stranded DNA. Sample 2 is single-stranded RNA. Sample 3 could be single-stranded or double-stranded DNA or RNA. Sample 4 is single-stranded DNA. Sample 5 is double-stranded DNA. Sample 6 is single-stranded or double-stranded RNA.

For more practice, see Problems 11, 12, 13, 14, 17, 19, and 21.

3. Describe the processes of gene expression.

Problems of this type ask you to recall the steps in gene expression and use that information to analyze DNA, RNA, or protein sequences. You may be given a DNA sequence and asked to predict the results of transcription into RNA and then translation of the RNA into protein. You may also be given amino acid sequences and asked to predict the corresponding RNA and DNA sequences.

Example Problem: Arrange the following list of events in order of their occurrence during gene expression: (a) tRNAs carry amino acids to the ribosome. (b) RNA polymerase transcribes DNA into mRNA. (c) The polypeptide is released from the ribosome. (d) RNA polymerase finds the promoter. (e) The ribosome locates the AUG codon. (f) RNA polymerase finds the terminator sequence. (g) Introns are removed from mRNA. (h) The ribosome locates a UAG codon. (i) The ribosome catalyzes peptide bond formation.

Solution Strategies	Solution Steps
Evaluate	
1. Identify the topic this problem addresses, and explain the nature of the requested answer.	1. This problem tests your understanding of the major steps of gene expression. The answer will be a list of gene expression processes ordered by when they occur during gene expression.
2. Identify the critical information given in the problem.	2. You are given nine steps of gene expression: (a) - (i)
Deduce	
3. Link each statement to a process in gene expression. **TIP:** Remember the central dogma of genetics. **TIP:** Remember that, in eukaryotes, introns are removed from mRNA after transcription but before translation.	3. Events b, d, and f refer to steps in transcription. Events a, c, e, h, and i refer to translation. Event g occurs after transcription but before translation.
4. Order the events concerning transcription relative to each other. Do the same for the events concerning translation. **TIP:** Transcription begins near the promoter and ends at the terminator. Translation begins at AUG and ends at a stop codon (UAA, UAG, or UGA).	4. The order of the events concerning transcription is d, b, and f. The order of the events concerning translation is e, a, i, h, and c. **TIP:** Steps a and i occur throughout translation; therefore, their order in this sequence could be reversed, reiterated, or both.

(continued)

Solution Strategies	Solution Steps
Solve	
5. Formulate the answer by ordering all the steps relative to each other. TIP: Remember to place step g after the last step in transcription and before the first step in translation.	**5.** **Answer:** The order of the events concerning steps in gene expression is d, b, f, g, e, a, i, h, and c.

For more practice, see Problems 14, 15, 16, 17, 19, and 20.

4. Describe the forces driving evolutionary change.

Problems of this type ask you to describe the evolutionary forces that cause changes in genotype frequencies in a population and ultimately lead to the formation of new species. They also may ask you to explain how these forces act to drive those changes. The forces described in Chapter 1 are natural selection, mutation, random genetic drift, and migration.

Example Problem: For each evolutionary force described in Chapter 1, indicate which ones can (a) introduce a new genotype into a population and which ones cannot. For those forces that cannot introduce new genotypes, indicate which ones can (b) drive evolutionary change based on the interaction of phenotype and environment and which ones cannot. If an evolutionary force cannot introduce a new genotype into a population and cannot drive evolutionary change based on the interaction of phenotype and environment, then (c) indicate what can it do and describe how.

Solution Strategies	Solution Steps
Evaluate	
1. Identify the topic this problem addresses, and explain the nature of the requested answer.	**1.** This problem tests your understanding of the forces that drive changes in genotype frequencies within populations. The answer will be a list of forces categorized first by those that can introduce new alleles and those that cannot, and then by which of the latter can drive change based on phenotype–environment interactions and which cannot. Finally, there should be a description of the last category of forces in detail.
2. Identify the critical information given in the problem.	**2.** You are asked to address the evolutionary forces described in Chapter 1, which are natural selection, mutation, random genetic drift, and migration.
Deduce	
3. Describe each evolutionary force in terms of how it drives changes in allele frequency.	**3.** Mutation refers to a heritable change in the genotype of an organism. Natural selection refers to the change in allele frequency due to differential reproduction among individuals in a population due to the interaction of their heritable phenotype with the environment. Random genetic drift is the change in allele frequency due to differential reproduction among individuals in a population due purely to chance. Migration causes changes in allele frequency due to the introduction of new individuals from a different population.

(continued)

Solve

4. Identify which forces meet criteria a and b of the question, and describe in detail the force(s) that meet criterion c.

4. (a) *Forces that can introduce new alleles* include mutation and migration.

    Forces that cannot introduce new alleles include natural selection and genetic drift.

    (b) Natural selection can alter allele frequency due to the interaction of phenotype and environment, whereas random genetic drift cannot.

    (c) Random genetic drift alters allele frequency due to differential reproduction of genotypes due purely to chance. The magnitude of the effect of random genetic drift increases with decreasing population size.

For more practice, see Problems 7, 8, and 9.

5. Construct and analyze phylogenetic trees.

Problems of this type ask you to apply your understanding of phylogenetic trees to the analysis of existing trees and the construction of new trees given specific data. You may be given a tree and the character data and asked to identify a shared derived characteristic that defines a clade or distinguishes one clade from another. You may be given a tree and asked to identify the most closely or most distantly related taxa. You may also be given data on the shared derived characteristics of multiple taxa and asked to apply the principle of parsimony to identify the most likely phylogenetic relationship among the taxa.

Example Problem: The table shows homologous DNA sequences of four taxa and the proposed sequence of their common ancestor. The accompanying figure shows three phylogenetic trees, each representing a different hypothesis for the evolutionary relationships among taxa 1, 2, and 3, with taxon 4 as the outgroup. Determine which of the trees is most likely by applying the principle of parsimony.

	Nucleotide Position						
	1	**2**	**3**	**4**	**5**	**6**	**7**
taxon 1	A	T	C	G	G	G	G
taxon 2	A	T	T	T	G	T	G
taxon 3	A	G	C	G	G	G	G
taxon 4	G	T	T	T	A	T	T
ancestor	C	T	T	T	T	T	T

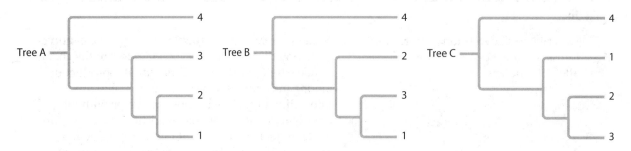

Solution Strategies	Solution Steps
Evaluate	
1. Identify the topic this problem addresses, and explain the nature of the requested answer.	1. This problem tests your understanding of parsimony. The answer will be the tree that requires the fewest proposed sequence changes.
2. Identify the critical information given in the problem.	2. The problem supplies the sequence of three ingroups (1, 2, and 3), one outgroup (4), and the sequence of their common ancestor.
Deduce	
3. Use the sequence data to label each tree with tick marks indicating the location of every evolutionary change, the character number of each change, and the new sequence character. Refer to the Phylogenetic Tree Key Analytical Tool for help with notation of evolutionary changes.	3.

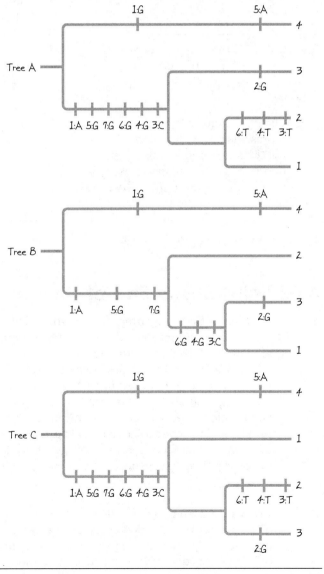

TIP: There may be more than one way to mark a tree that is consistent with the sequence data. For example, in trees A and C, an alternate to that shown would be to place the 3:C, 4:G, and 6:G tick marks on the lineages specific to taxa 1 and 3 and to remove the 3:T, 4:T, and 6:T tick marks in the lineage specific to taxa 2.

Solution Strategies	Solution Steps
	The changes at position 1 (to G) and 2 (to G) are specific to taxon 4, and the changes at position 5 (to A) are specific to taxon 2. These are placed at the same relative location in each tree. The changes at position 1 (to A), 5 (to G), and 7 (to G) are present in taxa 1, 2, and 3. These could have been present in their common ancestor and so are placed on the diagonal before the node representing their common ancestor. The changes at position 3 (to C), 4 (to G), and 6 (to G) are present in taxa 1 and 3, but not 2. For tree B, these could have been present in their common ancestor, and they can be placed on the diagonal before the node representing their common ancestor. For trees A and C, these changes could have been present in the common ancestor of taxa 1, 2, and 3 and therefore can be placed on the diagonal next to the changes at positions 1, 5, and 7. Since taxon 2 contains the ancestral T at these characters, changes from G back to T at positions 3, 4, and 6 need to be placed on the taxon-2-specific lineage in trees A and C.
4. Calculate the total number of changes for each tree.	4. Trees A and C each require 12 changes. Tree B requires only 9 changes.
Solve	
5. The phylogenetic tree that best represents the evolutionary relationships among the sequences under the principle of parsimony is the one that requires the fewest total number of evolutionary changes.	5. **Answer:** The most parsimonious tree is tree B.

For more practice, see Problem 18.

1.03 Solutions to End-of-Chapter Problems

1. *Evaluate > Deduce > Solve*: This is type 1. This problem tests your understanding of the ways that genetics affects your life. Start by considering which aspects of your health and well-being are inherited. Next, consider the food you eat and medications that you take, and investigate whether genetic technologies are involved in their production. Consider whether your pets or plants have been produced in part by genetic manipulation. Consider whether political issues requiring your attention are influenced by genetic technologies; for example, animal cloning, embryonic stem-cell research, or the ethical use of your genetic information by healthcare providers, insurers, and government agencies. Think broadly about the world around you now and throughout your life to consider how genetics affects you and your society, and strive to understand the basic principles of genetics so that you will be well armed to tackle these issues as an informed citizen.

2. *Evaluate > Deduce > Solve*: This is type 1. This problem tests your understanding of the impact that the discovery of DNA as the genetic material had on biological research. Remember that protein, not DNA, was the focus of efforts to understand heredity before this discovery and that only after this discovery was serious attention turned to understanding DNA structure and

function. Whereas the complexity of protein structure confounded thinking about mechanisms of inheritance, DNA structure provided profound insight into the mechanism of heredity, suggesting a simple, elegant mechanism for duplication (inheritance), change (mutation and evolution), and phenotype specification (coding). The finding that DNA is universal facilitated rapid progress because study results from all organisms were now directly related. This finding also fostered the development of recombinant DNA technologies in bacteria and bacteriophages, which led to the explosion of biological information that excites and confounds us today.

3.   *Evaluate > Deduce > Solve:* This is type 1. This problem tests your understanding of the overarching importance of genetics in the biological sciences. Genetics provides the principles and mechanisms for evolution, and evolution is the foundation of our understanding of biology. Dobzhansky said, "Nothing in biology makes sense except in the light of evolution" in an *American Biology Teacher* essay published in 1973, and Campbell stated that evolution is the "overarching theme of biology" in his 2008 textbook. Given that evolution can be understood only through principles of genetics, it is not unreasonable to place genetics alongside evolution as the two fundamental fields in the biological sciences.

4.   *Evaluate > Deduce > Solve:* This is type 1. This problem tests your understanding of the evolutionary argument for DNA as the genetic material of all living things (viruses are not considered "alive"). Evolution states that all life descended from a common ancestor, which passed its genes and the mechanisms by which those genes were used to its descendants. These mechanisms would include the structure of nucleotides, the structure of DNA, the enzymes that replicate and read DNA, the enzymes that translate mRNA into amino acid sequences, and many more. Any change to one of these components would be harmful, slowing or preventing reproduction, and would be removed by natural selection (mountains of experimental evidence demonstrate that mutations in genes encoding basic genetic machinery are lethal). Once established as the genetic material, DNA would be maintained as the genetic material by natural selection and therefore would be expected to be found as the genetic material in all existing organisms.

5.   *Evaluate > Deduce > Solve:* This is type 1. A *chromosome* is a single DNA molecule (plus associated proteins) that contains some or all of the genetic information of an organism. A chromosome includes genes and other DNA sequences that are not genes. The term *locus* refers to a "genetic locus," which is a specific location on a chromosome. A genetic locus is an arbitrary designation that does not imply size. A locus may contain no genes, part of a gene, an entire gene, or many genes. A *gene* is a genetic locus that contains all the information required for transcription a specific RNA. In many cases, the RNA is translated to produce a specific protein, therefore, the gene is said to "code" for a protein. An *allele* is a specific form of a gene or genetic locus. For example, the normal allele of a gene ("wild type") will code for the normal version of a protein, whereas an abnormal allele of that gene ("mutant") will code for an abnormal protein. Two chromosomes can contain the same set of genes but different alleles of one or more of these genes.

Street map analogy: The three levels of structure in the street map analogy, from largest to smallest, are street, type of building, and apartment floor plan. Assigning genetic terms from largest to smallest relates street to chromosome, type of building to gene, and apartment floor plan to allele. A street map is like a chromosome map: each address on the street map corresponds to a locus on the chromosome. There are different types of buildings on the street, each at a different location, just as there are different genes on the chromosome, each residing at a different locus. At any particular building, there is a specific floor plan for each apartment (or home, office, etc.), somewhat analogous to different alleles of a particular gene. The last part of this analogy is the weakest, since a building on a specific street has only one floor plan, whereas two copies of a gene in a cell can differ (different alleles). Nevertheless, there may be many copies of the building (identical structures built one after another along a street or on different streets) that differ only in the floor plan of an apartment, which is analogous to the different alleles of the same gene.

6. *Evaluate > Deduce > Solve:* This is type 1. *Genotype* refers to the genetic makeup of a cell or organism, whereas *phenotype* refers to observable characteristics of the cell or organism, such as appearance, physiology, and behavior. The genotype of an organism is part of what determines the phenotype of the organism; however, environment also plays a role in the organism's phenotype. The genotype is heritable and, therefore, its contribution to phenotype will be inherited. The aspects of phenotype that are due to environment are not heritable.

7. *Evaluate > Deduce > Solve:* This is type 4. *Natural selection* is differential reproduction among individuals in a population due to differences in heritable phenotypes, which are said to be adaptive. Individuals with the adaptive phenotype have a reproductive advantage (increased reproductive fitness); therefore, they contribute a disproportionately high number of progeny to the next generation. Since the phenotype is heritable, the adaptive phenotype and the underlying genotype will increase in frequency in the population over time. Evolution by natural selection is also called adaptive evolution. Natural selection differs from all three other evolutionary forces—migration, random genetic drift, and mutation—in that natural selection can explain only adaptive evolution, whereas the other three can account for nonadaptive evolution. Mutation is random, and therefore changes in genotype due to mutation occur regardless of whether the change is beneficial, detrimental, or neutral with respect to reproductive fitness. Mutation can also create new genotypes and phenotypes, whereas natural selection can act only on existing phenotypes. Random genetic drift (genetic drift) is the change in genotypic frequencies due purely to chance instead of due to a reproductive advantage. Population size influences the effect of genetic drift on genotype frequencies (smaller populations are more severely affected than larger ones), whereas population size does not necessarily influence natural selection. Migration changes the frequencies of genotypes in a population by adding new individuals from a different population that has different genotype frequencies.

8. *Evaluate > Deduce > Solve:* This is type 4. The modern synthesis of evolution is the reconciliation of Darwin's evolutionary theory with the findings of modern genetics. Darwin's theory proposed that all evolution was adaptive. Genetic studies on mutation and genetic recombination initially argued against the importance of natural selection as an agent for change because mutation and recombination were nonadaptive. The modern synthesis stated that evolution is due to the combined action of adaptive and nonadaptive evolutionary forces. In particular, the modern synthesis explained how mutation and recombination could provide the raw material (new genotypes and phenotypes) on which natural selection acts.

9. *Evaluate > Deduce > Solve:* This is type 4. The four processes of evolution are mutation, natural selection, random genetic drift, and migration. *Natural selection* is the differential reproduction among individuals in a population that is due to differences in heritable phenotypes and the environment. Because these phenotypes are said to be adaptive, natural selection thus can explain only adaptive evolution. *Random genetic drift* is also due to differential reproduction among individuals in a population, but it is due to chance and is independent of phenotype and environment. Because such changes are nonadaptive, random genetic drift is a mechanism that can explain nonadaptive evolution. *Mutation* is the random change in genetic information that occurs because of errors in DNA replication or DNA damage or chromosome alterations. Mutation is the ultimate source of genetic variation and thus provides the raw material on which natural selection and random genetic drift act. *Migration* is the addition of new individuals into a population. Migration will change the overall genetic makeup of a population if the added individuals differ in their genetic composition.

10a. *Evaluate > Deduce > Solve: Transcription* is the synthesis of RNA by RNA polymerase. The RNA is complementary to the strand of DNA that was used as the template for transcription.

10b. *Evaluate > Deduce > Solve:* An *allele* is a specific form of a gene or genetic locus.

10c. *Evaluate > Deduce > Solve:* The *central dogma of biology* originally stated that genetic information flows from DNA to RNA (by transcription) and from RNA to protein (by translation). A point of emphasis of this dogma was that information does not flow in the reverse direction and has been modified to account for reverse transcription.

10d. *Evaluate > Deduce > Solve: Translation* is the synthesis of a polypeptide using the information in an mRNA. *Translation* and *protein synthesis* are synonyms.

10e. *Evaluate > Deduce > Solve: DNA replication* is the process by which DNA is copied by DNA polymerase.

10f. *Evaluate > Deduce > Solve:* A *gene* is a segment of DNA that contains all the information necessary for its proper transcription, including the promoter, transcribed region, and termination signals.

10g. *Evaluate > Deduce > Solve:* A *chromosome* is a heritable molecule composed of DNA and protein that typically contains genes.

10h. *Evaluate > Deduce > Solve:* The term *antiparallel* refers to the orientation of the two strands of nucleic acid in a double-stranded nucleic acid (RNA or DNA). Each end of the double-stranded nucleic acid will contain the 5' end of one strand and the 3' end of the other.

10i. *Evaluate > Deduce > Solve: Phenotype* refers to the observable characteristics of an organism, which include morphology, physiology, and molecular composition. The *phenotype* of an organism is a product of the interaction between its genotype and its environment.

10j. *Evaluate > Deduce > Solve:* The term *complementary base pair* refers to the two nucleotides on opposite, antiparallel strands of a double-stranded nucleic acid, which are hydrogen bonded to each other. One nucleotide will contain a purine base that makes hydrogen bonds to the pyrimidine base that is part of the other nucleotide.

10k. *Evaluate > Deduce > Solve: Nucleic acid strand polarity* refers to the orientation of the nucleotides along a single strand of nucleic acid. One end of the strand terminates at the 3' hydroxyl group of a ribose (or deoxyribose) sugar, whereas the other strand terminates at the 5' phosphate group on the sugar. These are commonly referred to as the 3' and 5' ends of the strand.

10l. *Evaluate > Deduce > Solve: Genotype* refers to the genetic makeup of an organism. The genotype can refer to the organism's entire genetic makeup, or it can refer to the genetic information at only one or a few loci.

10m. *Evaluate > Deduce > Solve: Natural selection* is the process by which populations and species evolve and diverge through differential rates of survival and reproduction of members that are due to their inherited differences.

10n. *Evaluate > Deduce > Solve: Mutation* is the process that generates new genetic variety through change to existing alleles.

10o. *Evaluate > Deduce > Solve: Modern synthesis of evolution* is the term applied to the reconciliation of modern genetic analysis with Darwin's theory of evolution by natural selection.

11. *Evaluate:* This is type 2. Recall that thymine base-pairs with adenine, and cytosine base-pairs with guanine. *Deduce:* If thymine makes up 21% of the DNA nucleotides in the genome of a plant species, then adenine makes up 21%, and cytosine plus guanine make up the remainder, which is 58% (29% cytosine and 29% guanine). *Solve:* 21% adenine, 34% cysteine, and 34% guanine.

12. *Evaluate > Deduce:* This is type 2. The reactive chemical group at the 5' carbon of nucleotides is a phosphate group, and one at the 3' carbon is a hydroxyl group. The bond that joins adjacent nucleotides is a phosphodiester bond. The phosphodiester linkage that joins nucleotides is covalent. *Solve:* The 5' end is a phosphate group. The 3' end is a hydroxyl group. A phosphodiester bond is a covalent bond that joins nucleotides in a strand of nucleic acid.

13. *Evaluate > Deduce:* This is type 2. Two structural features distinguish DNA and RNA. One is the sugar, which is ribose in RNA and is deoxyribose in DNA. The other is the base that is complementary to adenine, which is uracil in RNA but is thymine in DNA. *Solve:* The sugar in DNA is deoxyribose, whereas the sugar in RNA is ribose. Thymine is a base in DNA, whereas uracil is the corresponding base in RNA.

14.   *Evaluate > Deduce > Solve:* This is type 3. The central dogma is a description of the flow of genetic information. The flow is unidirectional, from DNA to RNA to protein. The process by which information flows from DNA to RNA is called transcription. Transcription is the synthesis of RNA by RNA polymerase. The RNA is complementary to the strand of DNA that was used as the template for transcription. The flow of information from RNA to protein is called translation. Translation is the synthesis of a polypeptide using the information in an mRNA.

15.   *Evaluate:* This is type 3. The RNA sequence will contain the codons for all the amino acids in the peptide. The codon for Trp is UGG, for Lys is AA^A/$_G$, for Met is AUG, for Ala is GCN, and for Val is GUN. The template-strand DNA sequence is complementary to the RNA sequence. *Deduce:* The RNA encoding this peptide has the sequence 5'-UGGAA^A/$_G$AUGGCNGUN-3'. The corresponding template-strand DNA sequence is 5'-NACNGCCAT^T/$_C$TTCCA-3'. *Solve:* The RNA encoding this peptide has the sequence 5'-UGGAA^A/$_G$AUGGCNGUN-3'. The corresponding template-strand DNA sequence is 5'-NACNGCCAT^T/$_C$TTCCA-3'.

16a-b.   *Evaluate:* This is type 3. *Deduce:* The mRNA sequence will be complementary to the template sequence. The amino acid sequence will be determined by the order of the codons in the mRNA. *Solve:* (a) The mRNA sequence is 5'-UUCCAUGUC-3'. (b) The amino acid sequence is Phe His Val.

17a.   *Evaluate > Deduce:* This is type 2. There are 18 nucleosides in each strand of DNA, with one phosphodiester bond connecting each neighboring nucleotide. *Solve:* There are 18 − 1 = 17 phosphodiester bonds in each strand of DNA; therefore, there are 34 total phosphodiester bonds.

17b.   *Evaluate:* This is type 2. *Deduce:* There are 18 base pairs in the DNA molecule, 9 A-T base pairs and 9 G-C base pairs. There are 2 hydrogen bonds in each A-T base pair and 3 hydrogen bonds in each G-C base pair. *Solve:* The total number of hydrogen bonds is (9 × 2) + (9 × 3) = 45.

17c.   *Evaluate:* This is type 3. The bottom strand is the template for transcription. *Deduce:* The RNA transcribed from this DNA molecule is 5'-AUGCCAGUCACUGACUUG-3'. Although the correct reading frame is not specified, the RNA sequence begins an AUG start codon, which set the reading frame. *Solve:* The codons are AUG (Met), CCA (Pro), GUC (Val), ACU (Thr), GAC (Asp), and UUG (Leu). This polypeptide has 6 amino acids and 5 peptide bonds.

18a.   *Evaluate > Deduce > Solve:* This is type 5. The figure identifies six clades (vertebrate, tetrapod, therian, placental, mammalian, and primate).

18b.   *Evaluate > Deduce > Solve:* This is type 5. A backbone (they are all vertebrates).

18c.   *Evaluate:* This is type 5. The characteristics are (1) backbone, (2) four legs, (3) fur and milk, (4) live young, (5) placenta, (6) opposable thumbs. *Deduce:* mammals have a backbone, four legs, fur and produce milk, whereas humans also give birth to live young, have placenta, and have opposable thumbs. *Solve:* The mammalian clade and humans share these characteristics: they have backbones, have four legs, have fur and produce milk. They differ in that members of the human clade give birth to live young, have placenta and opposable thumbs.

19.   *Evaluate:* This is type 3. This problem tests your understanding of the relationships between DNA, mRNA, tRNA, and amino acids. The coding strand of DNA is the same sequence as the mRNA, except that T in DNA is replaced by U in mRNA. The template strand of DNA is complementary to the DNA coding strand, and the tRNA anticodon sequences are complementary to the mRNA codons. The correspondence

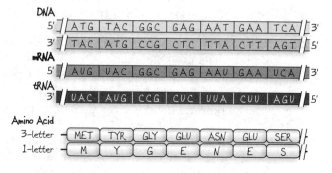

between mRNA codons and amino acids, as well as 3-letter and 1-letter amino acid designations, are given in the Genetic Code Analytical Tool (see page 3). *Deduce*: You are given complete information in one row for four of the seven positions (3-letter amino acid code MET, mRNA 5'-UAC-3', DNA coding strand 5'-GGC-3', and tRNA anticodon 3'-UUA-5'), which allows you to fill in the corresponding boxes in the other rows using your knowledge of the relationships between DNA, mRNA, tRNA, and amino acids. For example, the 1-letter amino acid designation for MET is M, the mRNA is 5'-AUG-3', the DNA coding strand is 5'-ATG-3', the DNA template strand is 3'-TAC-5', and the tRNA anticodon is 3'-UAC-5'. To fill in the remaining three positions, you must combine information from two or more rows. For example, in the sixth position, the 3-letter designation for E is GLU, which has two possible codons, 5'-GAA-3' and 5'-GAG-3'. The last position in the corresponding mRNA codon is an A; therefore, the mRNA codon sequence is 5'-GAA-3'. The corresponding boxes in the other rows can be solved for in the same way. *Solve:* See figure.

20. *Evaluate:* This is type 2. DNA will contain thymine (T) but not uracil (U) whereas RNA will contain U but not T. Double-stranded DNA will contain equal proportions of adenine (A) and T and equal proportions of guanine (G) and cytosine (C) whereas single-stranded DNA typically does not. Double-stranded RNA will contain equal proportions of A and U and equal proportions of G and C whereas single-stranded RNA typically does not. *Deduce:* Sample 1 contains T but not U and the percentage of A is the same as T, therefore, it is probably double-stranded DNA. Sample 2 contains U but not T and the percentage of A is not the same as U, therefore, it is single-stranded RNA. Sample 3 contains U but not T and the percentage of A is the same as U, therefore, it is probably double-stranded RNA. Sample 4 contains T but not U and the percentage of A is not the same as T; therefore, it is single-stranded DNA. *Solve:* Samples 1 and 3 are double-stranded, whereas samples 2 and 4 are single-stranded.

21. *Evaluate > Deduce > Solve:* This is type 5. Recall that monophyletic groups include a common ancestor and all of its descendants, whereas paraphyletic groups include a common ancestor but all of its descendants. Seed-eating finches include some ground finches and the sharp-beaked finches, whose common ancestor includes species that do not eat seeds. Thus seed-eating finches are a paraphyletic group. Cactus-flower-eating finches are a monophyletic group because their common ancestor and all of its descendants are cactus-flower eaters.

22. *Evaluate:* This is type 5. Recall that the outgroup is a taxon less closely related to any of the taxa being compared. *Deduce:* Reptiles are a paraphyletic group that includes birds; therefore, birds cannot be the outgroup. Fish and amphibians share a common ancestor with reptiles, whereas mammals are monophyletic and do not. *Solve:* Therefore, mammals would be the best choice as an outgroup for phylogenetic analysis of reptiles.

23. *Evaluate*: This is type 5. This problem tests your understanding of phylogenetics and parsimony and your ability to apply these principles to tree building. *Deduce > Solve*: See figure and the following explanation. Three pairs of taxa were identical to each other in amino acid sequence (*Pan troglodytes* and *Pan paniscus*, *Gorilla beringei* and *Gorilla gorilla*, and *Pongo pygmaeus* and *Pongo abelii*) and can be considered as one taxon each (*Pan*, *Gorilla*, and *Pongo*), leaving three taxa to analyze (*Homo*, *Hylobates*, and *Hoolock*). *Hoolock* has the same sequence as the common

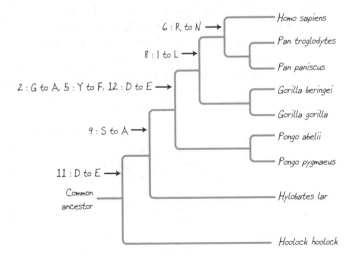

ancestor, and therefore is shown on a branch that has no evolutionary changes listed. *Hylobates* is most closely related to *Hoolock*, except that it shares the derived character E at position 11 with all other taxa; thus, a change of D to E at position 11 is proposed to have occurred after divergence of the lineages leading to *Hoolock* and *Hylobates*. *Homo, Pan, Gorilla,* and *Pongo* share the derived character, A, at position 9; thus, a change of S to A at position 9 is proposed to have occurred after the divergence of the lineages leading to *Hylobates* and these four taxa. *Homo, Pan,* and *Gorilla* share the derived characters A, F, and E at positions 2, 5, and 12, respectively; thus, changes to these characters are proposed to have occurred after the divergence of the lineages leading to *Pongo* and these three taxa. *Pan* and *Homo* share the derived character L at position 8; thus, a change of I to L at position 8 is proposed to have occurred after divergence of the lineages leading to *Gorilla* and these two taxa. N at position 6 is unique to *Homo*; thus, a change of R to N at position 6 is proposed to have occurred in the lineage specific to *Homo*.

1.04 Test Yourself

Problems

1. Consider the following statement: Mutation and genetic recombination, not natural selection, account for evolution. Modify this statement to reflect the understanding of the roles of mutation, recombination, and natural selection in evolution.

2. Label the arrows in the concept map shown in the figure with one of the following terms: transcription, translation, reverse transcription.

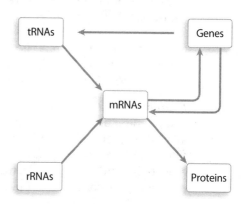

3. The table shows homologous DNA sequences of four taxa and the proposed sequence of their common ancestor. The accompanying figure shows three phylogenetic trees, each representing a different hypothesis for the evolutionary relationships among taxa 1, 2, and 3, with taxon 4 as the outgroup. Indicate the location of each evolutionary change on all three trees, and use that information to determine which of the trees is most likely by applying the principle of parsimony.

	Nucleotide Position						
	1	2	3	4	5	6	7
taxon 1	A	T	G	G	T	C	T
taxon 2	A	T	T	C	A	G	T
taxon 3	A	T	G	G	A	C	T
taxon 4	C	G	T	C	T	C	A
ancestor	C	G	T	C	T	C	T

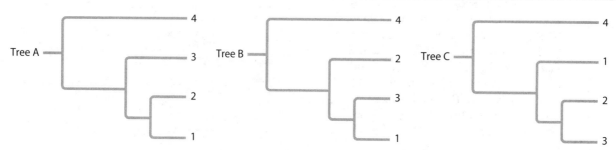

4. The figure shows an accepted phylogeny for the taxa tuna, lizard, monkey, hippopotamus, and whale. The identity of the characters corresponding to tick marks is shown in the key. Use this information to answer the following questions.
   a. What are the shared derived characters that identify the mammalian clade?
   b. What are the shared derived characters that identify the tetrapod clade?
   c. Which characters are proposed to represent homoplasy?

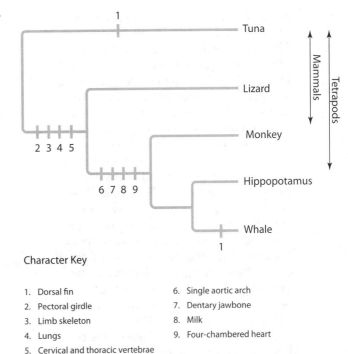

Character Key

1. Dorsal fin
2. Pectoral girdle
3. Limb skeleton
4. Lungs
5. Cervical and thoracic vertebrae

6. Single aortic arch
7. Dentary jawbone
8. Milk
9. Four-chambered heart

Solutions

1.   *Evaluate*: This is type 4. This statement reflects the sentiment of some experimental geneticists before the modern synthesis of evolution. They observed changes in the phenotype of organisms due to mutation and recombination and suggested that these forces alone could account for evolution. Furthermore, their observations suggested that evolution was nonadaptive. *Deduce*: The modern synthesis reconciled these findings with Darwinian evolution by natural selection by recognizing the importance of these genetic mechanisms in creating genetic diversity, upon which natural selection could act. *Solve*: Mutation and genetic recombination create genetic variation, which is used by natural selection to drive adaptive evolution.

2.   *Evaluate*: This is type 3. The updated central dogma of molecular biology states that information flows from DNA to RNA by transcription, from RNA to protein by translation, and from RNA back to DNA by reverse transcription. *Deduce*: tRNA, mRNAs, and rRNAs are produced by transcription of genes. tRNAs and rRNAs are involved in translation of mRNAs. Proteins are produced by translation of mRNAs. DNA is produced by reverse transcription of mRNAs. *Solve*: See figure.

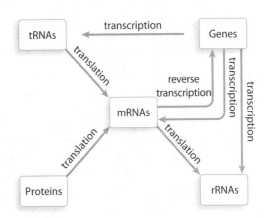

3. *Evaluate:* This is type 5. This problem tests your understanding of parsimony. The answer will be the tree that requires the fewest proposed sequence changes. *Deduce:* The changes in positions 1 (to A) and 2 (to T) are present in taxa 1, 2, and 3 and so could have been present in their common ancestor. The changes are placed on the diagonal line before the node corresponding to their common ancestor. The changes at positions 3 and 4 (to G) are present in taxa 1 and 3. For tree B, these changes could have been present in their common ancestor and can be placed on the diagonal before the node corresponding to their common ancestor. For trees A and C, these changes have occurred twice, once in the taxon-1-specific lineage and once in the taxon-3-specific lineage. The change at position 5 (to A) is present in taxa 2 and 3. For tree C, this change could have been present in their common ancestor and is placed on the diagonal before the node corresponding to the common ancestor. For trees A and B, this change could have occurred twice, once in the taxon-2-specific lineage and once in the taxon-3-specific lineage. The change at position 6 (to C) is specific to taxon 2 and at 7 (to A) is specific to taxon 4. Both changes are placed on the appropriate taxon specific lineage in each tree. *Solve:* Tree A requires 10 changes, tree B requires 8 changes, and tree C requires 9 changes. Tree B is the most likely under the principle of parsimony; however, which—if any—of these trees is correct is not known. *Solve:* Tree B is the most likely under the principle of parsimony.

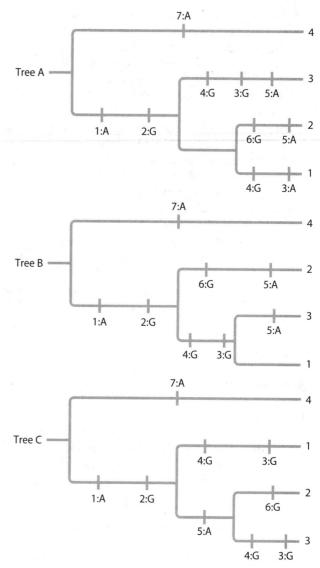

4a. *Evaluate:* This is type 5. The shared derived characteristics will be the characteristics that are present in the common ancestor to mammals but absent in the common ancestor to other clades. *Deduce:* The common ancestor to mammals have characteristics two through nine, but four of these characters—a single aortic arch, a dentary jawbone, a four-chambered heart, and production of milk—are specific to the mammalian clade. The other four characteristics are also present in tetrapods. *Solve:* The shared derived characteristics that distinguish the mammalian clade in this tree are a single aortic arch, a dentary jawbone, production of milk, and a four-chambered heart.

4b. *Evaluate:* This is type 5. The shared derived characteristics for tetrapods will be the characteristics present in the common ancestor to all tetrapods but absent in tuna (the outgroup). *Deduce:* The common ancestor to all tetrapods has four characteristics: a pectoral girdle, a limb skeleton, lungs, and cervical and thoracic vertebrae. *Solve:* The shared derived characteristics that distinguish tetrapods from tuna (the outgroup) are a pectoral girdle, a limb skeleton, lungs, and cervical and thoracic vertebrae.

4c. *Evaluate:* This is type 5. Homoplasmic characters are those that represent convergent evolution, and they will have evolved independently more than once. *Deduce:* The first characteristic, a dorsal fin, is shown to have evolved independently into separate lineages: the lineage specific to whale and the lineage specific to tuna (the outgroup). *Solve:* A dorsal fin is the only homoplasious character shown in the figure.

2

Transmission Genetics

2.01 Genetics Problem-Solving Toolkit

Key Terms and Concepts

Mendel's law of segregation: The two alleles of a single gene in parents segregate from each other during meiosis such that one-half of the gametes produced carry one allele and the other half carry the other allele. Random fusion of gametes from two parents results in progeny that contain allele combinations determined by chance.

Mendel's law of independent assortment: The alleles of different genes in a parent segregate independently of one another during meiosis, such that gametes are equally likely to carry all possible combinations of alleles.

Test cross: A procedure typically used to determine the genotype of an individual with a dominant phenotype. This is done by crossing that individual to an individual with a homozygous recessive genotype.

Key Analytical Tools

Chi-square formula: $\chi^2 = \sum \frac{(O - E)^2}{E}$

Binomial distribution: $(a + b)^n$

Punnett square

	$\frac{1}{2}A$	$\frac{1}{2}a$
$\frac{1}{2}A$	$\frac{1}{4}AA$	$\frac{1}{4}Aa$
$\frac{1}{2}a$	$\frac{1}{4}Aa$	$\frac{1}{4}aa$

Forked-line diagram

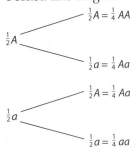

Key Genetic Relationships

Mendel's laws of segregation and independent assortment predict specific relationships between the genotypes of parents and the genotypic and phenotypic proportions of their progeny. You should be able to apply Mendel's laws using the Punnett square and the forked-line methods to determine the expected genotypic and phenotypic proportions of progeny given the parental genotypes, and work backward to deduce the possible parental genotypes given the proportions of progeny genotypes or phenotypes.

There are six different types of crosses involving one genetic locus. Use the Punnett squares below to fill in the progeny genotypes and proportions for all six crosses.

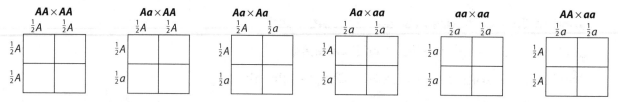

These crosses provide *six key relationships* that are as basic to genetics as multiplication is to math **(Table 2.01)**. Think of Table 2.01 as your "genetic multiplication table" for problems involving one genetic locus with two alleles, one dominant and one recessive. As you work through the problems in the textbook and study guide that deal with genetic relationships between parents and progeny, your first step will be to identify which of these relationships applies to the problem.

Table 2.01   Genetic Relationships for All Crosses Involving a Single Genetic Locus

	1	2	3	4	5	6
Parent 1						
Genotype	AA	Aa	Aa	Aa	aa	AA
Phenotype	dominant	dominant	dominant	dominant	recessive	dominant
Gametes	A	$\frac{1}{2}A, \frac{1}{2}a$	$\frac{1}{2}A, \frac{1}{2}a$	$\frac{1}{2}A, \frac{1}{2}a$	a	A
Parent 2						
Genotype	AA	AA	Aa	aa	aa	aa
Phenotype	dominant	dominant	dominant	recessive	recessive	recessive
Gametes	A	A	$\frac{1}{2}A, \frac{1}{2}a$	a	a	a
Progeny						
Genotypic ratio	AA	$\frac{1}{2}AA : \frac{1}{2}Aa$	$\frac{1}{4}AA : \frac{1}{2}Aa : \frac{1}{4}aa$	$\frac{1}{2}Aa : \frac{1}{2}aa$	aa	Aa
Phenotypic ratio	dominant	dominant	$\frac{3}{4}$ dominant : $\frac{1}{4}$ recessive	$\frac{1}{2}$ dominant : $\frac{1}{2}$ recessive	recessive	dominant

2.02 Types of Genetics Problems

1. Predict the outcome of crosses given parental genotypes.
2. Determine the genotype of individuals using information from a pedigree.
3. Calculate the probabilities of specific progeny types given information on parents.
4. Determine parental genotypes given progeny genotypes or phenotypes.
5. Test the goodness of fit between observed values and expected outcomes.
6. Calculate the probability of specific progeny distributions using binomial probability calculations.

1. Predict the outcome of crosses given parental genotypes.

Problems of this type provide you with the genotypes of parents and ask you for information concerning the expected genotypes or phenotypes of the progeny. You may be asked for the *ratio* of all possible progeny genotypes or phenotypes, the *proportion* (fraction) of progeny that will have a specific genotype or phenotype, or the expected *number* of progeny that have a specific genotype or phenotype given the total number of progeny analyzed.

Variations:

(a) Predict the outcome of crosses involving one genetic locus.

Table 2.01 provides information on the expected *ratios* and *proportions* of progeny genotypes and phenotypes. **TIP:** The expected *number* of progeny of a specific type is the expected *proportion* of that type multiplied by the *total number* of progeny analyzed. See Problem 2 for an example.

(b) Predict the outcome of crosses involving two or more genetic loci that assort independently.

The expected progeny proportions for each locus are provided in Table 2.01. **TIP:** Use the forked-line method to determine the progeny types and the multiplication rule to determine the proportions of each progeny type. See Problem 6 for an example.

Example Problem: In pea plants, the appearance of flowers along the main stem is a dominant phenotype called "axial" that is controlled by an allele T. The recessive phenotype, produced by an allele t, has flowers only at the end of the stem and is called "terminal." Pod form displays a dominant phenotype "inflated," controlled by an allele C, and a recessive "constricted" form, produced by the c allele. A pure-breeding axial, inflated plant is crossed to one that is pure-breeding terminal, constricted. The F_1 progeny of this cross are allowed to self-fertilize and produce 320 progeny. What genotypes are expected among the F_2, and how many plants of each genotype are expected?

Solution Strategies	Solution Steps
Evaluate	
1. Identify the topic this problem addresses, and explain the nature of the requested answer.	1. This problem tests your understanding of segregation and independent assortment. The answer will be a list of F_2 genotypes and the numbers of F_2 progeny of each genotype. The F_1 genotype must be determined in order to predict the F_2.
2. Identify the critical information given in the problem.	2. There are two traits determined by independently assorting genetic loci: flower position, where axial (T) is dominant to terminal (t); pod form, where inflated (C) is dominant to constricted (c). P_1 was pure-breeding axial, inflated. P_2 was pure-breeding terminal, constricted. Self-fertilizing the F_1 produced 320 F_2 progeny.

(continued)

Solution Strategies	Solution Steps
Deduce	
3. Determine which genetic relationships are relevant to this problem. Perform calculations if necessary to relate the given information to the list of standard genetic relationships. **TIP:** P_1 and P_2 are pure-breeding and therefore homozygous at both loci.	3. Genotypes of parents and F_1 can be deduced: Parents: P_1 is *TTCC*, and P_2 is *ttcc*; genetic relationship 6 of Table 2.01 is used for both loci to determine the F_1 genotype. F_1: *TtCc* dihybrids. • Self-fertilizing the F_1 is *TtCc* × *TtCc*. • Genetic relationship 3 of Table 2.01 is relevant for both loci: *Tt* × *Tt* and *Cc* × *Cc*.
Solve	
4. <u>Calculate</u> answers based on the identified genetic relationships. **TIP:** Use the forked-line method to combine the ratios predicted for the *T* and *C* loci to determine the proportion of F_2 genotypes.	4. The expected proportions of progeny genotypes for each locus from Table 2.01 are $Tt \times Tt$: $\frac{1}{4} TT; \frac{1}{2} Tt; \frac{1}{4} tt$ $Cc \times Cc$: $\frac{1}{4} CC : \frac{1}{2} Cc : \frac{1}{4} cc$ Combining both loci: $\frac{1}{16} TTCC, \frac{2}{16} TTCc, \frac{1}{16} TTcc, \frac{2}{16} TtCC, \frac{4}{16} TtCc, \frac{2}{16} Ttcc, \frac{1}{16} ttCC, \frac{2}{16} ttCc$, and $\frac{1}{16} ttcc$.
5. Apply the results of the calculations from the genetic relationships to derive the answer to this problem. **TIP:** The total number of progeny analyzed multiplied by proportions of each genotype equals the number of each genotype.	5. **Answer:** The expected numbers and types of progeny are $\frac{1}{16} TTCC \times 320$ progeny = 20 *TTCC*; $\frac{2}{16} TTCc \times 320 = 40$ *TTCc*; $\frac{1}{16} TTcc \times 320 = 20$ *TTcc*; $\frac{2}{16} TtCC \times 320 = 40$ *TtCC*; $\frac{4}{16} TtCc \times 320 = 80$ *TtCc*; $\frac{2}{16} Ttcc \times 320 = 40$ *Ttcc*; $\frac{1}{16} ttCC \times 320 = 20$ *ttCC*; $\frac{2}{16} ttCc \times 320 = 40$ *ttCc*; and $\frac{1}{16} ttcc \times 320 = 20$ *ttcc*.

2. Determine the genotype of individuals using information from a pedigree.

Problems of this type provide you with a pedigree showing a trait segregating in two or more generations and ask you to assign genotypes to individuals in the pedigree. First, determine whether the trait is dominant or recessive. Then, use the information in Table 2.01 to deduce the genotype of individuals in the pedigree.

Variations:

(a) Determine the genotype of individuals in a pedigree showing inheritance of a dominant trait. TIP: Dominant traits typically appear in every generation, and each individual with the trait has a parent with the trait. All individuals with the trait have at least one dominant allele. See Problem 4 for an example.

(b) Determine the genotype of individuals in a pedigree showing inheritance of a recessive trait. TIP: Recessive traits do not appear in every generation, and individuals with the trait typically have parents that do not have the trait. All individuals with the trait are homozygous recessive, and both their parents must carry at least one copy of the recessive allele. See Problem 15 for an example.

Example Problem: You are a genetic counselor. A patient (III-1 in the pedigree) who has been told that she has type 1 neurofibromatosis (NF1) comes to you and asks you to explain the inheritance of this disease in her family. You examined her siblings, her parents, her uncles and aunts, and her grandparents. You recorded your results in the following pedigree, where dark symbols indicate the presence of traits associated with NF1.

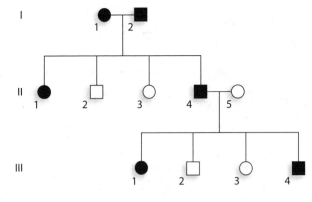

Use your pedigree to determine (a) whether the results are consistent with inheritance of NF1, a dominant trait, and (b) the likely genotype of each individual in the pedigree.

Solution Strategies	Solution Steps
Evaluate	
1. Identify the topic this problem addresses, and explain the nature of the requested answer.	1. This problem tests your understanding of pedigrees and dominance. The answer to (a) will be yes if the trait appears to be dominant and will be no if it does not. The answer to (b) will be a genotype for each individual in the pedigree. The dominance of the trait will be needed to determine the answers to (b).
2. Identify the critical information given in the problem.	2. The pedigree shows affected individuals in each generation. Affected individuals are I-1, I-2, II-1, II-4, III-1, and III-4. Unaffected individuals are II-2, II-3, II-5, III-2, and III-3.
Deduce	
3. Determine whether the trait is dominant or recessive, and identify individuals with dominant and recessive phenotypes.	3. The trait is dominant because it appears in every generation and each affected individual has an affected parent. Therefore, I-1, 1-2, II-1, II-4, III-1, and III-4 have the dominant phenotype; and II-2, II-3, II-5, III-2, and III-3 have the recessive phenotype.
4. Identify the key genetic relationships relevant to the pedigree. **TIP:** Use parent phenotypes to identify relationships (see Table 2.01) relevant to each mating. **TIP:** List expected outcomes of the matings based on each relevant relationship.	4. The two matings are I-1 × I-2 and II-4 × II-5. I-1 × I-2: genetic relationships 1, 2, and 3 (dominant × dominant) are potentially relevant. All progeny should have the dominant trait if either I-1 or I-2 is homozygous. Some progeny may have the recessive trait if both I-1 and I-2 are heterozygous. II-4 × II-5: genetic relationships 4 and 6 (dominant × recessive) are potentially relevant. All progeny will have the dominant trait if II-4 is homozygous. Some progeny may have the recessive trait if II-4 is heterozygous.

(continued)

Solution Strategies	Solution Steps
Solve	
5. Identify the most relevant genetic relationships, and determine the genotypes of all individuals. **TIP:** The genotype of some individuals with the dominant phenotype may remain ambiguous (*AA* or *Aa*).	5. **Answer:** II-2, II-3, II-5, III-2, and III-3 have the recessive phenotype and so are homozygous recessive (*aa*). I-1, I-2, and II-4 have progeny with recessive phenotype and so are heterozygous (*Aa*). II-1 has the dominant phenotype but no progeny, and it is either homozygous dominant or heterozygous (*AA* or *Aa*). III-1 and III-4 have the dominant phenotype and an unaffected mother and so are heterozygous (*Aa*).

3. Calculate the probabilities of specific progeny types given information on parents.

This problem differs from the first type in that it asks for the *probability* that progeny will have a specific genotype or phenotype. The probability of a specific progeny genotype or phenotype is the same as the *proportion* of progeny that are expected to have that genotype or phenotype. For example, in the cross $Aa \times Aa$, the proportion of progeny that are expected to be homozygous dominant (*AA*) is $\frac{1}{4}$. Therefore, the probability that any particular offspring will be *AA* is also $\frac{1}{4}$.

Variations:

(a) Calculate the probability of a specific progeny type given parental genotypes for a cross involving one genetic locus. Table 2.01 provides the information on the proportion of progeny types. See Problem 11 for an example.

(b) Calculate the probability of a specific progeny type given parental genotypes for a cross involving two or more genetic loci that assort independently. The expected progeny proportions for each locus are provided in Table 2.01. **TIP:** Combine the probabilities of specific types for each locus using the multiplication rule. See Problem 43 for an example.

(c) Calculate the probability of a specific progeny type given a pedigree where parental genotypes are unambiguous. This problem type differs from **3(a)** and **(b)** only in that you must first use the pedigree to deduce the parental genotypes. See Problem 34 for an example.

(d) Calculate the probability of a specific progeny type given a pedigree where parental genotypes are ambiguous. This problem differs from those above in that the genotype of at least one parent will be ambiguous. **TIP:** Typically, *only one possible parental genotype* will be relevant to the problem. Determine the probability of *that parental genotype* and the probability of the progeny type *assuming that parental genotype* and then combine the probabilities using the multiplication rule. See "Example Problem" below.

Example Problem: You are a genetic counselor. A woman (III-1) comes to you concerned that she may be a carrier of the recessive disorder, cystic fibrosis (CF), because her aunt has CF. You interview her and record the results of your interview in the pedigree in the accompanying figure. Assuming that II-5 is not a carrier, what is the probability that the woman is a carrier?

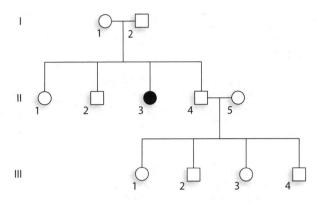

Solution Strategies	Solution Steps
Evaluate	
1. Identify the topic this problem addresses, and explain the nature of the requested answer.	1. The problem tests whether you can calculate the probability of an individual having a specific genotype given information from a pedigree. The probability that III-1 is a carrier will be a number between zero (cannot be heterozygous) and one (must be heterozygous). The probability that III-1 is a carrier will depend on the genotype of her parents and grandparents.
2. Identify the critical information given in the problem.	2. The pedigree and problem provide the following: CF is a recessive trait, II-3 is the only affected individual, and II-5 is not a carrier.
Deduce	
3. Identify the relevant key genetic relationships for the matings that are critical to the probability calculation. **TIP:** The genotypes of the individual's parents and grandparents affect the probability calculation.	3. III-1 is the daughter of II-4 and II-5 and the granddaughter of I-1 and I-2. II-4 × II-5: Genetic relationships 1 and 2 are relevant. All progeny will be homozygous dominant if II-4 is homozygous dominant. Some progeny may be heterozygous if II-4 is heterozygous. I-1 × I-2: Genetic relationships 1, 2, and 3 (dominant × dominant) are relevant. All progeny should have the dominant trait if either I-1 or I-2 are homozygous. Some progeny may have the recessive trait if both I-1 and I-2 are heterozygous.
4. Identify the most relevant genetic relationships, and determine the genotypes of the parents of each mating. **TIP:** If the genotypes of the individual's parents are unambiguous, then go to step 6; if the genotype of at least one of the individual's parents is ambiguous, then proceed to step 5.	4. I-1 × I-2: Genetic relationship 3 is applicable because they have an affected child. Both I-1 and I-2 are *Aa*. II-4 × II-5: Genotypes are not clear because they have no affected children and II-4's genotype is ambiguous, so it can be either *AA* or *Aa*.

(continued)

Solution Strategies	Solution Steps
Deduce	
5. Assign a genotype to the ambiguous parent that is required to meet the condition of the problem, and calculate the probability of that genotype. **TIP:** The probabilities obtained using Table 2.01 are based on proportions that consider all possible progeny types. II-4 cannot be *aa*; therefore, the probability that II-4 is *Aa* is a conditional probability calculation. **TIP:** Use the relevant genetic relationship to calculate the probability of that parent's genotype.	**5.** II-4 must be *Aa* in order for III-1 to have any chance of being a carrier (*Aa*). • II-4's parents are I-1 and I-2. • I-1 and I-2 are *Aa*; thus, genetic relationship 3 is used to calculate the probability of II-4 being *Aa*. • The probabilities taken directly from the information in Table 2.01 are $\frac{1}{4}$ probability of being *aa*, $\frac{1}{2}$ probability of being *Aa*, $\frac{1}{4}$ probability of being *AA*. • Since II-4 cannot be *aa*, then the total probability is $\frac{1}{2}$ *Aa* + $\frac{1}{4}$ *AA* = $\frac{3}{4}$. Therefore, II-4 has a $\frac{1}{2}$ out of $\frac{3}{4}$ = $\frac{2}{3}$ probability of being *Aa*, and $\frac{1}{4}$ out of $\frac{3}{4}$ = $\frac{1}{3}$ probability of being *AA*.
Solve	
6. Calculate answers based on the identified genetic relationships. **TIP:** Use the genotype assigned to II-4 to identify the relevant genetic relationship for II-4 × II-5.	**6.** The probability that II-4 is *Aa* is $\frac{2}{3}$. The probability that III-1 is *Aa* given that II-4 is *Aa* is $\frac{1}{2}$.
7. Apply the results of the calculations to the problem to derive the answer. **TIP:** Use the multiplication rule to combine the probabilities from step 5.	**7.** **Answer:** The overall probability that III-1 is a carrier is $\frac{2}{3} \times \frac{1}{2} = \frac{1}{3}$.

4. Determine parental genotypes given progeny genotypes or phenotypes.

Problems of this type provide information about the progeny of a cross and ask you to deduce the genotypes of their parents. You may be specifically asked to determine the parental genotypes, or you may need to determine the parental genotypes as part of the process of analyzing the genetic basis of a trait (e.g., whether the trait is dominant or recessive). These types of problems focus on a single trait controlled by a single genetic locus, and so the genetic relationships that are relevant to answering the questions are found in Table 2.01.

Example Problem: Determine the genetic basis of brown and black fur color in mice using the following information. A pure-breeding brown male mouse mated with a pure-breeding black female. All the F_1 were black. Matings between F_1 mice yielded F_2 progeny that were $\frac{3}{4}$ black and $\frac{1}{4}$ brown. Reciprocal crosses gave the same results.

Solution Strategies	Solution Steps
Evaluate	
1. Identify the topic this problem addresses, and explain the nature of the requested answer.	**1.** This problem tests your understanding of how genotypes and phenotypes are related. The answer should include the dominance relationship between black and brown, and the genotypes of the parents, F_1, and F_2 progeny.

Solution Strategies	Solution Steps
Evaluate	
2. Identify the critical information given in the problem.	2. The question supplies information on the phenotype of the parents, the F_1, and the F_2, and the proportions of each F_1 and F_2 phenotype in a cross and the reciprocal cross. Parents were pure-breeding black and pure-breeding brown, F_1 were all black, and F_2 were $\frac{3}{4}$ black and $\frac{1}{4}$ brown.
Deduce	
3. Determine the genotypes of parents and F_1 if possible.	3. Parents are pure-breeding and therefore must be homozygous. Each parent is homozygous for a different allele controlling fur color. The F_1 are heterozygous for these alleles (monohybrids).
4. Identify the key genetic relationship relevant to each cross, and list expected outcomes.	4. Parent cross: Genetic relationship 6. The progeny are heterozygotes and will express the dominant phenotype. F_1 cross: Genetic relationship 3. The progeny will be $\frac{3}{4}$ dominant, $\frac{1}{4}$ recessive.
Solve	
5. Apply the expected outcomes to the information provided to determine the dominance relationships and the genotypes of the parents, F_1, and F_2.	5. **Answers:** Black is dominant to brown, the parents are homozygous black (*BB*) and homozygous brown (*bb*); the F_1 are all heterozygotes (*Bb*); and the F_2 are $\frac{1}{4}$ black (*BB*) : $\frac{1}{2}$ black (*Bb*) : $\frac{1}{4}$ brown (*bb*).

5. Test the goodness of fit between observed values and expected outcomes.

Problems of this type typically provide you with the genotypes or phenotypes of parents *and* the types and numbers of progeny (*observed values*), and ask you to determine whether the observed values are a good statistical fit to the *expected outcomes* based on a specific *genetic hypothesis*. In Chapter 2, the genetic hypotheses are based on the laws of segregation and independent assortment, and the statistical test is the chi-square test.

Example Problem: A variety of pea plant called Blue Persian produces a tall plant with blue seeds. A second variety of pea plant called Spanish Dwarf produces a short plant with white seeds. One hypothesis to explain the difference between Blue Persians and Spanish Dwarfs is that they have different alleles at two independently assorting genes: one controls seed color and the other controls plant height. To test this hypothesis, the two varieties are crossed, and the resulting seeds are collected. All of the seeds are white,

F_2 Plant Phenotype	Number
Blue seed, tall plant	105
White seed, tall plant	350
Blue seed, short plant	35
White seed, short plant	110
	600

and when planted, they produce all tall plants. These tall F_1 plants are allowed to self-fertilize. The results for seed color and plant stature in the F_2 generation are as shown in the table. Examine the data in the table using the chi-square test, and determine whether they conform to expectations of the hypothesis.

Solution Strategies	Solution Steps
Evaluate	
1. Identify the topic this problem addresses, and explain the nature of the requested answer.	1. This problem tests your understanding of segregation, independent assortment, and your ability to test the goodness of fit between observed values and expected outcomes. The answer will be that the observed results are either consistent with the hypothesis or that they are statistically significantly different, so the hypothesis should be rejected. The answer will be based on the P value obtained from a chi-square analysis of the F_2 progeny. The P value will be determined using the chi-square value and 3 degrees of freedom in this analysis (4 phenotypic classes − 1).
2. Identify the critical information given in the problem.	2. Both parents are pure-breeding; the F_1 are white and tall; the F_2 are 105 blue and tall, 350 white and tall, 35 blue and short, 110 white and short. There are 600 total F_2 progeny, and the hypothesis is that two independently assorting genes are segregating in the cross: one controls plant height and one controls flower color.
Deduce	
3. Use the experimental hypothesis to assign the following: • Dominance relationships of the traits • Allele designations • Genotype of the parents and the F_1	3. According to the hypothesis, the parents are homozygous at both loci, so the F_1 are dihybrids. The F_1 are white and tall; therefore, white (*B*) is dominant to blue (*b*) and tall (*D*) is dominant to dwarf (*d*). The Blue Persian parent is *bbDD* and the Spanish Dwarf is *BBdd*. The F_1 are therefore *BbDd*.
4. Determine which key genetic relationships apply to the F_1 self-fertilization for each locus, and list the expected proportions of F_2 progeny phenotypes. **TIP:** Use the forked-line method to determine the types of F_2 progeny and the multiplication rule to determine the proportion of each type.	4. The F_2 are progeny of a dihybrid cross; therefore, genetic relationship 3 is relevant for both genetic loci. The expected proportion of progeny phenotypes for each phenotype is $\frac{3}{4}$ dominant : $\frac{1}{4}$ recessive. The expected proportion of progeny phenotype for both phenotypes is $\frac{3}{4}$ white $\times \frac{3}{4}$ tall = $\frac{9}{16}$ white, tall; $\frac{3}{4}$ white $\times \frac{1}{4}$ dwarf = $\frac{3}{16}$ white, dwarf; $\frac{1}{4}$ blue $\times \frac{3}{4}$ tall = $\frac{3}{16}$ blue, tall; $\frac{1}{4}$ blue $\times \frac{1}{4}$ dwarf = $\frac{1}{16}$ blue, dwarf.
Solve	
5. Calculate the expected number of F_2 progeny in each phenotypic class. **TIP:** Use the expected proportions of each F_2 class multiplied by the total number of F_2 progeny.	5. The expected numbers of each F_2 phenotype are $\frac{9}{6}$ white, tall $\times$ 600 = 337.5; $\frac{3}{16}$ white, dwarf $\times$ 600 = 112.5; $\frac{3}{16}$ blue, tall $\times$ 600 = 112.5; $\frac{1}{16}$ blue, dwarf $\times$ 600 = 37.5.

(continued)

Solution Strategies	Solution Steps
6. Calculate the chi-square value and the *P* value for 3 degrees of freedom. TIP: $\chi^2 = \sum \frac{(O-E)^2}{E}$ TIP: *P* value is taken from Table 2.4 in the textbook.	6. The chi-square value is 1.19, and there are 3 degrees of freedom. The *P* value is between 0.9 and 0.7. The calculations are $\left(\frac{[350-337.5]^2}{337.5}\right) + \left(\frac{[110-112.5]^2}{112.5}\right) + \left(\frac{[105-112.5]^2}{112.5}\right) + \left(\frac{[135-37.5]^2}{37.5}\right) = 1.19$ *P* value for $\chi^2 = 1.19$ and 3 df is between 0.9 and 0.7.
7. Apply the results of the calculations to the question to derive the answer to this problem. TIP: Reject the hypothesis only when the *P* value is less than 0.05.	7. **Answer:** The hypothesis cannot be rejected; therefore, the observed results are consistent with the hypothesis.

6. Calculate the probability of specific progeny distributions using binomial probability calculations.

This type of problem asks about the probability of specific combinations of progeny phenotypes where only two alternative phenotypes are possible. Classic examples include the probability of a certain number of boys and girls in a family or the probability of a certain number of affected and unaffected individuals in a family. In all cases, the probability of the two alternative phenotypes sums to 1, and the probability of each possible combination of phenotypes is determined by expanding the binomial equation.

Example Problem: Each of the seeds in bush bean pods are the product of an independent fertilization event. Green seed color is dominant to white seed color. If a heterozygous plant with green seeds self-fertilizes, what is the probability that five seeds in a single pod of the progeny plant will consist of two green and three white seeds?

Solution Strategies	Solution Steps
Solve 1. Identify the topic this problem addresses, and explain the nature of the requested answer.	1. This problem tests your understanding of binomial probability calculations. The answer will be a number indicated by a term in the expansion of the binomial $(a + b)^5$, where a = probability of green; b = probability of white.
2. Identify the critical information given in the problem.	2. The following information was provided: • Each seed in a pod is the result of an independent fertilization event. • Green is dominant to white. • The cross is a monohybrid self-cross.

(*continued*)

Solution Strategies	Solution Steps
Deduce	
3. Determine which genetic relationships are relevant to this question, and list the expected proportion of progeny phenotypes.	3. The cross is a self-fertilization of a monohybrid, therefore • Genetic relationship 3 is relevant to this question. • The expected proportion of green and white seeds is $\frac{3}{4}$ green : $\frac{1}{4}$ white.
Solve	
4. Indicate the term in the expanded binomial that is relevant to this question, and insert the values for a (probability of green) and b (probability of white) into the term. TIP: The probability of a phenotype is the same as the expected proportion of progeny from genetic relationship 3.	4. The expansion of the binomial $(a + b)^5$ is • $(a)^0(b)^5 + 5(a)(b)^4 + 10(a)^2(b)^3 + 10(a)^3(b)^2 + 5(a)^4(b) + (a)^5(b)^0$. • The term that applies to two green and three white seeds is $10(a)^2(b)^3$. • Plugging in $\frac{3}{4}$ for a and $\frac{1}{4}$ for b, this term is $10\left(\frac{3}{4}\right)^2\left(\frac{1}{4}\right)^3$.
5. Apply the results of the calculations to the question to derive the answer to this problem.	5. **Answer:** $10\left(\frac{3}{4}\right)^2\left(\frac{1}{4}\right)^3 = 0.088$; therefore, the probability of a pod from this plant having two green and three white seeds is 0.088 or 8.8 percent.

2.03 Solutions to End-of-Chapter Problems

1a. *Evaluate > Deduce > Solve: Dominant* and *recessive* refer to phenotypes affected by alternative alleles at a single genetic locus. Each phenotype is observed in an individual that is homozygous for one of the alleles, but only the dominant phenotype is observed in the heterozygote. In Mendel's pea plants, seeds with the genotype *GG* were yellow, *gg* were green, and *Gg* were yellow; therefore, yellow is dominant and green is recessive.

1b. *Evaluate > Deduce > Solve: Genotype* and *phenotype* are both properties of a cell or organism. Genotype refers to the genetic makeup of the organism. Phenotype is an observable property (or trait) of the organism, which is due to a combination of its genotype and its environment. We typically limit discussion of genotype and phenotype to only a few genes (genetic loci) and their associated phenotypes. The genotype indicates the alleles at those genetic loci, and the phenotype indicates the traits determined by those alleles.

1c. *Evaluate > Deduce > Solve: Homozygous* and *heterozygous* refer to the genotype of one genetic locus in a diploid organism. Homozygous individuals have two copies of the same allele, whereas heterozygotes have two different alleles.

1d. *Evaluate > Deduce > Solve:* In a *monohybrid cross*, both parents are heterozygous at a single genetic locus. The parents have the same genotype and are homozygous at all genes except for the single heterozygous locus. For a gene with two alleles that show simple dominance, the expected ratio of phenotypes in the progeny is expressed as fourths, with $\frac{3}{4}$ dominant and $\frac{1}{4}$ recessive. Two-thirds of the individuals showing the dominant phenotype will be heterozygous, and one-third are homozygous. In a *test cross*, one parent is homozygous for the allele associated with the recessive trait ("recessive allele"), and the other parent exhibits the dominant trait but its genotype at that locus is typically unknown. The expected ratio of phenotypes in the progeny is either 1 dominant : 1 recessive (unknown parent was heterozygous) or all dominant (unknown parent was homozygous).

1e. *Evaluate > Deduce > Solve:* Dihybrid and *trihybrid* crosses are similar to monohybrid crosses; both parents have the same genotype, and they are homozygous at almost all genetic loci. In a dihybrid cross, the parents are heterozygous at two genetic loci, whereas in a trihybrid cross they are heterozygous at three genetic loci. For a dihybrid cross in which the loci assort independently and both loci have two alleles that show simple dominance, the expected ratio of phenotypes in the progeny is expressed as sixteenths: $\frac{9}{16}$ show both dominant traits, $\frac{3}{16}$ show one dominant and one recessive trait, $\frac{3}{16}$ show the other dominant and recessive traits, and $\frac{1}{16}$ show both recessive traits. For the trihybrid cross, the ratio of phenotypes of the progeny will be expressed as sixty-fourths.

2. *Evaluate:* This is type 1a. The *BB* parent can make only *B* gametes. The *Bb* parent makes $\frac{1}{2}$ *B* and $\frac{1}{2}$ *b* gametes. *Deduce:* All the progeny will receive a *B* gamete from *BB*. One-half will receive a *B* gamete from *Bb*, and the other half will receive a *b* gamete from *Bb*. *Solve:* The genotypic ratio will be $\frac{1}{2}$ *Bb* and $\frac{1}{2}$ *BB*. The phenotypic ratio will be all dominant "B."

3. *Evaluate > Deduce:* This is type 1b. The *Aabb* parent makes two types of gametes, $\frac{1}{2}$ *Ab* and $\frac{1}{2}$ *ab*. The *aaBb* parent makes two types of gametes, $\frac{1}{2}$ *aB* and $\frac{1}{2}$ *ab*. *Deduce:* These gametes will combine as follows: $\frac{1}{2}$ *Ab* with $\frac{1}{2}$ *aB* to make $\frac{1}{4}$ *AaBb* progeny, $\frac{1}{2}$ *Ab* with $\frac{1}{2}$ *ab* to make $\frac{1}{4}$ *Aabb* progeny, $\frac{1}{2}$ *ab* with $\frac{1}{2}$ *aB* to make $\frac{1}{4}$ *aaBb* progeny, and $\frac{1}{2}$ *ab* with $\frac{1}{2}$ *ab* to make $\frac{1}{4}$ *aabb* progeny. *Solve:* The expected genotype ratio is $\frac{1}{4}$ *AaBb*, $\frac{1}{4}$ *Aabb*, $\frac{1}{4}$ *aaBb*, $\frac{1}{4}$ *aabb*. The expected phenotype ratio is $\frac{1}{4}$ "A" and "B," $\frac{1}{4}$ "A" and "b," $\frac{1}{4}$ "a" and "B," $\frac{1}{4}$ "a" and "b."

4. *Evaluate:* This is type 2a. I-1, I-2, II-2, and II-9 are black mice with white progeny. II-3 and II-5 are black mice with two black parents. II-6 is a black mouse with two black parents that has only black progeny. II-1, II-4, II-7, II-8, III-2, III-5, III-6, and III-13 are white. III-1, III-3, III-4, III-7, III-8, III-9, III-10, III-11, III-12, and III-14 are black mice with a white parent. *Deduce:* Black mice with white progeny must have a dominant and recessive allele, and so they are *Bb*. All white mice are homozygous recessive, so they must be *bb*. Black mice that did not mate must have at least one dominant allele and therefore are *B_*. II-6 mated with a white mouse and had only black offspring, suggesting but not proving that II-6 is *BB*, so she is listed at *B_*. Black mice with a white parent must have a dominant and recessive allele, so they are *Bb*. *Solve:* See figure.

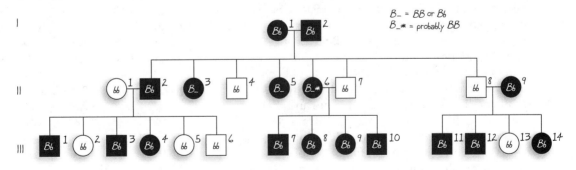

5. *Evaluate:* This is type 6. Each child will be either a boy or a girl, since the probability of either event is $\frac{1}{2}$. *Deduce:* Three sequences result in two girls and one boy; they are g-g-b, g-b-g, and b-g-g. *Solve:* The probability of each sequence is $\frac{1}{2} \times \frac{1}{2} \times \frac{1}{2}$, or $\left(\frac{1}{2}\right)^3$. The probability of two girls and one boy is therefore $3 \times \left(\frac{1}{2}\right)^3$, or 0.375. This problem could also be solved by expanding the binomial $(a + b)^n$ for $n = 3$, where *a* is the probability of a girl and *b* is the probability of a boy. The expansion is $\left(\frac{1}{2}\right)^3 + 3\left(\frac{1}{2}\right)^2\frac{1}{2} + 3\left(\frac{1}{2}\right)\left(\frac{1}{2}\right)^2 + \left(\frac{1}{2}\right)^3$. *Solve:* The formula for calculating the probability of 2 girls and 1 boy is $3\left(\frac{1}{2}\right)^2\left(\frac{1}{2}\right)^1 = 0.375$.

6a. *Evaluate:* This is type 1b. The parent genotypes are *AaBbCC* and *AABbCc*. *Deduce:* All the gametes from the *AaBbCC* parent will have the *C* allele. This parent is a dihybrid for the *A* and *B* loci, so it will produce four types of gametes; $\frac{1}{2}$ will carry *A* and $\frac{1}{2}$ will carry *a*. Among the gametes carrying *A*, $\frac{1}{2}$ will carry B $\left(\frac{1}{2} \times \frac{1}{2} = \frac{1}{4}$ of gametes will be ABC$\right)$ and $\frac{1}{2}$ will carry b $\left(\frac{1}{2} \times \frac{1}{2} = \frac{1}{4}$ of gametes will be *AbC*$\right)$. Among the gametes carrying *a*, $\frac{1}{2}$ will carry B $\left(\frac{1}{2} \cdot \frac{1}{2} = \frac{1}{4} aBC\right)$ and $\frac{1}{2}$ will carry *b* $\left(\frac{1}{2} \times \frac{1}{2} = \frac{1}{4}abc\right)$. All the gametes from the *AABbCc* parent will carry the *A* allele. This parent is a dihybrid for *B* and *C* loci, so it will produce four types of gametes; $\frac{1}{2}$ will carry *B* and $\frac{1}{2}$ will carry *b*. Among the gametes carrying *B*, $\frac{1}{2}$ will carry C $\left(\frac{1}{2} \times \frac{1}{2} ABC\right)$ and $\frac{1}{2}$ will carry c $\left(\frac{1}{2} \times \frac{1}{2} = \frac{1}{4} ABc\right)$. *Solve:* *AaBbCC* can produce four different gametes (*ABC*, *AbC*, *aBC*, and *abC*). *AABbCc* can produce four different gametes (*ABC*, *AbC*, *ABc*, and *Abc*).

6b. *Evaluate:* This is type 1b. The parent genotypes are *AaBbCC* and *AABbCc*. *Deduce:* Since both parents produce four types of gametes, the Punnett square should have 4 rows × 4 columns = 16 boxes to predict the progeny genotypes. The four gamete genotypes and frequencies from one parent are written to the left of the square, one for each row, and the four gamete genotypes and frequencies from the other parent are written across the top of the square, one for each column. The gamete genotypes and frequencies in each row are combined with the gamete genotypes and frequencies in each column using the multiplication rule to determine the progeny genotypes and their frequency. The progeny phenotypes are written below the genotypes in each box. *Solve:* See figure. There are 12 boxes corresponding to a phenotype of "ABC" $\left(\text{frequency} = \frac{12}{16}\right)$ and 4 boxes corresponding to "AbC" $\left(\text{frequency} = \frac{4}{16}\right)$.

6c. *Evaluate > Deduce:* This is type 1b. The parent genotypes are *AaBbCC* and *AABbCc*. *Solve:* See figure. Since both parents produce four gamete types, the forked-line drawing will have four gamete genotypes and their frequencies from one parent written in the first vertical column. Each of these genotypes has four lines forking to the right and listing the four gamete types and their frequencies from the other parent. This results in 16 lines, each line representing a different progeny genotype. The genotype and frequency of each progeny are determined by combining the gamete genotypes along each line and multiplying their frequencies. The progeny phenotype is listed to the right of each progeny genotype. There are 12 lines corresponding to a phenotype of "ABC" $\left(\text{frequency} = \frac{12}{16}\right)$ and 4 lines corresponding to a phenotype of "AbC" $\left(\text{frequency} = \frac{4}{16}\right)$.

7a. *Evaluate:* This is type 5. Table 2.4 shows the relationship between chi-square values and *P* values at different degrees of freedom. *Deduce:* For four degrees of freedom, a chi-square value of 7.83 is greater than 7.78 at the *P* value of 0.1, and it is less than 9.49 at the *P* value of 0.05. *Solve:* Therefore, the *P* value from this chi-square analysis is less than 0.1 but greater than 0.05.

7b. *Evaluate:* This is type 5. To reject the null hypothesis, the P value corresponding to the chi-square value must be less than 0.05. *Deduce > Solve:* Since the P value in this example is greater than 0.05, the null hypothesis cannot be rejected.

7c. *Evaluate:* This is type 5. To reject the null hypothesis, the chi-square value must correspond to a P value of less than 0.05. *Deduce > Solve:* The chi-square value must be greater than 14.07 to reject the null hypothesis for an experiment with 7 degrees of freedom.

8a. *Evaluate > Deduce > Solve:* False. The expected *phenotypic* ratio is $\frac{9}{16} : \frac{3}{16} : \frac{3}{16} : \frac{1}{16}$, assuming simple dominance and independent assortment of genetic loci ($AaBb \times AaBb$). There are nine different genotypes, and the genotypic ratio is $\frac{1}{16}$ ($AABB$) : $\frac{2}{16}$ ($AABb$) : $\frac{1}{16}$ ($AAbb$) : $\frac{2}{16}$ ($AaBB$) : $\frac{4}{16}$ ($AaBb$) : $\frac{2}{16}$ ($Aabb$) : $\frac{1}{16}$ ($aaBB$) : $\frac{2}{16}$ ($aaBb$) : $\frac{1}{16}$ ($aabb$).

8b. *Evaluate > Deduce > Solve:* True

8c. *Evaluate > Deduce > Solve:* True

8d. *Evaluate > Deduce > Solve:* False. The law of *independent assortment* is of primary importance in predicting the outcome of dihybrid and trihybrid crosses. The law of segregation is also necessary, but not sufficient.

8e. *Evaluate > Deduce > Solve:* False. Reciprocal crosses that produce identical results indicate that the traits being studied are autosomal.

8f. *Evaluate > Deduce > Solve:* False. The law of segregation predicts that she will produce two gamete genotypes with respect to her albinism gene at equal frequency.

8g. *Evaluate > Deduce > Solve:* True

8h-1. *Evaluate > Deduce > Solve:* True

8h-2. *Evaluate > Deduce > Solve:* False. There will be $\frac{1}{16}$ $AABB$, $\frac{1}{16}$ $AAbb$, $\frac{1}{16}$ $aaBB$, and $\frac{1}{16}$ $aabb$. All four genotypes will be true-breeding; therefore, $\frac{1}{4}$ of the progeny will be true-breeding.

8h-3. *Evaluate > Deduce > Solve:* False. Being "heterozygous at one or both loci" excludes only the progeny that are homozygous at both loci. In part b of this question, it was calculated that $\frac{1}{4}$ will be homozygous at both loci; therefore, $\frac{3}{4}$ will be heterozygous at one or both loci $\left(\frac{2}{16} AaBB, \frac{2}{16} AaBB, \frac{4}{16} AaBb, \frac{2}{16} Aabb, \text{ and } \frac{2}{16} aaBb\right)$.

9a. *Evaluate:* This is type 1a. Out of 38 progeny, 28 were purple $\left(0.74 \text{ or about } \frac{3}{4}\right)$ and 10 were white $\left(0.26 \text{ or about } \frac{1}{4}\right)$. *Deduce:* Since purple is dominant to white, purple-flowered plants are either WW or Ww. Self-fertilizing WW ($WW \times WW$) would give only purple-flowered progeny, whereas self-fertilizing Ww plants ($Ww \times Ww$) would give $\frac{3}{4}$ purple $\left(\frac{1}{4} WW \text{ and } \frac{1}{2} Ww\right)$ and $\frac{1}{4}$ white (ww). *Solve:* The results are consistent with the original purple-flowered plant having the genotype Ww.

9b. *Evaluate:* This is type 1a. Plants that "breed true" are homozygous at the genetic loci in question. Asking what proportion of the purple-flowered plants will breed true is asking what proportion of the plants are homozygous dominant (WW). *Deduce:* From part 9a, the purple flowered-plants would be $\frac{3}{4}$ of the progeny $\left(\frac{1}{4} WW \text{ and } \frac{1}{2} Ww\right)$. *Solve:* The proportion WW is $\frac{1}{4}$ out of $\frac{3}{4} = \frac{1}{3}$ would breed true.

10a. *Evaluate:* This is type 1a. All the F_1 of both crosses are mottled, and the leopard phenotype is not detected. The leopard phenotype reappears in the F_2. *Deduce:* Leopard behaves as a recessive phenotype. *Solve:* Therefore, mottled is dominant.

10b. *Evaluate:* This is type 1. Reciprocal crosses are crosses in which the genotypes of the male and female parents are reversed in one cross relative to the other. Reciprocal crosses are done to determine whether the trait is associated with a gene on a sex chromosome as opposed to an autosome. For autosomal inheritance, the results of the reciprocal crosses should be essentially identical. *Deduce:* In this problem, the first cross involved a leopard male and a mottled female, whereas the second cross involved a mottled male and a leopard female. *Solve:* The results from both crosses were the same, consistent with autosomal inheritance.

10c. *Evaluate:* This is type 1a. The F_1 are all mottled, and the F_2 show a 3:1 ratio of mottled to leopard. This is as expected for the F_2 progeny of a monohybrid cross. *Deduce:* If we assign L for mottled and l for leopard, the F_1 are Ll and the expected genotypic ratios among the F_2 are $\frac{1}{4} LL$, $\frac{1}{2} Ll$, and $\frac{1}{4} ll$. *Solve:* LL and ll are homozygous genotypes; therefore, $\frac{1}{4} + \frac{1}{4} = \frac{1}{2}$ are homozygous, and $\frac{1}{2}$ are heterozygous.

10d. *Evaluate:* This is type 1. Mottled frogs are either LL or Ll. So the crosses would be ones that distinguish LL from Ll. *Deduce:* A test cross is effective in revealing whether an individual with a dominant phenotype carries a recessive allele. If the mottled F_2 is homozygous dominant (genotype LL), then the test cross will yield all mottled progeny. If the mottled F_2 is heterozygous (Ll), then the test cross will produce $\frac{1}{2}$ mottled and $\frac{1}{2}$ leopard progeny. A backcross to one of its heterozygous (Ll) parents would also be revealing. If the mottled frog is LL, then the backcross will produce all mottled progeny. If the mottled frog is Ll, then the backcross will yield $\frac{3}{4}$ mottled and $\frac{1}{4}$ leopard. *Solve:* One cross would be a test cross and the other would be a backcross. The test cross would be the mottled F_2 to a true-breeding leopard individual. The backcross would be the mottled F_2 to one of its parents (both are heterozygotes).

11a. *Evaluate:* This is type 1a. The probability of the first offspring being pink is the same as the proportion of offspring expected to be pink. *Deduce > Solve:* Since black (P) is dominant to pink (p), only animals with the genotype pp will be pink. Crossing $Pp \times Pp$ yields $\frac{1}{4} PP$ (black), $\frac{1}{2} Pp$ (black), and $\frac{1}{4} pp$ (pink); therefore, $\frac{1}{4}$ are expected to be pink. If $\frac{1}{4}$ of the total are expected to be pink, then any individual pig, including the firstborn, has a $\frac{1}{4}$ probability of being pink.

11b. *Evaluate:* This is type 1a. The probability of the first and second offspring being black-skinned is the probability of the first being black multiplied by the probability of the second being black. *Deduce:* The probability of a black offspring is the same as the proportion of offspring expected to be black, which is $\frac{3}{4}$. *Solve:* The probability of the first and second offspring being black is $\frac{3}{4} \times \frac{3}{4} = \frac{9}{16}$.

11c. *Evaluate:* This is type 6. The probability of sequential events, two pink and one black, in any of the three possible orders (p-p-b, p-b-p, and b-p-p) can be found using the binomial equation. *Deduce:* Expand the binomial $(a + b)^3$, where a is the probability of being black $\left(\frac{3}{4}\right)$ and b is the probability of being pink $\left(\frac{1}{4}\right)$, and solve for the formula corresponding to two pink piglets and one black piglet. *Solve:* The formula for two pink and one black is $3\left(\frac{3}{4}\right)\left(\frac{1}{4}\right)^2$, which equals $\frac{9}{64}$.

12a. *Evaluate > Deduce > Solve:* The mode of fur color inheritance in these crosses is likely to be a single gene with two alleles controlling fur color, and black will be dominant to brown.

12b. *Evaluate:* This is type 4. There is one phenotype (fur color) and two variants (black and brown). The mode of inheritance should indicate how many genes are segregating and the dominance relationship for the alleles at each locus. *Deduce:* Assume that two alleles of a single gene are segregating in these crosses. If simple dominance applies, one of the phenotypes should be exhibited by individuals that are either homozygous dominant or heterozygous, and the other phenotype should be shown only in homozygous recessive individuals. *Solve:* The second cross is informative for dominance; since all the progeny of a black × brown are black, this indicates that black is dominant to brown. If we assign B to the black allele and b to the brown allele, the genotypes predicted by this hypothesis are bb for the brown male and BB for the black female in the second cross. Given that the brown male is bb and that the progeny from the first cross include both black and brown pups, the black female in that cross must be a Bb heterozygote. The expected ratio from a $bb × Bb$ cross is $\frac{1}{2}$ black: $\frac{1}{2}$ brown. The 9 black : 7 brown is close to this prediction (1.2:1).

13. *Evaluate:* This is type 5. The observed values were 55 round yellow, 51 round green, 49 wrinkled yellow, and 52 wrinkled green out of 207 progeny plants. The cross was a test cross of a dihybrid ($RrGr × rrgg$). The expected results assume R and G assort independently. *Deduce:* The chi-square value is calculated from the sum of (each observed value – the expected value)^2/expected value. The expected values for each phenotype are $\frac{1}{4}$ round, yellow ($RrGg$); $\frac{1}{4}$ round, green ($Rrgg$); $\frac{1}{4}$ wrinkled, yellow ($rrGg$); and $\frac{1}{4}$ wrinkled, green ($rrgg$). For 207 progeny, this is 51.75 for each class. *Solve:* The chi-square calculation is ($[55 – 51.75]^2$ / 51.75) + ($[51 – 51.75]^2$ / 51.75) + ($[49 – 51.75]^2$ / 51.75) + ($[52 – 51.75]^2$ / 51.75) = 0.362. The number of degrees of freedom, which is the number of progeny classes minus 1, is $4 – 1 = 3$. For 3 degrees of freedom, the chi-square value of 0.362 corresponds to a P value between 0.9 and 0.95; therefore, the hypothesis that R and G assort independently cannot be rejected.

14a. *Evaluate > Deduce > Solve:* The results suggest that the inheritance of color and fin shape are each due to segregation of two alleles at a single gene.

14b. *Evaluate:* Parts a and b are type 4. There are two phenotypes (color and fin shape) and two variants of each phenotype (black or gold color and single- or split-fin shape). The mode of inheritance should indicate how many genes are segregating and the dominance relationship for the alleles at each locus. *Deduce:* Assume that two alleles of a single gene are segregating in each of these crosses. If simple dominance applies, one of the phenotypes should be exhibited by individuals that are either homozygous dominant or heterozygous, and the other phenotype should be shown only in homozygous recessive individuals. *Solve:* Consider color first. Black × gold produced all gold, indicating that gold is dominant to black. Assigning b for the black allele and B for the gold allele, the black male would be bb. Since all the F_1 are gold, the female would be BB and all the F_1 Bb. The black male parent × F_1 gold female would be $bb × Bb$, which is predicted to yield black and gold at 1:1. This is very close to the observed results (34 black : 32 gold). For fin shape, single × split produced only split fin, indicating that split fin is dominant to single. Assigning s for single and S for split, the single-finned male parent would be ss. Since all the F_1 are split, the female parent would be SS and all the F_1 Ss. The single male × the split F_1 female would be $ss × Ss$, which is predicted to yield single and split at 1:1. This is very close to the observed results (41 split : 39 single).

14c. *Evaluate:* This is type 5. The observed values for the black male × gold F_1 were 34 black and 32 gold. The observed values for the single male × split F_1 were 41 split and 39 single. *Deduce:* The chi-square value is calculated from the sum of (each observed value – the expected value)^2/expected value. The expected values for gold and black are $\frac{1}{2}$ gold and $\frac{1}{2}$ black, which is 33 each. The expected values for single and split fin are $\frac{1}{2}$ single and $\frac{1}{2}$ split, which is 40 each. *Solve:* The chi-square calculation for color is $\left(\frac{[34 - 33]^2}{33}\right) + \left(\frac{[32 - 33]^2}{33}\right) = 0.061$. The number of degrees of freedom for color is 2 classes $– 1 = 1$. The P value for 0.061 and 1 degree of freedom is between 0.7 and 0.9. The chi-square calculation for fin shape is $\left(\frac{[41 - 40]^2}{40}\right) + \left(\frac{[39 - 40]^2}{40}\right) = 0.05$. The number of degrees of freedom for fin shape is also 1, and the P value for 0.05 and 1 degree of freedom is between 0.7 and 0.9. Neither P value is less than 0.05, which indicates that the one gene with two allele hypotheses for color and fin shape cannot be rejected.

15a.   *Evaluate:* This is type 2. The affected individuals have unaffected parents. Both males and females are affected. *Deduce:* The trait is recessive. The trait is most likely autosomal, since it appears in males and females at the same frequency. *Solve:* Autosomal recessive.

15b.   *Evaluate:* This is type 2. The affected individuals have unaffected parents. *Deduce:* The parents must be heterozygotes. Since albinism is recessive, we assign A to normal pigmentation and a to the albino allele, and Aa is the genotype for I-1, I-2, and I-3. *Solve:* Assuming autosomal recessive, all three individuals in generation I are Aa, where A corresponds to the allele for normal pigmentation and a is the allele for albinism.

15c.   *Evaluate:* This is type 6. The probability that I-1 and I-2 mate and have four children with *any* outcome other than 3 normal and 1 albino is the *sum of the probabilities of all possible outcomes* minus the *probability of 3 normal, 1 albino*. *Deduce:* The sum of the probabilities of all possible outcomes is 1. The probability of 3 normal and 1 albino is calculated by expanding the binomial $(a + b)^4$, where a is the probability of normal and b is the probability of albino, and then choosing the term corresponding to 3 normal and 1 albino. The expanded binomial is $(a)^4 + 4(a)^3(b) + 6(a)^2(b)^2 + 4(a)(b)^3 + (b)^4$, where $4(a)^3(b)$ corresponds to 3 normal and 1 albino.

Since the parents are both Aa, the probability of normal is $\frac{3}{4}$ and the probability of albino is $\frac{1}{4}$ *Solve:* The probability of 3 normal, 1 albino is $4\left(\frac{3}{4}\right)^3\left(\frac{1}{4}\right) = 0.422$. The probability of any outcome other than 3 normal, 1 albino is therefore $1 - 0.422 = 0.578$.

15d.   *Evaluate:* This is type 2. *Deduce:* I-3 is an unaffected parent of a child with a recessive trait *Solve:* The probability that I-3 is a heterozygous carrier for the albino allele is 1.0.

15e.   *Evaluate:* This is type 6. The probability that *any* of the other four children are carriers is the *sum of the probabilities of all possible outcomes* minus the *probability that none* of the other four children are carriers. *Deduce:* The sum of the probabilities of all possible outcomes is 1. The probability that none of the other four children are carriers is obtained by expanding the binomial $(a + b)^4$ and choosing the term corresponding to 0 carriers and 4 noncarriers. Since the pedigree shows that none of the four are albinos, the probability of being albino is zero, the probability of being a carrier is $\frac{2}{3}$, and the probability of being a noncarrier is $\frac{1}{3}$. *Solve:* The expanded binomial is $\left(\frac{2}{3}\right)^4 + 4\left(\frac{2}{3}\right)^3\left(\frac{1}{3}\right) + 6\left(\frac{2}{3}\right)^2\left(\frac{1}{3}\right)^2 + 4\left(\frac{2}{3}\right)\left(\frac{1}{3}\right)^3 + \left(\frac{1}{3}\right)^4$, where the term corresponding to zero carriers and four noncarriers is $\left(\frac{1}{3}\right)^4$. Using these numbers, the probability of the four being noncarriers is $\left(\frac{1}{3}\right)^4 = 0.012$. The probability that any of the four are carriers is therefore $1 - 0.012 = 0.988$.

16a.   *Evaluate:* This is type 1. One parent is a pure-breeding strain producing yellow, wrinkled seeds. Yellow is dominant and wrinkled is recessive. The other parent is a pure-breeding strain producing green, round seeds. Green is recessive and round is dominant. *Deduce:* Designate G for yellow, g for green, R for round, and r for wrinkled. The yellow, wrinkled plant's genotype is $GGrr$ and produces only one type of gamete, Gr. The green, round plant's genotype is $ggRR$ and produces only one type of gamete, gR. The F_1, therefore, will be $GgRr$ and will produce four types of gametes, GR, Gr, gR, and gr, each at a frequency of $\frac{1}{4}$. *Solve:* See figure. The Punnett square representing the self-fertilization of the F_1 is that of a dihybrid cross with four columns and four rows, one column and row for each of the four gamete

	$\frac{1}{4}$ GR	$\frac{1}{4}$ Gr	$\frac{1}{4}$ gR	$\frac{1}{4}$ gr
$\frac{1}{4}$ GR	$\frac{1}{16}$ GGRR ○ Yellow, round	$\frac{1}{16}$ GGRr ○ Yellow, round	$\frac{1}{16}$ GgRR ○ Yellow, round	$\frac{1}{16}$ GgRr ○ Yellow, round
$\frac{1}{4}$ Gr	$\frac{1}{16}$ GGRr ○ Yellow, round	$\frac{1}{16}$ GGrr ⊛ Yellow, wrinkled	$\frac{1}{16}$ GgRr ○ Yellow, round	$\frac{1}{16}$ Ggrr ⊛ Yellow, wrinkled
$\frac{1}{4}$ gR	$\frac{1}{16}$ GgRR ○ Yellow, round	$\frac{1}{16}$ GgRr ○ Yellow, round	$\frac{1}{16}$ ggRR ● Green, round	$\frac{1}{16}$ ggRr ● Green, round
$\frac{1}{4}$ gr	$\frac{1}{16}$ GgRr ○ Yellow, round	$\frac{1}{16}$ Ggrr ⊛ Yellow, wrinkled	$\frac{1}{16}$ ggRr ● Green, round	$\frac{1}{16}$ ggrr ⊛ Green, wrinkled

genotypes. The progeny genotypes will be the union of the parental genotypes in each column and row, and their frequency will be the product of the gamete frequencies.

16b. *Evaluate:* This is type 1a. The distribution of the dominant and recessive phenotypes at each locus will be as expected for a monohybrid cross. *Deduce:* $Gg \times Gg$ is expected to yield $\frac{1}{4}$ GG, $\frac{1}{2}$ Gg, and $\frac{1}{4}$ gg, which is $\frac{3}{4}$ yellow and $\frac{1}{4}$ green. $Rr \times Rr$ is expected to yield $\frac{1}{4}$ RR, $\frac{1}{2}$ Rr, and $\frac{1}{4}$ rr, which is $\frac{3}{4}$ round and $\frac{1}{4}$ wrinkled. *Solve:* Of the F_2 progeny, $\frac{3}{4}$ will have yellow seeds, $\frac{1}{4}$ will have green seeds, $\frac{3}{4}$ will have round seeds, and $\frac{1}{4}$ will have wrinkled seeds.

16c. *Evaluate:* This is type 1b. The F_2 are from a self-fertilization of the F_1, which is a dihybrid. The phenotypic distribution among the F_2 is the classic distribution for a dihybrid cross and can also be taken directly from the Punnett square (see answer to part a). *Deduce:* The phenotypic classes are in the ratio of 9:3:3:1, where $\frac{9}{16}$ has both dominant phenotypes (yellow and round), $\frac{3}{16}$ has one dominant and one recessive phenotype (yellow and wrinkled), $\frac{3}{16}$ has one dominant and one recessive phenotype (green and round), and $\frac{1}{16}$ has both recessive phenotypes (green and wrinkled). *Solve:* From the Punnett square, 9 of 16 boxes have genotypes that correspond to yellow, round $\left(\frac{9}{16}\right)$; three to yellow, wrinkled $\left(\frac{3}{16}\right)$; three to green, round $\left(\frac{3}{16}\right)$; and one to green, wrinkled $\left(\frac{3}{16}\right)$.

17. *Evaluate:* This is type 1b. The F_1 parent is $GgRr$. The green, round parent is true-breeding for both traits, so its genotype must be $ggRR$. *Deduce:* The F_1 parent produces four types of gametes: GR, Gr, gR, and gr, each at a frequency of $\frac{1}{4}$. The green, round parent produces only one type of gamete: gR. *Solve:* Place the gR gamete at the start of the diagram and assign it a frequency of 1.0 (the only gamete being produced), and draw four lines forking away from the gR gamete, with each line leading to one of the

F_1 parent's gametes. Assign each F_1 gamete a frequency of $\frac{1}{4}$. The genotype and frequency of each progeny are determined by combining the gamete genotypes along each line and multiplying their frequencies. The four possible combinations of gametes are as follows: 1 gR and $\frac{1}{4}$ $GR = \frac{1}{4}$ $GgRR$ (yellow, round); 1 gR and $\frac{1}{4}$·Gr $= \frac{1}{4}$ $GgRr$ (yellow, round); 1 gR and $\frac{1}{4}$ $gR = ggRR$ (green, round); and 1 gR and $\frac{1}{4}$ $gr = ggRr$ (green, round). The phenotypic distribution is $\frac{1}{4} + \frac{1}{4} = \frac{1}{2}$ yellow, round; and $\frac{1}{4} + \frac{1}{4} = \frac{1}{2}$ green, round.

18a. *Evaluate:* This is type 1b. The true-breeding axial, constricted parent has the genotype $TTcc$ and the true-breeding terminal, inflated parent has the genotype $ttCC$. *Deduce:* The F_1 are $TtCc$ dihybrids. According to Figure 2.1, flower pattern and pod form assort independently; therefore, self-fertilizing the F_1 gives the standard distribution of phenotypes for a dihybrid cross : $\frac{9}{16}$ dominant for both flower pattern and pod form : $\frac{3}{16}$ dominant for flower pattern and recessive for pod form : $\frac{3}{16}$ recessive for flower pattern and dominant for pod form : $\frac{1}{16}$ recessive for flower pattern and pod form. *Solve:* The answer is $\frac{9}{16}$ axial inflated : $\frac{3}{16}$ axial constricted : $\frac{3}{16}$ terminal inflated : $\frac{1}{16}$ terminal constricted.

18b. *Evaluate:* This is type 1b. The F_2 plants that have terminal flowers will be self-fertilized to produce the F_3 progeny. The result of self-fertilizing these plants will depend on the genotype of each plant and the relative abundance of each genotype. *Deduce:* The genotypic distribution of F_2 plants is $\frac{1}{16}$ $TTCC$: $\frac{2}{16}$ $TTCc$: $\frac{1}{16}$ $TTcc$: $\frac{2}{16}$ $TtCC$: $\frac{4}{16}$ $TtCc$: $\frac{2}{16}$ $Ttcc$: $\frac{1}{16}$ $ttCC$: $\frac{2}{16}$ $ttCc$: $\frac{1}{16}$ $ttcc$. The genotypes of F_2 with terminal flowers are $ttCC$, $ttCc$, and $ttcc$, which accounts for $\frac{4}{16}$ of the F_2. Among the F_2 progeny with terminal flowers, $\frac{1}{4}$ $\left(\frac{1}{16}\text{ out of }\frac{4}{16}\right)$ are $ttCC$, $\frac{1}{2}$ $\left(\frac{2}{16}\text{ out of }\frac{4}{16}\right)$ are $ttCc$, and $\frac{1}{4}$ $\left(\frac{1}{16}\text{ out of }\frac{4}{16}\right)$ are *ttcc*. When these genotypes are self-fertilized, (1) *ttCC* produces $\frac{1}{4}$ of the F_3, which are all terminal, inflated (*ttCC*) progeny; (2) *ttCc* produces $\frac{1}{2}$ of the F_3, of which $\frac{3}{4}$ are terminal, inflated $\left(\frac{1}{4}\ ttCC\text{ and }\frac{1}{2}\ ttCc\right)$ and $\frac{1}{4}$ are terminal, constricted (*ttcc*); (3) *ttcc* produces $\frac{1}{4}$ of the F_3, all of which are terminal, constricted (*ttcc*). **Solve:** The phenotypic distribution among the F_3 is $\frac{1}{4} + \left(\frac{1}{2} \times \frac{3}{4}\right) = \frac{5}{8}$ terminal inflated and $\left(\frac{1}{2} \times \frac{1}{4}\right) + \frac{1}{4} = \frac{3}{8}$ terminal, constricted.

18c.   *Evaluate:* This is type 1b. The F_1 from the original cross is a dihybrid, *TtCc*. A terminal, constricted plant has the genotype *ttcc*. **Deduce:** *TtCc* × *ttcc* is a test cross of a dihybrid, which yields $\frac{1}{4}$ *TtCc* (axial, inflated), $\frac{1}{4}$ *Ttcc* (axial, constricted), $\frac{1}{4}$ *ttCc* (terminal, inflated), and $\frac{1}{4}$ *ttcc* (terminal, constricted). **Solve:** The expected distribution is $\frac{1}{4}$ axial, inflated; $\frac{1}{4}$ axial, constricted; $\frac{1}{4}$ terminal, inflated; and $\frac{1}{4}$ terminal, constricted.

18d.   *Evaluate:* This is type 1b. The plants from part c that have terminal flowers will be self-fertilized. The result of self-fertilizing these plants will depend on the genotype of each plant and the relative abundance of each genotype. **Deduce:** $\frac{1}{2}$ of the progeny from the cross in part c have terminal flowers. $\frac{1}{2}$ of those are *ttCc* and $\frac{1}{2}$ are *ttcc*. If these are allowed to self-fertilize to produce a new generation, then (1) *ttCc* will produce $\frac{1}{2}$ of the progeny, of which $\frac{3}{4}$ will be terminal, inflated $\left(\frac{1}{4}\ ttCC\text{ and }\frac{1}{2}\ ttCc\right)$ and $\frac{1}{4}$ will be terminal, constricted (*ttcc*); and (2) *ttcc* will produce $\frac{1}{2}$ of the progeny, all of which will be terminal, constricted (*ttcc*). **Solve:** The phenotypic distribution of this generation is $\left(\frac{1}{2} \times \frac{3}{4}\right) = \frac{3}{8}$ terminal, inflated and $\left(\frac{1}{2} \times \frac{1}{4}\right) + \frac{1}{2} = \frac{5}{8}$ terminal, constricted.

19a.   *Evaluate:* The probability of each die showing any number is $\frac{1}{6}$, since there are six faces on each die. **Deduce:** Two dice faces can show a total of four spots in three ways: 1 and 3, 3 and 1, and two 2's. The probability of the 1 and 3 combination is $\left(\frac{1}{6}\right) \times \left(\frac{1}{6}\right)$, or $\frac{1}{36}$. The probability of 3 and 1 is also $\frac{1}{36}$, as is the probability of 2 and 2. **Solve:** The probability of either 1 and 3 or 3 and 1 or 2 and 2 is $\frac{1}{36} + \frac{1}{36} + \frac{1}{36} = \frac{3}{36} = 0.083$.

19b.   *Evaluate:* The probability of each die showing any number is $\frac{1}{6}$, since each die has six faces. **Deduce:** Two dice faces can show a total of seven spots in six ways: 1 and 6, 6 and 1, 2 and 5, 5 and 2, 3 and 4, and 4 and 3. **Solve:** Each combination has a probability of $\frac{1}{6} \times \frac{1}{6} = \frac{1}{36}$, so the probability that one of these combinations will occur is $6 \times \frac{1}{36} = \frac{6}{36} = 0.167$.

19c.   *Evaluate:* The probability that the dice will show more than 5 spots is the same as the probability that they will show 6, 7, 8, 9, 10, 11, or 12 spots. It is also the same as the probability that they won't show 2, 3, 4, or 5 spots. The second expression is simpler to calculate since it requires that we calculate those four probabilities (instead of the seven probabilities needed for the first expression) and then subtract the sum of those probabilities from 1. **Deduce:** There is only one way to total 2, two ways to total 3, three ways to total 4, and four ways to total 5. The sum of these probabilities is therefore $\left(\frac{1}{36}\right) + 2\left(\frac{1}{36}\right) + 3\left(\frac{1}{36}\right) + 4\left(\frac{1}{36}\right) = \frac{10}{36} = 0.28$. **Solve:** The probability of any total other than 2, 3, 4, or 5 is $1 - 0.28 = 0.72$.

19d.   *Evaluate:* The dice are just as likely to total an odd number as an even number. **Deduce:** Odd-number possibilities are 3 (two ways), 5 (four ways), 7 (six ways), 9 (four ways), and 11 (two ways). **Solve:** Thus the probability of an odd-number combination is $\frac{1}{36}$ (the probability of any two-dice combination) times the 18 ways of those combinations resulting in an odd number. $\frac{1}{36} \times 18 = \frac{18}{36} = \frac{1}{2}$. To check this, calculate the probability of an even number, which is also 0.5.

20.   *Evaluate:* This is type 5. The corn plant cross from Experimental Insight 2.1 is a monohybrid cross, *Ww* × *Ww*, where *W* is the allele for the dominant, yellow kernel color and *w* is the allele for the recessive, white kernel color. *Deduce:* The expected distribution of yellow to white in this cross is $\frac{3}{4}$ yellow $\left(\frac{1}{4}WW + \frac{1}{2}Ww\right)$ and $\frac{1}{4}$ white $\left(\frac{1}{4}ww\right)$. Out of 9882 total kernels, you would expect $\frac{3}{4} \times 9882$ = 7411.5 to be yellow and $\frac{1}{4} \times 9882$ = 2470.5 to be white. The chi-square calculation is, therefore, $(7506 - 7411.5)^2 / 7411.5 + (2376 - 2470.5)^2 / 2470.5 = 4.81$. There is 1 degree of freedom (2 phenotypic classes minus 1). In Table 2.4, in the row corresponding to 1 degree of freedom, 4.81 is a chi-square value that falls between *P* values of 0.05 and 0.01. *Solve:* A *P* value of less than 0.05 indicates that the observed data differ from expectations by a statistically significant amount, and the hypothesis should be rejected.

21a. *Evaluate:* This is type 2a. Assuming autosomal dominant inheritance, any member of the pedigree that does not show the trait is *bb*, and those showing the trait are either *BB* or *Bb*. *Deduce > Solve:* Homozygous recessive individuals are I-1, II-2, and II-4. I-2 shows the trait and therefore has at least one dominant allele. If I-2 was homozygous dominant, all of his children would show the trait; but this is not the case, so I-2 is *Bb*. Since II-1 is *bb*, everyone in generation II must have a *b* allele. The affected individuals in II (II-1 and II-3) must have a *B* allele and a *b* allele. Therefore, II-1 and II-3 are *Bb*.

21b. *Evaluate:* This is type 2b. Assuming autosomal recessive inheritance, any member of the pedigree that shows the trait is *bb*, and those that do not are either *BB* or *Bb*. *Deduce > Solve:* Homozygous recessive individuals are I-2, II-1, and II-3. I-1 does not show the trait, so she is either *BB* or *Bb*. If she was *BB*, then none of her children would show the trait, but two do. Therefore, I-1 must carry a *b* allele and is *Bb*. Since II-2 is *bb*, everyone in generation II must have a *b* allele. II-2 and II-4 do not show the trait, so they must have a *B* allele. Therefore, II-2 and II-4 are *Bb*.

21c. *Evaluate:* This is type 2. First, consider the two possible inheritance patterns: autosomal dominant and autosomal recessive. For autosomal dominant inheritance, II-4 must be *bb*. For autosomal recessive inheritance, II-4 must be *Bb*. In either case, the genotype of II-4's mate will be important to consider when predicting generation III phenotypes. *Deduce:* Assuming autosomal dominant inheritance and II-4's mate being unaffected, II-4 and II-5 would be homozygous recessive (*bb*) and so all their children will be unaffected. Assuming autosomal dominant inheritance and II-4's mate being affected and having the genotype *Bb*, then the mating is *bb* × *Bb* and $\frac{1}{2}$ of the progeny is expected to be affected. Assuming autosomal recessive inheritance and II-4's mate being an unaffected noncarrier (*BB*), then the mating is *Bb* × *BB* and none of the progeny will be affected. Assuming that II-4's mate is a carrier, then the mating is *Bb* × *Bb* and $\frac{1}{4}$ of the progeny are expected to be affected. Assuming that II-4's mate is affected, then the mating is *Bb* × *bb* and $\frac{1}{2}$ of the progeny are expected to be affected. *Solve:* This question involves several variations. Assuming autosomal dominant inheritance and that II-4's mate is unaffected, the pedigree would look like Solution 1 in the figure. Assuming autosomal dominant inheritance and that II-4's mate is affected *and* has the genotype *Bb*, then the pedigree might look like Solution 2. Assuming autosomal recessive inheritance and that II-4's mate is an unaffected noncarrier, the pedigree would look like Solution 1. If we assume that II-4's mate is a carrier, then the pedigree might look like Solution 3. If we assume that II-4's mate is affected, then the pedigree might look like that in Solution 2.

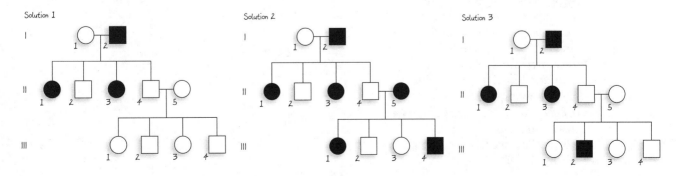

22a. *Evaluate:* This is type 6. The probability of the green plant producing 3 green and 3 white seeds is calculated by expanding the binomial $(a + b)^6$, where a is the probability of being green and b is the probability of being white. The cross is a monohybrid cross. *Deduce:* The expanded binomial is $(a)^6 + 6(a)^5(b) + 15(a)^4(b)^2 + 20(a)^3(b)^3 + 15(a)^2(b)^4 + 6(a)(b)^5 + (b)^6$, where $20(a)^3(b)^3$ corresponds to the probability of 3 green and 3 white. In a monohybrid cross, the probability of the dominant phenotype (green) is ¾ and the probability of the recessive phenotype (white) is $\frac{1}{4}$. *Solve:* The probability of 3 green and 3 white seeds is $20\left(\frac{3}{4}\right)^3\left(\frac{1}{4}\right)^3 = 0.132$.

22b. *Evaluate:* See part a. *Deduce:* The term in the binomial expansion in part a that corresponds to 6 green and 0 white is $(a)^6$. *Solve:* Since $a = \frac{3}{4}$, this term is 0.178.

22c. *Evaluate:* See part a. The probability of having *at least one white seed* is the same as the *sum of all probabilities* minus the probability of having *no white seeds*. *Deduce:* The probability of no white seeds is the same as the probability of all green seeds, which was calculated to be 0.178 in part b. *Solve:* The probability of at least one white seed is $1 - 0.178 = 0.822$.

23a. *Evaluate:* This is type 1b. Alleles at these loci assort independently. *Deduce:* All gametes produced by *AABbCcDd* will contain *A*. For each of the other three loci, half of the gametes will contain the dominant allele and half will contain the recessive allele. *Solve:* Create a forked-line diagram to obtain the following: *ABCD, ABCd, ABcD, ABcd, AbCD, AbCd, AbcD, Abcd.*

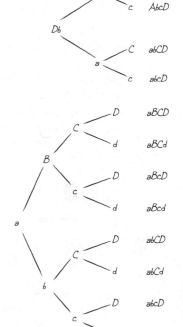

23b. *Evaluate:* This is type 1b. Alleles at these loci assort independently. *Deduce:* All gametes produced by *AabbCcDD* will contain *D* and *b*. For each of the other two loci, half of the gametes will contain the dominant allele and half will contain the recessive allele. *Solve:* Create a forked-line diagram to obtain the following: *AbCD, AbcD, abCD, abcD.*

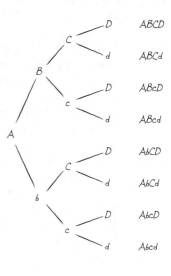

23c. *Evaluate:* This is type 1b. Alleles at these loci assort independently. *Deduce:* For each locus, half the gametes produced by *AaBbCcDd* will contain the dominant allele and half will contain the recessive allele. *Solve:* Create a forked-line diagram to obtain the following: *ABCD, ABCd, ABcD, ABcd, AbCD, AbCd, AbcD, Abcd, aBCD, aBCd, aBcD, aBcd, abCD, abCd, abcD, abcd.*

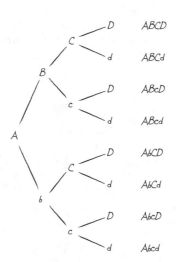

23d.   *Evaluate:* This is type 1b. Alleles at these loci assort independently. *Deduce:* In *AabbCCdd*, only the *A* locus is heterozygous. Therefore, half of the gametes will contain the *A* allele and half will contain the *a* allele. All gametes will contain the *b*, *C*, and *d* alleles. *Solve:* The two gamete genotypes will be *AbCd* and *abCd*.

24a.   *Evaluate:* This is type 1b. The probability of *A_B_C_D_* is the probability of *A_* × probability of *B_* × probability of *C_* × the probability of *D_*. *Deduce:* The probability of *A_* is 1.0 because one parent is *AA*, and so all its gametes contain the *A* allele. The probability of *B_* is $\frac{3}{4}$ because each parent is *Bb*, and in a *Bb* × *Bb* cross, $\frac{3}{4}$ are *B_*. The probability for *C_* and *D_* is the same as for *B_*. *Solve:* The probability of *A_B_C_D_* is $1 \times \frac{3}{4} \times \frac{3}{4} \times \frac{3}{4} = 0.422$.

24b.   *Evaluate:* This is type 1b. The probability of *AabbCcDd* is the probability of *Aa* × the probability of *bb* × the probability of *Cc* × the probability of *Dd*. *Deduce:* The probability of *Aa* is $\frac{1}{2}$, the probability of *bb* is $\frac{1}{4}$, the probability of *Cc* is $\frac{1}{2}$, and the probability of *Dd* is $\frac{1}{2}$. *Solve:* The probability of *AabbCcDd* is $\frac{1}{2} \times \frac{1}{4} \times \frac{1}{2} \times \frac{1}{2} = 0.0313$.

24c.   *Evaluate:* This is type 1b. The problem tests your understanding of how to find the probability of a phenotype that is identical to either parent. The probability that a progeny would have the same phenotype as either parent is the probability that the progeny will have the genotype *A_B_C_D_*. The probability of this genotype is the probability of *A_* × the probability of *B_* × the probability of *C_* × the probability of *D_*. *Deduce:* The probabilities are as follows: *A_* is 1, *B_* is $\frac{3}{4}$, *C_* is $\frac{3}{4}$, and *D_* is $\frac{3}{4}$. *Solve:* The probability of *A_B_C_D_* is $1 \times \frac{3}{4} \times \frac{3}{4} \times \frac{3}{4} = \frac{27}{64} = 0.422$.

24d.   *Evaluate:* This is type 1b. The probability of *A_B_ccdd* is the probability of *A_* × the probability of *B_* × the probability of *cc* × the probability of *dd*. *Deduce:* The probability of *A_* is 1, the probability of *B_* is $\frac{3}{4}$, the probability of *cc* is $\frac{1}{4}$, and the probability of *dd* is $\frac{1}{4}$. *Solve:* The probability of *A_B_ccdd* is $1 \times \frac{3}{4} \times \frac{1}{4} \times \frac{1}{4} = 0.0469$.

25.   *Evaluate:* This is type 2. I-1, I-2, and II-4 exhibit dominant forms of both traits. II-1 exhibits the recessive form of both traits. II-2 and II-3 exhibit the recessive form for thumb bending and the dominant form for earlobe attachment. *Deduce:* I-1 and I-2 show both dominant traits and therefore have a dominant allele at each locus. Since their daughter, II-1, shows both recessive traits, she must be homozygous recessive (*hhee*) and they must be heterozygous at both loci (*HhEe*). II-2 and II-3 show the recessive trait for thumb bending and so must be homozygous recessive for that locus (*ee*). They both show the dominant trait for earlobe attachment, and so are either homozygous dominant or heterozygous at that locus (*H_*). II-4 shows both dominant traits and so must have at least one dominant allele at both loci (*H_E_*). *Solve:* I-1 and I-2 are *HhEe*. II-1 is *hhee*. II-2 and II-3 are *hhE_*. II-4 is *H_E_*.

26.   *Evaluate:* This is type 1b. This is a dihybrid cross in which 3200 progeny are examined. *Deduce:* The expected distribution from a dihybrid cross is $\frac{9}{16}$ with both dominant phenotypes, $\frac{3}{16}$ with one dominant and the other recessive, $\frac{3}{16}$ with one recessive and one dominant, and $\frac{1}{16}$ with both recessive phenotypes. *Solve:* One expects $\frac{9}{16}$ of 3200 or 1800 to have full wings and gray bodies, $\frac{3}{16}$ of 3200 or 600 to have full wings and ebony bodies, $\frac{3}{16}$ of 3200 or 600 to have vestigial wings and gray bodies, and $\frac{1}{16}$ of 3200 or 200 to have vestigial wings and ebony bodies.

27a.   *Evaluate:* This is type 1b. The parents are true-breeding and so homozygous at each locus. The cross is $TTrrGG \times ttRRgg$. *Deduce:* The F_1 are trihybrids ($TtRrGg$). The distribution of the phenotypes among the F_2 can be determined using the forked-line method, where the probability of the dominant phenotype for each trait is $\frac{3}{4}$ and the probability of the recessive phenotype is $\frac{1}{4}$. *Solve:* See figure. The F_1 are all tall, round, and yellow. The F_2 are expected to be $\frac{27}{64}$ tall, round, yellow; $\frac{9}{64}$ tall, round, green; $\frac{9}{64}$

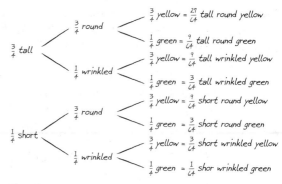

tall, wrinkled, yellow; $\frac{9}{64}$ short, round, yellow; $\frac{3}{64}$ tall, wrinkled, green; $\frac{3}{64}$ short, round, green; $\frac{3}{64}$ short, wrinkled, yellow; and $\frac{1}{64}$ short, wrinkled, green.

27b.   *Evaluate:* This is type 1b. The parents are true-breeding and so homozygous at each locus. The cross is $TTrrGG \times ttRRgg$. *Deduce:* The F_1 are heterozygous at each locus. Tall and yellow are dominant phenotypes, whereas wrinkled is recessive. The expected proportion of F_2 with a dominant phenotype at one locus is $\frac{3}{4}$, and the expected proportion with a recessive phenotype at one locus is $\frac{1}{4}$. *Solve:* The expected proportion with tall, wrinkled, and yellow is $\frac{3}{4} \times \frac{1}{4} \times \frac{3}{4} = \frac{9}{64}$. The expected proportion of $ttRRGg$ is the expected frequency of $tt \left(\frac{1}{4}\right) \times$ the expected frequency of $RR \left(\frac{1}{4}\right) \times$ the expected frequency of $Gg \left(\frac{1}{2}\right) = \frac{1}{32}$.

27c.   *Evaluate:* This is type 1b. The proportion of the round, green seed producers that breed true is the proportion with the genotypes $RRggTT$ or $RRggtt$. *Deduce:* The round seed producers are $\frac{1}{3}$ RR and $\frac{2}{3}$ Rr; therefore, $\frac{1}{3}$ of plants that produce round seeds will breed true for round seeds. Green seed producers are gg; therefore, all will breed true. *Solve:* Combining both loci, $\frac{1}{3} \times 1 = \frac{1}{3}$ of the round, green seed producers will breed true.

28a.   *Evaluate:* This is type 1. The Blue Persian and Spanish Dwarf *varieties* are assumed to be true-breeding for seed color and plant height; therefore, the resulting F_1 generation will be heterozygotes. *Deduce > Solve:* Since tall and white are expressed in the heterozygotes, they are the dominant traits. The reappearance of short and blue in the F_2 confirm that they are the recessive traits.

28b.   *Evaluate:* This is type 1. The F_1 are dihybrids. *Deduce:* The expected phenotypic distribution in the F_2 is $\frac{9}{16} : \frac{3}{16} : \frac{3}{16} : \frac{1}{16}$. *Solve:* Since tall and white are dominant, and short and blue are recessive, the expected distribution is $\frac{9}{16}$ tall, white; $\frac{3}{16}$ tall, blue; $\frac{3}{16}$ short, white; and $\frac{1}{16}$ short, blue.

28c.   *Evaluate:* This is type 2. The two varieties of pea plants differ in two traits, plant height and seed color. This experiment is testing the genetic basis of these differences. Which phenotype for each trait is dominant? How many genes are involved? How many alleles are at each locus? Do the genes assort independently of each other? *Deduce:* A straightforward hypothesis for two traits is that they are controlled by two independently assorting genes and that each gene has two alleles, one for the dominant trait and one for the recessive trait. *Solve:* Therefore, the hypothesis being tested in this experiment is that the two pea plant varieties differ at two independently assorting genetic loci, and two alleles show simple dominance at each locus.

28d.   *Evaluate:* This is type 5. There were 500 F_2 individuals. From part b, $\frac{9}{16}$ are expected to be tall and white, $\frac{3}{16}$ are expected to be tall and blue, $\frac{3}{16}$ are expected to be short and white, and $\frac{1}{16}$ are expected to be short and blue. *Deduce:* The expected outcomes are $\frac{9}{16} \times 500 = 281.25$ tall, white; $\frac{3}{16} \times 500 = 93.75$ tall, blue; $\frac{3}{16} \times 500 = 93.75$ short, white; and $\frac{1}{16} \times 500 = 31.25$ short, blue. *Solve:* The chi-square value for these data is 0.78. Since there are four phenotypic classes, there are $4 - 1 = 3$ degrees of freedom. In the 3 degrees of freedom row of Table 2.4, this chi-square value is greater than that for a P value of 0.9 but less than that for a P value of 0.7; therefore, the P value for these results is between 0.9 and 0.7. This P value is much greater than 0.05, so the hypothesis cannot be rejected (the results are consistent with the hypothesis).

29.   *Evaluate:* This is type 4. Consider each trait separately in each cross. In cross 1, the ratio of red to yellow fruit was 1:1 $\left(\frac{4}{8} : \frac{4}{8}\right)$, and the ratio of two-lobed to multilobed was 3:1 $\left(\frac{6}{8} : \frac{2}{8}\right)$. In cross 2, red to yellow was again 1:1 $\left(\frac{1}{2} : \frac{1}{2}\right)$, and two-lobed to multilobed was 1:1 $\left(\frac{1}{2} : \frac{1}{2}\right)$. *Deduce:* A ratio of 1:1 is expected for crosses of the type $Aa \times aa$; therefore, the parental genotypes at the R locus for both crosses was $Rr \times rr$, and the parental genotypes at the T locus in the second cross was $Tt \times tt$. A ratio of 3:1 is as expected for crosses of the type $Aa \times Aa$; therefore, the parental genotypes at the T locus for the first cross was $Tt \times Tt$. *Solve:* Cross 1 was $RrTt \times rrTt$. In cross 2, there are two possibilities, either $RrTt \times rrtt$ or $Rrtt \times rrTt$.

30a.   *Evaluate:* This is type 1b. This is a dihybrid cross ($CcPp \times CcPp$). *Deduce:* The expected phenotypic distribution is $\frac{9}{16}$ with dominant traits ($C\_P\_$), $\frac{3}{16}$ with one recessive trait ($ccP\_$), $\frac{3}{16}$ with the other recessive trait (C_pp), and $\frac{1}{16}$ with both recessive traits ($ccpp$). *Solve:* Therefore, $\frac{9}{16}$ will be expected to have neither CF nor PKU.

30b.   *Evaluate:* This is type 1b. The proportion that will have either PKU or CF includes those with only CF, those with only PKU, and those with both CF and PKU. *Deduce > Solve:* Using the information from part a, this is $\frac{3}{16} + \frac{3}{16} = \frac{3}{8}$.

30c.   *Evaluate:* This is type 1b. The proportion of children that will be carriers for one or both includes the genotypes $CcPP$, $CcPp$, $Ccpp$, $CCPp$, and $ccPp$. *Deduce:* The expected distribution of genotypes for a dihybrid cross is $\frac{1}{16} CCPP$, $\frac{2}{16} CCPp$, $\frac{1}{16} CCpp$, $\frac{2}{16} CcPP$, $\frac{4}{16} CcPp$, $\frac{2}{16} Ccpp$, $\frac{1}{16} ccPP$, $\frac{2}{16} ccPp$, and $\frac{1}{16} ccpp$. *Solve:* The expected proportion of children that will be carriers of one or both disorders is therefore $\frac{2}{16} CCPp + \frac{2}{16} CcPP + \frac{4}{16} CcPp + \frac{2}{16} Ccpp + \frac{2}{16} ccPp = \frac{12}{16} = \frac{3}{4}$.

31a.   *Evaluate:* This is type 5. The total number of boys observed was $(6 \times 9) + (63 \times 5) + (147 \times 4) + (204 \times 3) + (151 \times 2) + (56 \times 1) = 1927$. The total number of girls observed was $(63 \times 1) + (147 \times 2) + (204 \times 3) + (151 \times 4) + (56 \times 5) + (10 \times 6) = 1913$. The total number of children was $1927 + 1913 = 3840$. *Deduce:* The expected number of boys and girls is 1920 each. *Solve:* The chi-square value is 0.051. The number of degrees of freedom is (2 classes) $- 1 = 1$. For 1 degree of freedom, the chi-square of 0.051 corresponds to a P value between 0.9 and 0.7, which is much greater than 0.05. Therefore, the results are consistent with the expected ratio of 1:1 (you cannot reject the hypothesis of a boy-to-girl ratio of 1:1).

31b.   *Evaluate:* This is a combination of types 5 and 6. The observed values for distribution of boys and girls in 640 six-child families are given in the problem. The expected number is determined using the binomial for the distribution of boys and girls in a six-child family, which is $\left(\frac{1}{2}\right)^6 + 6\left(\frac{1}{2}\right)^5 \left(\frac{1}{2}\right) + 15\left(\frac{1}{2}\right)^4 \left(\frac{1}{2}\right)^2 + 20\left(\frac{1}{2}\right)^3 \left(\frac{1}{2}\right)^3 + 15 \left(\frac{1}{2}\right)^2 \left(\frac{1}{2}\right)^4 + 6 \left(\frac{1}{2}\right) \left(\frac{1}{2}\right)^5 + \left(\frac{1}{2}\right)^6$. *Deduce:* Solving each term in the expanded binomial yields $0.0156 + 0.0938 + 0.234 + 0.313 + 0.234 + 0.0938 + 0.0156$. Thus we can expect $0.0156 \times 640 = 10$, with six boys and no girls; $0.0938 \times 640 = 60$, with 5 boys and 1 girl; $0.234 \times 640 = 150$, with four boys and two girls; $0.313 \times 640 = 200$, with three boys and three girls; 150, with two boys and four girls; 60, with one boy and five girls; and 10, with no boys and six girls. *Solve:* The chi-square value is 0.663. The number of degrees of freedom is (7 classes) $- 1 = 6$. Using Table 2.4, the chi-square of 0.663 for 6 degrees of freedom corresponds to a P value greater than 0.95, indicating that the observed results fit the expected results; therefore, you cannot reject the hypothesis of a binomial distribution.

32a. *Evaluate:* This is type 5. The total number of children in these families was 480. The number of children with CF was $(52 \times 1) + (32 \times 2) + (18 \times 3) + (2 \times 4) = 178$; therefore, $480 - 178 = 302$ were normal. *Deduce:* The expected number of children with CF is $480 \times \frac{1}{4} = 120$; therefore, $480 - 120 = 360$ are expected to be normal. *Solve:* The chi-square value using these numbers is 37.4. The number of degrees of freedom is (2 classes) $- 1 = 1$. Using Table 2.4, the chi-square of 37.4 for 1 degree of freedom corresponds to a P value less than 0.001, which is much lower than 0.05, indicating that observed results are inconsistent with those expected and that you must reject the hypothesis that CF is inherited as an autosomal recessive trait based on these results.

32b. *Evaluate:* This is type 6. The observed distribution of children with CF in these families is given in the problem. The expected distribution is calculated by expanding the binomial $(a + b)^4$, where a = proportion of normal children = $\frac{3}{4}$ and b = proportion of children with CF = $\frac{1}{4}$. *Deduce:* The expanded binomial is $\left(\frac{3}{4}\right)^4 + 4\left(\frac{3}{4}\right)^3\left(\frac{1}{4}\right) + 6\left(\frac{3}{4}\right)^2\left(\frac{1}{4}\right)^2 + 4\left(\frac{3}{4}\right)\left(\frac{1}{4}\right)^3 + \left(\frac{1}{4}\right)^4$. *Solve:* Calculating each term gives $0.316 + 0.422 + 0.211 + 0.0469 + 0.00391$. Thus, in families where both parents are carriers and have four children, one expects $0.316 \times 120 = 38$ to have no CF children, $0.422 \times 120 = 50.6$ to have one child with CF, $0.211 \times 120 = 25.3$ to have two children with CF, $0.0469 \times 120 = 5.63$ to have three children with CF, and $0.00391 \times 120 = 0.469$ to have four children with CF.

32c. *Evaluate:* This is type 5. The observed values are given in the problem, and the expected values were calculated in the answer to part b. *Deduce:* The chi-square using these numbers is 47. The number of degrees of freedom is (5 classes) $- 1 = 4$. Using Table 2.4, the chi-square value of 47 for 4 degrees of freedom corresponds to a P value less than 0.001. *Solve:* This indicates that the difference between the observed and expected results is highly statistically significant and therefore is not consistent with that expected under binomial probability.

33a. *Evaluate:* This is type 1. The chance that her grandchild will inherit her D allele is the product of the probability that she will pass the allele to her child and the probability that her child will pass it on to her grandchild. *Deduce:* There is a $\frac{1}{2}$ chance that her child will inherit the dominant allele. If her child inherits the allele, then there is a $\frac{1}{2}$ chance that her child will pass it on to her grandchild. *Solve:* Therefore, the probability that her grandchild will inherit her dominant D allele is $\frac{1}{2} \times \frac{1}{2} = \frac{1}{4}$.

33b. *Evaluate:* This is type 1. The probability that two of her grandchildren will inherit her D allele is the square of the probability that one grandchild will inherit the allele. *Deduce:* As considered in part a, the probability that one grandchild will inherit her dominant allele is $\frac{1}{4}$. *Solve:* Therefore, the probability that two grandchildren inherit her dominant allele is $\frac{1}{4} \times \frac{1}{4} = \frac{1}{16}$.

33c. *Evaluate > Deduce > Solve:* See figure.

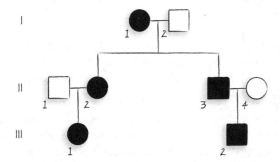

34. *Evaluate:* This is type 3. The probability that one or more of the children are carriers is the same as the sum of all probabilities minus the probability that none are carriers. *Deduce:* The probability that a child is not a carrier given that you know he or she is not homozygous recessive is $\frac{1}{3}$. The probability that all four are not carriers is $\left(\frac{1}{3}\right)^4 = 0.012$. *Solve:* The probability that one or more of the children is a carrier is $1 - 0.012 = 0.988$.

35a.   *Evaluate:* This is type 1b. The proportion of gametes carrying only dominant alleles is the same as the probability that a gamete is *ABCDE*. *Deduce:* The probability that a gamete is *ABCDE* is the product of the probabilities of a gamete carrying a dominant allele at each genetic locus. The probability of a dominant allele at one locus is $\frac{1}{2}$; therefore, the probability of all five alleles being dominant is $\left(\frac{1}{2}\right)^5 = 0.0313$. *Solve:* The proportion of gametes with only dominant alleles is 0.0313.

35b.   *Evaluate:* This is type 1b. The proportion of progeny with a genotype identical to the parent is the same as the probability of a progeny being *AaBbCcDdEe*. *Deduce:* The probability of this genotype is the product of the probabilities of *Aa*, *Bb*, *Cc*, *Dd*, and *Ee*. *Solve:* The probability of being heterozygous at one locus is $\frac{1}{2}$; therefore, the probability of being heterozygous at all five loci is $\left(\frac{1}{2}\right)^5 = 0.0313$.

35c.   *Evaluate:* This is type 1b. The parent's phenotype is dominant for each of the five traits. *Deduce:* The proportion of the progeny with dominant phenotypes for all five traits is the same as the probability of a progeny having at least one dominant allele at each locus (*A_B_C_D_E_*). *Solve:* The probability of having at least one dominant allele at one locus is $\frac{3}{4}$; therefore, the probability of having at least one dominant allele at all five loci is $\left(\frac{3}{4}\right)^5 = 0.237$.

35d.   *Evaluate:* This is type 1b. The proportion of gametes expected to be *ABcde* is the same as the probability that a gamete is *ABcde*. *Deduce:* Since the parent is heterozygous at all loci, the probability of a gamete carrying a particular allele is $\frac{1}{2}$ for each locus (e.g., the probability of a gamete carrying the *A* allele is $\frac{1}{2}$). *Solve:* The probability of any particular combination of five alleles is therefore $\left(\frac{1}{2}\right)^5 = 0.0313$.

35e.   *Evaluate:* This is type 1b. The proportion of the progeny expected to be *AabbCcDdE_* is the same as the probability of a progeny being *AabbCcDdE_*. *Deduce:* The probability of *Aa* is $\frac{1}{2}$, *bb* is $\frac{1}{4}$, *Cc* is $\frac{1}{2}$, *Dd* is $\frac{1}{2}$, and *E_* is $\frac{3}{4}$. *Solve:* Therefore, the probability of *AabbCcDdE_* is $\frac{1}{2} \times \frac{1}{4} \times \frac{1}{2} \times \frac{1}{2} \times \frac{3}{4} = 0.0234$.

36.   *Evaluate:* This is type 6. The probability of the couple having five children, with 0, 1, 2, 3, 4, or 5 of them affected by the disorder, can be calculated by expanding the binomial $(a + b)^5$, where a = probability of being unaffected and b = probability of being affected. *Deduce:* The probability of being unaffected is $\frac{3}{4}$ and the probability of being affected is $\frac{1}{4}$. The expanded binomial is $\left(\frac{3}{4}\right)^5 + 5\left(\frac{3}{4}\right)^4\frac{1}{4} + 10\left(\frac{3}{4}\right)^3\left(\frac{1}{4}\right)^2 + 10\left(\frac{3}{4}\right)^2\left(\frac{1}{4}\right)^3 + 5\left(\frac{3}{4}\right)\left(\frac{1}{4}\right)^4 + \left(\frac{1}{4}\right)^5$. *Solve:* The probability of the couple having five unaffected children is $\left(\frac{3}{4}\right)^5 = 0.237$, four unaffected and one affected is $5\left(\frac{3}{4}\right)^4\left(\frac{1}{4}\right) = 0.396$, three unaffected and two affected is $10\left(\frac{3}{4}\right)^3\left(\frac{1}{4}\right)^2 = 0.264$, two unaffected and three affected is $10\left(\frac{3}{4}\right)^2\left(\frac{1}{4}\right)^3 = 0.0879$, one unaffected and four affected is $5\left(\frac{3}{4}\right)\left(\frac{1}{4}\right)^4 = 0.0146$, and five affected is $\left(\frac{1}{4}\right)^5 = 0.000977$.

37.   *Evaluate:* This is type 5. The observed results are provided by the problem and are shown in the first two columns of the following table. *Deduce:* The expected frequency for each total value of the two dice is the frequency of the specific combination of dice $\left(\frac{1}{6} \times \frac{1}{6}\right)$, or $\left(\frac{1}{6}\right)^2$, and is given in the third column. The expected number of times that combination should appear in 300 rolls is the frequency × 300 and is given in the fourth column. *Solve:* The chi-square value using these observed and expected results is 5.78. The number of degrees of freedom is 11 classes – 1 = 10. Using Table 2.4, the chi-square value of 5.78 for 10 degrees of freedom corresponds to a P value between 0.9 and 0.7, which is much greater than 0.05. Therefore the student is incorrect because the observed results agree quite well with the expected results, and we cannot reject the hypothesis that random chance explains the results she observed.

Total Value on Two Dice	Number of Times Rolled	Expected Frequency	Expected Number of Times Rolled
2	7	$\left(\frac{1}{6}\right)^2 = 0.0278$	8.3
3	11	$2\left(\frac{1}{6}\right)^2 = 0.0556$	16.7
4	23	$3\left(\frac{1}{6}\right)^2 = 0.0833$	25
5	36	$4\left(\frac{1}{6}\right)^2 = 0.111$	33.3
6	42	$5\left(\frac{1}{6}\right)^2 = 0.139$	41.7
7	53	$6\left(\frac{1}{6}\right)^2 = 0.167$	50
8	40	$5\left(\frac{1}{6}\right)^2 = 0.139$	41.7
9	38	$4\left(\frac{1}{6}\right)^2 = 0.111$	33.3
10	30	$3\left(\frac{1}{6}\right)^2 = 0.0833$	25
11	12	$2\left(\frac{1}{6}\right)^2 = 0.0556$	16.7
12	8	$\left(\frac{1}{6}\right)^2 = 0.0278$	8.3
Sum	300	1.0	300

38.   *Evaluate:* This is type 1b. The experiments should test the hypothesis that short fur is dominant to long fur and brown fur color is dominant to white fur color. You have pure-breeding lines that, therefore, are homozygous at all loci. You can assume that fur length and fur color assort independently. *Deduce:* To test for dominance, you need to create strains heterozygous at each locus and determine the phenotype. The dominant phenotype will be apparent in the heterozygotes, and the recessive phenotype will be absent. The recessive phenotype should reappear in progeny of crosses between heterozygotes. You will need to confirm the heterozygote genotype by analyzing the phenotypic ratio among their progeny. *Solve:* <u>First Experiment:</u> Cross the true-breeding short, brown-furred and long, white-furred guinea pigs to create an F_1 population. According to the hypothesis, all of the F_1 should be dihybrids and have short, brown fur. To test that they are dihybrids, we will cross male F_1 with female F_1 to produce F_2 guinea pigs. According to the hypothesis, the expected fur length and color in the F_2 should be $\frac{9}{16}$ short, brown-furred; $\frac{3}{16}$ short, white-furred; $\frac{3}{16}$ long, brown-furred; and $\frac{1}{16}$ long, white-furred. Since larger numbers of progeny increase the likelihood that the observed phenotypic distribution will match that which is expected, cross both sets of pure-breeding parents to create 24 F_1 progeny and intercross all the F_1 (12 crosses) to produce 144 F_2. <u>Second Experiment:</u> Cross the F_1 with their long, white-furred parent to produce backcrossed guinea pigs. The expected fur length and color in these guinea pigs should be $\frac{1}{4}$ short, brown; $\frac{1}{4}$ short, white; $\frac{1}{4}$ long, brown; and $\frac{1}{4}$ long, white. To increase the numbers of backcrossed progeny, cross two of the F_1 males with their long, white-haired mother (one at a time; the mothers have now mated the maximum of three times), thus producing 24 progeny. Then cross the long, white-haired male guinea pig with all of the short, brown-haired female F_1, which would produce 144 progeny.

39a-b.   *Evaluate:* This is type 2. Amanda and Brice both have siblings with a recessive disease. *Deduce:* Amanda and Brice's parents are known to be carriers (Gg), and the expected genotypic distribution of their children is $\frac{1}{4}$ GG, $\frac{1}{2}$ Gg, and $\frac{1}{4}$ gg. Since neither Amanda nor Brice has galactosemia, they cannot be gg. Since they are not gg, they are either GG (probability of $\frac{1}{3}$) or Gg (probabilty of $\frac{2}{3}$). *Solve:* (a) See figure. (b) Amanda and Brice have the same probabilty of being a carrier, which is $\frac{2}{3}$.

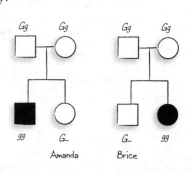

39c. *Evaluate:* This is type 1a. The probability of Amanda and Brice's first child having galactosemia is the probability that both are carriers multiplied by the chance that, given they are carriers, they will have a galactosemic child. Amanda and Brice each have a $\frac{2}{3}$ probability of being carriers (from part b). *Deduce:* The probability that both are carriers is $\frac{2}{3} \times \frac{2}{3} = \frac{4}{9}$. The probability that if they are carriers, they have a galactosemic child is $\frac{1}{4}$. *Solve:* The overall probability given what is known is therefore $\frac{4}{9} \times \frac{1}{4} = \frac{4}{36}$, or $\frac{2}{18}$.

39d. *Evaluate:* This is type 1. If their first child has galactosemia, then Amanda and Brice are both known carriers for galactosemia. *Deduce > Solve:* Given that they are carriers (Gg), there is a $\frac{1}{4}$ chance that any child they have will have galactosemia (gg).

40. *Evaluate:* This is type 1b. Pear shape and yellow color are recessive traits, so the desired plant must have the genotype *ffrr*. Axial flower position is a dominant trait, so the desired plant should have the genotype *TT*. You are starting with lines that are *FFrrtt* and *ffRRTT*. *Deduce:* The desired genotype is not present in the parents and will not be present in the F_1; therefore, you will need to create an F_2 population and determine which among the F_2 breed true for the desired traits. *Solve:* Cross the parents to produce (*FfRrTt*) trihybrids, and self-fertilize the F_1 to create an F_2 population. Among the F_2, $\frac{3}{64}$ will produce yellow, pear-shaped tomatoes with axial flowers (*ffrr*). To determine which of these are *TT*, self-fertilize them and identify the plants that breed true for axial flower position.

41a. *Evaluate:* This is type 5. According to the hypothesis, the F_1 plants have the genotype *Pp* and the F_2 are expected to be $\frac{3}{4}$ spicy (*P_*) and $\frac{1}{4}$ sweet (*pp*). Among 76 F_2, 56 spicy and 20 sweet were observed. *Deduce:* The expected distribution is $\left(\frac{3}{4}\right) \times 76 = 57$ spicy and $\left(\frac{1}{4}\right) \times 76 = 19$ sweet. *Solve:* The chi-square value for these results is 0.07, which corresponds to a P value between 0.9 and 0.7 for one degree of freedom. Therefore, Dr. Dopsis's hypothesis cannot be rejected.

41b. *Evaluate > Deduce > Solve:* The expected distribution among the F_2 is $\frac{1}{4}$ *PP*, $\frac{1}{2}$ *Pp*, and $\frac{1}{4}$ *pp*. The spicy plants make up $\frac{3}{4}$ of the F_2 plants. Among these, *PP* will be pure-breeding. $\frac{1}{4}$ out of $\frac{3}{4}$ are *PP*; therefore, $\frac{1}{3}$ will be pure-breeding.

42a. *Evaluate:* This is type 1. Alkaptonuria is a recessive disorder. Both couples had an affected child. *Deduce:* Unaffected parents with an affected child must be carriers; therefore, both sets of parents must be carriers. *Solve:* All four adults are heterozygous carriers of alkaptonuria (*Aa*).

42b. *Evaluate:* This is type 1a. Sara and James were informed that because their first child was affected, the chance of their next child having alkaptonuria was very low. Mary and Frank were informed that there was *no* chance of them having a child with alkaptonuria. *Deduce:* The probability that Sara and James's next child would be affected is independent of whether their first child was affected and is $\frac{1}{4}$. Although Frank's family history did not indicate that he was a carrier, he still has the same chance of anyone in the general population of being a carrier, which is $\frac{1}{1000}$. Before it was known that Mary was a carrier, the probability that she was a carrier would have been calculated as $\frac{2}{3}$ (both her parents are carriers). Therefore, the chance that she and Frank will have an affected child would have been calculated as $\frac{2}{3} \times \frac{1}{1000} \times \frac{1}{4} = \frac{1}{6000}$. *Solve:* For Sarah and James, the chance that their next child will have alkaptonuria is $\frac{1}{4}$, not *very low*; for Mary and Frank, the chance is $\frac{1}{6000}$, not *0*.

42c. *Evaluate:* This is type 1. They have one affected child. *Deduce > Solve:* They are both carriers. The probability that the next child of two carriers will be affected is $\frac{1}{4}$.

42d. *Evaluate:* They have one affected child. *Deduce > Solve:* They are both carriers. The probability that the next child of two carriers being unaffected is $\frac{3}{4}$.

42e.   *Evaluate:* This is type 1a. The probability that one of their children with alkaptonuria will have a child with alkaptonuria is dependent on the genotype of their child's mate. *Deduce > Solve:* If their child's mate has no family history of alkaptonuria, then there is a $\frac{4}{1000}$ chance that they will be a carrier. If the mate is a carrier, then there will be a $\frac{1}{2}$ chance that a grandchild will have alkaptonuria. The overall probability is therefore $\left(\frac{4}{1000}\right) \times \frac{1}{2} = \frac{1}{500}$. The probability that their affected child will have a child with alkaptonuria will increase dramatically if their child's mate has a close relative with the disorder. For example, if their child's mate's grandmother had alkaptonuria, then the mate will have at least a $\frac{1}{2}$ chance of being a carrier, and the overall probability that their first child will be affected is at least $\frac{1}{2} \times \frac{1}{2} = \frac{1}{4}$.

43a.   *Evaluate:* This is type 3b. The probability that a child will have the same phenotype as the parents is the probability that they will have the genotype $F\_E\_W\_H\_D\_$. The probability of that genotype is the product of the probabilities of $F\_$ and $E\_$ and $W\_$ and $H\_$ and $D\_$. *Deduce > Solve:* The probability of each of these genotypes is $\frac{3}{4}$; therefore, the probability that a child will have the same phenotype as the parents is $\left(\frac{3}{4}\right)^5 = 0.237$.

43b.   *Evaluate:* This is type 6. The probability of having any four dominant traits and one recessive trait is given by the appropriate term in the expanded binomial $(a + b)^5$, where a = probability of dominant and b = probability of recessive. *Deduce > Solve:* The probability of being dominant for a trait is $\frac{3}{4}$, and the probability of being recessive for a trait is $\frac{1}{4}$. The appropriate term in the expansion of the binomial is $5(a)^4(b)$. Therefore, the probability of having four dominant traits and one recessive trait is $5\left(\frac{3}{4}\right)^4\left(\frac{1}{4}\right) = 0.396$.

43c.   *Evaluate:* This is type 3b. The probability of being recessive for all five traits is the product of the probabilities of being recessive at each locus. *Deduce > Solve:* The probability of being recessive for one trait is $\frac{1}{4}$; therefore, the probability of being recessive for all five traits is $\left(\frac{1}{4}\right)^5 = 0.000977$.

43d.   *Evaluate:* This is type 3b. The probability of a child having the genotype $FfEEWwhhdd$ is the product of the probabilities of being Ff, EE, Ww, hh, and dd. *Deduce > solve:* The probability of Ff is $\frac{1}{2}$, $EE = \frac{1}{4}$, $Ww = \frac{1}{2}$, $hh = \frac{1}{4}$, and $dd = \frac{1}{4}$. Therefore, the probability of the genotype $FfEEWwhhdd$ is $\frac{1}{2} \times \frac{1}{4} \times \frac{1}{2} \times \frac{1}{4} \times \frac{1}{4} = 0.00391$.

44.   *Evaluate:* This is type 1b. The F_1 are dihybrids ($FfPp$) and produce F_2 that include chickens with feathered legs and single comb. These chickens are allowed to mate to produce the F_3. The genotypes of the mating F_2 chickens and the relative proportions of each genotype must be determined and then used to predict the types of gametes and their relative proportions of gamete types that contribute to the F_3. *Deduce:* Among the F_2 with feathered legs and single combs, $\frac{1}{16}$ will be $FFpp$ and $\frac{2}{16}$ will be $Ffpp$. Thus, $\frac{1}{3}$ of the mating population will be $FFpp$ and $\frac{2}{3}$ will be $Ffpp$. All the gametes from the $FFpp$ chickens will be Fp. Since $FFpp$ chickens make up $\frac{1}{3}$ of the mating population, $\frac{1}{3}$ of the gametes contributing to the F_3 will be Fp. Half of the gametes from the $Ffpp$ chickens will be Fp and half will be fp. Since $Ffpp$ chickens make up $\frac{2}{3}$ of the mating population, $\frac{2}{3} \times \frac{1}{2}$ or $\frac{1}{3}$ will be Fp and $\frac{2}{3} \times \frac{1}{2}$ or $\frac{1}{3}$ will be fp. *Solve:* Therefore, there will be two gamete types: $\frac{1}{3} + \frac{1}{3} = \frac{2}{3}$ will be Fp and $\frac{1}{3}$ will be fp. Using a Punnett square, $\frac{4}{9}$ of the F_3 will be $FFpp$, $\frac{4}{9}$ will be $Ffpp$, and $\frac{1}{9}$ will be $ffpp$ (see figure). The phenotypic ratio will be $\frac{8}{9}$ feathered legs and single comb, and $\frac{1}{9}$ no leg feathers and single comb.

	$\frac{2}{3}Fp$	$\frac{1}{3}fp$
$\frac{2}{3}Fp$	$\frac{4}{9}FFpp$	$\frac{2}{9}Ffpp$
$\frac{1}{3}fp$	$\frac{2}{9}Ffpp$	$\frac{1}{9}ffpp$

45.   *Evaluate:* This is type 3. The F_1 are monohybrids (Cc) and the F_2 are $\frac{1}{4}CC$, $\frac{1}{2}Cc$, $\frac{1}{4}cc$. Normal-winged males and females are mated. *Deduce:* $\frac{1}{3}$ of the normal-winged flies of both sexes are CC, and $\frac{2}{3}$ are Cc. The probability of a cross is the product of the probability of each parental genotype. *Solve:* There are two male genotypes and two female genotypes; therefore, there are four types of crosses: $\frac{1}{3} \times \frac{1}{3} = \frac{1}{9}$ will be male *CC* × female *CC*; $\frac{1}{3} \times \frac{2}{3} = \frac{2}{9}$ will be male *CC* × female *Cc*; $\frac{2}{3} \times \frac{1}{3} = \frac{2}{9}$ will be male *Cc* × female *CC*; $\frac{2}{3} \times \frac{2}{3} = \frac{4}{9}$ will be male *Cc* × female *Cc*.

2.04 Test Yourself

Problems

1. A trait called "creeper" reduces the size of the wings and legs in chickens. Chickens that do not have the creeper trait have normal, full-sized legs and wings. Creeper is due to a dominant allele (*C*) *that is lethal when homozygous.* *CC* chickens are never hatched; thus, all chickens that have the creeper trait are *Cc* heterozygotes. Chickens with normal, full-sized legs are *cc*. What is the expected genotypic and phenotypic proportion among the progeny of a cross between two creeper chickens?

2. You are studying a new strain of *Drosophila* that has curly wings. Crossing any curly-winged fly with any normal fly results in $\frac{1}{2}$ curly-winged and $\frac{1}{2}$ normal-winged progeny. Crossing any pair of curly-winged flies produces $\frac{2}{3}$ curly-winged progeny and $\frac{1}{3}$ normal progeny. Reciprocal crosses of both types give the same results. Use this information to determine the mode of inheritance of curly wings. Use *Cy* to refer to the dominant allele at the locus controlling wing shape in these flies.

3. The figure here shows a pedigree for a human trait. This trait is associated with the *A* locus, which has two alleles, *A* and *a*. Assume that II-5 is *AA*, but make no other assumptions. Use this figure to answer the following questions.
    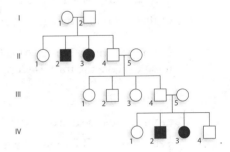
    a. Is the trait dominant or recessive? Explain the reasoning behind your answer.
    b. What is the probability that II-1 is *Aa*?
    c. What is the probability that II-4 is *Aa*?
    d. What is the probability that III-3 is *Aa*?
    e. If III-4 and III-5 have five more children, what is the probability that three of those children will have the trait?

4. Determine whether the observed results of 9 black mice and 7 brown mice from the first cross in Problem 12 is a good fit to the mode of inheritance you proposed as your answer to that question. How would the results of your goodness-of-fit analysis differ if the same 9:7 ratio was obtained, but the total number of progeny was 320 instead of only 16? What does this exercise illustrate about the use of a chi-square analysis to test a genetic hypothesis?

5. You have two strains of *Drosophila*. One is pure-breeding for normal wings and gray body color, and the other is pure-breeding for vestigial wings and ebony body color. The mode of inheritance of vestigial wings and ebony body color are described in Test Yourself Problem 1. Design a breeding experiment that will produce a pure-breeding line of flies that have vestigial wings and gray body color.

6. A male *Drosophila* that has curly wings and gray body is crossed to a female fly that has curly wings and gray body. Out of 240 progeny, 120 had curly wings and gray bodies, 40 had curly wings and ebony bodies, 60 had normal wings and gray bodies, and 20 had normal wings and ebony bodies. Use the information from Test Yourself Problems 1 and 2, and their answers, to solve the following questions.
    a. What were the genotypes of the parents?
    b. What proportion of the progeny with normal wings and gray bodies will be pure-breeding for these traits?

7. Travis and Kaylee are unrelated individuals whose parents are normal. They marry, and their first child has a rare autosomal recessive disorder. Travis's brother, Justin, marries Kaylee's sister, Katelyn.
    a. Assuming that only one of Justin's and Katelyn's parents were carriers, what is the probability that Justin and Katelyn's first child will have the same autosomal recessive disorder as Travis and Kaylee's child?

b. If Travis and Kaylee have six more children, what is the probability that only one of them will have the recessive disorder?

Solutions

1.   *Evaluate:* This is type 1a with a lethal genotype. The cross is $Cc \times Cc$. The genotype CC is lethal; therefore, CC chickens are never born. *Deduce > Solve:* The expected genotypic distribution would be $\frac{1}{4}\,CC$, $\frac{1}{2}\,Cc$, and $\frac{1}{4}\,cc$ except that CC is a lethal genotype, so no CC progeny are ever detected. The proportion of progeny genotypes is therefore $\frac{1}{2}$ out of $\frac{3}{4} = \frac{2}{3}\,Cc$ (creeper) and $\frac{1}{4}$ out of $\frac{3}{4} = \frac{1}{3}\,cc$ (normal).

2.   *Evaluate:* This is type 1a with a lethal genotype. Reciprocal crosses give the same results. Test crosses of all curly-winged flies yield $\frac{1}{2}$ curly-winged and $\frac{1}{2}$ normal-winged. Crosses in which both parents have curly wings yield $\frac{2}{3}$ curly-winged and $\frac{1}{3}$ normal-winged. *Deduce > Solve:* Cy is autosomal, and *all* curly-winged flies are heterozygotes. The absence of homozygous dominant $CyCy$ flies indicates that $CyCy$ is a lethal genotype. This is confirmed by the results of crosses in which both parents have curly wings. $Cycy \times Cycy$ is typically expected to result in $\frac{1}{4}\,CyCy$, $\frac{1}{2}\,Cycy$, and $\frac{1}{4}\,cycy$. If $CyCy$ is a lethal genotype, then $CyCy$ flies will never develop, and the ratio of the remaining two genotypes is $\frac{1}{2}$ out of $\frac{3}{4} = \frac{2}{3}\,Cycy$ and $\frac{1}{4}$ out of $\frac{3}{4} = \frac{1}{3}\,cycy$.

3a.   *Evaluate:* This is type 2. Affected individuals have parents that are not affected. Both males and females have the trait. *Deduce > Solve:* The trait is autosomal recessive.

3b.   *Evaluate:* This is type 3. II-1 has affected siblings and unaffected parents. *Deduce > Solve:* II-1's parents must be carriers (Aa). The probable genotypes for progeny of an $Aa \times Aa$ cross are $\frac{1}{4}\,AA$, $\frac{1}{2}\,Aa$, and $\frac{1}{4}\,aa$. II-2 cannot be homozygous recessive (she is unaffected); therefore, she has a $\frac{1}{4}$ out of $\frac{3}{4} = \frac{1}{3}$ chance of being AA and a $\frac{1}{2}$ out of $\frac{3}{4} = \frac{2}{3}$ chance of being Aa.

3c.   *Evaluate:* This is type 3. II-4 has siblings and a grandchild that are affected. *Deduce:* II-4's son, III-4 and his son's mate, III-5, must be carriers. For his son to be a carrier, either II-4 or II-5 would have to be a carrier. The problem states that II-5 is AA; therefore, II-4 must be Aa. *Solve:* The probability is 100%.

3d.   *Evaluate:* This is type 3. II-4 is a carrier and II-5 is not. *Deduce > Solve:* III-3 is the daughter of an $Aa \times AA$ cross. The probability that she is Aa is therefore $\frac{1}{2}$.

3e.   *Evaluate:* This is type 3. III-4 and III-5 are carriers. The probability that they have 3 affected and 2 unaffected children can be determined by expanding the binomial $(a + b)^5$, where a is the probability of having an affected child and b is the probability of having an unaffected child, and by calculating the value corresponding to 3 affected and 2 unaffected. *Deduce:* This is an $Aa \times Aa$ cross; therefore, the probability of affected is $\frac{1}{4}$ and the probability of unaffected is $\frac{3}{4}$. *Solve:* The expanded binomial is $\left(\frac{1}{4}\right)^5 + 5\left(\frac{1}{4}\right)^4\left(\frac{3}{4}\right) + 10\left(\frac{1}{4}\right)^3\left(\frac{3}{4}\right)^2 + 10\left(\frac{1}{4}\right)^2\left(\frac{3}{4}\right)^3 + 5\left(\frac{1}{4}\right)\left(\frac{3}{4}\right)^4 + \left(\frac{3}{4}\right)^5$. The term that applies to 3 affected and two unaffected is $10\left(\frac{1}{4}\right)^3\left(\frac{3}{4}\right)^2 = 0.088$.

4.   *Evaluate:* This is type 5. The mode of inheritance of mouse fur color was proposed to be controlled by an autosomal gene with black (B) dominant to brown (b). The first cross was proposed to be $bb \times Bb$, and the observed results were 9 black and 7 brown. If the same 9:7 ratio was obtained from 320 total progeny, then the observed results would have been 180 black and 140 brown. *Deduce:* $bb \times Bb$ is expected to produce $\frac{1}{2}$ black and $\frac{1}{2}$ brown progeny. The expected result for 16 progeny is 8 black and 8 brown. The expected result for 320 progeny is 160 black

and 160 brown. *Solve:* The chi-square value for the 9 plus 7 result is $\left(\frac{[9-8]^2}{8}\right) + \left(\frac{[7-8]^2}{8}\right) = 0.25$. For 1 degree of freedom, this corresponds to a *P* value between 0.7 and 0.5. This *P* value is greater than 0.05; therefore, the results are consistent with the hypothesis. The chi-square value for the 180 plus 140 result would be $\left(\frac{[180-160]^2}{160}\right) + \left(\frac{[140-160]^2}{140}\right) = 5.0$. For 1 degree of freedom, this corresponds to a *P* value between 0.05 and 0.01. This *P* value is less than 0.05; therefore, the hypothesis would be rejected. This exercise illustrates that small sample sizes are not as effective as large sample sizes in detecting deviations between observed and expected results when using the chi-square calculation. At a sample size of 16, there was no significant difference between a ratio of 9:7 and 1:1. This is also true for sample sizes of 160 or 240 (test this for yourself). A sample size of 320 is needed for chi-square to detect a statistically significant difference between the observed 9:7 ratio (180 black : 140 brown) and the expected 1:1 ratio (160 black : 160 brown).

5.   *Evaluate:* The available strains have the genotypes *VVEE* and *vvee*. You want to create a line of interbreeding male and female files that have the genotype *vvEE*. *Deduce > Solve:* Cross the pure-breeding normal-winged, gray flies to the vestigial-winged, ebony flies to create *VvEe* F_1. Allow the F_1 to intercross to create a F_2 population that includes vestigial-winged, gray flies of two genotypes, *vvEE* and *vvEe*. Mate male and female combinations of these flies to identify a combination that produces only vestigial, gray progeny. These matings identify parents that are either *vvEE* × *vvEE*, which represent a line of flies that is pure-breeding for vestigial wings and gray body, or *vvEE* × *vvEe*, which represents a line of flies that is not pure-breeding. Test-cross progeny from each line to determine which include *Ee* heterozygotes and which do not. The flies from the cross that did not produce heterozygotes will be pure-breeding for vestigial wings and ebony body.

6a.   *Evaluate:* This is type 4 with a lethal genotype. Curly wings is a dominant trait that is lethal when homozygous. Out of 240 progeny, 120 had curly wings and gray bodies, 40 had curly wings and ebony bodies, 60 had normal wings and gray bodies, and 20 had normal wings and ebony bodies. *Deduce:* The ratio of curly-winged to normal-winged progeny was 160:80, which is 2:1. This ratio is expected if both parents were *Cycy* heterozygotes. The ratio of gray-bodied to ebony-bodied progeny was 180:60, which is 3:1. This is expected if the parents were *Ee* heterozygotes. *Solve:* The parents were both *CycyEe*.

6b.   *Evaluate:* This is type 1b with a lethal genotype. The cross was *CycyEe* × *CycyEe*. Flies that are pure-breeding for normal wings and gray bodies have the genotype *cycyEE*. *Deduce > Solve:* The proportion of progeny with normal wings and gray bodies is $\frac{3}{12}$ ($\frac{1}{12}$ are *cycyEE* and $\frac{2}{12}$ are *cycyEe*). Therefore $\frac{1}{12}$ out of $\frac{3}{12}$, or $\frac{1}{3}$, of the normal-winged progeny will be pure-breeding for normal wings and gray body.

7a.   *Evaluate:* This is type 3. Both Justin and Katelyn have siblings who are carriers. *Deduce:* One of Justin's and one of Katelyn's parents are carriers for the trait; therefore, each has a $\frac{1}{2}$ probability of being a carrier. If both are carriers, then they have $\frac{1}{4}$ chance of their first child being affected. *Solve:* The probability of their first child being affected is therefore $\frac{1}{2} \times \frac{1}{2} \times \frac{1}{4} = \frac{1}{16}$.

7b.   *Evaluate:* This is type 6. Travis and Kaylee's first child was affected with an autosomal recessive disorder. The probability that they will have only one more affected child if they have six more children can be determined by expanding the binomial $(a + b)^6$ and solving for the term that corresponds to five unaffected and one affected. *Deduce:* Both Travis and Kaylee are carriers; therefore, this is an *Aa* × *Aa* mating. The probability of affected is $\frac{1}{4}$ and the probability of unaffected is $\frac{3}{4}$. The expanded binomial is $\left(\frac{3}{4}\right)^6 + 6\left(\frac{3}{4}\right)^5\left(\frac{1}{4}\right) + 15\left(\frac{3}{4}\right)^4\left(\frac{1}{4}\right)^2 + 20\left(\frac{3}{4}\right)^3\left(\frac{1}{4}\right)^3 + 15\left(\frac{3}{4}\right)^2\left(\frac{1}{4}\right)^4 + 6\left(\frac{3}{4}\right)\left(\frac{1}{4}\right)^4 + \left(\frac{1}{4}\right)^6$. *Solve:* The term corresponding to one affected and five unaffected is $6\left(\frac{3}{4}\right)^5\left(\frac{1}{4}\right) = 0.356$.

3

Cell Division and Chromosome Heredity

Section 3.01 Genetics Problem-Solving Toolkit

Section 3.02 Types of Genetics Problems

Section 3.03 Solutions to End-of-Chapter Problems

Section 3.04 Test Yourself

3.01 Genetics Problem-Solving Toolkit

Key Terms and Concepts

Sister chromatids: Genetically identical copies of a chromosome generated and attached to each other during S phase.

Homologous chromosomes: Chromosomes that contain the same genes but not necessarily the same alleles.

Mitosis: A single nuclear division that segregates sister chromatids to opposite poles of the cell, leading to a pair of genetically identical nuclei.

Meiosis: Two successive nuclear divisions, the first of which (Meiosis I) segregates homologous chromosomes and the second of which (Meiosis II) segregates sister chromatids.

Mendel's law of segregation: The segregation of *homologous chromosomes* and *sister chromatids* during meiosis.

Mendel's law of independent assortment: The segregation of *non-homologous chromosomes* and *non-sister chromatids* or recombination between *non-sister chromatids*.

Sex chromosomes: Chromosomes that differ in structure or number in males and females.

Sex-linked transmission: Transmission that occurs for genes that reside on sex chromosomes. The number of *sex-linked genes* differs in males and females.

Dosage compensation: Mechanisms that compensate for the difference in the number of genes on sex chromosomes in males and females.

Key Analytical Tools

Punnett squares for sex determination, reciprocal crosses, and the *forked-line* method.

Relationship between chromosome segregation and Mendelian ratios for segregation and independent assortment.

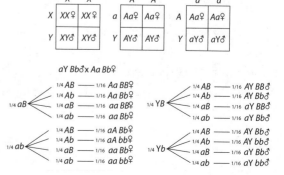

Key Genetic Relationships

Table 3.01   Genetic Relationships for Crosses Involving Sex-Linked Genes in XY Sex Determination

Genetic Relationship Number	Male Parent			Female Parent			Progeny	
	Genotype	Phenotype	Gametes	Genotype	Phenotype	Gametes	Genotypic ratio	Phenotypic ratio
1	AY	dominant	$\frac{1}{2}A, \frac{1}{2}Y$	aa	recessive	a	$\frac{1}{2}Aa : \frac{1}{2}aY$	all males recessive all females dominant
2	AY	dominant	$\frac{1}{2}A, \frac{1}{2}Y$	Aa	dominant	½ A, ½ a	$\frac{1}{4}AA,$ $\frac{1}{4}Aa,$ $\frac{1}{4}AY, \frac{1}{4}aY$	$\frac{1}{2}$ males recessive $\frac{1}{2}$ males dominant all females dominant
3	AY	dominant	$\frac{1}{2}A, \frac{1}{2}Y$	AA	dominant	A	$\frac{1}{2}AA, \frac{1}{2}AY$	all dominant
4	aY	recessive	$\frac{1}{2}a, \frac{1}{2}Y$	aa	recessive	a	$\frac{1}{2}aa, \frac{1}{2}aY$	all recessive
5	aY	recessive	$\frac{1}{2}a, \frac{1}{2}Y$	Aa	dominant	½ A : ½ a	$\frac{1}{4}Aa : \frac{1}{4}aa : \frac{1}{4}AY : \frac{1}{4}aY$	$\frac{1}{2}$ males dominant $\frac{1}{2}$ males recessive $\frac{1}{2}$ females dominant $\frac{1}{2}$ females recessive
6	aY	recessive	$\frac{1}{2}a, \frac{1}{2}Y$	AA	dominant	A	$\frac{1}{2}Aa : \frac{1}{2}AY$	all dominant

3.02 Types of Genetics Problems

1. **Relate mitosis and meiosis to segregation and independent assortment.**
2. **Analyze crosses or matings involving sex-linked transmission.**
3. **Describe the chromosomal basis of sex determination and the mechanism of X-linked gene dosage compensation in mammals.**

1. Relate mitosis and meiosis to segregation and independent assortment.

These problems ask you to integrate your understanding of Mendel's laws of segregation and independent assortment with your knowledge of the processes of mitosis and meiosis. You may be asked to draw chromosomes at various stages of mitosis or meiosis, assign alleles to chromosomes, and demonstrate how chromosome segregation explains Mendel's rules. You may also be asked to consider the molecular mechanisms underlying mitosis or meiosis and their role in cell division and chromosome segregation.

Example Problem: An alien organism is recovered from a remote region in the western United States. It is discovered that the alien is diploid and that sexual reproduction in this organism follows the law of segregation but not the law of independent assortment. A scientist on the genetics team charged with studying reproduction in the alien claims that the lack of independent assortment is due to the nonrandom alignment of non-homologous chromosomes on the meiotic I spindle. Defend the scientist's hypothesis based on your knowledge of meiosis. Assume that recombination does not occur during gamete formation in the alien.

Solution Strategies	Solution Steps
Evaluate	
1. Identify the topic this problem addresses, and explain the nature of the requested answer.	1. This problem is about understanding independent assortment and random alignment of non-homologous chromosomes on the meiotic I spindle. The answer will be an argument based on chromosome behavior during meiosis and the relationship between meiosis and Mendelian genetics.
2. Identify the critical information given in the problem.	2. The alien is diploid, and alleles at each genetic locus segregate during gamete formation according to Mendel's law, whereas alleles at different genetic loci do not assort independently. Recombination does not occur during meiosis.
Deduce	
3. Recall the relationship between chromosome movements during meiosis and Mendel's laws of segregation and independent assortment. **PITFALL:** Do not include recombination in your consideration of this problem.	3. In the absence of recombination, segregation of alleles occurs during meiosis I, and half of the gametes receive one allele and the other half receive the other allele. Independent assortment is due to the random alignment of non-homologous chromosomes on the meiotic spindle.
Solve	
4. Draw a figure illustrating the relationship between meiosis, segregation, and the lack of independent assortment of alleles on two chromosomes in the alien species.	

(*continued*)

Solution Strategies	Solution Steps
5. Use the information in the drawing to defend the hypothesis that nonrandom alignment during meiosis I explains allele segregation but not independent assortment.	5. **Answer:** Nonrandom alignment implies that non-homologous chromosomes align on the spindle such that a specific pair of alleles, for example *A* and *B* in the top half of the figure, are always on the same side. This determines that *A* and *B* and *a* and *b* always segregate together during meiosis I, creating only two gamete types instead of four gamete types as predicted by Mendel's law of independent assortment.

2. Analyze crosses or matings involving sex-linked transmission.

These problems ask you to apply dominance, segregation, and independent assortment to crosses or pedigrees that involve sex-linked traits. They are similar to problem types 1 through 5 in Chapter 3, except that one or more traits are controlled by sex-linked genes. You may be asked to predict the outcome of crosses given parental genotypes, determine the genotypes of individuals using information from a pedigree, calculate the probability of specific progeny types given information on parents, determine parental genotypes given progeny genotypes or phenotypes, or to test the goodness of fit between observed values and expected outcomes. You may also be asked to interpret pedigrees showing sex-linked inheritance or predict pedigree patterns for autosomal or sex-linked inheritance. Table 3.01 provides information on the expected *ratios* and *proportions* of progeny types for X-linked inheritance.

Variations:

(a) Analyze crosses or pedigrees involving only sex-linked traits.
The probability of a specific progeny type is the same as the expected proportion of that type.
See Problem 12 for an example.

(b) Analyze crosses or pedigrees in which sex-linked and autosomal traits are segregating.
Table 3.01 provides information on the expected ratios and proportions of progeny types for autosomal traits. **TIP:** See Key Analytical Tools to learn more about how the forked-line method is used to determine the progeny types and the multiplication rule to determine the proportions of each progeny type. See Problem 13 for an example.

Example Problem: Female *Drosophila* that are homozygous recessive for the X-linked *vermilion* (*v*) allele and male *Drosophila* hemizygous for *v* have bright red ("vermilion color") eyes. *Drosophila* homozygous for the autosomal recessive allele, *ebony* (*e*), have dark body color. Predict the phenotypic distribution of progeny from the cross *vvEe* × *VYEe*.

Solution Strategies	Solution Steps
Evaluate	
1. Identify the topic this problem addresses, and explain the nature of the requested answer.	1. This problem tests your understanding of crosses involving one X-linked gene and one autosomal gene. The answer will be a list of *expected progeny phenotypes* and the *proportions of each phenotype*.
2. Identify the critical information given in the problem.	2. Vermilion eye color is an X-linked recessive trait. Ebony body is an autosomal recessive trait. The cross is *vvEe* × *VYEe*.

(continued)

Solution Strategies	Solution Steps
Deduce	
3. Determine which genetic relationships are relevant to this problem.	3. For the X-linked *vermilion* gene, relationship 1 in Table 3.01 is the most relevant. For the autosomal ebony gene, relationship 3 in Table 3.01 is the most relevant.
Solve	
4. Calculate answers based on the identified genetic relationships.	4. The expected proportions of progeny phenotypes for eye-color gene is given by relationship 1 from Table 3.01 : Half of the progeny are wild-type females and half are vermilion males. The expected proportions of progeny phenotypes for body color is given by relationship 3 in Table 3.01: $\frac{3}{4}$ wild type, $\frac{1}{4}$ ebony.
5. Use the forked-line method to combine the ratios predicted for the vermilion and ebony phenotypes.	5. **Answer:** The expected phenotypic distribution is $\frac{3}{8}$ males with vermilion eyes and normal body color; $\frac{1}{8}$ males with vermilion eyes and ebony body color; $\frac{3}{8}$ females with wild-type eyes and body color; $\frac{1}{8}$ females with wild-type eyes and ebony body color.

3. Describe the chromosomal basis of sex determination and the mechanism of X-linked gene dosage compensation in mammals.

These problems ask you to relate the sex-chromosome content of cells or organisms to sex determination and sex-linked gene expression. You may be asked to predict the sex of individuals based on information on sex-chromosome composition. You may be asked to analyze crosses in organisms with a ZW sex-chromosome determination mechanism. You may also be asked interpret the pattern of X-linked gene expression in female mammals and relate that to the molecular mechanisms underlying dosage compensation in mammals. In solving these problems, it is important to remember that multiple mechanisms for sex determination and dosage compensation are known: The focus in this chapter is on fruit flies and mammals.

Example Problem: Calvin Bridges identified rare, unusual ("exceptional") flies among the progeny of his crosses of white-eyed females to red-eyed males. One exceptional type of fly was a sterile, red-eyed male and the other was a fertile, white-eyed female. If an exceptional white-eyed female was crossed with a normal red-eyed male by accident in an experiment, would the results of the cross alert a researcher familiar with Bridges's work to the exceptional nature of the white-eyed female? Assume that all gametes produced by both parents contain at least one sex chromosome and that the exceptional female produces gamete types in equal proportions.

Solution Strategies	Solution Steps
Evaluate	
1. Identify the topic this problem addresses, and explain the nature of the requested answer.	1. This problem tests your understanding of Bridges's experiments and your ability to predict the results of crosses using his exceptional flies and normal flies. The answer will be a comparison of the expected results from the two types of crosses and a statement of whether the results distinguish the crosses.

(continued)

Solution Strategies	Solution Steps
2. Identify the critical information given in the problem. TIP: Write the genotypes of Bridges's exceptional male and female flies.	**2.** The exceptional white-eyed female fly has the genotype $X^w X^w Y$, where X^w is an X chromosome carrying the mutant w allele. This female and the male produce gametes that contain at least one sex chromosome, and all gamete types are produced in equal proportions.

Deduce

3. Determine the types and proportions of gametes produced by each parent.	**3.** The gamete types and proportions are these: The $X^w X^w Y$ female produces four gamete types: $\frac{1}{4}$ $X^w X^w$, $\frac{1}{4} X^w$, $\frac{1}{4} Y$, and $\frac{1}{4} X^w Y$; a normal, white-eyed female produces two gamete types: $\frac{1}{2} X^w$ and $\frac{1}{2} Y$; a normal, red-eyed male produces $\frac{1}{2} X^{w+}$ and $\frac{1}{2} Y$.
4. Use a Punnett square to determine the progeny genotypes and their proportions for both crosses. TIP: Remember to account for the viability and sex of progeny with abnormal sex-chromosome composition.	**4.** 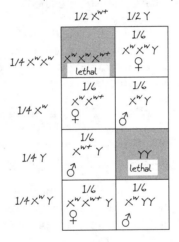

Solve

5. Using the Punnett square, determine the percentage of males and females with each eye color for each cross.	**5.** The cross involving the exceptional female will produce half males and half females; $\frac{2}{3}$ of the males and $\frac{1}{3}$ of the females will have white eyes. The rest of the males and females will have red eyes. The cross involving the normal female will produce males and females in equal proportions; all males will be white-eyed and all females red-eyed.

(continued)

Solution Strategies	Solution Steps
6. Use the expected results of the crosses to determine whether the results indicate the exceptional nature of the X^wX^wY female. TIP: Consider whether the results definitively identify the female's genotype.	6. **Answer:** Given that no red-eyed male progeny and no white-eyed female progeny are expected from a cross involving normal flies, the results do indicate something unusual about one or both parents. Bridges's "exceptional" white-eyed female flies should come to mind as an explanation of the unusual results.

3.03 Solutions to End-of-Chapter Problems

1a. *Evaluate:* This is type 1. *Deduce:* $2n = 6$; therefore, the organism contains three pairs of homologous chromosomes. The figure shows three chromosomes, each composed of a pair of sister chromatids, being pulled toward each pole. It also shows evidence of recombination between homologous chromosomes. *Solve:* This is anaphase I of meiosis. It cannot be anaphase of mitosis or anaphase II of meiosis because only three chromosomes are moving toward the poles, and sister chromatids are still attached.

1b. *Evaluate:* This is type 1. *Deduce:* $2n = 6$; therefore, the organism contains three pairs of homologous chromosomes. The figure shows six chromosomes aligned in the center of the spindle, each composed of a pair of sister chromatids attached to opposite poles. Homologs are not paired. *Solve:* This is metaphase of mitosis.

1c. *Evaluate:* This is type 1. *Deduce:* $2n = 6$; therefore, the organism contains three pairs of homologous chromosomes. The figure shows two small cells with six chromosomes each. Each chromosome is composed of a single chromatid. The cells also have a mitotic aster, not a spindle, and the nucleolus is detectable within the nucleus. *Solve:* This is telophase of mitosis.

1d. *Evaluate:* This is type 1. *Deduce:* $2n = 6$; therefore, the organism contains three pairs of homologous chromosomes. The figure shows six chromosomes, each composed of a single chromatid, being pulled toward opposite poles. *Solve:* This is anaphase of mitosis. It cannot be anaphase I or II of meiosis, because six chromosomes are being pulled toward opposite poles.

2a. *Evaluate:* This is type 1. *Deduce:* $2n = 48$; therefore, chimpanzees have 23 pairs of homologous autosomes and 1 pair of sex chromosomes. At the end of mitotic telophase, all sister chromatids will have separated from each other and will be located in two nuclei in two newly forming cells. *Solve:* Each cell will contain 48 chromosomes, 46 autosomes, and 2 sex chromosomes, and each chromosome will be composed of a single chromatid.

2b. *Evaluate:* This is type 1. *Deduce:* $2n = 48$; therefore, chimpanzees have 23 pairs of homologous autosomes and 1 pair of sex chromosomes. At meiotic metaphase I, homologous chromosomes will be paired, and the pairs will align in the center of the spindle. *Solve:* There will be 96 chromatids but only 48 chromosomes.

2c. *Evaluate:* This is type 1. *Deduce:* $2n = 48$; therefore, chimpanzees have 23 pairs of homologous autosomes and 1 pair of sex chromosomes. At the end of meiotic anaphase II, there will be four cells, each containing one copy of each homolog composed of a single chromatid. *Solve:* Each cell will contain 24 chromosomes (23 autosomes and 1 sex chromosome).

2d. *Evaluate:* This is type 1. *Deduce:* $2n = 48$; therefore, chimpanzees have 23 pairs of homologous autosomes and 1 pair of sex chromosomes. At early mitotic prophase, each chromosome will be composed of two sister chromatids. *Solve:* There will be 96 chromatids but only 48 chromosomes in the cell.

2e.   *Evaluate:* This is type 1. *Deduce:* $2n = 48$; therefore, chimpanzees have 23 pairs of homologous autosomes and 1 pair of sex chromosomes. At mitotic metaphase, there will be 48 chromosomes, and each chromosome will be composed of a pair of sister chromatids. *Solve:* There will be 96 chromatids but only 48 chromosomes in the cell.

2f.   *Evaluate:* This is type 1. *Deduce:* $2n = 48$; therefore, chimpanzees have 23 pairs of homologous autosomes and 1 pair of sex chromosomes. At early prophase I of meiosis, each chromosome is composed of a pair of sister chromatids. *Solve:* There will be 96 chromatids but only 48 chromosomes in the cell.

3.   *Evaluate:* This is type 3. This cross corresponds to genetic relationship 5 in Table 3.01, $aY \times Aa$, which predicts based on the law of segregation that there will be equal proportions of red- and white-eyed males and females. *Deduce:* The data indicate white-eyed flies are underrepresented in the progeny, thus deviating from expectations (261 red-eyed flies to 174 white-eyed flies). To perform a chi-square analysis for sex-linked inheritance, we will use the expected value of $\frac{1}{2}$ white-eyed flies being male and $\frac{1}{2}$ being female to correct for the deviant segregation ratio. *Solve:* The expected number of red-eyed males and red-eyed females is $\frac{1}{2}$ of 261, or 130.5 each. The expected number of white-eyed males and white-eyed females is $\frac{1}{2}$ of 174, or 87 each. The chi-square calculation is $\frac{(129 - 130.5)^2}{130.5} + \frac{(132 - 130.5)^2}{130.5} + \frac{(88 - 87)^2}{87} + \frac{(86 - 87)^2}{87} = 0.06$, which, for three degrees of freedom, gives a P value of greater than 0.95. Results are completely consistent with the hypothesis that white eye color is an X-linked trait.

4.   *Evaluate > Deduce > Solve:* This is type 1. Cohesion opposes the pulling forces attempting to separate sister chromatids until all pairs of sister chromatids are attached to microtubules from opposite poles of the spindle ("bipolar attachment"). Premature, as well as delayed, sister chromatid separation can cause sister chromatids to partition together instead of separating during anaphase, and lead to errors in chromosome segregation. Sister chromatid cohesion is due to cohesin, a protein complex that binds to sister chromatids and attaches them to each other. When all pairs of sister chromatids are under the tension generated by bipolar attachment and sister chromatid cohesion, the protease separase is activated. Separase cleaves a component of cohesin, which simultaneously ends cohesion on all pairs of sister chromatids and allows sister chromatids to be pulled toward opposite poles of the spindle.

5a.   *Evaluate:* This is type 1. G. *introductus* is a diploid with 18 pairs of homologous chromosomes ($2n = 36$) that contain 3 ng of DNA. *Deduce:* After S phase, each chromosome has been duplicated and is composed of a pair of identical sister chromatids. *Solve:* The chromosome number has not changed, but the DNA content has doubled; therefore, there are 36 chromosomes containing 6 ng of DNA.

5b.   *Evaluate:* This is type 1. Prophase of mitosis and prophase I of meiosis are the same with respect to chromosome and DNA content. *Deduce > Solve:* The beginning of prophase in mitosis and prophase I in meiosis both follow completion of S phase and G2; therefore, the number of chromosomes (36 in this case) and the amount of DNA (6 ng) is the same.

5c.   *Evaluate:* This is type 1. *Deduce:* At the end of telophase I, homologous chromosomes have been pulled to opposite poles of the meiotic spindle and cytokinesis is complete. Each of these daughter cells contains half the number of chromosomes and half the DNA content of a G2 phase cell. At the end of mitotic anaphase, sister chromatids of each chromosome have been pulled to opposite poles of the spindle but cytokinesis has not yet occurred. At the end of telophase II of meiosis, the sister chromatids of one set of homologs have been pulled to opposite poles of the meiotic spindle and cytokinesis is complete. Each of these daughter cells contains half the number of chromosomes and a fourth of the DNA content of a G2 phase cell. *Solve:* See table. Telophase I cells have 18 chromosomes containing 3 ng DNA. Mitotic anaphase cells have 72 chromosomes containing 6 ng DNA. Telophase II cells have 18 chromosomes containing 1.5 ng DNA.

End of Cell Cycle Stage	Number of Chromosomes	Amount of DNA
S phase	36	6 ng
G$_2$ of mitosis	36	6 ng
G$_2$ of meiosis	36	6 ng
Telophase I of meiosis (after cytokinesis)	18	3 ng
Anaphase of mitosis	72	6 ng
Telophase II of meiosis (after cytokinesis)	18	1.5 ng

6.   *Evaluate:* This is type 1. Recall that expected Mendelian proportions depend on the law of segregation and the law of independent assortment. **TIP:** Refer to Key Analytical Tools for relating Mendelian genetics to meiosis. *Deduce:* Alleles of the same locus segregate from each other during meiosis I because homologs pair at metaphase I. Segregation of alleles on non-homologous chromosomes is random because the relative attachment of non-homologous chromosomes to the metaphase I spindle is random. *Solve:* At metaphase I, homologous chromosomes can pair in one of two ways; with R$_1$ and T$_1$ on the same side of the spindle (top half of figure) or R$_1$ and T$_1$ on opposite sides of the spindle (bottom half of figure). For the former alignment, R$_1$ and T$_1$ will segregate together, as will R$_2$ and T$_2$. This generates four meiotic products, two R$_1$, T$_1$ and two R$_2$, T$_2$. For the latter metaphase alignment, R$_1$ and T$_2$ will segregate together, as will R$_2$ and T$_1$. This generates four meiotic products, two R$_1$, T$_2$ and two R$_2$, T$_1$. Both types of metaphase I alignments are equally likely, resulting in a gamete distribution that is 25% R$_1$, T$_1$; 25% R$_2$, T$_2$; 25% R$_1$, T$_2$; and 25% R$_2$, T$_1$.

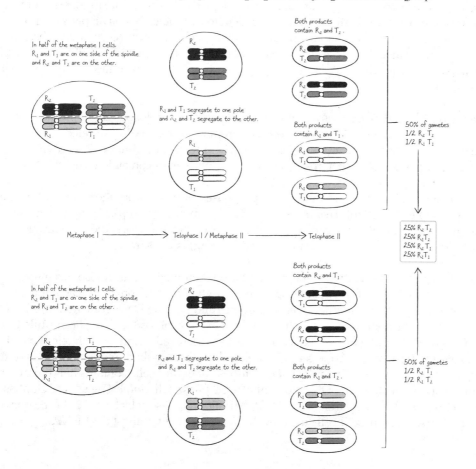

7. *Evaluate > Deduce > Solve:* This is type 1. Mendel's law of segregation states that D and d segregate during meiosis such that half the gametes will contain D and half will contain d. This is demonstrated by the pairing of homologous chromosomes during prophase I through metaphase I in meiosis and by the segregation of homologs to opposite poles of the spindle during anaphase I (assuming no crossovers between homologs). Half the gametes will be derived from the sister chromatids carrying D at one pole, and the other half will be derived from sister chromatids carrying d at the other pole.

8. *Evaluate:* This is type 1. *Deduce:* There are two possibilities for recombination: (1) recombination that occurs between the centromere and the D locus or (2) recombination that occurs between the D locus and the telomere. Mendel's law states that D and d segregate during meiosis such that half the gametes will contain D and half will contain d. The first recombination scenario can result in D and d gametes segregating together to both poles during anaphase I; however, D and d will segregate apart to separate gametes during anaphase II. In the second recombination scenario, D and d will segregate during anaphase I. *Solve:* There are two possible stages in M phase at which the alleles D and d could segregate: anaphase I or II.

9. *Evaluate:* This is type 1. *Deduce > Solve:* Crossovers between one or both pairs of homologs do not change the expected proportion of gamete types produced by $AaBb$ dihybrids. Crossovers can affect whether the alleles of one locus segregate during anaphase I or anaphase II; however, by the end of meiosis they will segregate from each other and will assort independently of alleles on non-homologous chromosomes. The figures here show that crossing over does not alter the outcome that half the gametes will contain the dominant allele and half will contain the recessive allele for each locus. The assortment of alleles on non-homologous chromosomes is determined by their relative attachment to the metaphase I and metaphase II spindles, which is random.

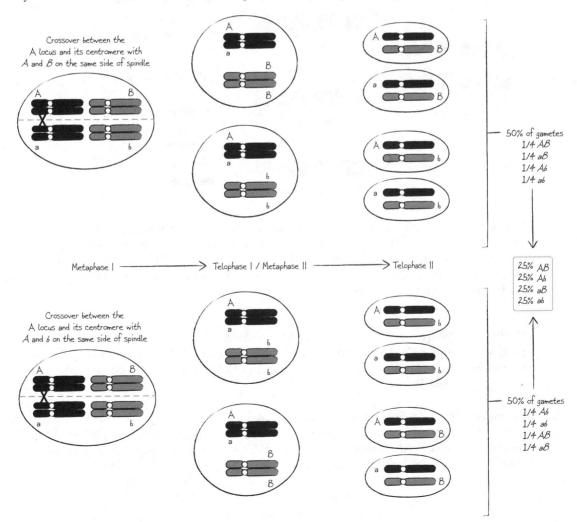

10.

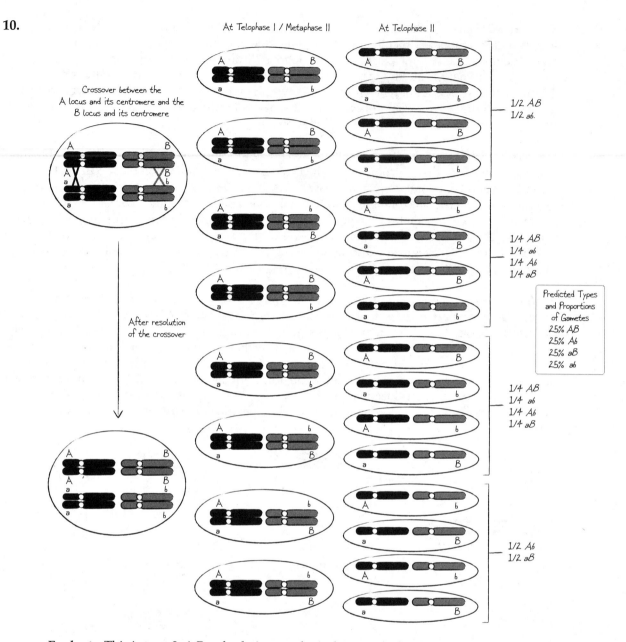

At Telophase I / Metaphase II   At Telophase II

Crossover between the
A locus and its centromere and the
B locus and its centromere

After resolution
of the crossover

1/2 AB
1/2 ab.

1/4 AB
1/4 ab
1/4 Ab
1/4 aB

Predicted Types
and Proportions
of Gametes
25% AB
25% Ab
25% aB
25% ab

1/4 AB
1/4 ab
1/4 Ab
1/4 aB

1/2 Ab
1/2 aB

Evaluate: This is type 3. A Barr body is a cytological structure that represents an inactive X chromosome. The somatic cells of normal human females have two X chromosomes, one of which is inactive. The somatic cells of a normal human male have one X chromosome, which is active. *Deduce > Solve:* A normal human female nucleus contains one Barr body, whereas a normal male nucleus contains no Barr bodies.

11a. *Evaluate > Deduce > Solve: Microtubules* are filaments in mitotic and meiotic spindles that are required for moving homologs to opposite sides of the poles during anaphase I of meiosis and for moving sister chromatids to opposite sides of the poles during anaphase II of meiosis and anaphase of mitosis.

11b. *Evaluate > Deduce > Solve: Cyclin-dependent kinases* are a family of related but distinct enzymes that control progression through the cell cycle by catalyzing the phosphorylation of specific proteins. Phosphorylation alters the function of those proteins, and the alteration of their function drives transitions in the cell cycle.

11c.   *Evaluate > Deduce > Solve: Kinetochores* are complex structures attached to the centromeres of chromosomes that bind to spindle microtubules. They are required for the segregation of chromosomes during anaphase of mitosis or meiosis.

11d.   *Evaluate > Deduce > Solve:* The *synaptonemal complex* is a layered structure located between paired homologs in prophase I of meiosis in many organisms. It is important for maintaining homologous chromosome pairing and plays a role in recombination.

12a.   *Evaluate:* This is type 2a. We know that the father has the recessive, X-linked disorder (dY), and that the mother is homozygous for the dominant allele (DD). *Deduce:* She had to inherit an X chromosome from each parent: a recessive d allele from her father, and one of her mother's dominant D alleles. *Solve:* She is Dd.

12b.   *Evaluate:* This is type 2a. We've determined that the woman has a Dd genotype. *Deduce:* Her son will inherit one of her two X chromosomes—one of which carries the recessive d allele—and a Y chromosome from his father. *Solve:* There is a $\frac{1}{2}$ or 50% chance that her son will have OTD.

12c.   *Evaluate:* This is type 2a. *Deduce:* Her daughter will inherit one of her mother's two X chromosomes, one of which carries the recessive allele for OTD, and her father's X chromosome, which carries the dominant allele. *Solve:* There is a 50% chance that her daughter will be Dd and a 50% chance that she will be DD. Therefore, there is no chance that her daughter will have OTD, but there is a $\frac{1}{2}$ or 50% chance that her daughter will be a carrier for OTD.

12d.   *Evaluate:* This is type 2a. *Deduce:* Her daughter would have to inherit an X chromosome carrying the recessive allele from both her mother and her father. *Solve:* The woman and a man could have a daughter with OTD if his genotype is dY and he has OTD.

12e.   *Evaluate:* This is type 2a. *Deduce:* The mother will give one half of her children the X chromosome with the dominant allele and half the X chromosome with the recessive allele. The father will give all his daughters the X chromosome with the recessive allele and all his sons his Y chromosome. *Solve:* Half the daughters will be Dd and half will be dd. Half the sons will be DY and half will be dY. Therefore, half of the daughters and half of the sons will have OTD.

13a.   *Evaluate:* This is type 2b. I-1 and II-5 have hemophilia, which is an X-linked recessive trait. II-1 and II-4 have albinism, which is an autosomal recessive trait. *Deduce:* I-1 and I-2 do not have albinism but have an albino son; therefore, both are carriers for albinism. I-3 and I-4 are normal but have a son with hemophilia and a daughter with albinism; therefore, both are carriers for albinism and I-4 is a carrier for hemophilia. *Solve:* I-1 is $hYAa$; I-2 is H_Aa; I-3 is $HYAa$; I-4 is $HhAa$.

13b-i. *Evaluate:* This is type 2a. *Deduce:* Clara's father had hemophilia, but she does not; therefore, she is a carrier for hemophilia. The probability that Clara's first child will be a boy is $\frac{1}{2}$. The probability that if the first child is a boy, he has hemophilia, is $\frac{1}{2}$. *Solve:* The probability that their first child is a boy with hemophilia is $\frac{1}{2} \times \frac{1}{2} = \frac{1}{4}$.

13b-ii. *Evaluate:* This is type 2b. *Deduce:* Clara's parents were both carriers for albinism (Aa); therefore, Clara has a $\frac{2}{3}$ probability of being a carrier for albinism. The same is true for Charles. If both Clara and Charles are carriers for albinism, they have a $\frac{1}{4}$ chance of having a child with albinism. The chance that they will have a girl is $\frac{1}{2}$. *Solve:* The chance that Clara and Charles's first child will have albinism is $\frac{2}{3} \times \frac{2}{3} \times \frac{1}{4} = \frac{4}{36} = \frac{1}{9}$. The chance that their first child will be a girl with albinism is $\frac{1}{9} \times \frac{1}{2} = \frac{1}{18}$.

13b-iii. *Evaluate:* This is type 2b. *Deduce:* The probability that their first child will be a girl is $\frac{1}{2}$. Since Charles does not have hemophilia, there is no chance that they will have a daughter with hemophilia. The probability that their first child will have albinism is $\frac{1}{9}$, as determined in part (b-ii); therefore, the probability that the child will be healthy is $1 - \frac{1}{9} = \frac{8}{9}$. *Solve:* The probability that their first child will be a healthy girl is $\frac{1}{2} \times \frac{8}{9} = \frac{8}{18} = \frac{4}{9}$.

13b-iv. *Evaluate:* This is type 2b. *Deduce:* There is a $\frac{1}{2}$ chance that their first child will be a boy. Clara's father had hemophilia; therefore, she is a carrier. If her first child is a boy, there is a $\frac{1}{2}$ chance that he will have hemophilia. Clara and Charles each have a $\frac{2}{3}$ chance of being carriers for albinism (see part b-ii). If they are both carriers, there is a $\frac{1}{4}$ chance that their child will have albinism. *Solve:* The probability that their first child will be a boy with both hemophilia and albinism is $\frac{1}{2} \times \frac{1}{2} \times \frac{2}{3} \times \frac{2}{3} \times \frac{1}{4} = \frac{4}{144} = \frac{1}{36}$.

13b-v. *Evaluate:* This is type 2b. *Deduce:* Since albinism is an autosomal trait, the probability that their first child will be a boy with albinism is the same as the probability that their first child will be a girl with albinism, which is calculated in part (b-ii). *Solve:* The chance that their first child will be a boy with albinism is $\frac{1}{9} \times \frac{1}{2} = \frac{1}{18}$.

13b-vi. *Evaluate:* This is type 2b. *Deduce:* Charles does not have hemophilia; therefore, he will give an X chromosome carrying the dominant H allele to his daughter. His daughter *cannot be* homozygous recessive for the h allele. *Solve:* The probability that their first child will be a girl with hemophilia is zero.

13c. *Evaluate:* This is type 2b. *Deduce:* If their first child has albinism, both Clara and Charles must be carriers. The chance that their next child will have albinism is therefore $\frac{1}{4}$. *Solve:* The probability that their next child will have albinism is $\frac{1}{4}$. This value is much higher than the probability calculated in part (b), for two reasons: (1) the calculation in part (b-ii) was for a girl and the one in part (b-v) was for a boy; and (2) the calculations in parts (b-ii and b-v) included the probability that both Clara and Charles were carriers, which was $\frac{2}{3} \times \frac{2}{3} = \frac{4}{9}$. Here, the probability is one.

14a. *Evaluate:* This is type 2b. Recall that X-linked and autosomal traits assort independently. Also note that miniature wings and purple eyes are absent in the parents but present in a proportion of the progeny. *Deduce:* Analyze the proportions of progeny for each trait individually to determine dominance and identify which trait shows X-linkage. Miniature wings and purple eyes are recessive traits; therefore, the female parent is heterozygous for both traits, and the male is heterozygous for the autosomal trait and hemizygous for the dominant allele of the X-linked trait. *Solve:* Among the male progeny, $\frac{3}{8} + \frac{1}{8} = \frac{4}{8} = \frac{1}{2}$ have full wings and $\frac{3}{8} + \frac{1}{8} = \frac{4}{8} = \frac{1}{2}$ have miniature wings; among the female progeny $\frac{3}{4} + \frac{1}{4}$, or all the progeny, have full wings. This indicates that the recessive trait, miniature wings, is X-linked. Among males, $\frac{3}{8} + \frac{3}{8} = \frac{6}{8} = \frac{3}{4}$ have red eyes and $\frac{1}{8} + \frac{1}{8} = \frac{2}{8} = \frac{1}{4}$ have purple eyes. This 3:1 ratio, typical of recessive autosomal inheritance, is also seen in the female progeny. Thus the male parent was hemizygous dominant for the X-linked, full-wing allele (M^+Y) and heterozygous at the eye-color locus (P^+p^-). The female parent was heterozygous at both loci (M^+m^- and P^+p^-).

14b. *Evaluate:* This is type 2b. Recall that miniature wings (m^-) and purple eyes (p^-) are recessive to full wings (M^+) and red eyes (P^+), and that miniature wings is X-linked and purple eyes is autosomal. The cross was M^+Y and $P^+p^- \times M^+m^-$ and P^+p^-. *Deduce:* Half the female progeny receive M^+ and P^+ from their father and half receive M^+ and p^-. One-fourth of the female progeny receive M^+ and P^+ from their mother, one-fourth receive M^+ P^+, one-fourth receive m^- and P^+, and one-fourth receive m^- and p^-. Half the male progeny receive Y and P^+ from their father and half receive Y and p^-. One-fourth of the males receive M^+ and P^+ from their mother, one-fourth receive M^+ P^+, one-fourth receive m^- and P^+, and one-fourth receive m^- and p^-. *Solve:* All female progeny with purple eyes are $p^- p^-$: Half are M^+m^- and half are M^+M^+. All males with miniature wings and purple eyes are m^-Y and p^-p^-.

15. *Evaluate:* This is type 2a. This problem asks you to determine the pattern of inheritance of the trait, discolored tooth enamel, based on the information provided, which is diagrammed in the figure. *Deduce:* The trait appears in all three generations, indicative of a dominant trait. Transmission occurs from females to both males and females.

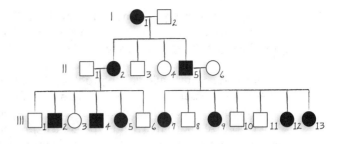

The only affected male who has children has only affected daughters, not sons. The transmission pattern from male to female only is consistent with an X-linked trait. *Solve:* Discolored tooth enamel appears to be an X-linked dominant trait in this family.

16. *Evaluate:* This is type 3. A Barr body is a cytologically detectable nuclear structure that represents an inactive X chromosome. All but one X chromosome in human somatic cells are inactivated. Normal human males have no Barr bodies in their somatic nuclei, whereas normal human females have one Barr body per somatic nucleus. *Deduce > Solve:* The number of X chromosomes per cell in an individual is determined by the number of X chromosomes in each of the gametes that fused to create that individual. The unusual males with two X chromosomes result from a normal egg fusing with a sperm containing an X and a Y chromosome (nondisjunction during meiosis I in the father) or fusion of a normal sperm with an egg containing two X chromosomes (nondisjunction during meiosis I or II in the mother). The unusual females with two Barr bodies result from fusion of a normal sperm with an egg containing two X chromosomes, or a normal egg with a sperm containing two X chromosomes.

17a. *Evaluate:* This is type 2. We know that brown males mated to orange females produce orange male and tortoiseshell female progeny, and that orange males mated to brown females produce brown male and tortoiseshell female progeny. *Deduce:* The transmission of brown and orange coat color from mothers to sons is indicative of an X-linked trait. The male progeny in these crosses will be hemizygous for the allele present in their mother, either brown or orange, and the female progeny will be heterozygous for brown and orange. *Solve:* Brown and orange are determined by alternate alleles of an X-linked gene. Males are hemizygous and are either brown or orange, depending on which allele is on their X chromosome. Brown females are homozygous for the brown allele, and orange females are homozygous for the orange allele. Tortoiseshell is the color of heterozygous females.

17b. *Evaluate:* This is type 3. *Deduce:* Females heterozygous for X-linked genes are mosaic for X-linked gene expression because of the dosage compensation mechanism called X-chromosome inactivation. Only one X chromosome is expressed in somatic cells; therefore, those cells will have the phenotype determined by the alleles on the "active" X chromosome. Females heterozygous for the X-linked brown and orange alleles are mosaic for fur color. *Solve:* Patches of their fur will be orange or dark brown, depending on which X chromosome is active in each patch of tissue. The mosaic pattern of brown and orange patches results in a coloration called tortoiseshell.

17c. *Evaluate:* This is type 3. *Deduce:* Tortoiseshell is the color of females heterozygous for the orange and brown alleles; therefore, the males must be orange/brown heterozygotes. Males normally have only one X chromosome; therefore, tortoiseshell male cats have an extra X chromosome. *Solve:* The tortoiseshell male cats have two X chromosomes (they are XXY) and are heterozygous for the brown and orange fur-color alleles. They result from a normal egg fusing with a sperm containing an X and a Y chromosome (nondisjunction during meiosis I in their father) or fusion of a normal sperm with an egg containing two X chromosomes (nondisjunction during meiosis I in their mother). They are sterile because not all genes on the X chromosome are silenced by X-chromosome inactivation. Some X-linked genes are meant to be expressed more highly in females than males. A high level of expression on these genes in XXY males causes sterility.

18. *Evaluate:* This is type 3. The syndrome is an X-linked, recessive trait. Therefore, males inheriting this mutation must express it (hemizygous), whereas it is possible for females to be heterozygous for the mutation (carrier females). *Deduce > Solve:* Because of random X inactivation, the tissue of heterozygous females will contain some cells that express the wild-type allele and some that express the mutant allele. The problem states that most female carriers do not show symptoms; therefore, the pattern of X-inactivation normally results in sufficient wild-type gene expression to promote normal development. In female carriers that show symptoms, X-inactivation must have occurred such that wild-type gene expression was insufficient for normal development. The symptoms are less severe in these symptomatic female carriers because they express some level of the wild-type allele.

19a. *Evaluate:* This is type 2b. Wild-type *Drosophila* have red eyes; red eyes are dominant. F_1 females from all the crosses have red eyes. Only F_1 males from the reciprocal crosses involving apricot or carnation eyes differ in eye color. F_1 males have red eyes when the dominant allele is donated by the female, whereas F_1 males have either apricot or carnation eyes, respectively, when the dominant allele is donated by the male. *Deduce:* X-linked recessive traits give different results in reciprocal crosses: They are transmitted from the mother to all male progeny but are not transmitted from the father to any progeny. *Solve:* Apricot and carnation are X-linked recessive traits because they show the expected pattern of transmission in reciprocal crosses. Because brown and purple show the same results in reciprocal crosses, they are autosomal recessive traits.

19b.

	Parents		F_1 Progeny		F_2 Progeny	
	Female	Male	Female	Male	Female	Male
A	apricot	red	red	apricot	$\frac{1}{2}$ red; $\frac{1}{2}$ apricot	$\frac{1}{2}$ red; $\frac{1}{2}$ apricot
B	brown	red	red	red	$\frac{3}{4}$ red; $\frac{1}{4}$ brown	$\frac{3}{4}$ red; $\frac{1}{4}$ brown
D	red	apricot	red	red	red	$\frac{1}{2}$ red; $\frac{1}{2}$ apricot
G	red	brown	red	red	$\frac{3}{4}$ red; $\frac{1}{4}$ brown	¾ red; ¼ brown

Evaluate: In part (a), apricot was determined to be an X-linked recessive trait whereas brown was determined to be autosomal recessive. *Deduce:* See table. For cross A, the F_1 males are aY and the F_1 females are aA, where a stands for the recessive apricot allele and A stands for the dominant wild-type allele. The cross is $Aa \times aY$. For cross B, the F_1 males and females are both bB, where b strands for the recessive brown allele and B for the wild-type allele. The cross is $Bb \times Bb$. For cross D, F_1 females are Aa and the F_1 males are AY. The cross is $Aa \times AY$. For cross G, the F_1 males and females are Bb. The cross is $Bb \times Bb$. *Solve:* For cross A, the F_2 females are $\frac{1}{2}$ red, $\frac{1}{2}$ apricot and the F_2 males are $\frac{1}{2}$ red, $\frac{1}{2}$ apricot. For cross B, the F_2 females and males are both $\frac{3}{4}$ red, $\frac{1}{4}$ brown. For cross D, the F_2 females are red and the F_2 males are $\frac{1}{2}$ red, $\frac{1}{2}$ apricot. For cross G, the F_2 females and males are both $\frac{3}{4}$ red, $\frac{1}{4}$ brown.

20. *Evaluate:* This is type 2b. *Deduce:* Since we are assessing likelihood, consider all four possible inheritance patterns for each pedigree. Assume that all traits are statistically rare; therefore, individuals from outside the family are unlikely to be carriers of a recessive trait, and traits appearing in all generations of a pedigree are more likely to be dominant than recessive. Also assume that dominant traits show complete penetrance; therefore, individuals with dominant traits must have at least one parent with that trait. Because X-linked traits are transmitted from mothers to sons and daughters and from fathers to daughters, X-linked recessive traits typically predominate in males; therefore, all children of males with X-linked dominant traits should show the trait. *Solve:* See table.

	Most Likely Inheritance Pattern	Transmitting Genotypes	Alternate, Possible Inheritance Pattern	Transmitting Genotypes	Impossible Inheritance Patterns	Reason
Pedigree A *Absence of trait in parents rules out dominance.* *This individual would have to be Aa for autosomal recessive inheritance.* *Both individuals would have to be unrelated carriers (Aa) for autosomal recessive inheritance.* (aY, Aa marked)	X-linked recessive	I–1 and II–6 are *Aa*.	autosomal recessive	I–1, I–2, II–5 and II–6 are *Aa*.	autosomal dominant	Neither parent in generation I is affected.
					X-linked dominant	III-3 is not affected.
Pedigree B *Rules out X-linked dominant* *This individual would have to be Aa for autosomal recessive inheritance.* *This individual would have to be aa for autosomal recessive inheritance.* *Rules out X-linked recessive* (Aa markings)	autosomal dominant	I–1, II–4, II–6, and III–4 are *Aa*.	autosomal recessive	I–1, II–3, and II–5 are *Aa* and I–2, II–4, II-6 and III–4 are *aa*.	X-linked recessive	III-7 is not affected.
					X-linked dominant	II-2 is not affected.
Pedigree C *One individual would have to be Aa.* *Rules out dominant inheritance and X-linked recessive.* (Aa, Aa, aa, aa markings)	autosomal recessive	I-1 or I-2 are *Aa*; II-5 and II-6 are *Aa*.	none		X-linked recessive	II-5 is not affected.
					X-linked dominant	II-6 is not affected.
					autosomal dominant	Neither II-5 nor II-6 is affected
Pedigree D *Trait appearing in three successive generations makes autosomal recessive inheritance highly unlikely. Unaffected parent in each mating would have to be Aa* *Would be AY for X-linked dominant* *Rules out X-linked recessive inheritance* (Aa markings)	autosomal dominant	I-1, II-6, II-7, III-8, III-9, and III-10 are *Aa*.	X-linked dominant	I-1 is *Aa*, II-6 is *Aa*, II-7 is *AY*.	Autosomal recessive	II-8 would have to be *Aa*.
					X-linked recessive	II-4 and III-7 are not affected.

21a.   *Evaluate:* This is type 2a.
Deduce > Solve: X-linked recessive traits are more common in males than females and often show up in consecutive generations of a pedigree even in the absence of inbreeding. Autosomal recessive traits are seen equally in males and females and most often occur only in one generation of a pedigree in the absence of inbreeding. See figure.

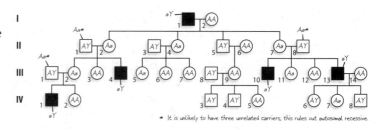

* It is unlikely to have three unrelated carriers; this rules out autosomal recessive.

21b.   *Evaluate:* This is type 2a.
Deduce > Solve: X-linked dominant traits are transmitted from affected fathers to all their daughters but none of their sons, whereas autosomal dominant traits are transmitted from fathers to $\frac{1}{2}$ of their sons and $\frac{1}{2}$ of their daughters. See figure.

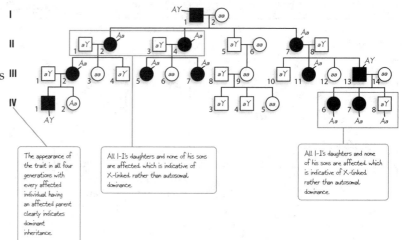

The appearance of the trait in all four generations with every affected individual having an affected parent clearly indicates dominant inheritance.

All I-1's daughters and none of his sons are affected which is indicative of X-linked rather than autosomal dominance.

All I-1's daughters and none of his sons are affected which is indicative of X-linked rather than autosomal dominance.

22.   *Evaluate:* This is type 3. The results for F_1 and F_2 are the same for both crosses. *Deduce:* The results of reciprocal crosses for X-linked traits differ, whereas reciprocal crosses for autosomal traits give the same results. Therefore, it is likely that horns are controlled by an autosomal gene. Because the data show more males than females with horns in both generations, it is likely that sex influences the phenotype. *Solve:* The presence of horns is a sex-influenced trait controlled by a single, autosomal gene "*H*" in these cattle. Males need only one *H* allele, whereas females need two. Cross I: Male is *HH*, female is *hh*; F_1 are $\frac{1}{2}$ *Hh* (horned males) and $\frac{1}{2}$ *Hh* (hornless females); F_2 males are $\frac{3}{4}$ horned ($\frac{1}{4}$ *HH* and $\frac{2}{4}$ *Hh*) and $\frac{1}{4}$ hornless ($\frac{1}{4}$ XY *hh*). F_2 females are $\frac{1}{4}$ horned ($\frac{1}{4}$ *HH*) and $\frac{3}{4}$ hornless ($\frac{2}{4}$ *Hh* and $\frac{1}{4}$ *hh*).

23a.   *Evaluate:* This is type 2 and 3. Recall that the sex-chromosome composition of chickens is ZZ for roosters and ZW for hens. Progeny are $\frac{1}{2}$ fast feathering and $\frac{1}{2}$ slow feathering regardless of sex. *Deduce:* The sex-chromosome composition of chickens is ZZ for roosters and ZW for hens. Assuming that *F*, for fast feathering, is dominant, the male can be either *FF* or *Ff*. The hen must be *f* W. *Solve:* The rooster was heterozygous (*Ff*) and the hen was hemizygous for *f* (*f* W). This would produce $\frac{1}{2}$ *Ff* and $\frac{1}{2}$ *ff* male progeny, and $\frac{1}{2}$ *FW* and $\frac{1}{2}$ *f* W female progeny.

23b.   *Evaluate:* This is type 2 and 3. All their male progeny are fast feathering, and all their female progeny are slow feathering. Recall that the sex-chromosome composition of chickens is ZZ for roosters and ZW for hens. *Deduce:* The Z chromosome of all females is paternally inherited, which is analogous to the X chromosome of mammalian males being maternally inherited. *Solve:* The rooster was homozygous (*ff*) and the hen was hemizygous for *F* (*FW*). This would produce all *Ff* male progeny, and all *f* W female progeny.

24a.   *Evaluate > Deduce > Solve:* This is type 3. Reciprocal crosses involving sex-linked traits give different results.

24b.   *Evaluate:* This is type 2 and 3. Reciprocal crosses involving sex-linked traits give different results. *Deduce:* Since we are not told which sex is heterogametic and which is homogametic, we can state the expected results of reciprocal crosses for a sex-linked recessive trait using these terms as follows. The trait will appear in all the heterogametic progeny and none of the homogametic progeny when the cross is a homogametic parent with the recessive trait crossed to heterogametic parent with the dominant trait. The trait will appear at equal frequencies in both the homogametic and heterogametic sexes in the reciprocal cross of a homogametic parent with the dominant trait to a heterogametic parent with the recessive trait. *Solve:* The results of Cross II suggest that nonspotted is recessive and the female is the heterogametic sex. The results of Cross I are consistent with that hypothesis. The female is the heterogametic sex. Black spot is dominant and nonspotted is recessive. Designating B for black-spot and b for nonspotted, the parents of Cross I are Bb male and bW female. Their progeny are $\frac{1}{4}$ black-spot males (Bb), $\frac{1}{4}$ nonspotted males (bb), $\frac{1}{4}$ black-spot females (BW), and $\frac{1}{4}$ nonspotted females (bW). For Cross II, the parents are a bb male and a BW female. Their progeny are approximately $\frac{1}{2}$ Bb males and $\frac{1}{2}$ bW females.

25a.   *Evaluate:* This is type 2a. In males, X-linked recessive disorders can be inherited only from their mothers. *Deduce:* A woman who has a brother with an X-linked recessive disorder has $\frac{1}{2}$ chance of being a carrier. If she is a carrier, there is a $\frac{1}{2}$ chance that her first son will have the disorder. *Solve:* The overall probability that her first son will have the disorder is $\frac{1}{2} \times \frac{1}{2} = \frac{1}{4}$.

25b.   *Evaluate:* This is type 2a. In males, X-linked recessive disorders can be inherited only from their mothers. *Deduce:* The woman is a carrier for the disorder. *Solve:* The probability that her next son will inherit the X chromosome with the recessive allele is $\frac{1}{2}$.

25c.   *Evaluate:* This is type 2a. In X-linked recessive disorders, a man cannot be a carrier. *Deduce:* The man in question is hemizygous for the wild-type allele because he does not have the disorder. *Solve:* There is no chance that this man's first son will have the disorder, assuming that he mates with an unrelated woman who is not a carrier.

26.   *Evaluate:* This is type 3. *Deduce:* Sex is determined by the presence (or absence) of the *SRY* gene, which is normally located on the Y chromosome. Presence of *SRY* directs development of male sex characteristics, whereas absence of *SRY* results in development of female sex characteristics. *Solve:* Rare sex-reversed males carry an altered X chromosome that contains a fragment of the Y chromosome including the *SRY* gene. Thus, they develop male sex characteristics yet lack a Y chromosome. Rare sex-reversed females carry an altered Y chromosome the lacks the *SRY* gene. Thus, they develop female sex characteristics even though they have a Y chromosome.

27.   *Evaluate:* This is type 1. *Deduce:* Recall that the chromosome theory of heredity states that genes are carried on chromosomes. Given that the genetic traits carried on chromosomes are encoded by DNA, one could restate Galton's law as "Each person inherits half his or her DNA from their parents, ¼ from their grandparents, and $\frac{1}{8}$ from their great grandparents." *Solve:* One should argue in favor of Galton's law, which accurately predicts the pattern of inheritance of chromosomal DNA, as shown in the figure. Individual II-2 inherits one copy of each of the four homologous chromosomes from each parent; thus, II-2 has $\frac{1}{2}$ of each parents' genetic information. During gamete formation, her chromosomes undergo recombination, resulting in each of her chromosomes containing about $\frac{1}{2}$ of the DNA sequence from

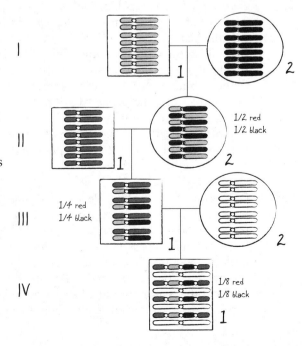

her father and $\frac{1}{2}$ of the DNA sequence from her mother. III-1 inherits one of each homolog from II-2, giving him $\frac{1}{2}$ of $\frac{1}{2}$ or $\frac{1}{4}$ of each of his grandparents' chromosomes. During gamete formation, his chromosomes undergo recombination, resulting in each of his chromosomes containing about $\frac{1}{2}$ of the DNA sequence from his father and $\frac{1}{2}$ of the DNA sequence from her mother. IV-1 inherits one of each homolog from III-1, giving him $\frac{1}{2}$ of $\frac{1}{4}$ or $\frac{1}{8}$ of his great grandparents' chromosomes.

28a. *Evaluate:* This is type 2b with autosomal genes also segregating. The sex-linked recessive trait is Echinus eyes (*ec*), and the autosomal recessive trait is vestigial wings (*v*). *Deduce:* Sex-linked and autosomal traits assort independently. The female was wild type for both traits, and the male was echinus but wild type for wings. Progeny include males and females that are wild type, echinus eyes and wild-type wings, vestigial wings and wild-type eyes, and echinus eyes and vestigial wings. *Solve:* Both normal-winged parents must be heterozygous for the recessive vestigial wing allele (call it *v*) if it appears in their progeny. The male must be hemizygous for echinus eyes (*ecY*) because he shows that trait. The female must be heterozygous for echinus eyes because both male and female progeny have echinus eyes. Therefore, the parents are *ECecVv* female and *ecYVv* male.

28b. *Evaluate:* This is type 2b. The sex-linked recessive trait is echinus eyes (*ec*), and the autosomal recessive traits are vestigial wings (*v*) and ebony body color (*e*). *Deduce:* Sex-linked and autosomal traits assort independently. The female and the male were wild type for all three traits. Progeny include individuals with vestigial, ebony, echinus, and all possible combinations of these traits. However, only males displayed the echinus phenotype. *Solve:* The male parent must be hemizygous for the wild-type echinus allele (*ECY*) since he has wild-type eyes. The female parent must be heterozygous for echinus eyes because she has male offspring with echinus eyes. Both parents must be heterozygous for the autosomal genes because they have offspring that show those recessive traits. Therefore, the parents are *ECecVvEe* female and *ECYVvEe* male.

28c. *Evaluate:* This is type 2b. The sex-linked recessive trait is echinus eyes (*ec*), and the autosomal recessive traits are vestigial wings (*v*) and ebony body color (*e*). *Deduce:* Sex-linked and autosomal traits assort independently. Ebony and vestigial are all expressed by at least some progeny; therefore, each parent carries at least one recessive allele for each autosomal gene. *Solve:* The male parent must be hemizygous for echinus (*ecY*) because he shows that trait. The female parent must be heterozygous for echinus (*ECec*) because both male and female progeny have echinus eyes. The female parent must be homozygous recessive for ebony (*ee*) because she shows that trait. Therefore, the parents are *ECecVvee* female and *ecYVvEe* male.

29a. *Evaluate:* This is type 2a. The number of progeny types, 324 females and 161 males, is a 2:1 ratio of females to males. *Deduce:* If the female fly is heterozygous for a recessive lethal mutation on X, then half her eggs will carry the mutant X chromosome and half will carry the wild-type X chromosome. The half with the mutant X chromosome will contribute to viable offspring only when combined with an X-bearing sperm (producing females): those that combine with a Y-bearing sperm will not produce viable offspring and will not be detected among the progeny. All of the female's eggs carrying the wild-type X chromosome will give rise to viable progeny, half of which will be male and half female. *Solve:* The progeny will include only half the expected numbers of males, causing the ratio of females to males to be 2:1. One possibility is that the female is heterozygous for an X-linked recessive lethal mutation.

29b. *Evaluate:* This is type 2a. *Deduce:* The hypothesis predicts that the half the female F_1 will be heterozygous for the recessive lethal mutation and half will be homozygous wild type. The F_1 females who are homozygous for the wild-type X chromosome will produce equal numbers of male and female progeny, whereas the F_1 females who are heterozygous for the recessive lethal mutation will produce females to males in a 2:1 ratio. *Solve:* Cross the female F_1 progeny to normal, wild-type males and determine the percentage of crosses that give 2:1 ratios of females to males. If the reason for the 2:1 segregation is a recessive lethal mutation on X, then about half of the F_1 females will produce this aberrant ratio and half will not.

30. *Evaluate:* This is type 2a. The children of the first mating, $Cc \times cY$, are expected to be $\frac{1}{2}$ color blind, $\frac{1}{2}$ normal, with equal proportions of male and female progeny in each class. The children of the second mating, $cc \times CY$, are expected to be $\frac{1}{2}$ color blind and $\frac{1}{2}$ normal, but all the color-blind individuals will be male and all the normal individuals will be female. *Deduce > Solve:* The reciprocal matings give different results, which is indicative of X-linked inheritance. The results suggest recessive inheritance because the mother in the first mating is not affected but has affected children. The results differ from those predicted for autosomal recessive inheritance because the second mating yields all males with one phenotype and all females with the other.

31. *Evaluate:* This is type 3. The explanation of the male tortoiseshell cat has two components: one is that it is tortoiseshell, and the other is that it is a male. *Deduce > Solve:* The former is explained by the fact that the cat has two X chromosomes and is heterozygous at a fur color locus on X. Because of the X-inactivation mechanism of dosage compensation in mammals, one of these two X chromosomes is randomly chosen for inactivation early in development in a cell-autonomous manner. Thus the animal is mosaic for X-linked gene expression, resulting in expression of one X-linked fur-color allele in some cells and the other in other cells. The second component, that the cat is male, is explained by the fact the cat has a Y chromosome, which contains the *SRY* gene, which determines male sexual development. Thus the XXY cat is a male that is mosaic for X-linked gene expression, resulting in a male tortoiseshell cat.

32a. *Evaluate > Deduce > Solve:* $A_1YB_1B_2C_1C_2D_1D_2$

32b. *Evaluate > Deduce > Solve:* This is type 1. At metaphase of mitosis, all chromosomes align on the spindle. Homologs do not pair. See figure.

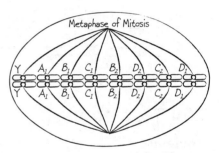

32c. *Evaluate > Deduce > Solve:* This is type 1. At metaphase I of meiosis, homologs pair and each pair aligns on the spindle. See figure.

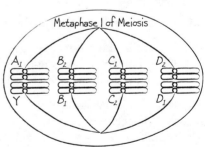

32d. *Evaluate:* This is type 1. *Deduce:* The chromosomes on the top of the figure segregate together during meiosis I, as do the chromosomes on the bottom. The products of the first meiotic division are $A_1A_1\ B_2B_2C_1C_1D_2D_2$ and $YYB_1B_1C_2C_2D_1D_1$. *Solve:* Sister chromatids segregate during meiosis II, resulting in formation of two $A_1B_2C_1D_2$ products and two $YA_2B_1C_2D_1$ products.

32e.   *Evaluate:* This is type 1. *Deduce:* Each pair of homologs can align in one of two possible ways in the spindle. Each pair of homologs aligns independently of the others; thus, the total number of alignments is the product of the number of ways that each pair can align. For each alignment, two different gamete types are produced. *Solve:* There are eight different types of alignments, each of which produces two types of gametes. The total number of gamete types is 8 × 2 = 16.

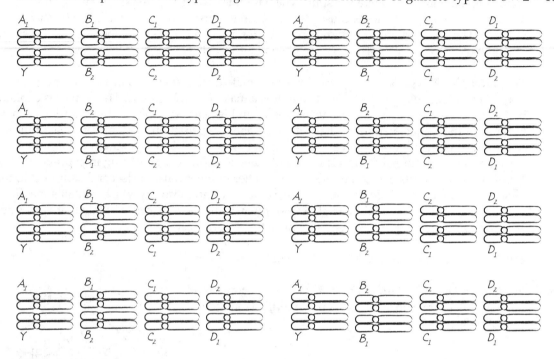

3.04 Test Yourself

Problems

1. Define *interphase*, and list the events that occur during interphase.

2. Two cells of the same type are in interphase of mitosis. One cell contains 3 picograms of chromosomal DNA and the other contains 6 picograms of chromosomal DNA. Use this information to determine the cell cycle stage of each cell.

3. Mitotic cell cycles during early embryonic development in organisms such as frogs and sea urchins lack detectable G_1 and G_2 phases. What is the consequence on cell size and chromosomal DNA content of mitotic cell cycles that lack G_1 and G_2 phases?

4. Name and describe the function of the three types of microtubules present in mitotic and meiotic spindles.

5. Describe the structure and function of the synaptonemal complex.

6. List the four cell cycle checkpoints and describe the function (purpose) of each.

7. Why are cyclins called cyclins, and why are cyclin-dependent kinases called cyclin dependent?

8. pRB and cyclin D1 are both involved in regulating entry into S phase, and entry into S phase is an event that occurs precociously in tumor cells. Why, then, is pRB referred to as a "tumor suppressor" gene whereas cyclin D1 is called a "proto-oncogene"?

9. Cohesin holds sister chromatids together, and the synaptonemal complex holds non-sister chromatids on homologous chromosomes together. What keeps homologous chromosomes attached at metaphase of meiosis I after the synaptonemal complex is removed?

Solutions

1.   *Evaluate:* This is type 1. *Deduce:* The cell cycle can be divided into interphase and M phase, where interphase is the period between successive M phases. The majority of the increase in cell mass and size occurs during interphase. *Solve: Interphase* is the period of the cell cycle after cell division is complete up to the moment before M phase begins. Interphase includes G_1, S phase, and G_2. Events that occur during interphase include preparation for DNA synthesis, DNA synthesis, preparation for M phase, and cell growth.

2.   *Evaluate:* This is type 1. *Deduce:* The cell cycle can be divided into interphase and M phase, where interphase is the period between successive M phases. Interphase includes G_1, S, and G_2 phases. *Solve:* The cell with 3 picograms of DNA is in G_1 phase, and the cell with 6 picograms of DNA is in G_2 phase.

3.   *Evaluate:* This is type 1. *Deduce:* Cells increase in size and mass during interphase and then divide during M phase. *Solve:* Mitotic cell cycles composed only of S and M phase result in an increase in the number of cells and a decrease in cell size.

4.   *Evaluate:* This is type 1. *Deduce > Solve:* Mitotic cell cycles composed only of S and M phase result in an increase in the number of cells and a decrease in cell size.*:* The three types of microtubules present in mitotic and meiotic spindles are kinetochore microtubules, polar microtubules, and astral microtubules. All three microtubule types emanate from the centrosome (in animals) and have their minus ends in the centrosome. Kinetochore microtubules terminate at the kinetochore of chromosomes, which contain microtubule motors that will be used to pull chromosomes along kinetochore microtubules toward the poles during anaphase. Polar microtubules terminate in the center of the spindles and are cross-linked to polar microtubules from the opposite pole by motor proteins that push the spindle poles apart during prometaphase and anaphase. Astral microtubules terminate at the periphery of the cytoplasm and are attached to the cell cortex by motor proteins that pull the spindle poles apart during anaphase.

5.   *Evaluate:* This is type 1. *Deduce > Solve:* The synaptonemal complex (SC) joins non-sister chromatids of homologous chromosomes during prophase I of meiosis. The SC is a layered, protein-based structure composed of lateral elements that bind to each non-sister chromatid and transverse filaments that connect the lateral elements to a central filament. Recombination nodules are embedded within the SC at sites that are presumed to be the site of crossing over, suggesting that the SC plays a role in recombination.

6.   *Evaluate:* This is type 1. *Deduce > Solve:* Four cell cycle checkpoints are described in the text; they are the G_1, S, G_2, and metaphase checkpoints. The G_1 checkpoint monitors factors affecting the decision to enter S phase (initiate DNA synthesis), including cell size, nutrient availability, and DNA damage. If any of these conditions indicate that the cell should not initiate S phase, then cyclin-dependent kinases required for S phase are inhibited. The S-phase checkpoint monitors completion of DNA synthesis and the presence of errors or damage to DNA. Completion of DNA synthesis and repair of damage is required before cells initiate processes that ready them for mitosis. The G_2 checkpoint monitors for completion of DNA synthesis and for DNA damage. If DNA synthesis is incomplete or DNA damage persists, cyclin-dependent kinases required for initiation of M phase are inhibited. The metaphase checkpoint monitors cells for the proper attachment of each pair of sister chromatids to microtubules emanating from opposite poles of the spindle. If all chromosomes are not attached properly, the separase enzyme is kept inactive; this prevents cleavage of cohesion, which in turn prevents initiation of anaphase.

7.   *Evaluate:* This is type 1. *Deduce > Solve:* The abundance of cyclin proteins varies in lockstep with the cell cycle. Since they accumulate and are destroyed in a cyclic fashion and are proteins, they were called cyclins. Cyclin-dependent kinases are kinase enzymes that require, or depend on, cyclin proteins for activity. Cyclins also control the substrate specificity of cyclin-dependent kinases.

8.   *Evaluate:* This is type 1. *Deduce:* Tumor suppressor genes are genes that encode proteins that normally inhibit cell cycle progression. The action of a tumor suppressor is to hold cell division in check. Proto-oncogenes are genes that encode proteins that stimulate cell cycle progression. The action of a proto-oncogene is to promote cell cycle progression, but it does so only in response to normal cell stimuli. *Solve:* pRB is an inhibitor of the transcription factor, E2F. E2F promotes transcription of genes that then drive entry into S phase. Therefore, pRB's normal function is to inhibit cell cycle progression, and mutations that inactivate pRB lead to inappropriate cell cycle progression. Cyclin D1 is a component of a cyclin-dependent kinase that phosphorylates and inhibits pRB function. Therefore, cyclin D1's normal function is to promote cell cycle progression, and mutations that hyperactivate cyclin D1 lead to inappropriate cell cycle progression.

9.   *Evaluate:* This is type 1. *Deduce:* Homologs pair and undergo crossing over during prophase of meiosis I. The crossovers connect the DNA molecules of one chromatid to one of its non-sister chromatids. *Solve:* Crossing over between non-sister chromatids on homologous chromosomes during prophase I of meiosis physically connects non-sister chromatids on homologous chromosomes. This physical connection holds homologs together at metaphase I of meiosis.

4

Gene Interaction

4.01 Genetics Problem-Solving Toolkit

Key Terms and Concepts

- **Codominance:** Both alleles are detectable in a heterozygote.
- **Incomplete dominance (partial dominance):** The phenotype of a heterozygote is between the phenotypes of the two homozygotes.
- **Haplosufficiency:** Individuals containing only one copy of an allele that have a wild-type phenotype.
- **Haploinsufficiency:** Individuals containing only one copy of an allele that have a mutant phenotype.
- **Gain-of-function mutation (neomorphic mutation):** Mutation that creates a mutant allele that has a function qualitatively different from the function of the wild-type allele.
- **Hypermorphic mutation:** Mutation that creates a mutant allele whose function is quantitatively greater than the wild-type allele.
- **Null mutation (amorphic mutation):** Mutation that creates a mutant allele with no measurable function. Null mutations are complete loss-of-function mutations.
- **Hypomorphic mutation (leaky mutation):** Mutation that creates a mutant allele with quantitatively decreased function as compared to the wild-type allele.
- **Genetic complementation:** Occurs when a cross between two individuals with the same mutant phenotype results in offspring having the wild-type phenotype—the mutations complement each other.
- **Complementation group:** Mutations in the same gene are part of the same complementation group, while mutations in different genes are in different complementation groups.
- **Epistatic interaction (epistasis):** The interaction of different genes in determining a phenotype, characterized by a ratio of phenotypes in the progeny of a dihybrid cross that differs from the expected 9:3:3:1 ratio.
- **Pleiotropy:** The effect of a mutation in one gene on more than one phenotype.
- **Genetic heterogeneity:** When a mutant phenotype is caused by mutations in more than one gene.
- **Variable expressivity:** Severity of a mutant phenotype varies among individuals with a mutant genotype.
- **Sex-limited trait:** A sex-influenced trait that is exhibited only by males or only by females, but not by both.

Key Analytical Tools

Chi-square formula: $\chi^2 = \sum \frac{(O-E)^2}{E}$

Key Genetic Relationships

Gene interaction:	None	Complementary	Duplicate	Dominant	Recessive epistasis	Dominant epistasis	Dominant supression
Phenotype ratio:	9:3:3:1	9:7	15:1	9:6:1	9:3:4	12:3:1	13:3
$\frac{1}{16}$ AABB / $\frac{2}{16}$ AaBB / $\frac{2}{16}$ AABb / $\frac{4}{16}$ AaBb	$\frac{9}{16}$ A–B–	$\frac{9}{16}$ A–B–	A–B–	$\frac{9}{16}$ A–B–	$\frac{9}{16}$ A–B–	A–B– ($\frac{12}{16}$)	$\frac{9}{16}$ A–B–
$\frac{1}{16}$ AAbb / $\frac{2}{16}$ Aabb	$\frac{3}{16}$ A–bb	A–bb	$\frac{15}{16}$ A–bb	A–bb ($\frac{6}{16}$)	$\frac{3}{16}$ A–bb	–bb	$\frac{3}{16}$ A–bb
$\frac{1}{16}$ aaBB / $\frac{2}{16}$ aaBb	$\frac{3}{16}$ aaB–	$\frac{7}{16}$ aaB–	aaB–	aaB–	aaB– ($\frac{4}{16}$)	$\frac{3}{16}$ aaB–	aaB– ($\frac{4}{16}$)
$\frac{1}{16}$ aabb	$\frac{1}{16}$ aabb	aabb	aabb	$\frac{1}{16}$ aabb	$\frac{1}{16}$ aabb	aabb	aabb

Genotype ratio (left column labels)

4.02 Types of Genetic Problems

1. Provide a hypothesis to explain the genetic basis of a phenotype.
2. Fit genetic data to a molecular model of a metabolic or developmental pathway.
3. Predict the outcome of a cross based on a genetic hypothesis.
4. Determine the genotypes of individuals based on a genetic hypothesis.
5. Interpret the results of or design a complementation test.

1. Provide a hypothesis to explain the genetic basis of a phenotype.

This type of problem provides you with information on the genetic basis of a phenotype and asks you to develop a hypothesis or model consistent with that information. The hypothesis typically addresses whether one or two genes are segregating in the cross and the dominance relationship of the alleles at each locus. For crosses in which two genes are segregating, you should also consider whether there is evidence for epistasis and, if so, which type of epistasis. You may also be asked to design crosses that will test your hypothesis or use chi-square statistics to test the goodness of fit of your hypothesis. See Problem 7 for an example.

Example Problem: A certain species of morning glories produces flowers that are blue, red, or purple. Two pure-breeding purple lines are crossed and produce F_1 progeny that all make blue flowers. The F_1 are allowed to self and produce 320 F_2 progeny with the following distribution: 185 blue, 115 purple, and 20 red. Use this information to deduce the genetic basis of flower color in this species of morning glories.

Solution Strategies	Solution Steps
Evaluate	
1. Identify the topic this problem addresses, and explain the nature of the requested answer.	1. This problem tests your ability to develop a genetic hypothesis based on the results of a cross. The answer will indicate the number of genes involved and the phenotype of all possible genotypes.
2. Identify the critical information given in the problem.	2. Flower color can be purple, blue, or red. Pure-breeding purple plants produce blue-flowering F_1, which in turn produce 185 blue-, 115 purple-, and 20 red-flowering F_2 plants.

(continued)

Solution Strategies	Solution Steps
Deduce	
3. Determine whether the F₁ progeny suggest segregation of one or two genes and what information on dominance is displayed.	3. For the F₁, the purple-flowered parents are pure-breeding (i.e., homozygous at genetic loci controlling flower color). Their progeny are blue, so the parents are homozygous recessive for mutations in two different genes.
4. Determine whether the F₂ confirm the number of genes segregating and whether they suggest epistasis.	4. The blue F₁ produce three types of F₂, including $\frac{20}{320} = \frac{1}{16}$ that produce red flowers. The $\frac{1}{16}$ red phenotype suggests two genes segregating. The appearance of three colors in the F₂, including a color not apparent in parents or F₁, suggests epistasis.
Solve	
5. Apply the genetic model of epistasis suggested from the analysis to the parents, F₁, and F₂.	5. Assuming two genetic loci controlling flower color, the results are these: **Parents** must be homozygous recessive at different loci: *AAbb* and *aaBB*. **F₁** would therefore be *AaBb*. **F₂** would therefore be $\frac{9}{16}$ *AaBb*, $\frac{3}{16}$ *A_bb*, $\frac{3}{16}$ *aaB*, and $\frac{1}{16}$ *aabb*. $\frac{9}{16}$ will be blue, $\frac{3}{16} + \frac{3}{16}$ would be purple, and $\frac{1}{16}$ would be red. Purple *AAbb* and *aaBB* and red *aabb* would be pure-breeding.
6. Assign phenotypes to all possible genotypes.	6. **Answer:** Purple-flowering plants are either *A_bb* or *aaB_*. Blue-flowering plants are *A_B_*. Red-flowering plants are *aabb*.

2. Fit genetic data to a molecular model of a metabolic or developmental pathway.

This type of problem provides you with information on the genetic basis of a developmental or metabolic pathway and asks you to infer the order of events in the pathway. The answer typically includes a description of the steps in the pathway and the identity of the gene that controls each step. See Problem 8 for an example.

Example Problem: Wild-type bacteria can grow on minimal medium. Four mutants that cannot grow on minimal medium but can grow on minimal medium supplemented with the nutrient "C" are isolated. It is suspected that metabolites A, B, and D are in the biochemical pathway for synthesis of C, so each mutant is tested for the ability to grow on minimal medium supplemented with these metabolites. Mutant 1 can grow on minimal medium supplemented with A, but not with B or D. Mutant 2 is unable to grow on minimal medium supplemented with A, B, or D. Mutant 3 is able to grow on minimal medium supplemented with B or A, but not D. Mutant 4 can grow on minimal medium supplemented with A, B, or D. Use this information to determine the order of metabolites A, B, and D in the C biosynthetic pathway, and indicate which step in the pathway is defective in each of the bacterial mutants.

Solution Strategies	Solution Steps
Evaluate	
1. Identify the topic this problem addresses, and explain the nature of the requested answer.	1. This problem tests your ability to interpret the results of a genetic analysis of a metabolic pathway. The answer will be a metabolic pathway showing the steps in the synthesis of "C" involving the metabolites A, B, and D, and will identify the step affected by each of the bacterial mutants.
2. Identify the critical information given in the problem.	2. All four mutants are unable to grow on minimal medium. The metabolites are A, B, and D, which are intermediates in the pathway for synthesis of C. Mutant 4 can grow on minimal medium supplemented with A, B, C, or D; mutant 3 can grow only if A, B, or C is supplied; mutant 1 can grow only if A or C is supplied; and mutant 2 can grow only if C is supplied.
Deduce	
3. Determine the relative order of the steps blocked in each mutant by ordering the mutants according to how many different supplements can be used to support their growth. **TIP:** The mutant defective for the earliest step in the pathway will be able to grow on the greatest number of supplemented minimal media.	3. Mutant 4 can grow on 4 different supplemented media, mutant 3 can grow on 3 media, mutant 1 can grow on 2 different media, and mutant 2 can grow on only 1 medium. The order of the steps blocked by mutants is mutant 4, 3, 1, and 2.
4. Determine the relative order of the metabolites by using the order of steps blocked in each mutant determined in part 3 and the information on the specific supplements that support growth of each mutant. **TIP:** A mutant can grow on medium supplemented with any metabolite that is part of the pathway *after* the step blocked in the mutant.	4. Mutant 4 blocks a step before all four metabolites; therefore, mutant 4 does not provide information on the order of the metabolites. Mutant 3 blocks a step before A, B, and C but not D; therefore, D is the earliest metabolite identified by this study. Mutant 1 blocks a step before A and C but not B or D; therefore, B occurs after D but before A and C. Mutant 2 blocks a step before C but not A, B, or D; therefore, A occurs after B and D but not before C.
Solve	
5. Draw the metabolic pathway, using arrows indicating steps from one metabolite to the next and placing the number of the mutant above the arrow, indicating the step blocked by each of the mutants.	5. **Answer:** See figure. $\xrightarrow{\text{4}}$ D $\xrightarrow{\text{3}}$ B $\xrightarrow{\text{1}}$ A $\xrightarrow{\text{2}}$ C

3. Predict the outcome of a cross based on a genetic hypothesis.

This type of problem provides you with a hypothesis or model for the genetic basis of a trait and the genotype of two parents and asks you to predict the outcome of the cross. The answer typically includes the phenotypic distribution among the progeny but may also include information on progeny genotypes. See Problem 25 for an example.

Example Problem: In rats, the genes B and D control fur color as follows: B_dd rats have black fur, $bbD\_$ have yellow fur, $B\_D\_$ have brown fur, and $bbdd$ rats have cream-colored fur. If a pure-breeding brown rat and a pure-breeding cream-colored rat mate to produce an F_1, and the F_1 are allowed to interbreed to produce an F_2, what is the expected phenotypic distribution among the F_2?

Solution Strategies	Solution Steps
Evaluate	
1. Identify the topic this problem addresses, and explain the nature of the requested answer.	1. This problem tests your ability to apply knowledge of epistasis to predict the outcome of a cross. The answer will be a list of predicted phenotypes and the expected proportion of each phenotype.
2. Identify the critical information given in the problem.	2. The B and D genes control fur color in this problem, such that B_dd rats are black, $bbD\_$ rats are yellow, $B\_D\_$ rats are brown, and $bbdd$ rats are cream-colored. A pure-breeding brown rat is crossed to a pure-breeding cream-colored rat to produce F_1 that interbreed to produce F_2.
Deduce	
3. Determine the genotype of the parents of each cross, and determine the genotypic frequency of the progeny of each cross. **TIP:** Pure-breeding strains are homozygous at all loci affecting fur color.	3. The genotypes are as follows: **Parents:** Brown is $BBDD$ and cream-colored is $bbdd$. **F_1:** All are $BbDd$. **F_2:** $\frac{9}{16}$ are $B\_D\_$, $\frac{3}{16}$ are B_dd, $\frac{3}{16}$ are $bbD\_$, and $\frac{1}{16}$ are $bbdd$.
4. Assign phenotypes to the genotypes of the F_2 and indicate the phenotypic frequency.	4. **Answer:** $\frac{9}{16}$ will be brown ($B\_D\_$), $\frac{3}{16}$ will be black (B_dd), $\frac{3}{16}$ will be yellow ($bbD\_$), and $\frac{1}{16}$ will be cream-colored ($bbdd$).

4. Determine the genotypes of individuals based on a genetic hypothesis.

This type of problem provides you with information on the genetic basis of a trait and the genotypes or phenotypes of progeny of a cross or mating, and it asks you to use that information to infer the parental genotypes. It may also ask you to match progeny to parents based on phenotypic information. See Problem 9 for an example.

Example Problem: In rats, the genes B and D control fur color as follows: B_dd rats have black fur, $bbD\_$ have yellow fur, $B\_D\_$ have brown fur, and $bbdd$ rats have cream-colored fur. A brown rat and a black rat have a litter of 24 pups, 9 with brown fur, 9 with black fur, 3 with yellow fur, and 3 with cream-colored fur. Use this information to determine the genotype of the parents.

Solution Strategies	Solution Steps
Evaluate	
1. Identify the topic this problem addresses, and explain the nature of the requested answer.	1. This problem tests your ability to apply knowledge of epistasis to determine parental genotypes. The answer will be a genotype for each parent.
2. Identify the critical information given in the problem.	2. The B and D genes control fur color in this problem, such that B_dd rats are black, $bbD\_$ rats are yellow, $B\_D\_$ rats are brown, and $bbdd$ rats are cream-colored. A brown rat and a black rate mate and have 9 brown pups, 9 black pups, 3 yellow pups, and 3 cream-colored pups.
Deduce	
3. Determine the possible phenotypes of the parents given the genetic model and then predict the progeny phenotypic distribution for each possible mating.	3. The possible genotypes of parents are brown $B\_D\_$ and black B_dd. The progeny phenotypic distribution would be the following: • $BBDD \times B\_dd$ would be all brown. • $BBDd \times B\_dd$ would be $\frac{1}{2}$ brown and $\frac{1}{2}$ black. • $BbDD \times BBdd$ would be all brown. • $BbDD \times Bbdd$ would be $\frac{3}{4}$ brown and $\frac{1}{4}$ yellow. • $BbDd \times BBdd$ would be $\frac{1}{2}$ brown and $\frac{1}{2}$ black. • $BbDd \times Bbdd$ would be $\frac{3}{8}$ brown, $\frac{3}{8}$ black, $\frac{1}{8}$ yellow, and $\frac{1}{8}$ cream-colored.
Solve	
4. Calculate the ratio of phenotypes among the progeny.	4. The distribution of the progeny was $\frac{9}{24} = \frac{3}{8}$ brown, $\frac{9}{24} = \frac{3}{8}$ black, $\frac{3}{24} = \frac{1}{8}$ yellow, and $\frac{3}{24} = \frac{1}{8}$ cream-colored.
5. Match the phenotypic distribution of progeny to the predicted progeny distribution for one of the possible parental combinations.	5. **Answer:** The actual progeny distribution matches the expected distribution for the cross $BbDd \times Bbdd$; therefore, the brown parent was $BbDd$ and the black parent was $Bbdd$.

5. Interpret the results of or design a complementation test.

This type of problem asks you to analyze the results of crosses between pure-breeding mutants with the same mutant phenotype to determine whether the mutants have mutations in the same gene or in different genes (a complementation test). You may also be asked how to determine whether two mutants harbor mutations in the same or different genes. See Problem 29 for an example.

Example Problem: Feather color in parakeets is controlled by multiple genes that encode enzymes in two metabolic pathways. Wild-type feather color is green. You have isolated 10 pure-breeding, recessive, white mutant lines and wish to determine how many different genes are identified by these mutants. You cross each mutant line with each other mutant line and assemble a table indicating the results, where + indicates that all progeny have green feathers and – indicates that all progeny have white feathers. The table is shown below. Use this information to determine how many genes are identified by these results and which mutants have mutations in the same gene.

Mutant	1	2	3	4	5	6	7	8	9	10
1	−									
2	+	−								
3	−	+	−							
4	−	+	−	−						
5	+	−	+	+	−					
6	+	+	+	+	+	−				
7	+	−	+	+	−	+	−			
8	−	+	−	−	+	+	+	−		
9	+	+	+	+	+	−	+	+	−	
10	+	+	+	+	+	−	+	+	−	−

Mutant

Solution Strategies	Solution Steps
Evaluate	
1. Identify the topic this problem addresses, and explain the nature of the requested answer.	1. This problem asks you to interpret the results of a complementation test. The answer will be a number of genes and lists of mutants corresponding to each gene.
2. Identify the critical information given in the problem.	2. Wild-type parakeets have green feathers, and mutant parakeets have white feathers. The table lists the results of crosses of each mutant line against each other mutant.
Deduce	
3. Identify the crosses that produce white parakeets. These mutants have mutations in the same gene.	3. Mutant 1 crossed with mutant 3, 4, or 8 produces white progeny; therefore, mutants 1, 3, 4, and 8 are in the same complementation group. Mutant 2 crossed with mutants 5 and 7 produces white progeny; therefore, mutants 2, 5, and 7 are in the same complementation group. Mutant 6 crossed with mutants 9 or 10 produces white progeny; therefore, mutants 6, 9, and 10 are in the same complementation group.
Solve	
4. List the mutants that are in the same complementation groups.	4. This study identifies three complementation groups: 1, 3, 4 and 8; 2, 5, and 7; and 6, 9, and 10.
5. Determine the number of genes identified by this study, and state which gene is mutant in each mutant parakeet line.	5. **Answer:** This study identifies three genes (let's call them *A*, *B*, and *C*): mutants 1, 3, 4, and 8 have mutations in gene *A*; mutants 2, 5, and 7 have mutations in gene *B*; and mutants 6, 9, and 10 have mutations in gene *C*.

4.03 Solutions to End-of-Chapter Problems

1.   *Evaluate:* This is type 1. *Incomplete penetrance* indicates that less than 100% of individuals with a mutant genotype have the mutant phenotyape. *Variable expressivity* indicates that the severity of a mutant phenotype differs among individuals with the same mutant genotype. *Deduce > Solve:* Incomplete penetrance and variable expressivity are distinguishable in that they describe different aspects of the relationship between genotype and phenotype. The terms are not mutually exclusive, however; a single mutation can show both incomplete penetrance *and* variable expressivity.

2.   *Evaluate:* This is type 1. *Epistasis* indicates that two or more genes interact to contribute to a particular phenotype. *Pleiotropy* indicates that one mutant allele contributes to two or more mutant phenotypes. ***Deduce > Solve:*** Epistasis and pleiotropy can be distinguished by inheritance patterns in pedigrees or from crosses. If the inheritance pattern of one phenotype indicates that more than one gene is segregating, then epistasis is occurring. If two or more phenotypes are inherited together in a pattern that indicates segregation of alleles of a single gene, then pleiotropy is occurring.

3.   *Evaluate:* This is type 5. *Deduce:* Crossing homozygous recessive mutants creates a hybrid whose phenotype is mutant if the two mutations are in the same gene. In contrast, the hybrid's phenotype will be wild type if the two mutations are in different genes. *Solve:* Cross the mutant lines, and examine the height of the F_1 hybrids. If the F_1 hybrids are short, the two mutations are in the same gene. If the F_1 hybrids are normal height (tall), then the two mutations are in different genes.

4a.   *Evaluate:* This is type 2. Complete medium contains a carbon source and all inorganic (salts) and organic (amino acids, vitamins, etc.) nutrients required for growth, whereas minimal medium lacks organic nutrients. *Deduce:* Bacteria that do not require any organic nutrients besides a carbon source are called prototrophs, whereas those that require a carbon source and one or more additional organic nutrients are called auxotrophs. *Solve:* Only 3 of the 15 colonies could not grow without a carbon source and one or more additional organic nutrients. The bacteria in the 12 colonies that grew on minimal medium are prototrophs, whereas those in the 3 colonies that could not grow on minimal medium were auxotrophs.

4b.   *Evaluate:* This is type 2. Bacteria that cannot grow on minimal medium but can grow on minimal medium plus serine are able to synthesize all organic nutrients except serine. *Deduce:* These bacteria require serine for growth because they carry mutations in one or more genes that encode enzymes required for serine biosynthesis. *Solve:* The bacteria in the three colonies that could not grow on minimal medium but could grow on minimal medium plus serine are serine-requiring auxotrophs. They carry mutations in one or more genes required for the biosynthesis of serine.

4c.   *Evaluate:* This is type 2. Serine-requiring auxotrophs can grow on minimal medium supplemented with serine or with any metabolic intermediate that is in the pathway after ("downstream of") the step catalyzed by the enzyme that they lack. *Deduce:* Mutant 1 cannot grow on minimal medium supplemented with any of the intermediates; therefore, that mutant is deficient for the enzyme that converts the final intermediate (3-PS) into serine. Mutant 2 can grow on minimal medium supplemented with the second (3-PHP) or third (3-PS) intermediate but not the first (3-phosphoglycerate); therefore, that mutant is deficient in the enzyme that converts 3-phosphoglycerate into 3-PHP. Mutant 3 can grow on minimal medium supplemented with the third intermediate (3-PS) but not the first two; therefore, that mutant is defective in the enzyme that converts 3-PHP into 3-PS. *Solve:* Mutant 1 carries a mutation in the gene coding for a component of enzyme C, mutant 2 carries a mutation in the gene coding for a component of enzyme A, and mutant 3 carries a mutation in the gene coding for enzyme B.

5a.   *Evaluate:* This is type 4. The information provided indicates that there are two genes, *Y* (yellow pigment) and *B* (blue pigment), involved in pigment synthesis and that the combination of both pigments results in a novel color (green). Recall the interaction between the *Drosophila vermilion* and *brown* genes. *Deduce:* Green birds produce both yellow and blue; they must have at least one dominant *Y* allele and one dominant *B* allele. Yellow birds produce only one pigment; they must have one dominant *Y* allele and be homozygous recessive for *b*. Blue birds produce only one pigment; they must have one dominant *B* allele and be homozygous recessive for *y*. Albino birds do not produce either pigment; they must be homozygous recessive for both *b* and *y*. *Solve:* Green is *Y_B_*; yellow is *Y_bb*; blue is *yyB_*; and albino is *yybb*.

5b. *Evaluate:* This is type 4. Pure-breeding birds are homozygous at all genetic loci.
Deduce > Solve: Pure-breeding green is *YYBB*, and pure-breeding albino is *yybb*.

5c. *Evaluate:* This is type 3. Recall from Chapter 3 that crossing two pure-breeding lines results in dihybrid progeny. *Deduce > Solve:* The F_1 are dihybrids with the genotype *YyBb*. *YyBb* birds are green.

5d. *Evaluate:* This is type 3. *YyBb* × *YyBb* is a dihybrid cross. Recall from Chapter 3 that the genotypic distribution of a dihybrid cross with simple dominance at both genes is $\frac{9}{16}$ Y_B_, $\frac{3}{16}$ Y_bb, $\frac{3}{16}$ yyB_, $\frac{1}{16}$ yybb. *Deduce > Solve:* The phenotypic distribution is $\frac{9}{16}$ green, $\frac{3}{16}$ yellow, $\frac{3}{16}$ blue, $\frac{1}{16}$ albino.

5e. *Evaluate:* This is type 4. One parent was green (Y_B_) and the other parent was yellow (Y_bb). Their offspring were 12 green (Y_B_), 4 blue (yyB_), 13 yellow (Y_bb), and 3 albino (yybb). *Deduce > Solve:* Because albino progeny were produced, each parent must have at least one recessive *y* allele and one recessive *b* allele. The green parent was *YyBb* and the yellow parent was *Yybb*.

6. *Evaluate:* This is type 4. Recall that blood type O indicates an $I^O I^O$ genotype. Blood type A indicates a genotype of $I^A I^A$ or $I^A I^O$. Blood type B indicates a genotype of $I^B I^B$ or $I^B I^O$. Blood type M indicates a genotype of *MM*. Blood type N indicates a genotype of *NN*. Blood type MN indicates a genotype of *MN*. *Deduce:* Child a has the genotype $I^B I^B MM$ or $I^B I^O MM$. Child b has the genotype $I^O I^O MM$. Child c has the genotype $I^A I^B MN$. Child d has the genotype $I^B I^B NN$ or $I^B I^O NN$. *Solve:* Child a's parents cannot be 2 or 4, because both parents in each pair do not have an *M* allele. Child b's parents could not be 2 or 4, because both parents in each pair do not have an *M* allele. Child b's parents also could not be 3, because the mother does not have an I^O allele. Child c's parents could not be 1 or 2, because none have an I^A allele. Child d's parents could not be 1, because they do not have an *N* allele. Child b's parents must be 1. Child a's parents could be 1 or 3; however, since child b's parents are 1, child a's parents must be 3. Child c's parents could be 3 or 4; however, since child a's parents are 3, child c's parents must be 4. Child d's parents could be 2, 3, or 4; however, since child a's parents are 3 and child c's parents are 4, child d's parents must be 2.

7. *Evaluate:* This is type 1. Pure-breeding black crossed to pure-breeding green resulted in all black F_1. Mating of the F_1 resulted in 179 black, 81 green, and 60 brown F_2. *Deduce:* $\frac{179}{320}$ were black, $\frac{81}{320}$ were green, and $\frac{60}{320}$ were brown, which is very close to a 9:4:3 ratio. A 9:4:3 ratio indicates that two genes are involved with epistasis, such that A_B_ are black, A_bb are brown, and aaB_ and aabb are green. *Solve:* One hypothesis to explain the results is that two genes are involved with 9:4:3 epistasis, such that A_B_ is black, A_bb is brown, and aaB_ and aabb are green.

8a. *Evaluate:* This is type 2. Recall that a dihybrid cross producing a phenotypic ratio of 9:6:1 indicates dominant gene interaction.
Deduce > Solve: At least one copy of each dominant allele results in dark blue, at least one copy of either dominant allele produces light blue, and the absence of either dominant allele produces white. See figure.

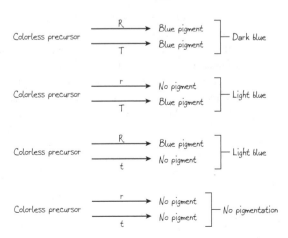

8b.   *Evaluate:* This is type 2. Recall that a dihybrid cross producing a phenotypic ratio of 12:3:1 indicates dominant epistasis. *Deduce > Solve:* At least one copy of both dominant alleles results in white; at least one copy of one of the dominant alleles also results in white, but at least one copy of the other dominant allele produces green; and the absence of either dominant allele produces yellow. See figure.

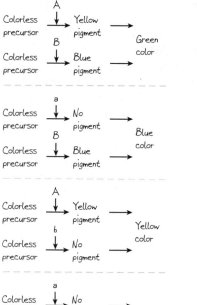

8c.   *Evaluate:* This is type 2. Recall that a dihybrid cross producing a phenotypic ratio of 9:3:3:1 indicates no gene epistasis. *Deduce > Solve:* At least one copy of both dominant alleles results in green, at least one copy of one dominant allele results in yellow, at least one copy of the other dominant allele results in blue, and the absence of either dominant allele results in white. See figure.

8d.   *Evaluate:* This is type 2. Recall that a dihybrid cross producing a phenotypic ratio of 9:7 indicates complementary gene interaction. *Deduce > Solve:* At least one copy of both dominant alleles is required for red pigment. The absence of either dominant allele results in white. See figure.

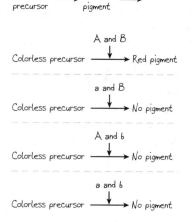

8e. *Evaluate:* This is type 2. Recall that a dihybrid cross producing a phenotypic ratio of 15:1 indicates duplicate gene interaction. *Deduce > Solve:* One copy of any dominant allele results in black. The absence of either dominant allele results in white. See figure.

Colorless precursor → (A or B) → Black pigment

Colorless precursor → (a or B) → Black pigment

Colorless precursor → (A or b) → Black pigment

Colorless precursor → (a or b) → No pigment

8f. *Evaluate:* This is type 2. Recall that a dihybrid cross producing a phenotypic ratio of 9:3:4 indicates recessive epistasis. *Deduce > Solve:* If both dominant alleles are present, the result is black. At least one copy of one specific dominant allele is required for gray. If that dominant allele is not present, the result is white, regardless of whether the other dominant allele is present. See figure.

Colorless precursor → (A) Gray pigment → (B) Black pigment → Black color

Colorless precursor → (a) No gray pigment → (B) No black pigment → No color

Colorless precursor → (A) Gray pigment → (b) No black pigment → Gray color

Colorless precursor → (a) No gray pigment → (b) No black pigment → No color

8g. *Evaluate:* This is type 2. Recall that a dihybrid cross producing a phenotypic ratio of 13:3 indicates dominant suppression. *Deduce > Solve:* The only way to produce green is to have at least one dominant allele at one specific locus and homozygous recessive alleles at the other locus. The second dominant allele functions as a suppressor; therefore, if both dominant alleles are present, it will suppress the expression of the first. See figure.

Colorless product → (A) ⊣ B → No green product → No color

Colorless product → (a) ⊣ B → No green product → No color

Colorless product → (A) ⊣ b → Green product → Green color

Colorless product → (a) ⊣ b → No green product → No color

9a. *Evaluate:* This is type 4. The child's blood type is A, Rh−, M; the mother's blood type is O, Rh−, MN; and the father's blood type is A, Rh+, M. Blood type A indicates a genotype of either I^AI^A or I^Ai. Blood type Rh− indicates a genotype of rr. Blood type M indicates a genotype of MM. The child's genotype must be I^A_rrMM. Blood type O indicates a genotype of ii. Blood type MN indicates a genotype of MN. The mother's genotype must be $iirrMN$. Rh+ genotype indicates a genotype of RR or Rr. The father's genotype must be $I^A\_$, $R\_$, and MM. *Deduce:* For the child to have Rh−, rr, the father must have the genotype Rr. We cannot determine whether the father has I^AI^A or I^Ai. *Solve:* The mother is $iirrMN$. The father is I^A_RrMM.

9b. *Evaluate:* This is type 3. To have B blood type, the child must receive an I^B allele from the father and an i allele from the mother. To be Rh−, the child must receive an r allele from both the father and the mother. To be MN, the child must receive an M allele from the father and an N allele from the mother. *Deduce:* Determine the probability of the child inheriting the required gametes from the father: $\frac{1}{2}(I^B) \times \frac{1}{2}(r) \times \frac{1}{2}(M) = \frac{1}{8}$. Determine the probability of the child inheriting the required gametes from the mother: $\frac{1}{2}(i) \times \frac{1}{2}(r) \times 1(N) = \frac{1}{4}$. **Solve:** The expected proportion of B, Rh–, MN children can be found by multiplying the probabilities: $\frac{1}{8} \times \frac{1}{4} = \frac{1}{32}$.

9c.   *Evaluate:* This is type 4. The child has the genotype *iirrMN* and must have received *irM* from the mother. *Deduce:* For the man in question to be the father, he would have to donate *irN*. **Solve:** Based on the father's phenotype, the genotype $I^B iRrN$ is possible. **Solve:** No. It is possible for the child to have inherited *irM* from the mother and *irN* from the man in question.

10a.   *Evaluate:* This is type 4. We know that *B_D_* is brown, *B_dd* is black, *bbD_* is yellow, and *bbdd* is cream. *Deduce:* The phenotypic ratio of the progeny is 1 brown (*B_D_*) : 1 black (*B_dd*) : 1 yellow (*bbD_*) : 1 cream (*bbdd*). Since there are *bbdd* progeny, each parent must have carried a recessive *b* and recessive *d* allele. For the B locus, there is a 1:1 ratio of *B_* to *bb*, consistent with a test cross of a *Bb* hybrid. The same is true for the D locus. **Solve:** At the B locus, one parent was *Bb* and the other was *bb*. At the D locus, one parent was *Dd* and the other was *dd*. This could have been *BbDd* (brown) × *bbdd* (cream) or *Bbdd* (black) × *bbDd* (yellow).

10b.   *Evaluate:* This is type 4. We know that *B_D_* is brown, *B_dd* is black, *bbD_* is yellow, and *bbdd* is cream. *Deduce:* The phenotypic ratio of the progeny is 3 brown (*B_D_*) : 3 yellow (*bbD_*) : 1 black (*B_dd*) : 1 cream (*bbdd*). Since there are *bbdd* progeny, both must have carried a recessive *b* and a recessive *d* allele. For the D locus, there was a 3:1 ratio of *D_* to *dd* consistent with a monohybrid cross. For the B locus, there was a 1:1 ratio of *B_* to *bb* consistent with a testcross of a monohybrid.

10c.   *Evaluate:* This is type 4. We know that *B_D_* is brown, *B_dd* is black, *bbD_* is yellow, and *bbdd* is cream. *Deduce:* The 9 black to 7 brown progeny is close to a 1 black : 1 brown ratio. Black rats must have a *B* and *D* allele, whereas brown rats must have a *B* allele and *dd* alleles. Since all progeny have a *B* allele, at least one of the parents must have been *BB*. Since half are black (*B_D_*) and half are brown (*B_dd*), one rat must have been *Dd* and the other *dd*. **Solve:** At the B locus, one parent was *BB* and the other parent could have had any genotype (*BB*, *Bb*, or *bb*). At the D locus, one parent was *Dd* and the other was *dd*.

11a.   *Evaluate:* This is type 3. This problem adds a third locus, C, to the model given in Problem 10. *C_* rats are able to express fur pigment and will be either brown (*B_D_C_*) or black (*B_ddC_*) or yellow (*bbD_C_*) or cream (*bbddC_*). The *cc* rats cannot express fur pigment and are albino regardless of their genotype at the B and D loci. *Deduce:* In the cross *BbDDCc* × *BbDdCc*, $\frac{3}{4}$ of the progeny will be *B_* and $\frac{1}{4}$ *bb*, all of the progeny will be *D_*, and $\frac{3}{4}$ will be *C_* and $\frac{1}{4}$ *cc*. Therefore, $\frac{3}{4} \times 1 \times \frac{3}{4}$ will be *B_D_C_* (brown), $\frac{1}{4} \times 1 \times \frac{3}{4}$ will be *bbD_C_* (yellow), $\frac{3}{4} \times 1 \times \frac{1}{4}$ will be *BbD_cc* (albino), and $\frac{1}{4} \times 1 \times \frac{1}{4}$ will be *bbD_cc* (albino). **Solve:** $\frac{9}{16}$ will be brown, $\frac{3}{16}$ will be yellow, and $\frac{4}{16}$ will be albino.

11b.   *Evaluate:* This is type 3. This problem adds a third locus, C, to the model given in Problem 10. The *C_* rats are able to express fur pigment and will be either brown (*B_D_C_*) or black (*B_ddC_*) or yellow (*bbD_C_*) or cream (*bbddC_*). The *cc* rats cannot express fur pigment and are albino regardless of their genotype at the B and D loci. *Deduce:* In the cross *BBDdcc* × *BbddCc*, all of the progeny will be *B_*, $\frac{1}{2}$ will be *Dd* and $\frac{1}{2}$ *dd*, and $\frac{1}{2}$ will be *Cc* and $\frac{1}{2}$ *cc*. Therefore, $\frac{1}{2} \times \frac{1}{2}$ will be *B_DdCc* (brown); $\frac{1}{2} \times \frac{1}{2}$ will be *B_Ddcc* (albino); $\frac{1}{2} \times \frac{1}{2}$ will be *B_ddCc* (black); $\frac{1}{2} \times \frac{1}{2}$ will be *B_ddcc* (albino); $\frac{1}{2} \times \frac{1}{2}$ will be *B_DdCc* (brown); $\frac{1}{2} \times \frac{1}{2}$ will be *B_Ddcc* (albino); $\frac{1}{2} \times \frac{1}{2}$ will be *B_ddCc* (black); and $\frac{1}{2} \times \frac{1}{2}$ will be *B_ddcc* (albino). **Solve:** $\frac{1}{8} + \frac{1}{8} = \frac{1}{4}$ will be brown, $\frac{1}{8} + \frac{1}{8} = \frac{1}{4}$ will be black, and $\frac{1}{8} + \frac{1}{8} + \frac{1}{8} + \frac{1}{8} = \frac{1}{2}$ will be albino.

11c.   *Evaluate:* This is type 3. This problem adds a third locus, C, to the model given in Problem 10. The C_ rats are able to express fur pigment and will be either brown (*B_D_C_*) or black (*B_ddC_*) or yellow (*bbD_C_*) or cream (*bbddC_*). The *cc* rats cannot express fur pigment and are albino regardless of their genotype at the B and D loci. *Deduce:* In the cross *bbDDCc × BBddCc*, all will be *Bb*, all will be *Dd*, $\frac{3}{4}$ will be *C_*, and $\frac{1}{4}$ will be *cc*. *Solve:* $\frac{3}{4}$ will be *BbDdCc* (brown) and $\frac{1}{4}$ will be *BbDdcc* (albino).

11d.   *Evaluate:* This is type 3. This problem adds a third locus, C, to the model given in Problem 10. The C_ rats are able to express fur pigment and will be either brown (*B_D_C_*) or black (*B_ddC_*) or yellow (*bbD_C_*) or cream (*bbddC_*). The *cc* rats cannot express fur pigment and are albino regardless of their genotype at the B and D loci. *Deduce:* In the cross *BbDdCC × BbDdCC*, $\frac{3}{4}$ will be *B_* and $\frac{1}{4}$ will be *bb*, $\frac{3}{4}$ will be *Dd* and $\frac{1}{4}$ will be *dd*, and all will be *CC*. Therefore, $\frac{3}{4} \times \frac{3}{4}$ will be *B_D_CC* (brown); $\frac{3}{4} \times \frac{1}{4}$ will be *B_ddCC* (black); $\frac{1}{4} \times \frac{3}{4}$ will be *bbD_CC* (yellow); and $\frac{1}{4} \times \frac{1}{4}$ will be *bbddCC* (cream). *Solve:* $\frac{9}{16}$ will be brown, $\frac{3}{16}$ will be black, $\frac{3}{16}$ will be yellow, and $\frac{1}{16}$ will be cream.

12a.   *Evaluate:* This is type 4. The phenotypic ratio of the progeny was $\frac{9}{16}$ brown (*B_D_C_*), $\frac{3}{16}$ black (*B_ddC_*), and $\frac{4}{16}$ albino (__ __*cc*). *Deduce:* Since there are black (*B_dd*) and albino (*cc*) progeny, both parents must have a *d* allele and *c* allele. Since there are no yellow progeny (*bbD_C_*), one parent must be *BB*. The ratio 9:3:4 is a modified 9:3:3:1, suggesting a dihybrid cross with epistasis. Since both parents must have a *d* and *c* allele, this suggests that the parents were both *DdCc*. Including the B locus, one parent must have been *BBDdCc* and the other was *DdCc* with all *B* genotypes possible. *Solve:* One parent was *BBDdCc* and the other was *DdCc* with any genotype possible at the B locus.

12b.   *Evaluate:* This is type 4. The phenotypic ratio of the progeny was $\frac{3}{8}$ black (*B_ddC_*), $\frac{3}{8}$ cream (*bbddC_*), and $\frac{2}{8}$ albino (__ __*cc*). *Deduce:* Since all progeny are *dd*, both parents are *dd*. Since cream and albino progeny were recovered, each parent must have a *b* and *c* allele. Since there is a 1:1 ratio of black (*B_dd*) to cream (*bbdd*), B must have been heterozygous in one parent and homozygous recessive in the other. Since there was a 3:1 ratio of pigmented (*C_*) to albino (*cc*), the C locus must have been heterozygous in both parents. *Solve:* The parents were *BbddCc × bbddCc*.

12c.   *Evaluate:* This is type 4. The phenotypic ratio of the progeny was $\frac{27}{64}$ brown (*B_D_C_*), $\frac{16}{64}$ albino (__ __*cc*), $\frac{9}{64}$ yellow (*bbD_C_*), $\frac{9}{64}$ black (*B_ddC_*), and $\frac{3}{64}$ cream (*bbddC_*). *Deduce:* Since there are albino and cream progeny, both parents must have *b*, *d*, and *c* alleles. The progeny proportions of sixty-fourths indicate a trihybrid cross; therefore, both parents were heterozygous at all three loci. *Solve:* Both parents were *BbDdCc*.

12d.   *Evaluate:* This is type 4. The phenotypic ratio of the progeny was $\frac{3}{4}$ brown (*B_D_*) and $\frac{1}{4}$ yellow (*bbD_*). *Deduce:* Since *bb* and *B_* progeny were observed, both parents must have a *b* allele. Since no black (*B_dd*) progeny were observed, at least one parent must have been *DD*. The 3:1 ratio of *B_* to *bb* indicates that both parents were heterozygous at the B locus. Since no albino progeny were observed, at least one parent must have been *CC*. *Solve:* The parents were both *Bb* at the B locus. Between the two parents, there was a *DD* and a *CC* locus, but these need not have been in the same parent. The other *D* and *C* loci could have any genotype.

13.   *Evaluate:* This is type 1. *Deduce:* Assign *R* to the functional receptor allele and *r* to the mutant receptor. The serum cholesterol level of *RR* homozygotes is 180–200 mg/100 ml, while *Rr* heterozygotes have 200–600 mg/100 ml, and *rr* homozygotes have 800–1000 mg/100 ml. The serum cholesterol levels of a heterozygote are between that of both homozygotes. *Solve:* The alleles of this cholesterol cell surface receptor gene are incompletely (or partially) dominant.

14.   *Evaluate:* This is type 3. The hypothesis for plant flower cover is that *An1* is fully functional and *An2* is nonfunctional (null), such that *An1An1* homozygotes produce the most pigment and have red flowers, *An1An2* produce half as much pigment and have pink flowers, and *An2An2* plants produce no pigment and have ivory flowers. *Deduce:* The F_1 is *An1An2*. The F_2 will be $\frac{1}{4}$ *An1An1*, $\frac{1}{2}$ *An1An2*, and $\frac{1}{4}$ *An2An2*. *Solve:* The F_2 will be $\frac{1}{4}$ red, $\frac{1}{2}$ pink, and $\frac{1}{4}$ ivory.

15a. *Evaluate:* This is type 1. A self-fertilization of this plant produces 622 viable and 204 nonviable seeds. *Deduce:* The production of 622 viable and 204 nonviable seeds is approximately a ratio of 3:1 (viable to nonviable). The 3:1 ratio is expected from a monohybrid self, corresponding to the genotypic distribution $\frac{1}{4}AA$, $\frac{1}{2}Aa$, and $\frac{1}{4}aa$. It could be that AA is a lethal genotype ($\frac{1}{4}AA : \frac{3}{4}Aa$ plus aa), or that aa is a lethal genotype ($\frac{1}{4}aa : \frac{3}{4}Aa$ plus AA). *Solve:* The plant is heterozygous at a locus A, where AA is a lethal genotype. Self-fertilization of heterozygotes produces seeds at a ratio of $\frac{1}{4}AA$, $\frac{1}{2}Aa$, and $\frac{1}{4}aa$. Since AA seeds are nonviable, this results in $\frac{3}{4}$ viable and $\frac{1}{4}$ nonviable seeds. Note: An alternative hypothesis is that aa is a lethal genotype.

15b. *Evaluate:* This is type 3. The hypothesis from part (a) predicts that $\frac{1}{3}$ of the viable seeds will be homozygous and $\frac{2}{3}$ will be heterozygous, and that selfing the homozygotes will produce all viable seeds, whereas selfing the heterozygotes will produce viable to nonviable seeds in a 3:1 ratio. *Deduce > Solve:* Allow the progeny from part (a) to self, and determine the percentage of plants that produce all viable seeds compared with those that produce 3:1 viable to nonviable. The prediction is that $\frac{1}{3}$ will produce all viable seeds, and $\frac{2}{3}$ will produce viable to nonviable in a 3:1 ratio.

16. *Evaluate:* This is type 3. *Dexter* is a dominant trait, in which *Dexter* heterozygotes have short stature and short limbs and *Dexter* homozygotes spontaneously abort. *Deduce:* Using D to refer to the dominant *Dexter* allele, DD is a lethal genotype, Dd cattle have short stature and short limbs, and dd cattle have normal stature and normal limbs. The cross in question is $Dd \times Dd$, which will yield $\frac{1}{4}DD$, $\frac{1}{2}Dd$, and $\frac{1}{4}dd$. *Solve:* The expected phenotypic proportions of this cross are $\frac{2}{3}$ ($\frac{1}{2} \times \frac{3}{4}$) short stature and limbs and $\frac{1}{3}$ ($\frac{1}{4} \times \frac{3}{4}$) normal stature and limbs.

17a. *Evaluate:* This is type 3. *Deduce:* $\frac{1}{2}$ of the progeny will be Ss and $\frac{1}{2}$ will be ss. $\frac{1}{2}$ will be R_1R_2 and $\frac{1}{2}$ will be R_2R_2. Therefore, $\frac{1}{2} \times \frac{1}{2}$ will be SsR_1R_2 (solid silver), $\frac{1}{2} \times \frac{1}{2}$ will be SsR_2R_2 (solid platinum), $\frac{1}{2} \times \frac{1}{2}$ will be ssR_1R_2 (spotted silver), and $\frac{1}{2} \times \frac{1}{2}$ will be ssR_2R_2 (spotted platinum). *Solve:* The phenotypic proportions of the progeny will be $\frac{1}{4}$ solid platinum, $\frac{1}{4}$ spotted platinum, $\frac{1}{4}$ solid silver, and $\frac{1}{4}$ spotted silver.

17b. *Evaluate:* This is type 3. *Deduce:* $\frac{3}{4}$ of the progeny will be $S\_$ and $\frac{1}{4}$ will be ss, and $\frac{1}{2}$ will be R_1R_1 and $\frac{1}{2}$ will be R_1R_2. Therefore, $\frac{3}{4} \times \frac{1}{2}$ will be $S\_R_1R_1$ (solid red), $\frac{3}{4} \times \frac{1}{2}$ will be $S\_R_1R_2$ (solid silver), $\frac{1}{4} \times \frac{1}{2}$ will be ssR_1R_1 (spotted red), and $\frac{1}{4} \times \frac{1}{2}$ will be ssR_1R_2 (spotted silver). *Solve:* The phenotypic proportions of the progeny will be $\frac{3}{8}$ solid red, $\frac{3}{8}$ solid silver, $\frac{1}{8}$ spotted red, and $\frac{1}{8}$ spotted silver.

17c. *Evaluate:* This is type 4. The genotype of the R loci of each parent is evident from their phenotype, since the R_1 and R_2 alleles show partial dominance. Thus, only the spotted versus solid phenotype of the progeny need to be analyzed to deduce the full genotype of the parents. *Deduce:* For Cross 1, the silver parent was R_1R_2 and the platinum parent was R_2R_2. Since both parents were solid, they each had an S allele. There were spotted minks among the progeny of Cross 1; therefore, both parents had an s allele. The S locus of both parents was Ss. For Cross 2, both parents were silver parent and therefore were R_1R_2. One parent was spotted and so was ss; the other parent was solid and so had an S allele. There were spotted minks among the progeny of Cross 2; therefore, the solid parent had an s allele and the S genotype of the parents was ss (spotted) and Ss (solid). *Solve:* For Cross 1, the solid silver parent was SsR_1R_2 and the solid platinum parent was SsR_2R_2. For Cross 2, the spotted silver parent was ssR_1R_2 and the solid silver parent was SsR_1R_2.

18a. *Evaluate:* This is type 4. $aaB\_$ petunias are red. *Deduce:* Pure-breeding red petunias will be homozygous at the A and B loci. *Solve:* Pure-breeding red petunias are $aaBB$.

18b. *Evaluate:* This is type 4. A_bb petunias are blue. *Deduce:* Pure-breeding blue petunias will be homozygous at the A and B loci. *Solve:* Pure-breeding blue petunias are $AAbb$.

18c.   *Evaluate:* This is type 3. From parts (a) and (b), the cross is *AAbb* × *aaBB*, which gives *AaBb* (purple) F_1. Recall from Chapter 3 that this is a dihybrid cross, which when self-crossed yields an F_2 ratio of 9:3:3:1. *Deduce > Solve:* The F_2 will be $\frac{9}{16}$ *A_B_* (purple), $\frac{3}{16}$ *A_bb* (blue), $\frac{3}{16}$ *aaB_* (red), and $\frac{1}{16}$ *aabb* (white).

19a.   *Evaluate:* This is type 4. The model involves two pathways and four genes (*A, B, C,* and *D*). A pure-breeding parakeet is homozygous for the A, B, C, and D loci, and a purple-colored parakeet results from a combination of red and blue colors. Red pigment from a colorless precursor requires an *A* allele. Blue pigment requires a *C* allele for production of a colorless intermediate and a *D* allele to convert the intermediate to blue. *Deduce:* To make red pigment, the parakeet must be *A_bb*. To make blue pigment, the parakeet must be *C_D_*. To fill in the blanks, we know that a pure-breeding parakeet is homozygous for the A, B, C, and D loci. *Solve:* The pure-breeding purple parakeet is *AAbbCCDD*.

19b.   *Evaluate:* This is type 4. A pure-breeding parakeet is homozygous at the A, B, C, and D loci and must make only yellow pigment. Yellow pigment requires an *A* allele for production of a colorless intermediate and a *B* allele to convert the intermediate to yellow. To ensure a colorless compound in pathway II, we need either *cc* or *dd* present. *Deduce:* The parakeet must be *A_B_*, and either *ccD_* or *C_dd* or *ccdd*, and given that it is pure-breeding, we can fill in the blanks. *Solve:* The pure-breeding yellow parakeet is either *AABBccDD* or *AABBCCdd* or *AABBccdd*.

19c.   *Evaluate:* This is type 3. From part (a), the genotype of a pure-breeding purple parakeet is *AAbbCCDD*. From part (c), the pure-breeding blue parakeet is *aaBBCCDD*. *Deduce:* The purple parent produces *AbCD* gametes and blue parent produces *aBCD*. *Solve:* The F1 will be *AaBbCCDD* and will be green.

19d.   *Evaluate:* This is type 3. *Deduce:* The F_1 are *AaBbCCDD*; therefore, the cross is *AaBbCCDD* × *AaBbCCDD*. All the progeny will be *CCDD*. $\frac{3}{4}$ of the progeny will be *A_* and $\frac{1}{4}$ will be *aa*. $\frac{3}{4}$ of the progeny will be *B_* and $\frac{1}{4}$ will be *bb*. Therefore, $\frac{3}{4} \times \frac{3}{4}$ will be *A_B_CCDD* (green), $\frac{3}{4} \times \frac{1}{4}$ will be *A_bbCCDD* (purple), $\frac{1}{4} \times \frac{3}{4}$ will be *aaB_CCDD* (blue), and $\frac{1}{4} \times \frac{1}{4}$ will be *aabbCCDD* (blue). *Solve:* The F_2 will be $\frac{9}{16}$ green, $\frac{3}{16}$ purple, and $\frac{4}{16}$ blue.

20a.   *Evaluate:* This is type 1. Recall that variable expressivity indicates that individuals with the same mutant genotype vary in severity of mutant phenotype. *Deduce:* Evidence of variable expressivity would be the appearance of some individuals with two thumbs affected and others with only one thumb affected. The pedigree shows both types of affected individuals. *Solve:* Yes, there is evidence of variable expressivity.

20b.   *Evaluate:* This is type 1. Recall that incomplete penetrance indicates that not all individuals with a mutant genotype show the mutant phenotype. *Deduce:* Evidence of reduced penetrance would be the appearance of an affected child who had unaffected parents. Individuals IV-4 and IV-5 are affected children of unaffected parents. *Solve:* Yes, there is evidence of incomplete penetrance.

21a.   *Evaluate:* This is type 1 and 4. The parents are pure-breeding for the albino trait and therefore are homozygous at all genes required for albino. *Deduce:* They have mutations in different genes, however, because when crossed, all F_1 progeny have wild-type coat color. If we designate the genes *A* and *B*, then the hypothesis is that a wild-type allele of both *A* and *B* is required for pigmentation: mice homozygous recessive for either *A* or *B* are albino. *Solve:* The parents are homozygous recessive for mutations in different genes (*A* and *B*) required for pigmentation. If one parent is designated *aaBB*, the other is *AAbb*. The F_1 are *AaBb*.

21b.   *Evaluate:* This is type 1 and 4. The F_1 from part (a) are *AaBb*; therefore, the F_2 are predicted to be $\frac{9}{16}$ *A_B_*, $\frac{3}{16}$ *A_bb*, $\frac{3}{16}$ *aaB_*, and $\frac{1}{16}$ *aabb*. *Deduce:* Given the genetic hypothesis from part (a), only the *A_B_* will have coat color; all other genotypes will be albino. This predicts $\frac{9}{16}$ colored and $\frac{7}{16}$ albino, which is consistent with the 31 pigmented and 25 albino F_2. *Solve:* The 31 pigmented F_2 are *A_B_*, and the 25 albino F_2 are composed of the genotypes *A_bb*, *aaB_*, and *aabb*. The

21b.   *Evaluate:* This is type 1 and 4. The F_1 from part (a) are *AaBb*; therefore, the F_2 are predicted to be $\frac{9}{16}$ *A_B_*, $\frac{3}{16}$ *A_bb*, $\frac{3}{16}$ *aaB_*, and $\frac{1}{16}$ *aabb*. *Deduce:* Given the genetic hypothesis from part (a), only the *A_B_* will have coat color; all other genotypes will be albino. This predicts $\frac{9}{16}$ colored and $\frac{7}{16}$ albino, which is consistent with the 31 pigmented and 25 albino F_2. *Solve:* The 31 pigmented F_2 are *A_B_*, and the 25 albino F_2 are composed of the genotypes *A_bb*, *aaB_*, and *aabb*. The genetic phenomenon that describes this phenotypic distribution is epistasis (specifically, complementary gene interaction).

22a.   *Evaluate > Deduce > Solve:* This is type 1. Recall that a disease caused by mutations in many different genes is said to express genetic heterogeneity.

22b.   *Evaluate:* This is type 5. Recall that mutant XP cell lines are deficient in repair of DNA damage, a phenotype that is measurable in cell culture. A complementation test using cell lines is performed by fusing cells from two lines together to make a heterokaryon. Recall that a heterokaryon expresses the genes in both genomes. If two mutant cell lines have recessive mutations in the same complementation group (gene), the heterokaryon formed by fusing those cell lines will *not* be capable of normal DNA damage repair. This would be indicated by a minus sign in the corresponding fusion cell line box. If two mutant cell lines have mutations in different complementation groups (genes), the heterokaryon formed by fusing those mutant cell lines *will* be capable of normal DNA damage repair. This would be indicated by a plus sign in the corresponding fusion cell line box. If either cell line has a dominant mutation, the heterokaryon formed by fusion will not be capable of normal DNA damage repair. *Deduce > Solve:* There are four complementation groups: Group 1 is defined by mutations 1, 3, and 7; Group 2 is defined by mutations 4 and 8; Group 3 is defined by mutations 5, 6, and 10; and Group 4 is defined by mutation 2. Mutation 9 cannot be assigned to a complementation group, because it is a dominant mutation.

23a.   *Evaluate:* This is type 1. The F_1 from each cross are yellow, indicating that yellow is dominant to green. *Deduce:* The F_2 of $G_1 \times Y$ were $\frac{3}{4}$ yellow and $\frac{1}{4}$ green, a 3:1 ratio typical of a monohybrid cross. The F_2 of $G_2 \times Y$ were $\frac{9}{16}$ yellow and $\frac{7}{16}$ green, a 9:7 ratio typical of 9:7 epistasis in a dihybrid cross. The F_2 of $G_3 \times Y$ were $\frac{27}{64}$ yellow and $\frac{37}{64}$ green, which suggests that three genes are segregating. The phenotypic distribution for a trihybrid cross with simple dominance and no epistasis is 27:9:9:9:3:3:3:1. A ratio of 27:37 suggests epistasis, where at least one dominant allele of all three genes is required for yellow. *Solve:* There is one gene segregating in the F_2 of $G_1 \times Y$, two genes segregating in the F_2 of $G_2 \times Y$, and three genes segregating in the F_2 of $G_3 \times Y$.

23b.   *Evaluate:* This is type 4. *Deduce:* The F_1 of each cross are hybrids, and the Y strain is homozygous dominant for all genes involved in seed color. The green parents are homozygous recessive for the gene(s) segregating in the F_2 of the cross. *Solve:* Y is *AABBDD*. G_1 is *aaBBDD*, and the F_1 of $G_1 \times Y$ are *AaBBCC*. G_2 is *aabbDD*, and the F_1 of $G_2 \times Y$ are *AaBbDD*. G_3 is *aabbdd*, and the F_1 of $G_3 \times Y$ are *AaBbDd*.

23c.   *Evaluate > Deduce > Solve:* This is type 1 and 4. For $G_1 \times Y$, there are 2 alleles of 1 gene segregating in the F_2. The alleles show simple dominance (*A* and *a*), with *A_* determining yellow. For $G_2 \times Y$, there are 4 alleles of 2 genes segregating in the F_2. The alleles show simple dominance (*Aa* and *Bb*). Yellow F_2 plants are *A_B_*. Green F_2 plants are *aaB_*, *A_bb*, and *aabb*. For $G_3 \times Y$, there are 6 alleles of 3 genes segregating in the F_2. The alleles show simple dominance (*Aa*, *Bb*, and *Dd*), and *A_B_D_* determines yellow.

23d.   *Evaluate:* This is type 3. *Deduce:* G_1 is *aaBBDD* and G_3 is *aabbdd*. G_1 produces one type of gamete, *aBD* and G_3 produces one type of gamete, *abd*, which are crossed to solve the problem. *Solve:* The F_1 will be green, *aaBbDd*.

23e.   *Evaluate:* This is type 3. *Deduce:* The gametes produced by *aaBbDd* are $\frac{1}{4}$ *aBD*, $\frac{1}{4}$ *aBd*, $\frac{1}{4}$ *abD*, and $\frac{1}{4}$ *abd*. Green seeds will include all genotypes that lack one dominant allele of at least one gene. *Solve:* All progeny will be homozygous for the *a* allele; therefore, all progeny will be green.

23f. *Evaluate:* This is type 3. *Deduce:* G_2 is *aabbDD* and G_3 is *aabbdd*. G_2 produces one type of gamete, *abD*. G_3 produces one type of gamete, *abd*. *Solve:* The F_1 will be green, *aabbDd*.

23g. *Evaluate:* This is type 3. *Deduce:* The gametes produced by *aabbDd* are $\frac{1}{2}$ *ab* and $\frac{1}{2}$ *abd*. Yellow seeds will include all genotypes that have at least one dominant allele of each gene. *Solve:* All the progeny will be homozygous for the *a* and *b* alleles; therefore, none of the F_2 will have yellow seeds.

24a. *Evaluate:* This is type 4. *Deduce:* Blue plants are *A_B_*, and purple plants are either *A_bb* or *aaB_*. Since both parents are pure-breeding purple plants, they must be *AAbb* or *aaBB*. *Solve:* The only combination of pure-breeding purple plants that will have blue-flowered progeny is *AAbb* and *aaBB*.

24b. *Evaluate:* This is type 3. Blue plants are *A_B_*, purple plants are *A_bb* or *aaB_*, and red plants are *aabb*. *Deduce:* The parents will produce four types of gametes: $\frac{1}{4}$ *AB*, $\frac{1}{4}$ *Ab*, $\frac{1}{4}$ *aB*, and $\frac{1}{4}$ *ab*. The genotypic distribution will be $\frac{1}{16}$ *AABB*, $\frac{2}{16}$ *AABb*, $\frac{1}{16}$ *AAbb*, $\frac{2}{16}$ *AaBB*, $\frac{4}{16}$ *AaBb*, $\frac{2}{16}$ *Aabb*, $\frac{1}{16}$ *aaBB*, $\frac{2}{16}$ *aaBb*, and $\frac{1}{16}$ *aabb*. *Solve:* The phenotypic distribution will be $\frac{9}{16}$ blue (*A_B_*), $\frac{6}{16}$ purple (*A_bb* + *aaB_*), and $\frac{1}{16}$ red (*aabb*).

24c. *Evaluate:* This is type 3. The F_1 is *AaBb* and the parents are either *AAbb* or *aaBB*. *Deduce:* The F_1 produces four types of gametes: $\frac{1}{4}$ *AB*, $\frac{1}{4}$ *Ab*, $\frac{1}{4}$ *aB*, and $\frac{1}{4}$ *ab*. The *AAbb* parent produces *Ab* gametes, and the *aaBB* parent produces *aB* gametes. A cross of the F_1 to the *AAbb* parent produces $\frac{1}{4}$ *AABb*, $\frac{1}{4}$ *AAbb*, $\frac{1}{4}$ *AaBb*, and $\frac{1}{4}$ *Aabb*. A cross of the F_1 to the *aaBB* parent produces $\frac{1}{4}$ *AaBB*, $\frac{1}{4}$ *AaBb*, $\frac{1}{4}$ *aaBB*, and $\frac{1}{4}$ *aaBb*. *Solve:* The progeny of the cross of the F_1 to the *AAbb* parent will be $\frac{1}{2}$ blue (*AABb* and *AaBb*) and $\frac{1}{2}$ purple (*AAbb* and *Aabb*) progeny. The progeny of the cross of the F_1 to the *aaBB* parent will be $\frac{1}{2}$ blue (*AaBB* and *AaBb*) and $\frac{1}{2}$ purple (*aaBB* and *aaBb*).

25a. *Evaluate:* This is type 4. The parents have blue flowers and therefore are *A_B_*. The progeny are $\frac{3}{4}$ blue (*A_B_*) and $\frac{1}{4}$ purple (either *aaB_* or *A_bb*). *Deduce:* Since the progeny are either *aa* or *bb*, both parents must carry a recessive *a* or *b* allele. The 3:1 phenotypic ratio suggests a monohybrid cross; therefore, both parents have the same genotype. *Solve:* The blue parents are either *AaBB* or *AABb*.

25b. *Evaluate:* This is type 4; see part (a). The parents have purple flowers and therefore are either *aaB_* or *A_bb*. The progeny are $\frac{1}{4}$ blue (*A_B_*), $\frac{1}{2}$ purple (*aaB_* or *A_bb*), and $\frac{1}{4}$ red (*aabb*). *Deduce:* Since there are red-flowered progeny, both parents must be homozygous recessive at one locus and heterozygous at the other. The 1:2:1 genotypic ratio suggests a monohybrid cross; therefore, the parents have the same genotype. *Solve:* Both purple parents are either *aaBb* or *Aabb*.

25c. *Evaluate:* This is type 4; see part (a). One parent makes red flowers (*aabb*) and the other makes blue flowers (*A_B_*). The progeny are $\frac{1}{4}$ blue (*A_B_*), $\frac{1}{2}$ purple (*aaB_* or *A_bb*), and $\frac{1}{4}$ red (*aabb*). *Deduce:* Since there are red progeny, the blue parent must have recessive *a* and *b* alleles. *Solve:* The blue parent is *AaBb* and the red parent is *aabb*.

25d. *Evaluate:* This is type 4; see part (a). One parent is purple (either *aaB_* or *A_bb*) and the other is red (*aabb*). The progeny are $\frac{1}{2}$ purple (*aaB_* or *A_bb*) and $\frac{1}{2}$ red (*aabb*). *Deduce:* Since there are red progeny, the purple parent must be heterozygous at the A or B locus. *Solve:* The purple parent is either *aaBb* or *Aabb*, and the red parent is *aabb*.

25e. *Evaluate:* This is type 4; see part (a). One parent is blue (*A_B_*) and the other is purple (either *A_bb* or *aa B_*). The progeny were $\frac{3}{8}$ blue (*A_ B_*), $\frac{1}{2}$ purple (either *A_bb* or *aaB_*), and $\frac{1}{8}$ red (*aabb*). *Deduce:* Since there are red progeny, the blue parent must have a recessive *a* and *b* allele and the purple parent must be heterozygous at the A or B locus. *Solve:* The blue parent is *AaBb* and the purple parent is either *Aabb* or *aaBb*.

26a.   *Evaluate:* This is type 1. *Deduce:* $Y_1 \times G_1$ produces all yellow F_1, which produce $\frac{3}{4}$ yellow and $\frac{1}{4}$ green F_2. Green appears to be recessive, and the 3:1 ratio suggests segregation of alleles of a single gene. $Y_2 \times G_1$ produces all green F_1, which produces $\frac{3}{4}$ green and $\frac{1}{4}$ yellow. Here, yellow appears recessive and, again, the 3:1 ratio suggests segregation of alleles at a single gene. $Y_1 \times Y_2$ produces all yellow F_1, which produces $\frac{13}{16}$ yellow and $\frac{3}{16}$ green. The sixteenths in the F_2 suggests alleles at two genes are segregating. The two genes segregating in the third cross correspond to the genes segregating in the first two crosses. *Solve:* The results of these crosses indicate that two genes control squash fruit color.

26b.   *Evaluate:* This is type 4. Part (a) indicates that two genes are involved, which we will call *A* and *B*. *Deduce:* The third cross is most informative, indicating a 13-to-3 epistasis, where yellow color is determined by one of three genotypes, *A_B_*, *A_bb*, and *aabb* and green is determined by *aaB_*. (Note that *A* and *B* are interchangeable; therefore, yellow could also be designated *aaB_* and green *A_bb*.) The $Y_1 \times G_1$ F_2 progeny show that alleles of one locus are segregating; therefore, the yellow F_1 are heterozygous at one locus. Yellow squash that are heterozygous at one locus are *AaBB*, making Y_1 *AABB* and G_1 *aaBB*. The $Y_2 \times G_1$ F_2 progeny show alleles of one locus segregating; therefore, the green F_1 are heterozygous at one locus. Green squash heterozygous at one locus are *aaBb*, making Y_2 *aabb* (G_1 is *aaBB*). $Y_1 \times Y_2$ is *AABB* × *aabb*, producing *AaBb* dihybrids, which yield $\frac{9}{16}$ *A_B_*, $\frac{3}{16}$ *A_bb*, $\frac{3}{16}$ *aaB_*, and $\frac{1}{16}$ *aabb*. *Solve:* In this model, a Y_1 is *AABB* × G_1 *aaBB*, their yellow F_1 progeny are *AaBB*, and their F_2 are $\frac{3}{4}$ yellow (*A_BB*) and $\frac{1}{4}$ green (*aaBB*). Y_2 is *aabb*, G_1 is *aaBB*, their green F_1 are *aaBb*, and their F_2 are $\frac{3}{4}$ green (*aaB_*) and $\frac{1}{4}$ yellow (*aabb*). The F_1 of $Y_1 \times Y_2$ are *AaBb* and their F_2 are $\frac{9}{16}$ yellow (*A_B_*), $\frac{3}{16}$ yellow (*A_bb*), $\frac{3}{16}$ green (*aaB_*), and $\frac{1}{16}$ yellow (*aabb*).

26c.   *Evaluate:* This is type 3. This problem asks you to use your answers from parts (a) and (b) to predict the outcome of crossing the F_1 of Crosses I and II. *Deduce:* The cross is *AaBB* × *aaBb*. The progeny will be $\frac{1}{2}$ *Aa* and $\frac{1}{2}$ *aa* at the A locus, and $\frac{1}{2}$ *BB* and $\frac{1}{2}$ *Bb* at the B locus, resulting in $\frac{1}{2} \times \frac{1}{2}$ *AaBB*, $\frac{1}{2} \times \frac{1}{2}$ *AaBb*, $\frac{1}{2} \times \frac{1}{2}$ *aaBB*, and $\frac{1}{2} \times \frac{1}{2}$ *aaBb*. *Solve:* The progeny will be $\frac{1}{2}$ yellow (*AaBB* and *AaBb*) and $\frac{1}{2}$ green (*aaBB* and *aaBb*).

27.   *Evaluate > Deduce > Solve:* This is type 1. Recall that variation in the symptoms of a disease, such as Marfan syndrome, resulting from mutations in one gene is called variable expressivity.

28a.   *Evaluate:* This is type 2. *Deduce:* Wild-type yeast strains grow well at both 37°C and 25°C. Five of the yeast mutants grow normally at 25°C but do not grow at 37°C. Two of the mutants grow well at 25°C and grow slowly at 37°C. *Solve:* All the mutants are temperature-sensitive mutants, carrying a mutant allele of a gene required for normal growth at 37°C. The typical mechanistic explanation of this is that the gene is required for growth at all temperatures, and the mutant allele is functional at 25°C but not 37°C (an alternative explanation is that the gene is required only at 37°C). This explains the five mutants that cannot grow at 37°C. For the two mutants that grow slowly at 37°C, the alleles probably retain partial function at 37°C.

28b.   *Evaluate:* This is type 5. A complementation test using haploid yeast mutants is done by mating all possible combinations of yeast mutants to create diploids and then testing the growth of each diploid at 37°C. *Deduce:* Diploids derived from mutants A and D, A and F, and F and D cannot grow at 37°C; therefore, mutants A, D, and F define one complementation group. Diploids derived from mutants B and G grow slowly at 37°C; therefore, mutants B and G define one complementation group. Diploids derived from mutants C and E cannot grow at 37°C; therefore, mutants C and E define one complementation group. *Solve:* This study identifies three complementation groups: Group 1 is defined by A, D, and F, Group 2 is defined by B and G, and Group 3 is defined by C and E.

29.   *Evaluate:* This is type 1. Out of 812 progeny from a cross of flies showing the phenotype, 598 exhibit the phenotype and 214 do not. *Deduce:* Assuming that the original fly line is homozygous at all genetic loci, the phenotype appears to be a dominant phenotype that shows reduced penetrance. The dominant mutation causes some but not all flies to become paralyzed at 30°C and to die if they are held at 30°C for too long. *Solve:* The phenotype is a dominant, temperature-sensitive phenotype that shows reduced penetrance.

30a. *Evaluate:* This is type 1. The results were 55 blue, 22 purple, and 23 white. *Deduce:* Using P for the partially dominant allele and p for the partially recessive allele, the hypothesis predicts a 1:2:1 ratio of purple (PP) : blue (Pp) : white (pp). For 100 progeny, the expected number and types of progeny are 50 blue, 25 blue, and 25 white. A chi-square test is used to evaluate the fit between the observed results and those predicted by a genetic hypothesis. *Solve:* The chi-square value is $\frac{(22-25)^2}{25} + \frac{(23-25)^2}{25} + \frac{(55-50)^2}{50} = 1.02$. There are three phenotypic classes; therefore, there are two degrees of freedom in this calculation. A chi-square value of 1.02 with 2 df gives a p value between 0.5 and 0.7; therefore, the results are consistent with a 1:2:1 hypothesis, and that hypothesis cannot be rejected.

30b. *Evaluate:* This is type 1. The results were 55 blue, 22 purple, and 23 white. *Deduce:* Using B and b for one gene and P and p for the other, the hypothesis predicts a 9:3:4 ratio of blue ($P\_B\_$) : purple ($ppB\_$) : white (P_bb and $ppbb$). For 100 progeny, the expected number and types of progeny are 56.25 blue, 18.75 purple, and 25 white. A chi-square test is used to evaluate the fit between the observed results and those predicted by a genetic hypothesis. *Solve:* The chi-square value is $\frac{(22-18.75)^2}{18.75} + \frac{(55-56)^2}{55} + \frac{(23-25)^2}{25} = 0.75$. There are three phenotypic classes; therefore, there are two degrees of freedom in this calculation. A chi-square value of 0.75 with 2 df gives a p value between 0.5 and 0.7; therefore, the results are consistent with a 9:4:3 hypothesis, and that hypothesis cannot be rejected.

30c. *Evaluate > Deduce > Solve:* Neither hypothesis can be rejected based on the chi-square analysis.

30d. *Evaluate:* This is type 1. The predicted genotypes for blue, purple, and white plants differ under the two hypotheses. *Deduce:* All purple (PP) plants will be true-breeding under the 1:2:1 hypothesis, whereas only a portion of purple ($ppB\_$) plants will be true-breeding under the 9:4:3 hypothesis. *Solve:* Self-fertilize all the purple progeny and determine the proportion that breed true. If 1:2:1 is correct, all purple plants will breed true. If 9:4:3 is correct, then $\frac{2}{3}$ of the purple plants will not breed true (i.e., they will produce some whites).

31. *Evaluate:* This is type 3. Bombay phenotype is an apparent O blood type that is due to a genotype hh regardless of the genotype of the I locus. *Deduce:* $I^A I^B Hh \times I^A I^B Hh$ results in $\frac{3}{4}$ of the progeny with $H\_$ and $\frac{1}{4}$ with hh. All the hh progeny will have type O blood. Among the $\frac{3}{4}$ $H\_$ individuals, $\frac{1}{4}$ will be $I^A I^A$, $\frac{1}{2}$ will be $I^A I^B$, and $\frac{1}{4}$ will be $I^B I^B$. Thus, $\frac{3}{4} \times \frac{1}{4}$ will be $H\_I^A I^A$, $\frac{3}{4} \times \frac{1}{4}$ will be $I^A I^B$, and $\frac{3}{4} \times \frac{1}{4}$ will be $H\_I^B I^B$. *Solve:* For this cross, $\frac{1}{4}$ will have type O blood, $\frac{3}{16}$ will have type A blood, $\frac{3}{8}$ will have type AB blood, and $\frac{3}{16}$ will have type B blood.

32. *Evaluate:* This is type 1. Strain 1 × strain 2 gives only albino progeny, which when intercrossed also give only albino progeny. *Deduce:* Thus, strains 1 and 2 are homozygous recessive for mutations in the same albino gene (we'll call the gene A, and strains 1 and 2 are aa). Crosses strain 1 × strain 3 and strain 2 × strain 3 each yield pigmented F_1 progeny that produce F_2 with a ratio of pigmented to unpigmented of 9:7. These results indicate that strain 3 is homozygous recessive for mutations in a gene different from strains 1 and 2 (we'll call the gene B, so strain 3 is bb). Cross A is $aaBB \times aaBB$, which yields $aaBB$ (albino) F_1 that produce $aaBB$ (albino) F_2. Cross B is $aaBB \times AAbb$, which yields $AaBb$ (pigmented) F_1 that produce $\frac{9}{16}$ $A\_B\_$ (pigmented), $\frac{3}{16}$ A_bb (albino), $\frac{3}{16}$ $aaB\_$ (albino), and $\frac{1}{16}$ $aabb$ (albino). Cross C is $aaBB \times AAbb$, which yields $AaBb$ (pigmented) F_1 that produce $\frac{9}{16}$ $A\_B\_$ (pigmented), $\frac{3}{16}$ A_bb (albino), $\frac{3}{16}$ $aaB\_$ (albino), and $\frac{1}{16}$ $aabb$ (albino). *Solve:* Strains 1 and 2 are homozygous for mutations in the same gene, A, that causes albinism. Strain 3 is homozygous for a mutation in a different gene, B, which causes albinism. Strains 1 and 2 are $aaBB$, as are the F_1 and F_2 of cross A. Strain 3 is $AAbb$. The F_1 of Cross B and Cross C are $AaBb$. The F_2 of Cross B and Cross C are $\frac{9}{16}$ $A\_B\_$ (pigmented), $\frac{3}{16}$ A_bb (albino), $\frac{3}{16}$ $aaB\_$ (albino), and $\frac{1}{16}$ $aabb$ (albino).

33.   *Evaluate:* This is type 1. Given this model, the ratio among the F_2 should be 1 blue : 2 purple : 1 red, which for 320 progeny, would be 80 blue, 160 purple, and 80 red. The observed results were 59 blue, 182 purple, and 79 red. *Deduce > Solve:* The chi-square value is $\frac{(59-80)^2}{80} + \frac{(182-160)^2}{160} + \frac{(79-80)^2}{80} = 8.55$. For two degrees of freedom, this chi-square value gives a P value between 0.05 and 0.01, which indicates that the difference between the observed results and those expected under the single-gene, partial dominance model is statistically significant, thus arguing against that genetic model.

4.04 Test Yourself

Problems

1. You have isolated two pure-breeding plant lines, each of which produces white flowers. When you cross the two lines, the progeny produce red flowers. When you allow those flowers to self-pollinate, you obtain 320 plants, 170 of which produce red flowers and 150 of which produce white flowers. Use this information to propose a biochemical model for red flower pigment production that is based on two genes controlling red flower pigment synthesis, and test whether the results are a good statistical fit for that model.

2. Chickens have four different shapes of combs: single, rose, pea, and walnut. Interbreeding pure-breeding rose chickens with pure-breeding pea chickens produced all walnut progeny. Intercross of the walnut progeny produced the following distribution of comb types: $\frac{9}{16}$ walnut, $\frac{3}{16}$ pea, $\frac{3}{16}$ rose, and $\frac{1}{16}$ single.
    a. Use this information to derive a genetic model for control of comb shape in these chickens.
    b. Use your genetic model from part (a) to predict the phenotype of the F_1 of a cross between pure-breeding walnut comb strain and a pure-breeding single comb strain and the F_2 generated by interbreeding the F_1.
    c. Determine what proportion of the F_2 from part (b) will be pure-breeding for their respective comb shape.
    d. A chicken with a rose-shaped comb is crossed to a chicken with a pea-shaped comb and produces $\frac{1}{4}$ walnut, $\frac{1}{4}$ rose, $\frac{1}{4}$ pea, and $\frac{1}{4}$ single-comb shaped progeny. What were the genotypes of the pea and rose parents?

3. Wild-type bacteria can grow on minimal medium. Five mutants are isolated that cannot grow on minimal medium but can grow on medium supplemented with the metabolite X. It is believed that the metabolites Y, W, and Z are intermediates in the biosynthetic pathway for X, but the details of the pathway are not known. You find that mutant 3 can grow on minimal medium supplemented with Z or X but not with Y and W. Mutants 1 and 4 can grow on minimal medium supplemented with Y, Z, or X but not with W. Mutant 2 can grow on medium supplemented with X, Y, W, or Z. Mutant 5 can grow on medium supplemented with X but not with Y, W, or Z. Use this information to describe the biosynthetic pathway for X, and identify the step in the pathway blocked by each mutant.

4. The figure shows a biosynthetic pathway for the $\xrightarrow{\;3\;}$ R $\xrightarrow{\;2\;}$ Q $\xrightarrow{\;1\;}$ P $\xrightarrow{\;4\;}$ M
    metabolite M. You have isolated four bacterial mutants, each mutant being defective for one of the enzymes shown (1, 2, 3, and 4). Assuming that synthesis or uptake of M is required for growth and that the bacterium can utilize each of the metabolites shown if they are supplied in the medium, predict whether each mutant can grow on the following medium: minimal medium, minimal medium supplemented with M, minimal medium supplemented with P, minimal medium supplemented with Q, and minimal medium supplemented with R.

5. Two genes, A and B, interact to produce flower color. The dominant A allele performs a function in the pigment pathway and the recessive allele, a, is null. The dominant allele B also performs a function in that pathway and the recessive allele, b, is null. If a pure-breeding green plant ($AAbb$) is crossed with a pure-breeding white plant ($aaBB$), the F_1 are green and the F_2 are $\frac{13}{16}$ green and $\frac{3}{16}$ white. Use this information to determine the genotypes of white parent plants whose progeny were $\frac{3}{4}$ white and $\frac{1}{4}$ green.

6. The table shows the results of a complementation test on 10 pure-breeding white-flowered mutants (wild-type flower color is blue). A "+" indicates that a cross between mutants resulted in production of progeny that produced blue flowers, and a "−" indicates that the cross resulted in production of progeny that produced white flowers. Use this information to assign each mutant to a complementation group and to determine the number of genes required for blue pigment production that were identified by this study.

	1	2	3	4	5	6	7	8	9	10
1	−									
2	+	−								
3	+	+	−							
4	+	+	+	−						
5	+	+	−	+	−					
6	+	+	+	+	+	−				
7	+	−	+	+	+	+	−			
8	+	−	+	+	+	+	−	−		
9	+	+	+	−	+	+	+	+	−	
10	−	+	+	+	+	+	+	+	+	−

Mutant (rows) / Mutant (columns)

7. Biosynthesis of flower pigment in a plant requires five enzymes (enzymes 1 through 5), each of which is encoded by a different gene (gene 1 through gene 5). You have been given 10 pure-breeding, recessive mutants that fail to make pigment. You were told that mutants A and B have null mutations in gene 1, mutants C and D have null mutations in gene 3, mutants E and F have null mutations in gene 5, mutants G and H have null mutations in gene 2, and mutants I and J have null mutations in gene 4. To confirm this information, you conduct a complementation test, crossing each mutant to all nine other mutants. If the information you were provided is correct, then predict the results of the complementation test.

Solutions

1. *Evaluate:* This is type 1. *Deduce:* The F_1 from the cross of the pure-breeding lines would be a dihybrid ($AaBb$, for example); therefore, the F_2 would be $\frac{9}{16}$ $A\_B\_$; $\frac{3}{16}$ A_bb; $\frac{3}{16}$ $aaB\_$, and $\frac{1}{16}$ $aabb$. Since the F_2 have only two phenotypes, then the genes show epistasis. 170 red-flowering plants out of 320 total is $\frac{8.5}{16}$, and 150 white-flowering plants out of 320 is $\frac{7.5}{16}$. Among the models given for epistasis, 8.5 to 7.5 is closest to 9:7, which would predict that 180 out of 320 progeny would produce red flowers and 140 would produce white flowers. The chi-square calculation using the observed and expected values is $\frac{(170-180)^2}{180} + \frac{(150-140)^2}{140}$. *Solve:* The fact that only $A\_ B\_$ F_2 produce red flowers indicates that at least one functional allele of both genes is required for red pigment synthesis. The chi-square value based on that model is 1.27, which for 1 degree of freedom corresponds to a p value between 0.3 and 0.2. This p value indicates that the model is a reasonable statistical fit for the experimental data. Therefore, A model proposing that two genes control red pigment synthesis such that one functional allele of both genes is required for pigment synthesis would fit the model (9:7 epistasis). The chi-square value using that model to determine the expected values is 1.27—which, for 1 degree of freedom, corresponds to a p value between 0.2 and 0.3.

2a. *Evaluate:* This is type 1. *Deduce:* The original pea and rose lines are pure-breeding and therefore are homozygous at all genetic loci controlling comb shape. Their walnut F_1 progeny, therefore, are heterozygous at one or more loci controlling comb shape. Since the phenotypic proportions of the F_2 are in sixteenths, it appears that the alleles of two genes are segregating and the F_1 was a dihybrid. The F_2, therefore, are $\frac{9}{16}$ $A\_B\_$, $\frac{3}{16}$ A_bb, $\frac{3}{16}$ $aaB\_$, and $\frac{1}{16}$ $aabb$. *Solve:* Walnut is $A\_B\_$, pea is A_bb, rose is $aaB\_$, and single is $aabb$.

2b. *Evaluate:* This is type 3. Walnut is $A\_B\_$, pea is A_bb, rose is $aaB\_$, and single is $aabb$. *Deduce:* The pure-breeding walnut strain is $AABB$, and the pure-breeding single strain is $aabb$. The F_1 are $AaBb$, and the F_2 are $\frac{9}{16} A\_B\_$, $\frac{3}{16} A\_bb$, $\frac{3}{15} aaB\_$, and $\frac{1}{16} aabb$. *Solve:* The F_1 will be all walnut comb shape, and the F_2 will be $\frac{9}{16}$ walnut, $\frac{3}{16}$ pea, $\frac{3}{16}$ rose, and $\frac{1}{16}$ single.

2c. *Evaluate:* This is type 3. *Deduce:* Pure-breeding strains are homozygous at all loci controlling comb shape. Among the $\frac{9}{16} A\_B\_$ walnut F_2, $\frac{1}{9}$ are $AABB$, $\frac{1}{9}$ are $AAbb$, $\frac{2}{9}$ are $AABb$, $\frac{2}{9}$ are $AaBB$, and $\frac{4}{9}$ are $AaBb$. Among the $\frac{3}{16}$ pea F_2, $\frac{1}{3}$ are $AAbb$ and $\frac{2}{3}$ are $Aabb$. Among the $\frac{3}{16}$ rose F_2, $\frac{1}{3}$ are $aaBB$ and $\frac{2}{3}$ are $aaBb$. Among the $\frac{1}{16}$ F_2 singles, all are $aabb$. Therefore, $\frac{9}{16} \times \frac{1}{9} = \frac{9}{144} = 0.0625$ will breed true for walnut comb shape, $\frac{3}{16} \times \frac{1}{3} = \frac{3}{48} = 0.0625$ will breed true for pea comb shape, $\frac{3}{16} \times \frac{1}{3} = \frac{3}{48} = 0.0625$ will breed true for rose comb shape, and $\frac{1}{16} \times 1 = \frac{1}{16} = 0.0625$ will breed true for single comb shape. *Solve:* The proportion of F_2 that breed true for comb shape is $0.0625 \times 4 = 0.25$.

2d. *Evaluate:* This is type 4. *Deduce:* Pea comb shape corresponds to A_bb and rose comb shape corresponds to $aaB\_$. Therefore, the pea parent is either $AAbb$ or $Aabb$ and the rose parent is either $aaBB$ or $aaBb$. Since single-comb-shaped progeny are obtained, and single comb is $aabb$, then each parent must have a recessive a and b allele. *Solve:* The pea-comb-shaped parent was $Aabb$, and the rose-comb-shaped parent was $aaBb$.

3. *Evaluate:* This is type 2. This problem asks you to interpret the results of a genetic analysis of an uncharacterized biochemical pathway. *Deduce:* Mutant 2 can grow on medium supplemented with X or any of the metabolites, W, Y, or Z, indicating that mutant 2 blocks at a step before all of these intermediates. Mutants 1 and 4 can grow on mediums supplemented with any metabolite except W; therefore, mutants 1 and 4 are blocked in a step after W but before Y, Z, and X. Mutant 3 can grow on medium supplemented with Z or X but not W or Y; therefore, mutant 3 blocks at a step after W and Y but before Z and X. Mutant 5 can grow on medium supplemented with X but not with W, Y, or Z; therefore, mutant 5 blocks after W, Y, and Z but before X. *Solve:* The order of mutants is 2, 1 and 4, 3, and 5. The order of metabolites is W, Y, Z, and X. See figure. $\xrightarrow{\;2\;}$ W $\xrightarrow{\;1\text{ and }4\;}$ Y $\xrightarrow{\;3\;}$ Z $\xrightarrow{\;5\;}$ X

4. *Evaluate:* This is type 2. This problem requires you to predict the results of a genetic analysis given the details of a metabolic pathway. *Deduce:* Mutants defective for an enzyme will be able to use any metabolite that is after that step in the pathway, but not metabolites that are before that step in the pathway. *Solve:* None of the mutants can grow on minimal medium. Mutant 1 can grow on minimal medium supplemented with P or M but not with R or Q. Mutant 2 can grow on minimal medium supplemented with Q, P, or M but not with R. Mutant 3 can grow on minimal medium supplemented with R, Q, P, or M. Mutant 4 can grow on minimal medium supplemented with M but not with R, Q, or P.

5. *Evaluate:* This is type 4. *Deduce:* The F_2 indicate that control of flower color in this plant by the A and B genes shows a 13:3-type epistasis. Plants produce white flowers only if they are homozygous recessive at the A locus and have at least one dominant B allele ($aaB\_$). To get white and green progeny ($aaB\_$), both white parents must have at least one recessive b allele. *Solve:* The white parents were both $aaBb$.

6. *Evaluate:* This is type 5. *Deduce:* Mutant 1 crossed to mutant 10 produces only white flowers; every other cross with mutants 1 or 10 produces blue flowers. Mutant 2 crossed to mutants 7 or 8 produces only white flowers; every other cross involving mutants 2, 7, or 8 produces blue flowers. Mutant 3 crossed with mutant 5 produces white flowers; every other cross involving mutants 3 and 5 produces blue flowers. Mutant 4 crossed with mutant 9 produces only white flowers; every other cross involving 4 or 9 produces blue flowers. All crosses (other than self-crosses) involving mutant 6 produce blue flowers. *Solve:* This study identified five complementation groups (genes). Mutants 1 and 10 define one gene; mutants 2, 7, and 8 define a second gene; mutants 3 and 5 define a third gene; mutants 4 and 9 define a fourth gene; and mutant 6 defines a fifth gene.

7. *Evaluate:* This is type 5. **Deduce:** A cross between mutants that have mutations in the same gene will yield progeny with unpigmented flowers, whereas a cross between mutants with mutations in different genes will yield progeny that produce pigmented flowers. *Solve:* See table.

Mutant	A	B	C	D	E	F	G	H	I	J
A	−									
B	−	−								
C	+	+	−							
D	+	+	−	−						
E	+	+	+	+	−					
F	+	+	+	+	−	−				
G	+	+	+	+	+	+	−			
H	+	+	+	+	+	+	−	−		
I	+	+	+	+	+	+	+	+	−	
J	+	+	+	+	+	+	+	+	−	−

Mutant

5

Genetic Linkage and Mapping in Eukaryotes

5.01 Genetic Problem-Solving Toolkit

Key Terms and Concepts

Genetic linkage: Genetic loci that do not assort independently during meiosis.

Synteny: The conserved order of genes together on a chromosome in species that share a common ancestor.

Crossing over (recombination): The breakage and reunion of homologous chromosomes that results in reciprocal recombination.

Parental-type progeny: Progeny that do not undergo crossing over.

Recombinant-type progeny: Progeny that do undergo crossing over.

Recombination frequency (% recombinants, _r_): The rate of recombination for a given pair of linked genes.

Genetic linkage mapping: Process for creating maps of genes based on their linkage relationships to other genes.

Chromosome interference: A crossover event at one location that can prevent additional crossovers from occurring nearby.

Key Analytical Tools

Genetic distance:

If you know the number of crossovers that occurred in a specified number of meioses, genetic distance is calculated as

$$\frac{1}{2} \times \left(\frac{\text{number of crossovers}}{\text{number of meiosis}} \right) \times 100 = \text{distance in centimorgans (cM)}.$$

If you know the number of recombinant gametes (or progeny) among the total number of gametes (or progeny), genetic distance is calculated as

$$\left(\frac{\text{number of recombinants}}{\text{total number of gametes}}\right) \times 100 = \text{distance in recombination frequency (\%).}$$

If you know the types and numbers of tetrads from a *S. cerevisiae* or *N. crassa* cross, genetic distance is calculated as

$$\left(\frac{\frac{1}{2}(\text{tetratype}) + 3(\text{nonparental ditype})}{\text{total tetrads}}\right) \times 100 = \text{distance in map units (m.u.).}$$

If you know the number of second-division segregation asci among the total number of asci from a *N. crassa* cross, genetic distance from the centromere is calculated as

$$\left(\frac{\frac{1}{2}(\text{second-division segregation asci})}{\text{total number of asci}}\right) \times 100 = \text{distance in map units (m.u.).}$$

5.02 Types of Genetics Problems

1. **Relate chromosome structure and recombination during meiosis to genetic linkage.**
2. **Make predictions or test hypotheses based on genetic linkage data.**
3. **Calculate genetic distance from the results of crosses.**
4. **Estimate genetic distance or calculate probabilities using pedigrees or lod scores.**

1. Relate chromosome structure and recombination during meiosis to genetic linkage.

These problems test your understanding of the relationship between the behavior of chromosomes during meiosis and the phenomenon of genetic linkage. You may be given genetic maps and asked to predict the outcome of specific types of recombination events or to comment on the relationship between recombination frequencies and genetic map distance. Alternatively, you may be asked to infer or draw genetic maps given linkage data.

Example Problem: The order of the syntenic genes *A*, *B*, and *D* in a plant is unknown. The results of genetic analysis support linkage of *A* to *B* at $r = 0.20$, and linkage of *B* to *D* at $r = 0.20$, but do not support linkage of *A* to *D*. Is the independent assortment of *A* and *D* surprising given the linkage between *A* and *B* and *B* and *D*? Explain.

Solution Strategies	Solution Steps
Evaluate	
1. Identify the topic this problem addresses, and explain the nature of the requested answer.	1. This problem tests your understanding of the relationship between recombination frequency and genetic linkage. The answer will be a yes or no and an explanation for that response.
2. Identify the critical information given in the problem.	2. The *A*, *B*, and *D* genes are on the same chromosome. The recombination frequency for *A* and *B* is 0.2. The recombination frequency for *B* and *D* is 0.2. *A* and *D* assort independently.

(continued)

Solution Strategies	Solution Steps
Deduce	
3. Draw the chromosome and indicate the genetic distance between each gene pair. TIP: What recombination frequency indicates independent assortment?	3. • *A* is 20 recombination units from *B*. • *B* is 20 recombination units from *D*. • *A* and *D* assort independently; therefore, they recombine at 50%. • *B* is in the middle.
4. Account for the relationship between recombination frequency and genetic distance as genetic distance increases. TIP: Recombination frequency underestimates genetic distances when genes are more than 15 map units apart.	4. The genetic distance between *A* and *B* and between *B* and *D* is likely to be greater than 20 map units.
Solve	
5. Use the genetic map and the relationship between recombination frequency and genetic distance to predict the recombination frequency between *A* and *D*.	5. If *A* to *B* and *B* to *D* were exactly 20 map units each, the distance from *A* to *D* would be 40 map units. Since the 20 map unit distances are likely to be underestimates, the distance from *A* to *D* is likely to be more than 40 map units.
6. Answer the question using the logic from step 4.	6. **Answer:** Given that *A* and *D* are likely to be more than 40 map units apart and that the maximum recombination frequency that can be observed for two genes is 50%, it is not surprising that *A* and *D* assort independently, since they may be separated by 50 map units or more.

2. Make predictions or test hypotheses based on genetic linkage data.

These problems test your ability to use genetic linkage data to predict the outcome of crosses or matings. You may be given genetic linkage information and asked to predict the types and proportions of gametes that would be produced by an individual or the types and proportions of progeny that would result from a specified mating. You may be given hypothetical genetic linkage information and asked to design crosses that would test the linkage hypothesis.

Example Problem: The order of genes on a plant chromosome is *A, B* and *D*, where *A* and *B* are separated by a recombination frequency of 0.1, and *B* and *D* are separated by a recombination frequency of 0.2. A pure-breeding plant that has the dominant *A* and *B* phenotypes with a recessive *d* phenotype is crossed to a pure-breeding plant that has the recessive *a* and *b* phenotypes with the dominant *D* phenotype. The resulting hybrid is crossed to a plant that has all three recessive phenotypes. Predict the types and proportions of progeny that will result from this cross under the assumption that the value for interference in this region is 0.

Solution Strategies	Solution Steps
Evaluate	
1. Identify the topic this problem addresses, and explain the nature of the requested answer.	1. This problem tests your ability to predict the outcome of a test cross of a trihybrid plant. The answer will be a list of progeny phenotypes and the relative proportion of each progeny phenotype.
2. Identify the critical information given in the problem.	2. The parents of the trihybrid have the genotypes *ABd/ABd* and *abD/abD*. The trihybrid is crossed to a strain with the genotype *abd/abd*. The value for interference in this region is 0.
Deduce	
3. Determine the genotype of the trihybrid and the expected types of progeny. State the corresponding phenotype. **TIP:** A test cross of a trihybrid produces two parental and four or six recombinant progeny classes. The two parental classes represent the meiotic products when **no crossovers** occur in the *A*-to-*D* interval. The four recombinant classes represent the reciprocal meiotic products of two different **single-crossover** events (between *A* and *B* or between *B* and *D*). The last two recombinant classes represent the meiotic products produced by **double crossovers** (between *A* and *B* and between *B* and *D*). Double crossovers will be detected if the value for interference (*i*) is less than 1.0.	3. The genotypes (and phenotypes) of the F_1 and test cross progeny are the following: F_1: *ABd/abD* (dominant *A*, *B*, and *D*) **No crossovers:** *ABd/abd* (dominant *A* and *B*, recessive *d*) and *abD/abd* (recessive *a* and *b*, dominant *D*) **Single crossover between *A* and *B*:** *AbD/abd* (dominant *A* and *D*, recessive *b*) and *aBd/abd* (dominant *B*, recessive *a* and *d*) **Single crossover between *B* and *D*:** *ABD/abd* (dominant *A*, *B*, and *D*) and *abd/abd* (recessive *a*, *b*, and *d*) **Double crossovers:** *Abd/abd* (dominant *A*, recessive *b* and *d*) and *aBD/abd* (recessive *A*, dominant *B* and *D*)
Solve	
4. Calculate the relative proportions of each progeny genotype by plugging the recombination frequencies given in the problem into the appropriate formula: The frequency of each parental progeny type is $\frac{1}{2}$ (the frequency of *no* recombination between *A* and *B*) (*no* recombination between *B* and *D*). **TIP:** The frequency of *no* recombination in a genetic interval is (1 – frequency of recombination).	4. Recombinant progeny proportions are these: Each progeny type due to no recombination in either interval is $\frac{1}{2} \times (0.9)(0.8) = .36$ Each progeny type due to a single crossover between *A* and *B* is $\frac{1}{2} \times (0.1)(0.8) = .04$ Each progeny type due to a single crossover between *B* and *D* is $\frac{1}{2} \times (0.9)(0.2) = .09$ Each progeny type due to a double crossover is $\frac{1}{2} \times (0.1)(0.2) = .01$

(continued)

Solution Strategies	Solution Steps
The frequency of each type of progeny due to a single crossover in a genetic interval is $\frac{1}{2}$ (the frequency of recombination in that interval) (the frequency of *no* recombination in the other interval).	
The frequency of each type of double-crossover progeny is $\frac{1}{2}$ (the product of the recombination frequencies of both genetic intervals).	
5. Answer by listing the F_1 and progeny phenotypes and the proportions of each.	**5. Answer:** **F_1:** Dominant *A*, *B*, and *D* **Progeny:** 0.36 dominant *A* and *B*, recessive *d* 0.36 recessive *a* and *b*, dominant *D* 0.04 dominant *A* and *D*, recessive *b* 0.04 dominant *B*, recessive *a* and *d* 0.09 dominant *A*, *B*, and *D* 0.09 recessive *a*, *b*, and *d* 0.01 dominant *A*, recessive *b* and *d* 0.01 dominant *B* and *D*, recessive *a*

3. Calculate genetic distance from the results of crosses.

These problems test your ability to interpret the results of crosses to detect genetic linkage and, if linkage is detected, to estimate genetic distance. You may be given the results of a test cross and asked to statistically analyze whether the data supports linkage. You may be given the results of a test cross and asked to *deduce* the linkage phase of the alleles in the hybrid parent, order the genes on the chromosome, determine genetic distances, draw a genetic map, or determine the value for interference.

Example Problem: The order of the syntenic genes *A*, *B*, and *D* in a plant is unknown. A plant heterozygous for all three genes is crossed to a pure-breeding plant with all three recessive phenotypes. The following results were obtained.

Phenotype	Number of Progeny
Dominant *A*, dominant *B*, dominant *D*	2
Dominant *A*, dominant *B*, recessive *d*	20
Dominant *A*, recessive *b*, dominant *D*	402
Recessive *a*, dominant *B*, dominant *D*	73
Dominant *A*, recessive *b*, recessive *d*	70
Recessive *a*, dominant *B*, recessive *d*	409
Recessive *a*, recessive *b*, dominant *D*	23
Recessive *a*, recessive *b*, recessive *d*	1

Use these results to determine the order of the three genes, the recombination frequency between adjacent genes, and the value for crossover interference (*I*).

Solution Strategies	Solution Steps
Evaluate	
1. Identify the topic this problem addresses, and explain the nature of the requested answer.	1. This problem tests your ability to analyze the results of a trihybrid test cross. The answer will be a statement of the gene order, the recombination frequency between adjacent genes, and the value for interference.
2. Identify the critical information given in the problem.	2. The parents of the trihybrid have the genotypes *AbD/AbD* and *aBd/aBd*. The trihybrid is crossed to a strain with the genotype *abd/abd*. The results of the test cross are shown in the table.
Deduce	
3. Infer the genotype of the trihybrid plant from the genotypes of its parents.	3. The trihybrid's genotype is *AbD/aBd* (no particular gene order implied).
4. Determine the order of the genes by comparing the progeny classes representing the parental types (no crossovers) and the double-crossover types, which differ only in the alleles of the gene that lies in the middle. Restate the genotype of the trihybrid, showing the correct gene order. **TIP:** The two predominant progeny types correspond to the parental types. The two least abundant classes correspond to double-crossover types.	4. The two predominant progeny types are dominant *A* and *D* with recessive *b* (402), and dominant *B* with recessive *a* and *d* (409). The two least represented classes are dominant *A*, *B*, and *D* (2) and recessive *a*, *b*, and *d* (1). The differences between the predominant and least represented classes are the alleles of the *B* locus, indicating that *B* is in the middle. The trihybrid's genotype was *AbD/aBd*.
5. Identify the number of progeny that provide evidence for a crossover in each genetic interval during meiosis in the trihybrid parent. **TIP:** Use the trihybrid genotype from step 4 to predict the phenotype of progeny resulting from trihybrid gametes that undergo a crossover between *A* and *B*, a crossover between *B* and *D*, or crossovers between *A* and *B and* between *B* and *D*. Identify these phenotypes among the progeny.	5. Recombinant progeny that represent the following: • A single crossover between *A* and *B* are dominant *A* and *B* with recessive *d* (20) and recessive *a* and *b* with dominant *D* (23). • A single crossover between *B* and *D* are dominant *A* with recessive *b* and *d* (70) and recessive *a* with dominant *B* and *D* (73). • A crossover between *A* and *B and* between *B* and *D* are dominant *A*, *B*, and *D* (2) and recessive *a*, *b*, and *d* (1).

(continued)

Solution Strategies	Solution Steps
Solve	
6. Calculate the genetic distance between the adjacent genes. **TIP:** The recombination frequency between *A* and *B* will be the number of progeny representing a crossover between *A* and *B* divided by the total. **PITFALL:** Do not forget to include the double-crossover progeny when calculating the recombination frequency for each interval.	**6.** The genetic distances in this interval are the following: • From *A* to *B* is $\frac{20+23+2+1}{1000} = 0.046$ • From *B* to *D* is $\frac{70+73+2+1}{1000} = 0.146$
7. Calculate the coefficient of coincidence (*c*) for use in calculating the interference value: $$c = \frac{\text{observed double crossover}}{\text{expected double crossover}}$$ **TIP:** The expected number of double crossovers = (recombination frequency 1)(recombination frequency 2) × (total number of progeny observed).	**7.** Expected number of double-crossover progeny is $0.046 \times 0.146 \times 1000 = 6.7$. Observed number of double-crossover progeny is 3. Plug into the formula to get: $$c = \frac{3}{6.7} = 0.45$$
8. Give the gene order (step 3), the recombination frequency for each interval (step 5) and the interference value (*I*). **TIP:** $I = 1 - c$.	**8. Answer:** The gene order is *A-B-D*. The *A* and *B* are separated by a recombination frequency of 0.046. *B* and *D* are separated by a recombination frequency of 0.146. The value for interference is 0.55.

4. Estimate genetic distance or calculate probabilities using pedigrees or lod scores.

These problems test your ability to apply knowledge of linkage and recombination to pedigrees and the statistical tests associated with linkage analysis using pedigrees. You may be asked to calculate genetic distance using information from a pedigree or interpret the lod scores generated by linkage analysis using pedigree information. You may also be asked to use genetic linkage data and pedigree information to determine the probability of outcomes of specified matings.

Example Problem: Neurofibromatosis type 1 (NF1) is an autosomal dominant disorder that can cause the formation of multiple benign tumors throughout the nervous tissue of affected individuals. Although the penetrance of NF1 is essentially 100%, its expressivity is highly variable, such that some individuals with the NF1 mutation have no detectable tumors and only a few unusual freckles called café au lait spots. Because of the extreme variation in phenotype, children with an affected parent are sometimes uncertain as to whether they carry the mutation. The pedigree shows a family in which the NF1 mutation

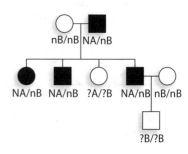

(*N* allele) is segregating. It also shows segregation of a two-allele RFLP (*A* and *B*) that is linked to the *NF1* gene at a recombination frequency of 0.2. Individuals II-3 and III-1 have not been tested for NF1 and have asked you to determine the probability that they carry the NF1 mutation (the *N* allele).

Solution Strategies	Solution Steps
Evaluate	
1. Identify the topic this problem addresses, and explain the nature of the requested answer.	1. This problem tests your ability to use genetic linkage and pedigree information to determine the probability that certain individuals have a specific genotype. The answer will be a probability between 0 (no chance) and 1.0 (it is certain).
2. Identify the critical information given in the problem.	2. NF1 is an autosomal dominant disorder with highly variable expressivity. II-3 and III-1 are children of an affected parent. A two-allele RFLP is linked to the *NF1* gene at $r = 0.2$. The linkage phase for the RFLP and the *NF1* gene is known for the affected parents of II-3 and III-1.
Deduce	
3. Assign II-3 and III-1 genotypes assuming that they carry the NF1 mutant allele (*N*). Determine whether the gamete from the affected parent that would have provided the *N* allele was parental or recombinant.	3. Consider II-3 and III-1 one a time. • II-3 has an *A* and a *B* allele. The *A* allele had to have come from I-2. If she carries the *N* allele, she is *NA/nB* and, therefore, inherited *NA* from I-2. *NA* is a parental gamete type. • III-1 has two *B* alleles and therefore inherited a *B* allele from II-4. If he carries the *N* allele, he is *NB/nB* and, therefore, inherited the *NB* from II-4. *NB* is a recombinant gamete type.
Solve	
4. Calculate the probability of the II-3 and III-1 genotypes assuming they carry the *N* allele. TIP: Since recombination frequency for *N* and the RFLP is 0.2, the probability that a gamete is parental for *N* and the RFLP is 0.8 and the probability that a gamete is recombinant is 0.2.	4. Calculate the probability for II-3 and III-1 one at a time • For gametes from I-2 that carry the *A* allele, the probability that the *N* allele is present (parental type) is 0.8. • For gametes from II-4 that carry the *B* allele, the probability that the *N* allele is present (recombinant type) is 0.2.
5. The probability calculated in step 4 is the probability that II-3 and III-1, respectively, carry the dominant *N* allele.	5. **Answer:** The probability that II-3 carries the *N* allele is 0.8. The probability that III-1 carries the *N* allele is 0.2.

5.03 Solutions to End-of-Chapter Problems

1a–e.   *Evaluate:* This is type 1. Recall that alleles on the same side of the slash are on the same homolog, whereas alleles on the opposite sides of the slash are on opposite homologs. *Deduce > Solve:* See figure.

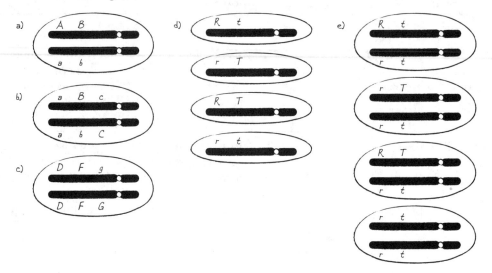

2a.   *Evaluate:* This is type 2. Recall from the text the meaning of *syntenic*, and that 1 m.u. = 1% recombination. *Deduce:* 18 map units predicts 18% recombinant gametes, 9% of each type, and 82% nonrecombinant (parental) gametes, 41% of each type. *DR* and *dr* are parental gametes, and *Dr* and *dR* are recombinant. *Solve:* Parental: 41% *DR*, 41% *dr*; recombinant: 9% *Dr*, 9% *dR*.

2b.   *Evaluate:* This is type 2. *Deduce:* 18 map units predicts 18% recombinant gametes, 9% of each type, and 82% nonrecombinant (parental) gametes, 41% of each type. *Dr* and *dR* are parental, and *DR* and *dr* are recombinant. *Solve:* Parental: 41% *Dr*, 41% *dR*; recombinant: 9% *DR*, 9% *dr*.

3a.   *Evaluate:* This is type 2. From Problem 2, we know that the parents are *Dr/Dr* × *dR/dR*. *Deduce:* The F_1 progeny are *Dr/dR*. The F_1's gametes are 41% *Dr*, 41% *dR*, 9% *DR*, and 9% *dr*. The F_1 is crossed to a *dr/dr* plant, which produces only *dr* gametes. *Solve:* The progeny genotype and proportions are, therefore, 41% *Dr/dr* (tall oval), 41% *dR/dr* (short round), 9% *DR/dr* (tall round), and 9% *dr/dr* (short oval).

3b.   *Evaluate:* This is type 2. The type and frequency of gametes produced by *Dr/ dR* plants are 41% *Dr*, 41% *dR*, 9% *DR*, and 9% *dr*. *Deduce:* The Punnett square shows these gamete frequencies and

	0.41 Dr	0.41 dR	0.09 DR	0.09 dr
0.41 Dr	0.1681 Dr/Dr	0.1681 Dr/dR	0.0369 Dr/DR	0.0369 Dr/dr
0.41 dR	0.1681 dR/Dr	0.1681 dR/dR	0.0369 dR/DR	0.0369 dR/dr
0.09 DR	0.0369 DR/Dr	0.0369 DR/dR	0.0081 DR/DR	0.0081 DR/dr
0.09 dr	0.0369 dr/Dr	0.0369 dr/dR	0.0081 dr/DR	0.0081 dr/dr

the types and frequencies of F_2 progeny produced by all combinations of gametes. The light-shaded squares are short round plants, and the dark-shaded squares are tall round plants. The total fraction of short round plants is obtained by adding the frequencies in all the light-shaded boxes and the fraction of tall round plants is obtained by adding the frequencies in the dark-shaded boxes. *Solve:* There will be 0.2419 short round and 0.5081 tall round.

4.   *Evaluate:* This is type 1. *Deduce:* A single crossover between *E* and *H* in an individual with the genotype *EH/eh* results in production of 1 *EH* : 1 *eh* : 1 *Eh* : 1 *eH* gametes. *Solve:* *E* and *H* are not genetically linked, because a single crossover in every meiosis results in production of equal proportion of *EH*, *Eh*, *eH*, and *eh* gametes. *EH* and *eh* are parental gametes, whereas *Eh* and *eH* are recombinant gametes. The percentage of parental gametes is the same as the percentage of recombinant gametes, which is 50%.

5a. *Evaluate:* This is type 2. Recall that 1 m.u. = 1% recombination. *Deduce: A* and *H* are 16 map units apart, indicating that 16% of gametes produced will be recombinant, and 84% will be parental. *C* is not linked to *A* or *H* and therefore will assort independently of both genes. *A_* are purple, *aa* are green, *H_* are hairy, *hh* are hairless, *C_* are cut, and *cc* are potato. The parents were purple (*A_*), hairy (*H_*), cut (*C_*), and green (*aa*), hairless (*hh*), potato (*cc*). The predominant progeny classes were purple, hairy and green, hairless, indicating that the *AH* and *ah* were parental gametes. *Solve:* The genotype of hairy, purple, cut individuals is *AH/ahCc*. The genotype of hairless, green, potato individuals is *ah/ahcc*. The genotypes of hairy, green, cut individuals is *aH/ahCc*. The genotype of hairless, purple, cut individuals is *Ah/ahCc*. The genotype of hairy, green, potato individuals is *aH/ahcc*. The genotype of hairless, purple, potato individuals is *Ah/ahcc*. The genotype of hairless, green, cut individuals is *ah/ahCc*. The genotype of purple, hairy, potato individuals is *AH/ahcc*.

5b. *Evaluate:* This is type 2. *Deduce > Solve:* Since the parents were *AH/ahCc* and *ah/ahcc*, the progeny phenotypes will be determined by the gametes produced by the purple, hairy, cut parent. The purple, hairy, cut parent will produce 42% *AH*, $\frac{1}{2}$ of which will be *C* and $\frac{1}{2}$ *c*, resulting in 21% *AHC* gametes (purple, hairy, cut progeny) and 21% *AHc* gametes (purple, hairy, potato progeny). That parent will also produce 42% *ah*, $\frac{1}{2}$ of which will be *C* and $\frac{1}{2}$ of which will be *c*, resulting in 21% *ahC* gametes (green, hairless, cut progeny) and 21% *ahc* gametes (green, hairless, potato progeny). The remaining gametes produced by the purple, hairy, cut parent will be 8% *Ah* and 8% *aH*, and $\frac{1}{2}$ of each will be *C* and $\frac{1}{2}$ will be *c*, resulting in 4% *AhC* (purple, hairless, cut progeny), 4% *Ahc* (purple, hairless, potato progeny), 4% *aHC* (green, hairy, cut progeny), and 4% *aHc* (green, hairy, potato progeny).

6a. *Evaluate:* This is type 1. Recall that syntenic genes separated by less than 50 map units will show genetic linkage. *Deduce > Solve:* Yes, the *y* and *w* genes are expected to show linkage because they are less than 50 map units apart.

6b. *Evaluate:* This is type 1. Recall that syntenic genes separated by 50 map units or more will assort independently. *Deduce > Solve:* Yes, *y* is expected to assort independently of *f* because *y* and *f* are more than 50 map units apart. The same applies to *w* and *f*.

6c. *Evaluate:* This is type 2. Recall that 1 m.u. = 1% recombination. *Deduce:* The male is hemizygous for recessive alleles; therefore, the progeny phenotypes will be determined by the female's gametes. Since *y* and *w* are separated by 1.5 m.u., 1.5% of her gametes will be recombinant for *y* and *w* (0.75% y^+w and 0.75% yw^+) and 98.5% will be parental (49.25% y^+w^+ and 49.25% yw). Since *f* is unlinked to *y* and *w*, $\frac{1}{2}$ of each of these gamete types will be f^+ and $\frac{1}{2}$ will be *f*. *Solve:* The female's gametes will be $\frac{1}{2} \times 0.75 = 0.375$ y^+wf^+, $\frac{1}{2} \times 0.75 = 0.375$ y^+wf, $\frac{1}{2} \times 0.75 = 0.375$ yw^+f^+, $\frac{1}{2} \times 0.75 = 0.375$ yw^+f, $\frac{1}{2} \times 49.25 = 24.625$ $y^+w^+f^+$, $\frac{1}{2} \times 49.25 = 24.625$ y^+w^+f, $\frac{1}{2} \times 49.25 = 24.625$ ywf^+, and $\frac{1}{2} \times 49.25 = 24.625$ ywf. There will be 24.625% of each of the following progeny types: gray with red eyes and forked bristles, gray with red eyes and normal bristles, yellow with white eyes and forked bristles, and yellow with white eyes and normal bristles. There will be 0.375% of each of the following types: gray with white eyes and forked bristles, gray with white eyes and normal bristles, yellow with red eyes and forked bristles, and yellow with red eyes and normal bristles.

6d. *Evaluate:* This is type 2. *Deduce > Solve:* The female is heterozygous at *y*, *w*, and *f*; the male is hemizygous recessive. Consider the linked *y* and *w* loci first. The *y* and *w* genes are 1.5 map units apart, so the females will make 0.4925 of each parental gamete type, which are y^+w^+ and y^-w^-, and 0.0075 of each recombinant gamete type, which are y^+w^- and y^-w^+. The *f* gene is unlinked to *y* and *w*, so its alleles assort independently of *y* and *w*, with 0.50 of each y^-w^- genotype receiving an f^+ allele and 0.50 receiving an f allele. The fraction of $y^+w^+f^+$ is $0.4925 \times 0.5 = 0.24625$. The same is true for y^+w^+f, $y^-w^-f^+$, and y^-w^-f. The fraction of $y^+w^-f^+$ is $0.0075 \times 0.50 = 0.00375$. The same is true for y^+w^-f, $y^-w^+f^+$, and $y^-w^+f^-$.

7.   *Evaluate:* This is type 2. *Deduce:* This is a three-factor test cross; therefore, the progeny types will be determined by the types of gametes produced by the $a^+b^+c^-/a^-b^-c^+$ parent. There will be eight classes of gametes; 2 parental, 2 with a single crossover between a and b, 2 with a single crossover between b and c, and 2 with crossovers between a and b and between b and c. A recombination frequency of 0.08 between a and b predicts that 0.08 of the gametes will have undergone a recombination event between a and b and that 0.92 will not. Similarly, 0.24 of the gametes will have undergone a recombination event between b and c and 0.76 will not. Since there is no interference, double crossovers occur at a frequency equal to the frequency of recombination between a and b multiplied by the frequency of recombination between b and c, which is $0.08 \times 0.24 = 0.0192$. The frequency of each gamete type can be calculated by multiplying the frequencies together. Parental-type gametes will occur at a frequency of 0.92 (no crossover between a and b) multiplied by 0.76 (no crossover between b and c), which is 0.6992. There are two parental types ($a^+b^+c^-$ and $a^-b^-c^+$), each occurring at equal frequency; therefore, each parental type occurs at $0.6992 \times \frac{1}{2} = 0.3496$. Recombinant gametes resulting from a single crossover between a and b will occur at a frequency of 0.08 multiplied by 0.76, which is 0.0608, with $\frac{1}{2}$ being $a^+b^-c^-$ and the other half $a^-b^+c^+$. Recombinant gametes resulting from a single crossover between b and c will occur at a frequency of 0.92 multiplied by 0.24, which is 0.2208, with $\frac{1}{2}$ being $a^+b^+c^+$ and the other half $a^-b^-c^-$. Recombinant gametes resulting from double crossovers, one between a and b and the other between b and c, will occur at a frequency of 0.08 multiplied by 0.24, which is 0.0192, with $\frac{1}{2}$ being $a^+b^-c^+$ and the other half being $a^-b^+c^-$. *Solve:* See table.

Gamete	Frequency	Gamete Type
a^+b^+c	$\frac{1}{2} \times 0.92 \times 0.76 = 0.3496$	Parental
abc^+	$\frac{1}{2} \times 0.92 \times 0.76 = 0.3496$	Parental
a^+bc^+	$\frac{1}{2} \times 0.08 \times 0.76 = 0.0304$	Single crossover between a and b
ab^+c	$\frac{1}{2} \times 0.08 \times 0.76 = 0.0304$	Single crossover between a and b
$a^+b^+c^+$	$\frac{1}{2} \times 0.92 \times 0.24 = 0.1104$	Single crossover between b and c
abc	$\frac{1}{2} \times 0.92 \times 0.24 = 0.1104$	Single crossover between b and c
a^+bc	$\frac{1}{2} \times 0.08 \times 0.24 = 0.0096$	Double crossover (one between a and b and one between b and c)
ab^+c^+	$\frac{1}{2} \times 0.08 \times 0.24 = 0.0096$	Double crossover (one between a and b and one between b and c)

8a.   *Evaluate:* This is type 1. *Deduce:* Since the order of the genes is not given, G or R could be in the middle. If G is in the middle, then T is on one side and R is on the other side. If R is in the middle, then G is on one side and T is on the other. T cannot be in the middle, because the distance from T to G is greater than the distance from R to G. *Solve:* See figure.

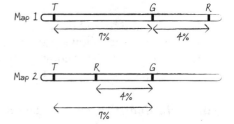

8b.   *Evaluate:* This is type 2. *Deduce:* The two maps differ in the order of the genes; therefore, a three-factor test cross can be used to determine which map is correct. For example, a cross between a strain that is TRG/trg and trg/trg will produce 8 progeny classes, two of which correspond to double-crossover progeny. The double-crossover progeny will be *TgR/tgr* and *tGr/tgr* if the order is *TGR*, whereas they will be *TrG/trg* and *tRg/trg* if the order is *TRG*. *Solve:* Perform a test cross of a *TRG/trg* trihybrid to determine the gene order. If the order is *TGR*, then the least abundant progeny classes will be *TgR/tgr* and *tGr/tgr*. If the order is *TRG*, the least abundant progeny classes will be *TrG/trg* and *tRg/trg*.

9a. *Evaluate:* This is type 2. *Deduce:* Since *A* and *E* recombine at a frequency of 0.28, 0.72 of the progeny will be parental ($\frac{1}{2}$ *Ae/ae* and $\frac{1}{2}$ *aE/ae*) and 0.28 will be recombinant ($\frac{1}{2}$ *AE/ae* and $\frac{1}{2}$ *ae/ae*). *Solve:* 360 *Ae*, 360 *aE*, 140 *AE*, and 140 *ae*

9b. *Evaluate:* This is type 1. *Deduce:* One measure is from experiments detecting recombination between *A* and *E*, whereas the other is from detecting recombination between four smaller genetic intervals between *A* and *E*. Recombination frequency is an accurate measure of genetic distance over short genetic intervals but underestimates genetic distance over longer intervals. Since genetic distance is additive, the sum of the recombination frequencies between *A* and *B*, *B* and *C*, *C* and *D*, and *D* and *E* is the more accurate measure of the genetic distance between *A* and *E*. *Solve:* The recombination frequency measured between *A* and *E* underestimates the actual genetic distance because double crossovers are occurring and not being counted. By measuring recombination frequency over several smaller genetic intervals from *A* to *E*, more of the crossovers that are occurring are counted, which gives a more accurate estimate of the true genetic distance.

10. *Evaluate:* This is type 1. Recall that syntenic genes are genes that are on the same chromosome and, although physically linked, can assort independently if separated by more than 50 m.u. *Solve:* Syntenic genes that are separated by 50 map units or more will assort independently because there will be one or more crossovers between them per meiosis.

11. *Evaluate:* This is type 1. *Deduce:* Genes on different chromosomes assort independently, resulting in a recombinant gamete frequency of 50%. *Solve:* The maximum recombination frequency is 50% because it is the frequency of recombinant gametes produced by dihybrids when the genes are on different chromosomes.

12a. *Evaluate:* This is type 2. *Deduce:* A test cross of a dihybrid is the most convenient way to measure recombination frequency. The dihybrid can have the trans (*y⁺l/yl⁺*) or cis (*y⁺l⁺/yl*) configuration. The progeny will be 72% parental type (36% *y⁺l* and 36% *yl⁺* if trans is used, or 36% *y⁺l⁺* and 36% *yl* if cis is used) and 28% recombinant type (14% *y⁺l⁺* and 14% *yl* if trans is used, or 14% *y⁺l* and 14% *yl⁺* if cis is used). *Solve:* To measure distance between *Y* and *L*, cross a *yl/y⁺l⁺* female to a *yl/Y* male. The progeny should be 36% yellow lozenge, 36% gray normal eyes, 14% yellow normal eyes, and 14% gray lozenge eyes. To measure the distance between *L* and *F*, cross an *lf/l⁺f⁺* female to an *lf/Y* male. The progeny will be 340 lozenge forked bristles, 340 normal eyes and normal bristles, 160 normal eyes with forked bristles, and 160 with lozenge eyes and normal bristles.

12b. *Evaluate:* This is type 1. Recall that syntenic genes separated by more than a 50% recombination are by definition unlinked and therefore cannot be demonstrated to be genetically linked. *Deduce > Solve:* No cross can demonstrate genetic linkage between genes *Y* and *F* because the recombination frequency between genes *Y* and *L* plus that between *L* and *F* is greater than 50%; therefore, the percent recombination between *Y* and *F* in any cross will be 50%.

12c. *Evaluate:* This is type 1. Recall that *independent assortment* is a term that applies to genes that show a recombination frequency of 50%. *Deduce > Solve:* A cross measuring recombination frequency between *Y* and *F* will determine that they are recombining at a frequency of 50%, which is the definition of independent assortment.

13a. *Evaluate:* This is type 3. *Deduce:* Linkage is indicated when the recombination frequency between genes is less than 50%. *Solve:* The data show that the predominant progeny classes were the parental types (colored, full, waxy and colorless, shrunken, starchy) and that all recombinant progeny classes were rare, which is consistent with genetic linkage between each pair of genes.

13b. *Evaluate:* This is type 3. *Deduce:* The expected distribution of progeny of the test cross of the trihybrid, under the assumption that all three genes assort independently, is eight classes, each present at a frequency of $\frac{1}{8}$. For 6708 progeny, $\frac{1}{8}$ is 838.5. *Solve:* The chi-square calculation is

$$\frac{(116 - 838.5)^2}{838.5} + \frac{(601 - 838.5)^2}{838.5} + \frac{(2538 - 838.5)^2}{838.5} + \frac{(4 - 838.5)^2}{838.5} + \frac{(2708 - 838.5)^2}{838.5}$$
$$+ \frac{(2 - 838.5)^2}{838.5} + \frac{(113 - 838.5)^2}{838.5} + \frac{(626 - 838.5)^2}{838.5} = 10{,}649$$

Solve: The chi-square value is 10,649. For seven degrees of freedom, this corresponds to a P value below 0.01, which indicates that the difference between the expected and observed results is highly statistically significant.

13c. *Evaluate:* This is type 3. Recall that the order of genes is determined by comparing the progeny genotypes corresponding to the nonrecombinant parental gametes and the genotype corresponding to double-crossover gametes. *Deduce:* These progeny will differ by only one of the traits, and that trait is in the middle. *ClShWx* and *clshwx* are the progeny genotypes corresponding to the nonrecombinant parental gametes. *ClshWx* and *clShwx* are progeny genotypes corresponding to the double-crossover gametes. They differ only in the *Sh* locus. *Solve:* The gene order is *Cl-Sh-Wx*.

13d. *Evaluate:* This is type 3. Recall that the recombination frequency is the number of progeny showing a crossover between genes divided by the total. *Deduce:* The progeny showing a crossover between *Cl* and *Sh* are *Clshwx*, *clShWx*, *ClshWx*, and *clShwx*. The progeny showing a crossover between *Sh* and *Wx* are *ClShwx*, *clshWx*, *ClshWx*, and *clShwx*. The progeny showing a crossover between *Cl* and *Wx* are *Clshwx*, *clShWx*, *ClShwx*, *clshWx*, *ClshWx*, and *clShwx*. **TIP:** Each *ClshWx* and *clShwx* progeny represent two crossovers between *Cl* and *Wx*. *Solve:* For *Cl* and *Sh*, the recombination frequency is $\frac{116 + 113 + 2 + 4}{6708} = .035$. For *Sh* and *Wx*, the recombination frequency is $\frac{601 + 626 + 2 + 4}{6708} = .184$. For *Cl* and *Wx*, the recombination frequency is $\frac{601 + 626 + 116 + 113 + 2 \times (2 + 4)}{6708} = .219$.

13e. *Evaluate:* This is type 3. Recall that the interference value is determined by subtracting the coefficient of coincidence from 1.0. The coefficient of coincidence is determined by dividing the number of observed double-crossover progeny by the number of expected progeny, assuming no interference. *Deduce:* The expected number of double crossovers is $0.035 \times 0.184 \times 6708 = 43.2$. The observed number was 6. The coefficient of coincidence is $6 \div 43.2 = 0.139$. *Solve:* The interference value is $1 - 0.139 = 0.861$.

14a. *Evaluate:* This is type 4. Nail–patella syndrome (NPS) is present in all three generations in the pedigree, and every affected individual has an affected parent. *Deduce:* Recall that traits apparent in every generation, where every affected individual has an affected parent, are usually dominant traits. *Solve:* Nail–patella syndrome is a dominant trait.

14b. *Evaluate:* This is type 4. Recall that genetic linkage is indicated if the frequency of recombinant progeny is less than 50%. *Deduce > Solve:* Yes, NPS appears to segregate with blood type A in this pedigree, indicating genetic linkage between these traits.

14c. *Evaluate:* This is type 8. *Deduce:* Since NPS is dominant, individuals with NPS have a dominant allele (N) and normal individuals are homozygous for the wild-type allele (n). Individuals with O blood type have the genotype $I^O I^O$. Individuals with A blood type have the genotype $I^A I^A$ or $I^A I^O$. Individuals with B blood type have the genotype $I^B I^B$ or $I^B I^O$. Individuals with AB blood type have the genotype $I^A I^B$. *Solve:* I-1 is $I^O n / I^O n$. I-2 is either $I^A N / I^O n$ or $I^A n / I^O N$. Since all three children who inherited N from I-2 also inherited I^A and the two children that inherited n from

I-2 also inherited I^O, we'll assume that I-2 is I^AN/I^On. II-2 is I^On/I^On, II-4 is I^AN/I^On, II-6 is I^AN/I^On, II-7 is I^On/I^On, and II-9 is I^AN/I^On.

14d.   *Evaluate:* This is type 4. *Deduce:* NPS segregated with the I^A allele in most cases in this pedigree, but not in the case of III-6. *Solve:* III-6 is I^ON/I^On and III-8 is I^On/I^On. Even though III-6 and III-8 are both *O* blood type, III-6 has NPS because he inherited the recombinant chromosome I^ON from his mother, whereas III-8 inherited the nonrecombinant I^On from her mother.

14e.   *Evaluate:* This is type 4. *Deduce > Solve:* The genotypes of III-11 and III-12 cannot be unambiguously determined, because they have type A blood and both of their parents are I^AI^O. Thus, either or both could I^AI^A or I^AI^O. For this reason, it is not clear whether III-11 and III-12 are parental-type or recombinant-type progeny.

15a.   *Evaluate:* This is type 3. Recall that genetic linkage will be apparent if the percentage of recombinant progeny is less than 50% for any two traits. In three-factor test crosses, linkage of all three genes results in two predominant progeny types, which represent the linkage phase for all three genes on the two parental chromosomes. Linkage between only two genes results in four progeny types that are equally predominant; those progeny types will show the parental chromosomes' linkage phase for the two linked genes. *Deduce:* Among the eight progeny types, four are the most highly represented (tunicate, glossy, liguled; tunicate, glossy, liguleless; nontunicate, nonglossy, liguless; and nontunicate, nonglossy, liguleless). These progeny represent the tunicate, glossy/nontunicate, nonglossy linkage phase in the parental chromosomes, indicating that tunicate and glossy are linked. *Solve:* The results indicate that tunicate and glossy are linked.

15b.   *Evaluate:* This is type 3. Independent assortment will be apparent if the percentage of recombinant progeny is 50% for any two traits. In three-factor test crosses, independent assortment of all three genes results in eight, equally represented progeny types. If one gene assorts independently of the other two, then four progeny types will predominate and be present at equivalent levels. Those progeny types will show the parental chromosomes' linkage phase for the two linked genes but not for the third, unlinked gene. *Deduce:* Among the eight progeny types, four are the most highly represented (tunicate, glossy, liguled; tunicate, glossy, liguleless; nontunicate, nonglossy, liguled; and nontunicate, nonglossy, liguleless). These progeny represent the tunicate, glossy/nontunicate, nonglossy linkage phase in the parental chromosomes, indicating that tunicate and glossy are linked and that liguled is unlinked to tunicate and glossy. *Solve:* The results indicate that liguled assorts independently of tunicate and glossy.

15c.   *Evaluate:* This is type 3. Recall that the alleles for syntenic genes are written on opposite sides of a slash (/) to indicate the linkage phase, for example, *AB/ab*. The alleles of genes whose chromosome location is not known are written side by side without a slash, for example, *AaBb*. *Deduce:* The results indicate that tunicate (*T*) and glossy (*G*) are linked (see 15a) and that the linkage phase in the F_1 was cis. Liguled (*L*) assorts independently of *T* and *G*, and its chromosome location is not known (see 15b). *Solve:* Tunicate (*T*) and glossy (*G*) alleles should be written on opposite sides of a slash and liguled (*L*) alleles should be written side by side. The F_1 trihybrid was *TG/tgLl*. Of the F_2, 102 were *TG/tgLl*; 106 were *TG/tgll*; 18 were *Tg/tgLl*; 20 were *Tg/tgll*; 22 were *tG/tgLl*; 23 were *tG/tgll*; 99 were *tg/tgLl*; and 110 were *tg/tgll*.

15d.   *Evaluate:* This is type 3. *Deduce:* The results indicate that tunicate (*T*) and glossy (*G*) show linkage (see 15a) and that liguled (*L*) assorts independently of both *T* and *G* (see 15b); therefore, recombination frequency needs to be calculated for *T* and *G*. *Solve:* The parental type progeny for *T* and *G* are tunicate, glossy and nontunicate, nonglossy. There are 102 + 106 tunicate, glossy and 99 + 110 nontunicate, nonglossy for a total of 417 parental-type progeny. The recombinant progeny for *T* and *G* are tunicate, nonglossy and nontunicate, glossy. There are 18 + 20 tunicate, nonglossy and 22 + 23 nontunicate, glossy for a total of 83 recombinant progeny. The formula for recombination frequency is $83/(417 + 83) = 0.166$.

15e.   *Evaluate:* This is type 3. *Deduce > Solve:* Yes. *T* and *G* are genetically linked, and therefore are syntenic. *L*, although not genetically linked to *T* or *G*, may or may not be on the same chromosome. If L is syntenic with T and G, then it is separated from both of them by a recombination frequency of 50% (a genetic distance of 50 cM or greater).

16a.   *Evaluate:* This is type 3. Recall that the order of genes in a three-point test cross is determined by comparing the parental-type progeny to the double-crossover progeny. These will differ by only one gene, and that gene is in the middle. *Deduce:* The parental types were *GLT* and *glt*, and the double-crossover progeny were *GLt* and *glT*. *Solve:* The difference between the parental and double-crossover progeny is the *T* gene; therefore, the *T* gene is in the middle. The order is *G-T-L*.

16b.   *Evaluate:* This is type 3. The question asks for frequencies between each gene pair, which means you are asked to consider only two genes at a time and not consider the evidence of double-crossover progeny that is provided in the table. *Deduce:* Recombinant progeny for *G* and *T* are *Ggtt* (109 + 3) and *ggTt* (103 + 7). The distance between *G* and *T* is $\frac{109 + 103 + 3 + 7}{1600} = 0.139$. Recombinant progeny for *T* and *L* are *Ttll* (64 + 7) and *ttLl* (67 + 3). The genetic distance between *T* and *L* is $\frac{64 + 67 + 3 + 7}{1600} = 0.088$. The recombinant progeny for *G* and *L* are *Ggll* (64 + 109) and *ggLl* (67 + 103). The genetic distance between *G* and *L* is $\frac{64 + 109 + 103 + 67}{1600} = 0.214$. *Solve:* The recombination frequencies are 0.139 between *G* and *T*, 0.088 between *T* and *L*, and .214 between *G* and *L*.

16c.   *Evaluate:* This is type 3. *Deduce > Solve:* The recombination frequency for *G* and *L* is less than for *G* and *T* plus *T* and *L* because the double-crossover progeny do not appear to be recombinant for *G* and *L* and, therefore, are not counted.

16d.   *Evaluate:* This is type 3. Recall that interference is equal to 1 minus the coefficient of coincidence. *Deduce:* The coefficient of coincidence is the observed number of double-crossover progeny divided by the expected number, assuming no interference. The expected number of double-crossover progeny is 0.139 × 0.088 × 1600 = 19.6. The observed number was 10. *Solve:* The coefficient of coincidence is 10 ÷ 19.6 = 0.51, and the interference value (*l*) is 1 − 0.51 = 0.49.

16e.   *Evaluate:* This is type 3. *Deduce:* A positive value for interference indicates that a crossover in one genetic interval in a meiotic cell has a negative effect on the likelihood that a second crossover will occur nearby in that same meiotic cell. *Solve:* The meaning of *I* = 0.49 is that only half of the expected number of double-crossover progeny were observed. This indicates that a meiotic cell undergoing a crossover between *G* and *T* is about half as likely to also have a crossover occur between *T* and *L*. Similarly, a crossover between *T* and *L* reduces the likelihood of a crossover between *G* and *T*.

17a.   *Evaluate:* This is type 2. Recall that the recombination frequency indicates the percent of recombinant gametes produced for linked genes. *Deduce:* *D* and *E* are separated by 8% recombination. *De* and *dE* are parental gametes, and *DE* and *de* are recombinant gametes. *Solve:* There will be $\frac{1}{2} \times 8 = 4\%$ of each recombinant gamete, and $\frac{1}{2} \times (100 - 8) = 46\%$ of each parental gamete. Therefore, the results are 46% *De*, 46% *dE*, 4% *DE*, 4% *de*.

17b.   *Evaluate:* This is type 2. Recall that the recombination frequency indicates the percent of recombinant gametes produced for linked genes and that recombination frequency over long genetic intervals underestimates genetic distance because of double crossovers. **TIP:** With no interference, the frequency of double crossovers is the product of the single-crossover frequencies. *Deduce:* *A* and *D* are separated by 28% recombination. *Ad* and *aD* are parental gametes, and *AD* and *ad* are recombinant gametes. *Solve:* The expected frequency of recombinant gametes for *A* and *D* is the recombination frequency minus the expected double-crossover frequency. The double-crossover frequency will be 0.1 × 0.18 = 0.018. The observed recombination frequency between *A* and *D* will be 0.28 − 0.018 = 0.262. There will be $\frac{1}{2} \times 26.2 = 13.1\%$ of each recombinant gamete and $\frac{1}{2} \times (100 - 26.2) = 36.9\%$ of each parental gamete. Therefore, the results are 36.9% *Ad*, 36.9% *aD*, 13.1% *AD*, 13.1% *ad*.

17c.   *Evaluate:* This is type 2. Recall that the recombination frequency indicates the percent of recombinant gametes produced for linked genes and that double-crossover gametes from a trihybrid are detectable. **TIP:** With no interference, the frequency of double crossovers is the product of the single-crossover frequencies. *Deduce:* D and E are separated by 8% recombination, and E and F are separated by 24% recombination. DeF and dEf are parental, DEf and deF are from a single crossover between D and E, Def and dEF are from a single crossover between E and F, and DEF and def are from a double crossover. *Solve:* The frequency of DeF and dEf will each be $\frac{1}{2}(0.92 \times 0.76) = 0.3496$. The frequency of DEf and deF gametes will each be $\frac{1}{2}(.08 \times 0.76) = 0.0304$. The frequency of Def and dEF will each be $\frac{1}{2}(0.92 \times 0.24) = 0.1104$. The frequency of DEF and def will each be $\frac{1}{2}(0.08 \times 0.24) = 0.0096$. Therefore, the results are 34.96% DeF, 34.96% dEf, 3.04% DEf and 3.04% deF, 11.04% Def and 11.04% dEF, 0.96% DEF and 0.96% def.

17d. *Evaluate:* This is type 2. Recall that the recombination frequency indicates the percent of recombinant gametes produced for linked genes and that recombination frequency over long genetic intervals underestimates genetic distance because of double crossovers. **TIP:** With no interference, the frequency of double crossovers is the product of the single-crossover frequencies. *Deduce:* BdE/bde is a dihybrid (the organism is dd; therefore, all gametes will contain a d allele). B and E are separated by 26% recombination. BdE and bde are parental gametes, and Bde and bdE are recombinant gametes. *Solve:* The parental gametes BdE and bde are expected at a frequency of $\frac{1}{2} \times (.82)(.92) = .378$. The double crossover gametes are also BdE and bde, and they are each expected at a frequency of $\frac{1}{2} \times (.18)(.08) = .007$. Single crossover between B and D produces the recombinant gametes Bde and bdE, each at an expected frequency of $\frac{1}{2} \times (.18)(.92) = .083$. Single crossover between D and E produces the same Bde and bdE crossover gametes. Each of these is expected at a frequency of $\frac{1}{2} \times (.82)(.08) = .033$. Therefore, the expected results are .385 Bde, .385 bde, .116 Bde, and .116 bdE.

18a. *Evaluate:* This is type 4. Recall that an lod score of 3.0 or higher at any θ value is strong statistical support for linkage at that genetic distance. *Deduce > Solve:* Yes, the data provides strong support for linkage between Rh and eliptocytosis because the maximum lod score supporting linkage is above 3 (it is about 5.5 for linkage at a θ value just over 0.1).

18b. *Evaluate:* This is type 4. Recall that Z_{max} is the highest lod score in the results. *Deduce > Solve:* The maximum lod score is about 5.5 for linkage at a θ value just over 0.1.

18c. *Evaluate:* This is type 4. Recall that an lod score of 3.0 or higher at any θ value is strong statistical support for linkage. *Deduce > Solve:* The results support linkage at θ values from just under 0.05 to about 0.30.

19a. *Evaluate:* This is type 4. *Deduce:* Tom inherited r and E from his mother and R and e from his father, making his genotype rE/Re. Terri inherited R and E from her father and r and e from her mother, making her genotype RE/re. Given that R and E recombine at a frequency of 4%, Tom's gamete types will be 0.48 rE, 0.48 Re, 0.02 re, and 0.02 RE. Terri's gamete types will be 0.48 RE, 0.48 re, 0.02 Re, and 0.02 rE. There are three ways they can have a child with Rh– blood type and eliptocytosis (assuming that homozygous dominant E is viable): rE from Tom and re from Terri, which occurs at a frequency of $0.48 \times 0.48 = 0.2304$; rE from Tom and rE from Terri, which occurs at a frequency of $0.48 \times 0.02 = 0.0096$; and re from Tom and rE from Terri, which occurs at a frequency of $0.02 \times 0.02 = 0.0004$. *Solve:* The probability that Tom and Terri's first child will have Rh– blood type and elliptocytosis is $0.2304 + 0.0096 + 0.0004 = 0.2404$.

19b. *Evaluate:* This is type 4. *Deduce:* Tom is rE/Re. Terri is RE/re. Tom's gamete types will be 0.48 rE, 0.48 Re, 0.02 re, and 0.02 RE. Terri's gamete types will be 0.48 RE, 0.48 re, 0.02 Re, and 0.02 rE. The probability that a child with Rh+ blood type will have eliptocytosis is a conditional probability, calculated as the (probability of Rh+ and eliptocytosis)/(probability of Rh+). *Solve:* The probability of Rh+ and eliptocytosis is $(0.48)(0.48) + (0.48)(0.48) + (0.48)(0.02) + (0.48)(0.02) + (0.48)(0.02) + (0.02)(0.48) + (0.02)(0.02) + (0.02)(0.48) + (0.02)(0.02) = 0.5096$. The probability of Rh+ blood type is 0.75. $0.5096 \div 0.75 = 0.679$.

20a.   *Evaluate:* This is type 1. *Deduce:* The distance between a gene and its centromere is calculated as $\frac{1}{2}\left(\frac{\text{(number of second division asci)}}{\text{(total number of asci)}}\right) \times 100 = cM$. There were 500 total asci analyzed, of which $23 + 27 + 29 + 21 = 100$ were second-division asci. *Solve:* The distance between a and its centromere is $\frac{1}{2}\left(\frac{100}{500}\right) \times 100 = 10\ cM$.

20b.   *Evaluate:* This is type 1. Like alleles will be located together on one side of the ordered ascus when there is no crossover between the gene and its centromere during meiosis. *Deduce:* The drawing shows meiosis occurring without a crossover between the centromere and the a locus. *Solve:* See figure.

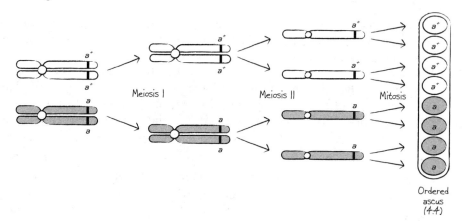

20c.   *Evaluate:* This is type 1. Like alleles will be located on both sides of the ordered ascus when there is a single crossover between the gene and its centromere during meiosis. *Deduce:* The drawing shows meiosis occurring with a single crossover occurring between the centromere and the a locus. *Solve:* See figure.

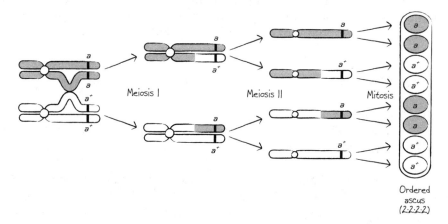

21a.   *Evaluate:* This is type 2. Recall that recombination frequency is the same as the proportion of gametes that will be recombinant. *Deduce:* if $r = 0.2$, then 0.2 (20%) of the gametes will be recombinant and 0.8 (80%) will be parental. Rt and rT are parental gametes. RT and rt are recombinant gametes. *Solve:* The gametes will be 40% Rt, 40% rT, 10% RT, and 10% rt.

21b.   *Evaluate:* This is type 1. A two-stranded double crossover results in formation of four parental gametes. *Deduce > Solve:* Two Rt gametes and two rT gametes.

21c.   *Evaluate:* This is type 1. A three-stranded double crossover produces two parental gametes and two recombinant gametes. *Deduce > Solve:* One Rt, one rT, one RT, and one rt.

21d.   *Evaluate:* This is type 1. A four-stranded double crossover produces four recombinant gametes. *Deduce > Solve:* Two RT and two rt.

22. *Evaluate:* This is type 3. *Deduce:* There were 1541 parental-type and 900 recombinant progeny out of 2441 analyzed. Independent assortment predicts 1220.5 parental type and 1220.5 recombinant type. *Solve:* The chi-square calculation is $\left(\frac{(1541-1220.5)^2}{1220.5}\right) + \frac{(900-1220.5)^2}{1220.5} = 168.3$. For 3 degrees of freedom, this corresponds to a *P* value of less than 0.001.

23a. *Evaluate:* This is type 3. *Deduce:* The two predominant progeny types are parental types and correspond to the linkage phase of the alleles in the trihybrid. The two least abundant classes correspond to double-crossover progeny. The difference between the parental-type progeny and the double-crossover progeny is the allelic pair of the gene in the middle. *Solve:* The parental types are pale (*lTR/ltr*) and oval, short (*Ltr/ltr*). The pure-breeding recessive plant was *ltr/ltr*; therefore, the trihybrid was *lTR/Ltr*. The double-crossover progeny were pale, short (*ltR/ltr*) and oval (*LTr/ltr*). The difference between the parental and double-crossover progeny are the alleles of the *T* locus. The trihybrid's genotype was *lTR/Ltr*, and the gene order is *L-T-R*.

23b. *Evaluate:* This is type 3. *Deduce:* Progeny that indicate a crossover occurred between *L* and *T* are wild type; pale, short, oval; pale, short; and oval. Progeny that indicate a crossover occurred between *T* and *R* include pale, oval; short; pale, short; and oval. *Solve:* The recombination frequency between *L* and *T* is $\frac{98+102+10+6}{1630} = 0.13$. The recombination frequency between *T* and *R* is $\frac{84+64+10+6}{1630} = 0.10$.

23c. *Evaluate:* This is type 3. *Deduce:* The interference value is 1 minus the coefficient of coincidence. The coefficient of coincidence is the number of observed double recombinants divided by the number expected assuming no interference. The number of double-crossover progeny expected is $0.10 \times 0.13 \times 1630 = 21.19$. The number observed was 16. *Solve:* The interference value is $1-(16 \div 21.19) = 0.25$.

24a. *Evaluate:* This is type 2. *Deduce:* Linkage can be detected by crossing pure-breeding brown-eyed and pure-breeding short-bristled lines to create a dihybrid. *Solve:* Assuming that *c* and *d* are syntenic, the pure-breeding brown-eyed fly line is cd^+/cd^+, the pure-breeding short-bristled line is c^+d/c^+d, and these can be crossed to create the F_1 dihybrid, which will be cd^+/c^+d.

24b. *Evaluate:* This is type 2. *Deduce:* The tester flies should be homozygous recessive at both genetic loci. *Solve:* The cross should be a test cross of the F_1 hybrid. The test-cross strain would be *ccdd*. The test cross will yield four progeny categories whose phenotypes will be determined by the dominant or recessive alleles contributed by the F_1 dihybrid.

24c. *Evaluate:* This is type 2. *Deduce:* A map distance of 28 m.u. corresponds to a recombination frequency of 28% and predicts the cross will produce 28% recombinant progeny and 72% parental-type progeny. *Solve:* The progeny will be 36% cd^+, 36% c^+d, 14% cd, and 14% c^+d^+.

24d. *Evaluate:* This is type 2. Recall that genes that are not linked assort independently. *Deduce > Solve:* The progeny will be 25% cd^+, 25% c^+d, 25% cd, and 25% c^+d^+.

25a. *Evaluate:* This is type 3. *Deduce:* The two predominant progeny types correspond to the linkage phase of the alleles in the trihybrid. The two least abundant classes correspond to double-crossover progeny. The difference between the nonrecombinant progeny and the double-crossover progeny is the allelic pair of the gene in the middle. The nonrecombinant progeny are chocolate, short, straight and white, long, curly. The double-crossover progeny are white, short, straight and chocolate, long, curly. The difference between these progeny are the alleles at the fur-color locus (w^+ and w), indicating that the *w* gene is in the middle. *Solve:* The chromosomes in the F_1 were c^+w^+s/cws^+, and the gene order is *c-w-s*.

25b. *Evaluate:* This is type 3. *Deduce:* Progeny that indicate a crossover occurred between *c* and *w* are chocolate, short, curly; white, long, straight; white, short, straight; and chocolate, long, straight. Progeny that indicate a crossover occurred between *w* and *s* are chocolate, straight, long; white, curly, short; white, short, straight; and chocolate, long, straight. *Solve:* The recombination frequency between c and w is $\frac{79 + 82 + 13 + 13}{1400} = 0.134$. The recombination frequency between w and s is $\frac{165 + 162 + 13 + 13}{1630} = 0.252$.

25c.   *Evaluate:* This is type 3. The interference value is 1 minus the coefficient of coincidence. The coefficient of coincidence is the number of observed double recombinants divided by the number expected assuming no interference. *Deduce:* The number of double-crossover progeny expected is $0.134 \times 0.252 \times 1400 = 47.28$. The number observed was 26. *Solve:* The interference value is $1 - (26/47.28) = 0.45$.

26a.   *Evaluate:* This is type 3. *Deduce:* The two predominant progeny types correspond to the linkage phase of the alleles in the trihybrid. The two least abundant classes correspond to double-crossover progeny. The difference between the nonrecombinant progeny and the double-crossover progeny is the allelic pair of the gene in the middle. The nonrecombinant progeny are compound leaves and intercalary leaflets with green fruits. The double-crossover progeny include compound leaves with green fruits and intercalary leaflets. The difference between these progeny is the alleles at the fruit-color locus, indicating that the g gene is in the middle. *Solve:* See figure.

26b.   *Evaluate:* This is type 3. *Deduce:* Progeny that indicate a crossover occurred between c and g genes were these: compound leaves with intercalary leaflets and green fruits; wild type; compound leaves with green fruits; and intercalary leaflets. Progeny that indicate a crossover occurred between g and I genes were these: green fruits; compound leaves with intercalary leaflets; compound leaves with green fruits; and intercalary leaflets. The recombination frequency between c and g is $\frac{51 + 49 + 5 + 3}{815} = 0.133$. The recombination frequency between g and i is $\frac{32 + 42 + 5 + 3}{815} = 0.101$. *Solve:* See figure.

26c.   *Evaluate:* This is type 3. The interference value is 1 minus the coefficient of coincidence. The coefficient of coincidence is the number of observed double recombinants divided by the number expected assuming no interference. *Deduce:* The number of double-crossover progeny expected is $0.133 \times 0.101 \times 815 = 10.95$. The number observed was 8. *Solve:* Eight of the progeny are produced by double crossover. The interference value is $1 - (8/10.95) = 0.27$.

27a.   *Evaluate:* This is type 2. Pure-breeding plants are homozygous at all genetic loci and therefore make only one type of gamete. *Deduce:* Pure-breeding tall, peach fuzz, round plants are TpR/TpR. Pure-breeding dwarf, smooth, oblong plants are tPr/tPr. *Solve:* The tall, peach fuzz, round plant will produce TpR gametes. The pure-breeding dwarf, smooth, oblong plant will produce tPr gametes.

27b.   *Evaluate:* This is type 2. *Deduce > Solve:* The F_1 are tall, smooth, round and have the genotype TpR/tPr.

27c.   *Evaluate:* This is type 2. *Deduce:* The types of gametes that will be produced include those resulting from meioses with no crossovers (TpR and tPr), with a crossover between T and P (TPr and tpR), with a crossover between P and R (Tpr and tPR) and with crossovers between T and P and P and R (TPR and tpr). *Solve:* The frequency of TpR and tPr gametes will each be $\frac{1}{2} \times 0.96 \times 0.82 = 0.3936$. The frequency of TPr and tpR gametes will each be $\frac{1}{2} \times 0.04 \times 0.82 = 0.0164$. The frequency of Tpr and tPR gametes will each be $\frac{1}{2} \times 0.96 \times 0.18 = 0.0864$. The frequency of TPR and tpr gametes will each be $\frac{1}{2} \times 0.04 \times 0.18 = 0.0036$. Therefore, the gamete types and frequencies will be 39.36% TpR, 39.36% tPr, 1.64% TPr, 1.64% tpR, 8.64% Tpr, 8.64% tPR, 0.36% TPR, and 0.36% tpr.

27d. *Evaluate:* This is type 2. The type of progeny will be determined by the gamete derived from the F_1 parent, since the other parent is homozygous recessive at all three loci. *Deduce:* The number of each progeny type will be the frequency of the gamete from part 27c multiplied by 1000. *Solve:* There will be 393.6 tall, peach fuzz, round; 393.6 dwarf, smooth, oblong; 16.4 tall, smooth, oblong; 16.4 dwarf, peach fuzz, round; 86.4 tall, peach fuzz, oblong; 86.4 dwarf, smooth, round; 3.6 tall, smooth, round; and 3.6 dwarf, peach fuzz, oblong.

28a. *Evaluate:* This is type 4. *NF1* is a dominant trait, so assign *N* to the mutant allele and *n* to the wild-type allele. *Deduce:* I-1 is heterozygous for *N* and the RFLP, but the linkage phase is unknown; therefore, it could be *N1/n2* or *N2/n1*. *Solve:* I-1 is either *N1/n2* or *N2/n1*. I-2 is *n2/n2*. II-1 is *N1/n2*. II-2 is *n2/n2*. III-1 is *N1/n2*. III-2 is *N1/n2*. III-3 is *n2/n2*. III-4 is *N1/n2*. III-5 is *n2/n2*. III-6 is *N2/n2*. III-7 is *N1/n2*. III-8 is *n2/n2*.

28b. *Evaluate:* This is type 4. **TIP:** Recombinants will be individuals that have a different linkage phase between *N* and the DNA marker when compared with their parents. *Deduce:* The linkage phase of II-1 is *N* with 1 and *n* with 2 (*N1/n2*). *Solve:* The only generation III individual with a different linkage phase is III-6, who has *N* with 2. Therefore, III-6 is a recombinant. Her genotype indicates that the marker allele 2 is on the same chromosome as the *NF1* allele, unlike the allele arrangement in her mother (II-1).

28c. *Evaluate:* This is type 4. The recombination frequency will be the number of recombinant progeny divided by the sum of the number of recombinant and parental progeny. *Deduce:* There are 7 parental-type progeny and 1 recombinant progeny. *Solve:* Therefore, *r* will be $\frac{1}{8}$.

29a. *Evaluate:* This is type 4. An lod score of 3 or more provides statistical support for linkage. *Deduce > Solve:* The lod scores for marker *A* but not marker *B* show statistically significant odds in favor of linkage to *SCA5*. All the θ values for maker *A* have an lod score greater than 3, whereas none of the θ values for *B* have an lod score greater than 3.

29b. *Evaluate:* This is type 4. *Deduce > Solve:* The maximum lod value for marker *A* is 12.26. The maximum lod value for marker *B* is 1.07.

29c. *Evaluate:* This is type 4. Marker *A* is linked to *SCA5* at a distance of 0.05 recombination frequency (θ). Linkage at $\theta = 0.05$ is supported by an lod score of 12.26, which is the highest lod score calculated for this data set and is much higher than 3.0, which is the established cutoff for statistical support for linkage. *Deduce > Solve:* An lod score of 12.26 indicates that the chance of obtaining these data by chance (if *SCA5* is unlinked to marker *A*) is 4×10^{12} to 1.

29d. *Evaluate:* This is type 4. *Deduce > Solve:* The data do not indicate whether marker *B* is linked to *SCA5*. Although the maximum lod score is 1.07 at $\theta = 0.10$, which indicates a 12:1 chance of obtaining these data by chance if *B* is not linked to *SCA5*, this is not strong enough to provide statistically significant evidence in favor of linkage.

30a. *Evaluate:* This is type 3. *Deduce:* The two predominant progeny types correspond to nonrecombinant progeny and show the linkage phase of the alleles in the trihybrid. The two least abundant classes correspond to double-crossover progeny. The difference between the nonrecombinant progeny and the double-crossover progeny is the allelic pair of the gene in the middle. The nonrecombinant progeny are echinus (*e*) and scute (*s*), crossveinless (*c*). The double-crossover progeny are wild type and echinus, scute, crossveinless. The difference between these progeny are the alleles at the echinus locus, indicating that the echinus gene (*e*) is in the middle. *Solve:* The gene order is scute, echinus, crossveinless. The allelic phase in the trihybrid is *+e+/s+c*.

30b.   *Evaluate:* This is type 3. Progeny that indicate a crossover occurred between *s* and *e* genes are these: crossveinless; scute, echinus; scute, echinus, crossveinless; and wild type. Progeny that indicate a crossover occurred between *e* and *c* genes are these: echinus, crossveinless; scute; scute, echinus, crossveinless; and wild type. Progeny that indicate a crossover occurred between *S* and *C* are these: scute; crossveinless; echinus, scute; and echinus, crossveinless. *Solve:* The recombination frequency between *s* and *e* is $\frac{716 + 681 + 4 + 1}{20,785} = 0.067$. The recombination frequency between *e* and *c* is $\frac{977 + 1002 + 4 + 1}{20,785} = 0.095$. The recombination frequency between *s* and *c* is $\frac{977 + 1002 + 716 + 681 + 4 + 1}{20,785} = 0.162$.

30c.   *Evaluate:* This is type 3. Recall that recombination frequencies are not always additive across genetic intervals, because double-crossover events can go unrecognized and reduce the apparent genetic distance between relatively weakly linked genes. *Deduce:* Double-crossover progeny were recognized and counted in calculation of recombination frequencies. *Solve:* There are no discrepancies across this genetic interval.

30d.   *Evaluate:* This is type 3. *Deduce:* The expected distribution of progeny of the test cross of a trihybrid under the assumption that all three genes assort independently is eight classes, each present at a frequency of $\frac{1}{8}$. *Solve:* For 20,764 progeny, this is 2598. The chi-square calculation is $\frac{(8576 - 2595.5)^2}{2595.5} + \frac{(977 - 2595.5)^2}{2595.5} + \frac{(716 - 2595.5)^2}{2595.5} + \frac{(681 - 2595.5)^2}{2595.5} + \frac{(8808 - 2595.5)^2}{2595.5} + \frac{(4 - 2595.5)^2}{2595.5} + \frac{(1002 - 2595.5)^2}{2595.5} + \frac{(1 - 2595.5)^2}{2595.5} = 38,592$. The chi-square value is 38,592, which corresponds to a *P* value well below 0.01, which indicates that the results of this experiment are not due to independent assortment.

31a.   *Evaluate:* This is type 1. *Deduce > Solve:* See figure.

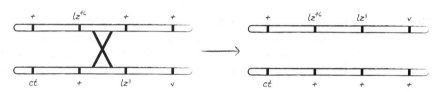

31b.   *Evaluate:* This is type 1. *Deduce:* There are two recombinant female gametes, one that is *ct+++* and the other that is $+lz^{46}lz^8v$. Male progeny will be hemizygous for these alleles. *Solve:* The phenotype of male progeny will be $\frac{1}{2}$ cut wings with wild-type eyes and $\frac{1}{2}$ with vermilion-colored, compound lozenge eyes.

32a.   *Evaluate:* This is type 3. *Deduce:* Independent assortment would result in 25% of each possible progeny phenotype. In cross 1, 25% of 2260 is 565. In cross 2, 25% of 845 is 211.25. The chi-square value for cross 1 is $\frac{(310 - 565)^2}{565} + \frac{(858 - 565)^2}{565} + \frac{(781 - 565)^2}{565} + \frac{(311 - 565)^2}{565} = 463.8$. The chi-square value for cross 2 is $\frac{(340 - 211.25)^2}{211.25} + \frac{(115 - 211.25)^2}{211.25} + \frac{(92 - 211.25)^2}{211.25} + \frac{(298 - 211.25)^2}{211.25} = 225$.

Solve: The chi-square values for both sets of data correspond to *P* values well below 0.01 and therefore indicate that the results significantly deviate from expectation based on independent assortment. This supports linkage of the colorless and waxy genes.

32b.   *Evaluate:* This is type 3. *Deduce:* In cross 1, the dihybrid is *ClWx/clwx*, which produces *ClWx* and *clwx* parental-type gametes and *Clwx* and *clWx* recombinant gametes. *Solve:* The recombination frequency is $\frac{310 + 311}{2260} = 0.27$. In cross 2, the dihybrid is *Clwx/clWx*, which produces *Clwx* and *clWx* parental gametes and *ClWx* and *clwx* recombinant gametes. The recombination frequency was $\frac{115 + 92}{845} = .24$.

32c. *Evaluate:* This is type 3. *Deduce:* Both sets of data support linkage because the recombination frequency from each set was less than 50%. The cross 1 and cross 2 results differed slightly in their estimate of recombination frequency (i.e., compare 27% to 24%). *Solve:* Yes, both sets of data are compatible with the hypothesis of genetic linkage, although the recombination frequencies of the two sets differed slightly.

32d. *Evaluate:* This is type 3. *Deduce:* The total number of progeny is 3105, and the total number of recombinant progeny is 828. *Solve:* The recombination frequency using combined data is $\frac{828}{3105} = 0.27$.

5.04 Test Yourself

Problems

1. The yeast genes x and y are syntenic, but it is not known whether they are linked. You mate an α mating type x^+y^- strain to an a mating type x^-y^+ strain to make an a/α diploid with the genotype x^+y^-/x^-y^+. You allow diploid cells to sporulate and produce asci, and you analyze 50 asci to determine the xy genotype of each ascospore. Your results are that there are 40 *PD*, 8 *TT*, and 2 *NPD*. Do these results support linkage of x and y? If so, what is the recombination frequency?

2. The yeast genes e and f are syntenic, but it is not known whether they are linked. You mate an α mating type e^+f^+ strain to the a mating type e^-f^- strain to make an a/α diploid with the genotype e^+f^+/e^-f^-. You allow diploid cells to sporulate and produce asci, and you analyze 50 asci to determine the ef genotype of each ascospore. Your results are shown in the table below. Do these results support linkage of e and f? If so, what is the recombination frequency?

Number of Asci	Spore 1	Spore 2	Spore 3	Spore 4
62	e^+f^+	e^+f^+	e^-f^-	e^-f^-
10	e^+f^+	e^+f^-	e^-f^+	e^-f^-
3	e^+f^-	e^-f^+	e^-f^+	e^+f^-

3. In yeast, the a and b genes are tightly linked to their centromeres but on different chromosomes. On average, a crossover occurs between a and its centromere in 2% of meioses. Recombination never occurs between b and its centromere. Given this information, indicate the number of *PD*, *NPD*, and *TT* asci that are expected if 100 diploid yeast cells with the genotype a^+ab^+b undergo meiosis.

4. The s gene in *Neurospora crassa* is said to be 10 map units from its centromere. If you were to mate an s^+ strain to an s^- strain and examine 100 asci, how many would exhibit second-division segregation of s^+ and s^-?

5. The genetic length of every chromosome in a newly discovered organism is 50 cM. What does this suggest about the frequency of double crossovers in the meiotic cells of this organism?

6. The yeast genes x and y are syntenic. What kind of ascus would be produced by meiosis of a diploid yeast with the genotype x^+y^+/x^-y^- under each of the following circumstances?
   a. There is a single crossover between x and y.
   b. There is a four-stranded double crossover between x and y.
   c. There is a three-stranded double crossover between x and y.
   d. There is a two-stranded double crossover between x and y.

7. The pedigree shows segregation of a dominant trait (*D*) and a linked RFLP. Individual I-1 is DR_1/dR_2. Use this information to estimate the recombination frequency between *D* and the RFLP locus.

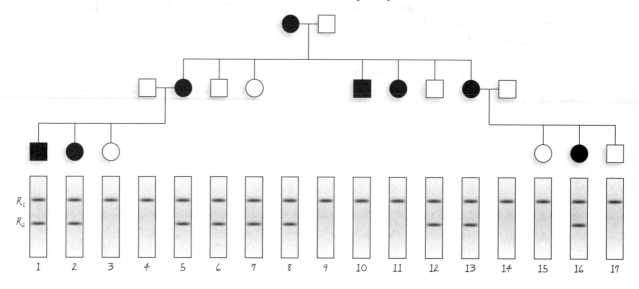

8. The table shows lod scores calculated using linkage data for three RFLPs, *A*, *B*, and *C*. Analyze the results to determine whether the analysis supports linkage for any pair of genes; if so, state the best estimate for θ.

θ values	0.05	0.10	0.15	0.20	0.25	0.30	0.35	0.40
Lod for *A* and *B*	1.0	2.5	3.0	3.25	3.0	2.0	1.0	0.5
Lod for *B* and *C*	0.5	0.75	1.0	1.25	1.5	1.2	0.75	0.5
Lod for *A* and *C*	0.1	0.8	0.4	0.2	0.15	0.12	0.10	0.75

Solutions

1.   *Evaluate:* This is type 3. *Deduce:* For linked genes, the number of *PD* > number of *TT* > number of *NPD*. Genetic distance can be estimated by $r = \frac{(\frac{1}{2} \times TT) + 3(NPD)}{\text{total asci}} \times 100$. *Solve:* The number of *PD* is much greater than the number of *NPD*, indicating linkage. The genetic distance is $r = \frac{(\frac{1}{2} \times 8) + 3(2)}{50} \times 100 = 20$ cM.

2.   *Evaluate:* This is type 3. Recall that, in yeast linkage problems, you must identify ascus types corresponding to *PD*, *NPD*, and *TT*. *Deduce:* For linked genes, the number of *PD* > number of *TT*. Genetic distance can be estimated by $r = \frac{(\frac{1}{2} \times TT) + 3(NPD)}{\text{total asci}} \times 100$. There are three ascus types: 62 *PD*, 10 *TT*, and 3 *NPD*. *Solve:* The number of *PD* (62) is much greater than the number of *TT* (10), indicating linkage. The genetic distance is $r = \frac{(\frac{1}{2} \times 10) + 3(3)}{75} \times 100 = 18.7$ cM.

3.   *Evaluate:* This is type 2. Recall that if recombination does not occur between two centromere-linked genes on different chromosomes, then 50% of the asci will be *PD* and 50% will be *NPD*. *Deduce:* Each meiosis in which no crossover occurs between *b* and its centromere and one crossover occurs between *a* and its centromere, a *TT* ascus will result. *Solve:* If one crossover occurs between *a* and its centromere in 2 out of 100 meioses, then there will be 2 *TT* and the other 98 are expected to be half *PD* (49) and half *NPD* (49).

4.   *Evaluate:* This is type 2. Recall that second-division segregation results when a single crossover occurs between a gene and its centromere. *Deduce:* Genetic distance from the centromere is calculated as $\frac{1}{2} \times \frac{\text{(number of second-division segregation asci)}}{\text{(total number of asci)}} \times 100 = $ distance in cM. *Solve:* $\frac{1}{2} \times \frac{\text{(number of second-division segregation asci)}}{100} \times 100 = 10$ cM; therefore 20 = (number of second-division segregation asci 20% (20 asci) would exhibit second-division segregation.

5.   *Evaluate:* This is type 1. Recall that double crossovers (and other classes of multiple crossovers) between genes decreases their apparent genetic distance. **TIP:** See Figure 5.15 in the textbook. *Deduce:* In the absence of double crossovers, the relationship between recombination frequency and map distance is linear. If each pair of homologs participates in one crossover during each meiosis, all chromosomes will be 50 cM in genetic length. *Solve:* The information suggests that double crossovers do not occur in this organism and that there is one crossover between homologs during each meiosis.

6a.   *Evaluate:* This is type 1. *Deduce:* A single crossover between x and y would result in two parental-type ascospores and two recombinant ascospores. *Solve:* A tetratype ascus.

6b.   *Evaluate:* This is type 1. *Deduce:* A four-stranded double crossover between x and y results in four recombinant ascospores. *Solve:* A nonparental ditype ascus.

6c.   *Evaluate:* This is type 1. *Deduce:* A three-stranded double crossover between x and y results in two parental-type ascospores and two recombinant ascospores. *Solve:* A tetratype ascus.

6d.   *Evaluate:* This is type 1. *Deduce:* A two-stranded double crossover between the x and y will result in four parental-type ascospores. *Solve:* A parental ditype ascus.

7.   *Evaluate:* This is type 4. Recall that recombination frequency will be the number of recombinant progeny divided by the number of recombinant plus parental progeny. *Deduce:* II-2 and II-8 are recombinant types and II-3, II-4, II-5, II-6, II-7, III-1, III-2, III-3, III-4, III-5, and III-6 are parental types. *Solve:* The recombination frequency is about $\frac{2}{13} = 0.154$.

8.   *Evaluate:* This is type 4. An lod score of 3.0 or more is necessary for statistical support for linkage between any pair of genes. Lod scores below 3.0 and above –3.0 do not argue for or against linkage. *Deduce:* The results support the hypothesis that the A and B genes are linked and indicate linkage at $\theta = 0.20$. The results do not argue for or against linkage of A with C or B with C. *Solve:* The results support linkage of A and B at $\theta = 0.20$. The results do not argue for or against linkage of A with C or B with C.

6

Genetic Analysis and Mapping in Bacteria and Bacteriophage

Section 6.01 Genetics Problem-Solving Toolkit

Section 6.02 Types of Genetics Problems

Section 6.03 Solutions to End-of-Chapter Problems

Section 6.04 Test Yourself

6.01 Genetics Problem-Solving Toolkit

Key Terms and Concepts

Bacterial plasmids: Circular DNA molecules that exist separately from the bacterial chromosome.

F plasmids (or F factors): Large plasmids that code for proteins that catalyze their own replication and transfer to another bacterial cell.

Hfr cell: Forms when F plasmids integrate into the bacterial chromosome. The F plasmid of Hfr cells can be excised from the chromosome. Imperfect excision creates an F′ plasmid that contains chromosomal DNA in addition to F plasmid DNA.

Time of entry: When chromosomal DNA from an Hfr strain enters a recipient cell; it is often used to map the relative position of genes in the chromosome.

Partial diploids: A cell containing two copies of a gene, one on the chromosome and one on the F′ plasmid, which is created by the transfer of *F′ DNA*. These can be used to perform complementation analysis.

Transformation: A process in which DNA released by a cell after lysis is taken up by other cells.

Bacteriophage: Viruses that infect and multiply in bacteria, typically resulting in lysis of the infected host bacterial cell.

Transducing phage: Bacterial viruses that have packaged host chromosomal DNA into phage particles. There are both generalized transducing phage and specialized transducing phage.

Cotransduction: Two bacterial sequences transduced by the same transducing phage particle. The likelihood of occurrence is inversely proportional to the distance between genes. Genes that are separated by more DNA than can fit into a single transducing phage are never cotransduced.

Key Genetic Relationships

Bacterial Conjugation	Outcome	
	Exconjugant converted to donor state?	Donor bacterial genes transferred to exconjugant?
$F^+ \times F^-$	Yes; $F^- \rightarrow F^+$	No
Hfr $\times F^-$	No	Yes
$F' \times F^-$	Yes; $F^- \rightarrow F'$	Yes

6.02 Types of Genetics Problems

1. Compare and contrast bacterial genetic approaches.
2. Describe, plan, or analyze bacterial conjugation experiments.
3. Describe, plan, or analyze bacterial transduction experiments.
4. Describe, plan, or analyze bacterial transformation experiments.
5. Describe or analyze bacteriophage fine-structure mapping experiments.

1. Compare and contrast bacterial genetic approaches.

This type of problem tests your understanding of principles and concepts of bacterial genetics. Common to all the methods are the use of growth media to determine the phenotype of bacterial strains, transfer of DNA from a donor cell to a recipient cell, and stable maintenance and expression of the transferred DNA after it enters the recipient cell. Those principles that distinguish methods include the mechanism of transfer, the amount of DNA transferred, and the mechanism by which transferred DNA is maintained in the recipient cell. See Problem 11 for an example.

Example Problem: You have isolated a new variant of *E. coli* that you believe was created by an inversion of a large region of the chromosome. The normal and variant *E. coli* are otherwise identical and can grow on minimal medium without any supplements added. You wish to generate a map of this variant *E. coli* chromosome and have at your disposal reagents to perform conjugation mapping, generalized transduction mapping, and mapping by transformation. Which is the most appropriate method for this purpose? Explain your answer.

Solution Strategies	Solution Steps
Evaluate	
1. Identify the topic this problem addresses, and explain the nature of the requested answer.	1. This problem tests your understanding of the methods of mapping a variant *E. coli* chromosome and ability to defend your choice. The answer will be conjugation, transduction, or transformation.
2. Identify the critical information given in the problem.	2. The difference between the normal *E. coli* chromosome and this variant is a large inversion. Both strains can grow on unsupplemented minimal medium.

(continued)

Solution Strategies	Solution Steps
Deduce	
3. Compare and contrast the three methods with respect to their utility in mapping chromosomes.	3. Conjugation mapping can map an entire chromosome if a few appropriate Hfr strains are available.
	Transduction and transformation are useful for mapping the distance and order of a few genes at a time. They also do not require new strains to be created, which is often an advantage.
4. Describe the differences between the *E. coli* chromosomes in terms of their genetic maps.	4. An inversion of a large segment of the *E. coli* chromosome will change the order of and distance between genes at the boundaries of the inversion but not within the inverted or normal portions.
Solve	
5. Identify the key characteristic required by a potential method to map the novel chromosome.	5. Since only a few genes will change in terms of linkage or order, the mapping strategy employed must be capable of scanning the entire chromosome to find the boundaries of the inverted segment.
6. Decide on the best method, and explain your choice.	6. Conjugation mapping is the best choice because it can determine gene order over large segments of the chromosome. It allows for rapid scanning across the entire chromosome to identify the boundaries of the inversion. The only disadvantage of this technique is that Hfr strains must be created before mapping can begin.
	Mapping by transduction or transformation would require many separate experiments, since only small segments of the chromosome are mapped in each experiment.

2. Describe, plan, or analyze bacterial conjugation experiments.

This type of problem tests your knowledge and understanding of the principles and applications of bacterial conjugation: description of events during conjugation, definitions of terminology related to conjugation, and design of conjugation experiments, including the definition of bacterial media that should be used.

Variation

Problems may provide results of interrupted mating experiments and ask for a genetic map of the region. See Problem 15 for an example.

Example Problem: You have four strains of bacteria in a box in the freezer, but the labels have worn off. You remember that the four bacteria include one Hfr strain, one F⁺ strain, one F′ strain, and one F⁻ strain. You also remember that the Hfr, F⁺, and F′ strains are streptomycin-sensitive prototrophs, whereas the F⁻ strain is auxotrophic for methionine, leucine, and threonine and is streptomycin resistant.

Your lab notebook indicates that the F′ strain was derived from the Hfr strain by an aberrant excision of the F plasmid and that the Hfr strain transfers *thr+*, *leu+*, and *met+* but not *str^S* 20 minutes after the start of conjugation. How would you determine which stock corresponds to which bacterial strain?

Solution Strategies	Solution Steps
Evaluate	
1. Identify the topic this problem addresses, and explain the nature of the requested answer.	1. This problem tests your understanding of the relationship between genotype and stock. The answer will be experiments that you would perform in order to positively identify each bacterial stock.
2. Identify the critical information given in the problem.	2. The F– strain is auxotrophic for methionine, leucine, and threonine and is streptomycin resistant, whereas the other three are streptomycin-sensitive prototrophs. The F′ strain was derived from the Hfr strain. The Hfr strain transfers *thr$^+$*, *leu$^+$*, and *met$^+$* but not *str^S* by 20 minutes.
Deduce	
3. Consider whether testing growth on specific bacterial media would provide information on the identity of the strains.	3. Test the growth of all four strains on three media: (a) minimal medium; (b) minimal medium plus methionine, leucine, and threonine; (c) minimal medium plus methionine, leucine, threonine, and streptomycin. The F⁻ strain will grow on (b) and (c), whereas the other strains will grow on (a) and (b). This step positively identifies the F⁻ strain.
4. Design mating experiments that would provide information on the identity of the strains. TIP: Only the F⁻, streptomycin-resistant strain can serve as a recipient in conjugation experiments. TIP: Hfr and F′ strains differ from F⁺ strains in their ability to transfer bacterial genes. TIP: Only the F′ strain will create an exconjugate that will be able to subsequently transfer F′ plasmid to another F⁻ strain.	4. Mate each of the remaining three unknown strains to the F⁻ strain. Interrupt the mating after 20 minutes, and plate the exconjugates on minimal medium plus streptomycin. Isolate an exconjugate from each mating, mate each of them with the F⁻ strain for 20 minutes, and plate the exconjugates on minimal medium containing streptomycin.

(continued)

Solution Strategies	Solution Steps
Solve	
5. Predict the results for each of the three experiments in step 4. **TIP:** The Hfr strain cannot transfer the entire F plasmid sequence in 20 minutes. **TIP:** The F′ strain will create F′ exconjugates identical to itself. **TIP:** The F⁺ strain can transfer only F plasmid DNA, which will not create prototrophic exconjugates.	5. The predicted results are these: **Phenotype 1:** Prototrophic exconjugate colonies that cannot act as donors in the second experiment **Phenotype 2:** Prototrophic exconjugate colonies that can act as donors in the next experiment **Phenotype 3:** No prototrophic exconjugate colonies
6. Answer the question by matching the predicted results to the appropriate strain.	6. **Answer:** The strain with phenotype 1 is the Hfr strain. The strain with phenotype 2 is the F′ strain. The strain with phenotype 3 is the F⁺ strain and the strain that could grow on medium (b) and (c) but not (a) in step 3 is the F⁻ strain.

3. Describe, plan, or analyze bacterial transduction experiments.

This type of problem tests your understanding of the principles and applications of bacterial transduction: descriptions of events during transduction, definitions of terms related to transduction and cotransduction frequency, and design of transduction experiments, including the definition of bacterial media that should be used.

Variation

Problems may provide the results of cotransduction experiments and ask for a genetic map of the region. See Problem 19 for an example.

Example Problem: The genes *met*, *thr*, and *leu* are tightly linked on the *E. coli* chromosome. *met⁻* mutants cannot synthesize methionine, *thr⁻* mutants cannot synthesize threonine, and *leu⁻* mutants cannot synthesize leucine. Design a cotransduction experiment that could determine the order of the three genes on the *E. coli* chromosome. Your answer should include the type of medium necessary, the genotype of donor strain(s), the genotype of recipient strains, what should be done, and the expected results if *leu* is located equidistant between *met* and *thr*.

Solution Strategies	Solution Steps
Evaluate	
1. Identify the topic this problem addresses, and explain the nature of the requested answer.	1. This problem tests your understanding of how to design a cotransduction experiment that could determine the order of genes. The answer will include the media and *E. coli* strains needed, the steps of the experiments, and the results expected assuming that *leu* is equidistant between *met* and *thr*.
2. Identify the critical information given in the problem.	2. *met, thr,* and *leu* are genes required for the synthesis of methionine, threonine, and leucine, respectively. The genes are tightly linked, but the order of the genes is not known.
Deduce	
3. Indicate the genotype of the donor and recipient strains. **TIP:** The donor should carry dominant (functional) alleles, and the recipient should carry recessive (nonfunctional) alleles.	3. The genotypes should be *met⁺, thr⁺, leu⁺* (donor) and *met⁻, thr⁻, leu⁻* (recipient).
4. Indicate the selected marker for each experiment, and describe the media needed. **TIP:** Each dominant allele should serve as the selected marker in one experiment. The selective medium should contain everything the recipient strain requires for growth except for the component related to expression of the selective marker.	4. Each marker in the donor strain will serve as a selected marker for one experiment: *met⁺*: by selection of transductants on minimal medium containing threonine and leucine *thr⁺*: by selection of transductants on minimal medium containing methionine and leucine *leu⁺*: by selection of transductants on minimal medium containing threonine and methionine
5. Indicate the unselected markers for each experiment, and describe the media needed. **TIP:** For each experiment, one marker is the selected marker and the other two are unselected markers. **TIP:** Unselected markers are scored using medium that contain all required nutrients except for that being scored.	5. There will be two unselected markers in each experiment: • For *met⁺* transductants, the unselected markers are *thr⁺* and *leu⁺*. *thr⁺* will be scored using minimal medium plus leucine, and *leu⁺* will be scored on minimal medium plus threonine. • For *thr⁺* transductants, the unselected markers are *met⁺* and *leu⁺*. *met⁺* will be scored using minimal medium plus leucine, and *leu⁺* will be scored on minimal medium plus methionine. • For *leu⁺* transductants, the unselected markers are *thr⁺* and *met⁺*. *thr⁺* will be scored using minimal medium plus methionine, and *met⁺* will be scored on minimal medium plus threonine.

(continued)

Solution Strategies	Solution Steps

Solve

6. Describe the steps of the transduction mapping experiment using the answers to steps 4 and 5.

6. The steps to the experiments are these: (a) Generate generalized transducing phage using the *met+*, *thr+*, *leu+* strain; (b) infect the *met−*, *thr−*, *leu−* strain; (c) select transductants on three media—minimal medium plus threonine and leucine, minimal medium plus methionine and leucine, and minimal medium plus threonine and methionine; (d) score transductant from medium 1 on minimal medium plus leucine and minimal medium plus threonine; (e) score transductants from medium 2 on minimal medium plus leucine and minimal medium plus methionine; and (f) score transductants from medium 3 on minimal medium plus threonine and minimal medium plus methionine.

7. Tabulate results under the assumption that *leu* is equidistant between *met* and *thr*.

7. Answer

1	*met+*	(a) *thr− leu−* cotransductants will be the most common.
		(b) *thr− leu+* cotransductants will be less common than a.
		(c) *thr+ leu+* cotransductants will be less common than b.
		(d) *thr+ leu−* cotransductants will be rare.
2	*thr+*	(e) *met− leu−* cotransductants will be the most common.
		(f) *met− leu+* cotransductants will be less common than e.
		(g) *met+ leu+* cotransductants will be less common than f.
		(h) *met+ leu−* cotransductants will be rare.
3	*leu+*	(i) *met− thr−* cotransductants will be the most common.
		(j) *met− thr+* cotransductants will be less common than i.
		(k) *met+ thr−* cotransductants will be as common as j.
		(l) *met+ thr+* cotransductants will be less common than j but not rare.

4. Describe, plan, or analyze bacterial transformation experiments.

This type of problem tests your understanding of the principles and applications of bacterial transformation: description of events during transformation, definitions of terms related to transformation, and design of transformation experiments, including the definition of bacterial media that should be used.

Variation

Problems may provide results of cotransformation mapping experiments and ask for a genetic map of the region. See Problem 25 for an example.

Example Problem: You have attempted to map the order of the genes *gal, lac,* and *pro* on the bacterial chromosome by using cotransformation mapping. *gal⁻* mutants cannot use galactose as a carbon source. *trp⁻* mutants cannot synthesize tryptophan. *pro⁻* mutants cannot synthesize proline. You isolated chromosomal DNA from *gal⁺, trp⁺, pro⁺* bacterium and added it to *gal⁻ trp⁻ pro⁻* bacteria under conditions that promote uptake of DNA. The results of your analysis are shown in the table below. Column A indicates the primary medium used to select transformants in each experiment. Column B indicates the secondary medium used to assay transformants for the unselected markers. Column C shows the percent of selected transformants that could grow on the secondary medium. Use this information to determine the gene order.

A	B	C
minimal medium plus proline and tryptophan with galactose as carbon source	minimal medium plus proline with galactose as carbon source	0.01
	minimal medium plus tryptophan with galactose as carbon source	0.005
	minimal medium with galactose as carbon source	0.0016
minimal medium plus tryptophan with glucose as carbon source	minimal medium with glucose as carbon source	0.0016
	minimal medium plus tryptophan with galactose as carbon source	0.005
	minimal medium with galactose as carbon source	0.0016
minimal medium plus proline with glucose as carbon source	minimal medium with glucose as carbon source	0.0016
	minimal medium plus proline with galactose as carbon source	0.01
	minimal medium with galactose as carbon source	0.0016

Solution Strategies	Solution Steps
Evaluate	
1. Identify the topic this problem addresses, and explain the nature of the requested answer.	1. This problem tests your understanding of gene order and distance. The answer will be a gene order and the relative distance between the genes expressed as cotransformation frequency.
2. Identify the critical information given in the problem.	2. The table indicates the percentage of transformants (column C) selected on the primary medium (column A) that can also grow on secondary media (column B).
Deduce	
3. Determine the selected marker for each of the primary media. TIP: The primary (selective) medium selects for the growth of a specific allele of one gene (the selected marker) while allowing the growth of either allele at all other genes.	3. There are three selective media and therefore three selected markers: • Minimal medium plus proline and tryptophan using galactose as carbon source selects for *gal⁺* • Minimal medium plus tryptophan using glucose as carbon source selects for *pro⁺* • Minimal medium plus proline using glucose as carbon source selects for *trp⁺*

(continued)

Solution Strategies	Solution Steps
4. Determine which cotransformation frequency corresponds to each value in column C.	4. The cotransformation frequencies when selecting for *gal⁺* are 0.01% for *gal⁺* and *trp⁺*, 0.005% for *gal⁺* and *pro⁺*, 0.0016% for *gal⁺* and *trp⁺* and *pro⁺*. The cotransformation frequencies when selecting for *pro⁺* are 0.0016% for *pro⁺* and *trp⁺*, 0.005% for *pro⁺* and *gal⁺*, and 0.0016 for *pro⁺* and *trp⁺* and *gal⁺*. The cotransformation frequencies when selecting for *trp⁺* are 0.0016 for *trp⁺* and *pro⁺*, 0.01% for *trp⁺* and *gal⁺*, 0.0016% for *trp⁺* and *pro⁺* and *gal⁺*.
Solve	
5. Draw all three possible gene orders and predict the class of cotransformants that will be the least frequent. **TIP:** There are three possible gene orders. **TIP:** The least frequent class will be the one that requires four crossover events.	5.
6. Identify the data representing the lowest cotransformation frequencies, and select the gene order that is completely consistent with the data.	6. **Answer:** See figure.

5. Describe or analyze bacteriophage fine-structure mapping experiments.

This type of problem tests your understanding of the principles and applications of bacteriophage genetics. Problems may provide information on complementation tests or deletion-mapping experiments and ask you to draw a map showing the location of mutations within the *rII* locus of bacteriophage T4.

Variation

Given a map, you may be asked to predict the results of complementation or recombination mapping experiments. See Problem 23 for an example.

Example Problem: You have received a new T4 *rII* mutant, *m1*, from a colleague who also sent you a description of the results of her complementation and mapping experiments. She said that when she coinfected *E. coli* K12 cells with *m1* phage and a phage point mutant in region *A5* (see Figure 6.28 of the textbook), wild-type lysis did not occur, although she noted that a few plaques did form and that they contained wild-type virus. When she used *m1* and a point mutant in region *B5*, wild-type lysis occurred. She said that when she coinfected *m1* and deletion mutant *PB242* or *221*, a few plaques formed, and they contained wild-type virus. In contrast, coinfection with *m1* and deletion mutant *PT1* or *PB28* yielded no plaques. Use this information to map the mutation *m1* to a region of the T4 *rII* locus.

Solution Strategies	Solution Steps
Evaluate	
1. Identify the topic this problem addresses, and explain the nature of the requested answer.	1. This problem tests your analysis of the results of complementation and deletion mapping experiments and ability to determine the location of a new mutation. The answer will be a region of the *rII* locus as indicated in Figure 6.28 of the textbook.
2. Identify the critical information given in the problem.	2. Coinfection of mutant *m1* and a mutation in *rIIA* does not produce wild-type lysis, whereas coinfection of *m1* and an *rIIB* mutant does. Coinfection of mutation *m1* deletion mutants *PB242* and *221* produces a few plaques that contain wild-type virus, whereas coinfection with deletion mutants *PT1* or *PB28* does not.
Deduce	
3. Use the results of the complementation analysis to determine which cistron of *rII* is mutant in mutant *m1*. TIP: Coinfection using mutants in the different cistrons results in wild-type levels of cell lysis.	3. Mutation *m1* complements a mutant in *rIIB* but not *rIIA*. The mutation is in the *rIIA* locus.
4. Use the results of the deletion-mapping experiment to determine the region with *rIIA* that contains the *m1* mutation. TIP: Mutations that are able to recombine with a deletion mutant to produce wild-type virus must lie outside the deleted region.	4. Mutation *m1* can recombine with deletion mutants *PB242* and *221* to yield wild-type virus. Mutation *m1* cannot recombine with deletion mutants *PT1* and *PB28* to produce wild-type virus. The *m1* mutation is in a region that overlaps *PT1* and *PB28* but not *PB242* or *221*, which corresponds to *A4g* and *A4f*.
Solve	
5. Determine the region of the *rII* locus that must contain mutation *m1* using Figure 6.28 in the textbook.	5. Mutation *m1* is in region *A4g* or *A4f*.

6.03 Solutions to End-of-Chapter Problems

1a.   *Evaluate:* This is type 2. *Deduce > Solve:* F⁺ means the F plasmid is present. Hfr means that the F plasmid is part of the chromosome. F′ indicates that the F plasmid is present and contains a portion of the chromosome. F⁻ means that the F plasmid is not present.

1b.   *Evaluate:* This is type 2. Recall that donors are bacteria that can transfer DNA to another cell. Recipients are cells that receive DNA from a donor. *Deduce > Solve:* F⁺, F′ and Hfr are donors. F⁻ is a recipient.

1c.   *Evaluate:* This is type 2. Recall that a donor cell must contain the entire F plasmid. *Deduce > Solve:* F⁺ and F′ bacteria can convert recipient cells into donor cells.

1d.   *Evaluate:* This is type 2. *Deduce > Solve:* Hfr and F′ transfer donor genes to the recipient cell.

1e.   *Evaluate:* This is type 2. *Deduce > Solve:* If a recipient genotype does not change except for becoming a donor, the donor was an F$^+$. If the recipient genotype changes in addition to becoming a donor, then the donor was an F′. If the recipient genotype changes but it does not become a donor, the donor was an Hfr.

1f.   *Evaluate:* This is type 2. *Deduce > Solve:* A partial diploid is a bacterium that has two copies of some genes but only one copy of most genes. It is formed when a recipient cell receives an F′ plasmid from a donor.

2.   *Evaluate:* This is type 2. *Deduce > Solve:* (1) Transfer of an entire F$^+$ plasmid from an F$^+$ cell converts an F$^-$ cell to a F$^+$ cell. (2) Integration of the F plasmid into the host chromosome converts an F$^+$ cell to an Hfr cell. (3) Precise excision of the F plasmid from the chromosome of an Hfr cell converts an Hfr cell to an F$^+$ cell. (4) Excision of the F plasmid plus some host DNA from the chromosome of an Hfr cell converts an Hfr cell into an F′ cell.

3.   *Evaluate:* This is type 2. *Deduce:* An Hfr cell is a bacterium that contains an F plasmid inserted into its chromosome. *Solve:* Transfer of the entire chromosome of the Hfr cell and recombination of the entire F plasmid DNA sequence into the recipient chromosome could convert an F$^-$ cell into an Hfr cell.

4.   *Evaluate:* This is type 1. *Deduce:* All three mechanisms can involve homologous recombination of the transferred DNA into the recipient chromosome; however, only transduction always involves homologous recombination. In all three mechanisms, if the DNA entering the cell is linear or does not contain an origin of replication, then recombination of the DNA into the recipient cell chromosome or episome is required for the DNA to be stably maintained. If the DNA entering the recipient is circular and contains sequences required for replication and maintenance, then recombination into the recipient cell chromosome or an episome is not required. *Solve:* These three mechanisms differ in how DNA is transferred from one cell to another. Only conjugation requires genetic information for transfer in the donor cell (F plasmid DNA) and physical contact between the donor and recipient cell. Transduction is characterized by infection of the donor cell by a bacteriophage. On the other hand, transformation does not require particular genes in the donor or the help of phage: DNA is released from the "donor" cell due to cell lysis and enters the "recipient" cell via DNA transporters.

5a.   *Evaluate:* This is type 2. *Deduce > Solve:* The *origin of transfer* is the site in F plasmid DNA where binding and single-strand cleavage of F DNA by the relaxosome occurs, and it is the site where rolling circle replication of F plasmid DNA is initiated. Part of the origin is also the first sequence transferred out of the donor cell and into the recipient cell.

5b.   *Evaluate:* This is type 2. *Deduce > Solve:* The *conjugation pilus* makes a physical connection between the donor and recipient cell.

5c.   *Evaluate:* This is type 2. *Deduce > Solve: Homologous recombination* is important for stable maintenance and expression of donor chromosomal DNA in the exconjugate cell when transfer is from an Hfr cell.

5d.   *Evaluate:* This is type 2. *Deduce > Solve:* The *relaxosome* is a protein complex that binds to the F origin of transfer and initiates transfer of DNA by cleaving one strand of origin DNA. Relaxosome components also bind to the 5′ end of the cleaved DNA and help it interact with a coupling protein, which is required for feeding the DNA strand into the conjugation pilus.

5e.   *Evaluate:* This is type 2. *Deduce > Solve: Relaxase* is the relaxosome component that binds to the 5′ end of the F origin after it is cleaved and that binds to coupling protein as part of the process of feeding the DNA into the conjugation pilus.

5f.   *Evaluate:* This is type 2. *Deduce > Solve:* The *T strand* is the strand of F-origin DNA that is cleaved, bound by relaxase, and fed into the conjugation pilus.

5g.   *Evaluate:* This is type 2. *Deduce > Solve: Pilin* is the protein component that assembles into the conjugation pilus.

6.   *Evaluate > Deduce > Solve:* Lysis of an infected bacterial host cell, and the release of progeny phage particles, is the end result of the lytic cycle of bacteriophage. Lysogeny involves the integration of the phage chromosome (known as a prophage once integrated) by site-specific recombination into a specific site (DNA sequence) in the bacterial chromosome. Once integrated, the prophage can replicate along with the rest of the bacterial chromosome until conditions induce excision of the prophage and resumption of the lytic cycle.

7.   *Evaluate:* This is type 3. *Deduce:* Lysogeny is an alternative bacteriophage life cycle that involves integration of the bacteriophage chromosome into the bacterial chromosome and repression of bacteriophage replication and cell lysis genes. The integrated phage genome is referred to as the prophage, and the bacterial cell is referred to as a lysogen. Specialized transducing phages are bacteriophage particles that contain specific host DNA sequences. *Solve: Site-specific recombination* refers to recombination between a specific site (DNA sequence) on a bacteriophage chromosome and a specific site on a bacterial chromosome. Site-specific recombination creates a lysogen containing a prophage genome. Inexact site-specific recombination in a lysogen releases the bacteriophage genome plus adjacent bacterial DNA. These aberrant chromosomes are packaged into a specialized transducing phage that will always carry certain bacterial DNA sequences, as opposed to a generalized transducing phage that can carry any bacterial chromosomal DNA.

8.   *Evaluate:* This is type 3. *Deduce > Solve:* A *prophage* is a bacteriophage genome that is part of the host cell chromosome. It is formed by integration of a bacteriophage chromosome into the host cell chromosome by site-specific recombination.

9.   *Evaluate:* This is type 3. *Deduce:* Cotransduction frequency is the fraction of transductants for one gene (called the selected marker) that are also transductants for a second gene (the unselected marker). Given that two genes are close enough to be included on a single DNA fragment that can be incorporated into a transducing particle, the frequency of cotransduction of two genes increases as the distance between the genes decreases. *Solve:* The drawing should show the position of three genes—*a*, *b*, and *c*—and three types of double recombination events

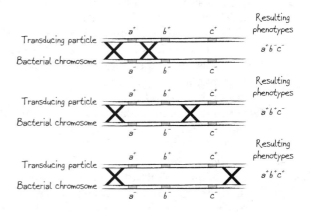

that select for *a*⁺. Whereas *b* is always contransduced along with *c*, *c* is not always cotransduced along with *b*. *Solve:* Given that two genes are close enough to be included on a single DNA fragment that can be incorporated into a transducing particle, the frequency of cotransduction of two genes increases as the distance between the genes decreases. The figure shows the position of three genes—*a*, *b*, and *c*. If a is the selected marker, then the cotransduction frequency of *a* and *b* will be greater than the cotransduction frequency of *a* and *c*.

10.   *Evaluate:* This is type 5. *Deduce:* In studies of bacteriophage complementation or recombination, mutant bacteriophage are isolated that are able to infect and reproduce in permissive bacterial strains but not in nonpermissive bacterial strains. Complementation and recombination between different mutant viruses is detected when they simultaneously infect a nonpermissive strain and are able to reproduce and lyse the strain. *Solve:* In genetic complementation, bacterial lysis occurs because the two viruses have mutations in different genes. This is analogous to complementation analysis in eukaryotes. In recombination, bacterial lysis occurs because, although the viruses have mutations in the same gene, rare homologous recombination events produce recombinant wild-type viruses. Complementation and recombination can be differentiated by the frequency of bacterial lysis after simultaneous infection with two mutant viruses: lysis is frequent in the case of complementation (many plaques are formed), whereas it is rare in the case of recombination.

11.   *Evaluate:* This is type 1. *Deduce:* Conjugation transfers chromosomal DNA from Hfr and F′ donors to F– recipients. Transformation transfers fragments of chromosomal DNA after lysis and partial degradation of donor chromosomal DNA. Transduction transfers fragments of donor DNA by incorporation into defective phage particles. *Solve:* Conjugation by an Hfr strain is capable of transforming the largest amount of donor chromosomal DNA because it can, in theory, transfer the entire *E. coli* chromosome. Transduction generally transfers the smallest amount of chromosomal DNA because it is limited to fragments of the chromosome that can fit into a transducing particle.

12.   *Evaluate:* This is type 5. *Deduce:* Simultaneous infection of a nonpermissive bacterium with a deletion mutant virus and a point mutant virus results in wild-type recombinant virus only if the deletion mutation does not overlap the site of the point mutation. Deletion mutant 1 includes the sites of point mutations *a* and *d*. Deletion 3 includes point mutation *a* and *e*; therefore, deletion 3 and 1 overlap at *a*, which is between *d* and *e*. Deletion 4 includes point mutations *b*, *d*, *a*, and *e*; therefore, deletion 4 includes the *d*, *a*, *e* region and places *b* next to *d* or *e*. Deletion 6 includes point mutations *c*, *b*, and *d*, which places *b* next to *d* and *c* on the other side of *b*. Deletions 2 and 5 each include only one gene, *e* and *c*, respectively. Deletion 7 includes point mutations *c*, *b*, *d*, *a*, and *c*, which indicates deletion 7 spans the entire region. *Solve:* Draw a genetic map as shown in the figure.

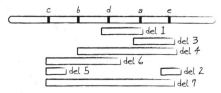

13.   *Evaluate:* This is type 5. *Deduce:* Simultaneous infection of a nonpermissive bacterium with a deletion mutant virus and a point mutant virus results in wild-type recombinant virus only if the deletion mutation does not overlap the site of the point mutation. For Series I deletion mutants, deletions *1272*, *1241*, and *J3* overlap the mutation but *PT1*, *PB242*, *A105*, and *638* do not. This places the mutation in the *rIIA* region A3. For Series II deletions, the smallest deletion that overlaps the point mutation is *250*, which includes A3h and A3g. Since *C33* includes A3h but does not overlap the point mutation, this places the point mutation at A3g. All other results with Series II deletion mutants are consistent with the A3g location. *Solve:* The point mutation is in region A3g.

14.   *Evaluate:* This is type 5. *Deduce:* Simultaneous infection of a nonpermissive bacterium with a deletion mutant virus and a point mutant virus results in wild-type recombinant virus only if the deletion mutation does not overlap the site of the point mutation. For Series I deletion mutants, deletions *1272* and *1241* include A2h2 and therefore, no wild-type recombinant virus would be detected using these mutants. *J3*, *PT1*, *PB242*, *A105*, and *638* do not overlap A2h2; therefore, wild-type recombinant virus would be detected using these deletions. For Series II deletion mutants, *EM66*, *1695*, *1993*, and *PT153* overlap A2h2; therefore, no wild-type recombinant viruses would be detected. All other Series II deletion mutants do not overlap A2h2; therefore, wild-type recombinant virus would be detected using these deletions. *Solve:* See table.

	Deletion Mutation	1272	1241	J3	PT1	PB242	A105	638		
Series I	Result	−	−	+	+	+	+	+		
	Deletion Mutation	1364	EM66	386	168	1993	1695	PT153	1231	C33
Series II	Result	+	−	+	+	−	−	−	−	+

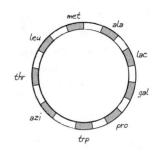

15.   *Evaluate:* This is type 2. *Deduce:* The order of gene entry is the same as the order of the genes in the bacterial chromosome. Chromosome segments can be transferred into the recipient cell in either direction; the direction depends on the orientation of the F plasmid sequence in the bacterial chromosome in the Hfr strain. *Solve:* Hfr1 sets the gene order *met-ala-lac-gal*. Hfr2 also transfers *met* first, but includes genes on the side of met opposite to *ala*. This gives the gene order *azi-thr-leu-met-ala-lac-ga*l. Hfr3 transfers *gal* first and *azi* last, indicating that the chromosome is circular in the order *azi-thr-leu-met-ala-lac-gal-pro-trp-azi* (*thr-leu-...*). Hfr5 adds the *trp* gene to the map, indicating that it's to the left of *thr*. Draw a map similar to the one shown in the figure.

16a.   *Evaluate:* This is type 2. *Deduce:* The time of entry corresponds to the X-intercept on a graph plotting "frequency of Hfr markers among selected conjugates vs. conjugation time" for each Hfr strain. The first gene that enters plateaus at the highest frequency, the next gene plateaus at the next highest frequency, and so on. *Solve:* Draw a time-of-entry profile similar to those shown in the figure.

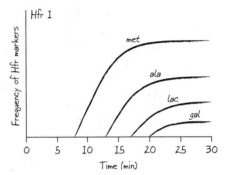

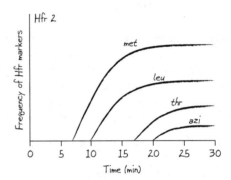

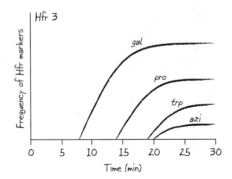

16b.   *Evaluate:* This is type 2. *Deduce:* The difference between the time of entry of two adjacent genes is the map distance, in minutes, between the genes. *Solve:* Draw a map similar to the figure shown.

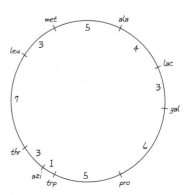

16c. *Evaluate:* This is type 2. *Deduce:* The map distance in the answer to 16b is in minutes and indicates the difference in time of entry of consecutive genes in the chromosome. *Solve: azi* is the last gene in Hfr2 to enter because the origin of transfer is 20 minutes away from the *azi* gene in Hfr2. The next gene to enter would be *trp*, which would require at least 21 minutes of conjugation to be detected (the graph of *trp* would intersect the x-axis at 21 minutes in the middle graph in the answer to 16a. *Solve:* The next gene to enter be *trp*, which would require at least 21 minutes of conjugation to be detected.

16d. *Evaluate:* This is type 2. *Deduce:* The map distance in the answer to 16b is in minutes and indicates the difference in time of entry of consecutive genes in the chromosome. The first gene to enter will be *oriT*, followed by the genes, in order, from the table of Problem 15. The time of entry of *oriT* and the first bacterial gene is arbitrary; however, the rest is determined from the genetic map. *Solve:* See table.

Hfr strain 4	oriT	leu	met	ala	lac
Duration (min)	2	4	7	12	16
Hfr strain 5	oriT	trp	thr	leu	met
Duration (min)	2	3	7	14	17

16e. *Evaluate:* This is type 2. *Deduce:* The total length of the donor strain chromosome is determined by adding the distances between each gene on the map. *Solve:* The total chromosome length is 37 min.

17a. *Evaluate:* This is type 2. *Deduce:* The selected marker is the dominant marker (allele) gene in the donor strain that must be incorporated into the exconjugate's chromosome in order for exconjugate cells to grow on the bacterial medium. Only cys^+ str^R bacteria can grow on minimal medium plus leucine, methionine, and streptomycin. Only leu^+ str^R bacteria can grow on minimal medium plus cysteine, methionine, and streptomycin. Only met^+ str^R bacteria can grow on minimal medium plus leucine, cysteine, and streptomycin. *Solve:* See table.

	Supplements added to minimal medium		
Selected marker	leucine, methionine, and streptomycin	cysteine, methionine, and streptomycin	cysteine, leucine, and streptomycin
Genotype of bacteria that can grow on medium indicated	cys^+ met^+ leu^+ str^R	leu^+ met^+ cys^+ str^R	met^+ leu^+ cys^+ str^R
	cys^+ met^+ leu^- str^R	leu^+ met^+ cys^- str^R	met^+ leu^+ cys^- str^R
	cys^+ met^- leu^+ str^R	leu^+ met^- cys^+ str^R	met^+ leu^- cys^+ str^R
	cys^+ met^- leu^- str^R	leu^+ met^- cys^- str^R	met^+ leu^- cys^- str^R
Counter-selected marker	str^S	str^S	str^S

17b. *Evaluate:* This is type 2. *Deduce > Solve:* The exconjugate must incorporate the selected marker and not incorporate the counter-selected marker into its chromosome in order to grow on the bacterial medium. cys^+ str^R bacteria with any combination of *leu* and *met* genotypes can grow on minimal medium plus leucine, methionine, and streptomycin. leu^+ str^R bacteria with any combination of *cys* and *met* genotypes can grow on minimal medium plus cysteine, methionine, and streptomycin. met^+ str^R bacteria with any combination of *leu* and *cys* genotypes can grow on minimal medium plus leucine, cysteine, and streptomycin.

17c. *Evaluate:* This is type 2. *Deduce:* Only str^R bacteria can grow on medium containing the antibiotic, streptomycin. The donor strain is str^S and the recipient strain is str^R. str^S is likely to be transferred much later than *cys+*, *met+*, and *leu+* in this Hfr; therefore, selection against *str^S^* (for *str^R^*) ensures that only exconjugate bacteria and not donor bacteria will grow on the media. *Solve:* *str^S^* is the counter-selected marker in this experiment. Adding it to each selection medium ensures that only exconjugate bacteria and not donor bacteria will grow on the media.

17d. *Evaluate:* This is type 2. *Deduce:* The presence of colonies on a medium at any sampling time indicates that the selected marker had entered the recipient cell at that time. The earliest sampling time resulting in colonies on medium 1 was 15 minutes, medium 2 was 9 minutes, and medium 3 was 18 minutes. The selected marker for medium 1 was *cys+*; medium 2, *leu+*; medium 3, *met+*. *Solve:* The order of gene transfer was *leu-cys-met*.

17e. *Evaluate:* This is type 2. *Deduce:* *cys+* and *met+* are each selected markers in this experiment for that medium; therefore, both *cys+* and *met+* must have entered the recipient cell at the time of sampling in order for colonies to grow on the medium. *met+* transfers later than *cys+*; therefore, the time of entry of *met+* will be the earliest sampling time yielding colonies on this medium. *Solve:* 18 minutes

18a. *Evaluate:* This is type 3. *Deduce:* *met+* is determined by the ability to grow on medium lacking methionine. *phe+* is determined by the ability to grow on medium lacking phenylalanine. *ara+* is determined by the ability to grow on medium containing arabinose as the sole source of carbon (the only sugar present). *Solve:* Selection for *met+* was done on minimal medium containing glucose and phenylalanine. *met+* transductants were assayed for cotransduction of *phe+* using minimal medium containing glucose. *met+* transductants were assayed for cotransduction of *ara+* using minimal medium containing arabinose. Selection of *phe+* transductants was done on minimal medium containing glucose and methionine. *phe+* transductants were assayed for cotransduction of *met+* on minimal medium containing glucose. *phe+* transductants were assayed for cotransduction of *ara+* on minimal medium containing arabinose. *met+ phe+* transductants were selected on minimal medium containing glucose. *met+ phe+* transductants were assayed for cotransduction of *ara+* on minimal medium containing arabinose. *ara+* transductants were selected on minimal medium containing arabinose, phenylalanine, and methionine. *ara+* transductants were assayed for cotransduction of *met+* on minimal medium containing arabinose and phenylalanine. *ara+* transductants were assayed for cotransduction of *phe+* on minimal medium containing arabinose and methionine.

18b. *Evaluate:* This is type 3. *Deduce:* Genes that are close together will have a higher cotransduction frequency than genes that are farther apart. Selection for the outer two genes will give close to 100% cotransduction of the middle gene. The cotransduction frequency of *met+* and *phe+* is higher than *met+* and *ara+*, placing *ara* between *phe* and *met*. *ara* and *phe* are closer to each other than either is to *met*. Simultaneous selection for *met+* and *phe+* gives 79% cotransduction of *ara+*, again consistent with the location of *ara* between *phe* and *met*. *Solve:* The gene order is *phe - ara - met* or *met - ara - phe*.

19a. *Evaluate:* This is type 3. *Deduce:* *ara1−* and *ara2−* are different mutants, each unable to use arabinose as a carbon source. If *ara1−* and *ara2−* are mutations at the same chromosomal location, then neither could be used as a donor to transduce the other to *ara+*. If they are in different locations, then each could act as a donor to transduce the other to *ara+*. *Solve:* Since *ara1−* can act as a donor to transduce *ara2−* to *ara+*, then, *ara1−* and *ara2−* are mutations at different locations.

19b. *Evaluate:* This is type 3. *Deduce:* Cotransduction will be more frequent when the mutations are closer together than when they are farther apart. Cotransduction with *ara1* and *ara2* is the least frequent; with *ara2* and *ara3*, it is more frequent; and with *ara1* and *ara3*, it is the most frequent. *Solve:* The order is *ara1-ara3* and *ara2* (or *ara2-ara3-ara1*).

20a. *Evaluate:* This is type 5. *Deduce:* Mutations in the same gene will not complement each other, whereas mutations in different genes will complement each other. Mutants 1, 5, and 8 do not complement each other. Mutant 2 complements all the others. Mutants 3 and 7 do not complement each other. Mutants 4 and 6 do not complement each other. *Solve:* There are 4 genes represented.

20b. *Evaluate:* This is type 5. *Deduce:* Mutations in the same gene will not complement each other, whereas mutations in different genes will complement each other. See 20a. *Solve:* Mutations 1, 5, and 8 are in one gene. Mutation 2 is in a second gene. Mutations 3 and 7 are in a third gene. Mutations 4 and 6 are in a fourth gene.

20c. *Evaluate:* This is type 5. *Deduce > Solve:* Complementation resulted in the formation of many plaques on each plate (the lysis of many different bacteria) due to coinfection by bacteriophage with mutations in different genes. The vast majority of these phages are mutants and cannot singly infect and lyse bacteria. Recombination between mutations in the same gene results in rare plaques (very few bacteria lyse); however, all the virus particles produced are wild type and can infect and lyse bacteria.

20d. *Evaluate:* This is type 5. *Deduce:* Mutations that cannot complement either gene are mutant for both genes. Mutations in the same gene that can recombine to give wild-type viruses are in different sites of that gene. *Solve:* Mutation 9 is a deletion that inactivates two genes. It overlaps mutations 1 and 7, but not 3, 5, or 8.

20e. *Evaluate:* This is type 5. *Deduce:* Mutations that cannot complement either gene are mutant for both genes. Mutations in the same gene that can recombine to give wild-type viruses are in different sites of that gene. *Solve:* Mutation 10 is a deletion mutation that inactivates two genes. It overlaps mutations 4 and 8, but not 1, 5, 6, or 9.

20f. *Evaluate:* This is type 5. *Deduce:* Mutations 1 and 7 are overlapped by deletion mutation 9. Since 7 is in the same gene as 3, the order is 3, 7, and 1. Since 1, 5, and 8 are in the same gene, the order is 3, 7, 1, 8, and 5 or 3, 7, 1, 5, and 8. Because mutations 4 and 8 are overlapped by deletion mutation 10, the order is 3, 7, 1, 5, 8, and 4. Since mutation 6 is in the same gene as 4, the order is 3, 7, 1, 5, 8, 4, and 6. That leaves mutation 2, which is on an end, so the final order is 3, 7, 1, 5, 8, 4, 6, and 2. *Solve:* The mutation order is 3, 7, 1, 5, 8, 4, 6, and 2.

21a. *Evaluate:* This is type 2. *Deduce:* Conjugation involving an F′ plasmid produces a partial diploid. A partial diploid with a dominant, functional copy of a gene and a recessive nonfunctional copy of the same gene will have the wild-type phenotype. *Solve:* The *his2⁻* mutation is in the *hisJ* gene.

21b. *Evaluate:* This is type 2. *Deduce:* Conjugation involving an F′ plasmid produces a partial diploid. A partial diploid with a dominant, functional copy of a gene and a recessive, nonfunctional copy of the same gene will have the wild-type phenotype. *Solve:* The exconjugate from the *his2⁻* recipient has the genotype *hisJ⁻* F′ *hisJ⁺*.

22a. *Evaluate:* This is type 3. *Deduce: leu⁺* is determined by the ability to grow on medium lacking leucine. *phe⁺* is determined by the ability to grow on medium lacking phenylalanine. *ala⁺* is determined by the ability to grow on medium lacking alanine. *Solve:* In experiment A, transductants were selected on minimal medium containing phenylalanine and alanine and then were tested on minimal medium, minimal medium containing phenylalanine, and minimal medium containing alanine. In experiment B, transductants were selected on minimal medium containing leucine and alanine and then were tested on minimal medium, minimal medium containing leucine, and minimal medium containing alanine. In experiment C, transductants were selected on minimal medium containing leucine and phenylalanine and then were tested on minimal medium, minimal medium containing leucine, and minimal medium containing phenylalanine.

22b. *Evaluate:* This is type 3. *Deduce:* Essentially all bacteria that are cotransductants for both outside genes should also be cotransductants for the middle gene. *Solve:* Among the transductants selected for *ala⁺* that are also *phe⁺*, all are *leu⁺*. Likewise, among the transductants selected for *phe⁺* that are also *ala⁺*, all are *leu⁺*. None of the other combinations (*leu⁺* that are also *ala⁺*, *ala⁺* that are also *leu⁺*, *leu⁺* that are also *phe⁺*, and *phe⁺* that are also *leu⁺*) meet this criterion. This places *leu* in the middle. Therefore, the order is *phe-leu-ala*.

22c.   *Evaluate:* This is type 3. *Deduce:* Two recombination events were required to incorporate the transduced genes into the recipient cell's chromosome. In experiment A, *leu+* was the selected marker; therefore, all recombination events must include one on either side of *leu*. *Solve:* Draw crossover events similar to those shown in the figure.

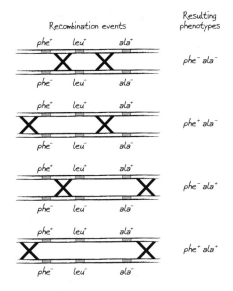

22d.   *Evaluate:* This is type 3. We are given the results of cotransduction mapping of three genes. *Deduce:* Two recombination events were required to incorporate the transduced genes into the recipient cell's chromosome. In experiment B, *phe+* was the selected marker; therefore, all recombination events must include one on either side of *phe*. *Solve:* Selection for *phe+* requires a recombination event to the left of *phe*. For this to also result in *leu−* and *ala+*, three more recombination events are necessary. Transductants resulting from quadruple crossovers are rare.

23.   *Evaluate:* This is type 5. *Deduce:* A "−" in the table indicates that the deletion and the point mutation overlap. *Solve:* Draw the length and endpoints of each deletion as shown in the figure.

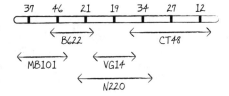

24a.   *Evaluate:* This is type 5. *Deduce:* A "−" result indicates that the deletion overlaps the point mutation. *Solve:* Mutation 55 is within M12, to the left of C19. Mutation 67 is in the region where L36, R22, and W42 overlap. Mutation 74 is in the region where R22 and W42 overlap but to the right of L36. Mutation 82 is in the region where C19 and M12 overlap but to the left of L36. Mutation 85 is within R22, to the right of W42. Mutation 91 is in the region of overlap between L36 and C19. *Solve:* Draw the point mutations on the chromosome as shown in the figure.

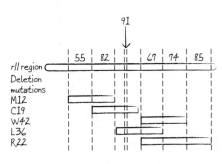

24b.   *Evaluate:* This is type 5. *Deduce:* A "+" indicates that the deletion mutation does not inactivate the gene; a "−" indicates that the deletion mutant does inactivate the gene. L36 is the only deletion that inactivates both genes; therefore, L36 overlaps both genes. C19 and M12 inactivate only *rIIB*, and R22 and W42 inactivate only *rIIA*. *Solve:* Draw the separation between *rIIA* and *rIIB* on the RII region as shown in the figure.

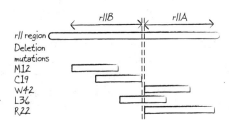

24c.   *Evaluate:* This is type 5. *Deduce:* If a mutation maps to the *rIIA* gene, then it will complement *rIIB* mutants but not *rIIA* mutants. If a mutation maps to the *rIIB* gene, then it will complement *rIIA* mutants but not *rIIB* mutants. *Solve:* Create a table similar to the one shown here.

Mutant	Complemented by	
	rIIA	*rIIB*
55	+	−
67	−	+
74	−	+
82	+	−
85	−	+

24d.   *Evaluate: Deduce > Solve:* This is type 5. The result for mutant 91 will be "−" with *rIIA* and "+" with *rIIB*.

25a.   *Evaluate:* This is type 4. *Deduce:* Genes that are close together will have a higher cotransformation frequency than genes that are farther apart. *Solve:* *zap* and *rag* are closest together, *zap* and *pip* are farthest apart, and *rag* is closer to *pip* than it is to *zap*. The order is *zap - rag - pip* (or *pip - rag - zap*).

25b.   *Evaluate:* This is type 4. *Deduce > Solve:* See Answer 25a. *zap* and *pip* are the most closely linked.

25c.   *Evaluate:* This is type 3. *Deduce:* The cotransformation frequency for unlinked genes is the product of their individual transformation frequencies. If the two genes are linked, then the cotransformation frequency will be significantly greater than the product of their individual frequencies. *Solve:* The cotransformation frequency if unlinked would be $0.00018 \times 0.00012 = 2.2 \times 10^{-8}$, which is the same as 0.0002%. Therefore, the results do not support genetic linkage.

26a.   *Evaluate:* This is type 2.
Deduce: The gene closest to *oriT* is the first gene to enter the recipient during conjugation. Genes transferred by more than one Hfr strain indicate the extent of overlap and the

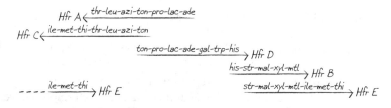

relative orientation of transfer of different Hfr strains. *Solve:* Lines representing the Hfr DNA as it enters the recipient are drawn with the arrow indicating direction of movement, and the first chromosomal gene to enter is directly behind the arrow. Hfr A overlaps Hfr C and Hfr D, and Hfr A and Hfr C transfer in the same direction, opposite to that of Hfr D. Hfr D overlaps with Hfr B, which overlaps with Hfr E, all of which transfer in the same direction. Hfr E overlaps with Hfr C, which transfers in the opposite direction. *Solve:* Draw the Hfr strains as shown to identify overlap.

26b.   *Evaluate:* This is type 2. *Deduce:* The map from Answer 26a shows overlap between Hfr E with both Hfr C and Hfr B, which indicates the circular nature of the chromosome. *Solve:* Draw a map similar to the one shown in the figure.

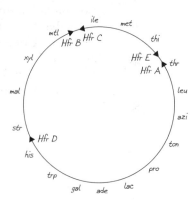

6.04 Test Yourself

Problems

1. A student in your laboratory was performing transformation experiments for the purpose of mapping three genes—*met, thi,* and *thr*—on the *E. coli* chromosome. Her results indicate that the gene order is *met-thi-thr*, with *met* and *thi* cotransformed at a frequency of 0.001%, and *thi* and *thr* cotransformed at a frequency of 0.005%. She left the lab without analyzing one set of transformants, which you find as ~200,000 colonies on several unlabeled petri plates in the refrigerator. You check the *met, thi,* and *thr* phenotypes of these colonies to find out which marker was the selected marker in this transformation and whether the results were consistent with the genetic map developed by your student. The results were that 100% of the colonies could grow on minimal medium plus thiamine and threonine, 0.001% could grow on minimal medium plus methionine and threonine, and none could grow on minimal plus methionine and thiamine. What was the selected marker, and do these results fit with the existing genetic mapping data?

2. You wish to perform transformation experiments to map the *gal, lac,* and *pro* genes. The donor strain is *gal⁺, lac⁺,* and *pro⁺*. The recipient strain is *gal⁻, lac⁻,* and *pro⁻*. *gal⁺* is required for galactose utilization. *lac⁺* is required for lactose utilization. *pro⁺* is required for proline synthesis. What media would you use to measure the *gal⁺* transformation frequency and the *gal⁺ pro⁺* cotransformation frequency? Can you directly measure the *gal⁺ lac⁺* cotransformation frequency on one type of medium? Why or why not?

3. A particular mutant strain of bacterium is able to be used as a recipient for Hfr mating and generalized transduction mapping, but it cannot be used in transformation experiments. Use your understanding of these mapping procedures and the information in Figure 6.11 in the textbook to create a hypothesis for the nature of the mutation in this bacterium.

4. You are given a new T4 *rII* mutant, *m2*, and are asked to map the site of the mutation. You find that *m2* does not complement known *rIIA* or *rIIB* mutations and that it can produce wild-type recombinants with deletion mutants *W833* and *1993* but not with deletion mutants *1695* or *196*. Use this information to determine the type and location of the *m2* mutation.

5. You have two Hfr strains, Hfr A and Hfr B. *str^S* (streptomycin sensitivity) is a useful counter-selected marker for both strains. *leu* is the first gene transferred by Hfr strain A, and *gal* is transferred late. *met* is the first gene transferred by Hfr strain B, and *mal* is transferred late. Use the genetic map to indicate the site and orientation of the F-plasmid in two Hfr strains and to determine the time of entry of each gene between the *oriT* and the counter-selected marker for each Hfr strain. Assume that the first gene to enter in both Hfr strains enters at 2 minutes.

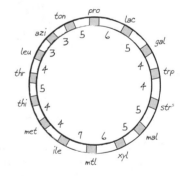

6. The figure shows a map of a bacterial chromosome and indicates the location and orientation of the F-plasmid DNA in each of four different Hfr strains (A, B, C, and D). Assume that each gene in all four Hfr strains is the wild-type, functional allele. You would like to use one of these strains to create a bacterial strain that is diploid for the *gal* gene. Indicate which strain you would use, and describe how you would use it to create a bacterial diploid for the *gal* gene.

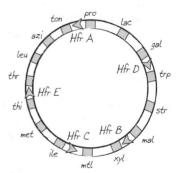

7. You have isolated a new *E. coli* mutant that requires methionine for growth. When you use this *met⁻* strain as a recipient in conjugation with a *met⁺* Hfr strain, you are able to isolate *met⁺* exconjugates. However, when you use the *met⁻* strains as a recipient with a *met⁺* F′ strain derived from the Hfr strain, which carries *met⁺* on the F plasmid, all exconjugates are *met⁻*. Explain.

8. In Charles Yanofsky's cotransduction experiments presented in the textbook, he mapped the relative order of the *trp* operon genes *trpB* and *trpC* using a gene required for cysteine synthesis (*cys*). The *trp* genes are located at 28.4 minutes on the *E. coli* chromosome. The identity and location of some *cys* genes is listed below. Considering that the *E. coli* chromosome is 100 minutes in length and that the generalized transducing virus, P2, can incorporate a segment of chromosome equal to about 2% the size of the *E. coli* chromosome, which of the *cys* genes most likely corresponds to the *cys* gene used by Yanofsky?

Gene	Chromosomal location in minutes
cysA	54.7
cysA	28.7
cysC	61.9
cysB	61.9
cysE	81.5
cysG	75.3

9. The table shows the results of three cotransduction experiments designed to determine the order of *ara*, *ilv*, and *leu* on the *E. coli* chromosome. *ara⁻* mutants cannot utilize the sugar, arabinose. *ilv⁻* mutants cannot synthesize isoleucine. *leu⁻* mutants cannot synthesize leucine. The transducing particles were created using a *leu⁺ ara⁺ ilv⁺* strain and allowed to infect a *leu⁻ ara⁻ ilv⁻* strain. In experiment 1, the transductants were selected on minimal medium plus leucine and isoleucine with arabinose as the carbon source. In experiment 2, transductants were selected on minimal medium plus isoleucine with glucose as carbon source. In experiment 3, transductants were selected on minimal medium plus leucine with glucose as carbon source. Use this information to determine the order of the genes.

Experiment 1		Experiment 2		Experiment 3	
leu⁺ ilv⁺	12.5%	*ilv⁺ ara⁺*	10%	*leu⁺ ara⁺*	12.5%
leu⁺ ilv⁻	25%	*ilv⁺ ara⁻*	20%	*leu⁺ ara⁻*	25%
leu⁻ ilv⁺	0%	*ilv⁻ ara⁺*	20%	*leu⁻ ara⁺*	0%
leu⁻ ilv⁻	62.5%	*ilv⁻ ara⁻*	50%	*leu⁻ ara⁻*	62.5%

10. The order of genes on a bacterial chromosome is *a-b-c*. A donor strain with the genotype $a^+b^+c^+$ and a recipient strain with the genotype a^-b^-c- were used in a mapping experiment. The resulting genotypes after DNA transfer included $a^+b^+c^+$, $a^+b^+c^-$, $a^+b^-c^-$, $a^-b^-c^-$, $a^-b^+c^+$, and $a^-b^-c^+$. No bacteria with the genotype $a^+b^-c^+$ were isolated. Do these results indicate the method used to perform the transfer/mapping experiments? Explain.

Solutions

1.   *Evaluate:* This is type 4. *Deduce:* The selected marker is the gene in the donor strain that must be taken up and incorporated into the recipient chromosome in order for transformants to grow. Growth on medium lacking methionine indicates transformation of *met⁺*, growth on medium lacking thiamine indicates transformation by *thi⁺*, and growth on medium lacking threonine indicates transformation by *thr⁺*. *Solve:* Since 100% of the colonies grew on medium lacking methionine, *met⁺* was the selected marker. 0.001% of *met⁺* transformants that are also *thi⁺* indicates 0.001% cotransformation for *met* and *thi*, which is also consistent with the map. The lack of any *met⁺ thr⁺* colonies out of 200,000 total is also consistent with the greater than 0.001% cotransformation frequency expected for *met* and *thr* given the map.

2. *Evaluate:* This is type 4. *Deduce:* gal^+ transformants are selected by plating on medium that contains galactose as the sole carbon source. lac^+ transformants are selected by plating on medium that contains lactose as the sole carbon source. pro^+ bacteria are selected by plating on medium that lacks proline. *Solve:* gal^+ transformants are selected using minimal medium containing proline with galactose as the sole carbon source. gal^+ pro^+ cotransformants are selected using minimal medium containing galactose as carbon source. gal^+ lac^+ cotransformation frequency cannot be determined by selecting gal^+ lac^+ on a single medium, because galactose and lactose in medium makes neither the sole carbon source. Either gal^+ transformants must be selected first and then scored for lac^+, or vice versa.

3. *Evaluate:* This is type 1. *Deduce:* The key difference in these three genetic methods is the mechanism by which donor DNA enters the recipient cell. *Solve:* The mutant is defective for uptake or processing of exogenous DNA.

4. *Evaluate:* This is type 5. *Deduce:* The mutation does not complement *rIIA* or *rIIB* mutants; therefore, the mutation inactivates both genes. Mutation *m2* cannot recombine with *1695* or *196* and therefore overlaps both mutants. Since *1695* and *196* do not overlap each other, then *m2* must be either a double mutant or a deletion that spans the distance between the endpoints of *1695* and *196*. *Solve:* Mutation *m2* is either a double point mutant, with one located in region A5c1 to A5c2 and the other located in region B5 to B7. Alternatively, mutation *m2* is a deletion whose endpoints are located in the regions just indicated.

5. *Evaluate:* This is type 2. *Deduce:* An arrow can be used to indicate the location and orientation of the origin of transfer. The arrow points in the direction of entry into the recipient cell, and genes behind the arrow follow in order. The time of entry of the first gene is 2 minutes, and the time of entry of each following gene is determined by summing the total distance between it and the first gene and adding 2. For example, for Hfr A, *leu* enters at 2 minutes; *leu* and *azi* are separated by 3 minutes, so *azi* enters at 5 minutes. *Solve:* See figure.

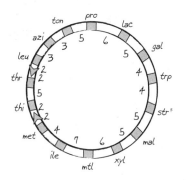

	gene	*leu*	*azi*	*ton*	*pro*	*lac*	*gal*	*trp*	*str*^{S}
Hfr strain A	time of entry	2	5	8	13	19	24	28	32
Hfr strain B	gene	*met*	*ile*	*mtl*	*xyl*	*mal*	*str*^{S}		
	time of entry	2	6	13	19	24	29		

6. *Evaluate:* This is type 2. This problem provides a genetic map that shows the location and orientation of four Hfr strains and asks which of these could be used to create a strain diploid for the *gal* gene. *Deduce:* Partial diploids can be created by mating an F' strain that carries the gene of interest on the F plasmid and a normal recipient strain. The *gal* gene is closest to the F plasmid DNA in Hfr strain D. *Solve:* Isolate an F' strain from Hfr strain D by looking for rare donors that produce donor exconjugates after mating. Determine whether the F' carries the gal^+ gene by mating the F' strain to a gal^-, F$^-$ recipient and confirming that the exconjugates are F' donors that are now gal^+. Mating this F' strain to any strain that carries a *gal* gene creates a partial diploid with two copies of *gal*.

7. *Evaluate:* This is type 2. *Deduce:* Stale maintenance and expression of genes donated from an Hfr strain requires integration of those genes into the recipient chromosome, which replaces the recipient alleles at each locus with the alleles from the donor. Genes donated from an F' strain are stably maintained on the F' episome, creating a strain that is diploid for all genes carried on the F' plasmid. *Solve:* The met^- mutation is dominant to the wild-type met^+ allele. The recessive met^+ allele from the Hfr donor replaces the dominant met^- allele in the recipient, so the recipient has a *met⁺* genotype and a *met⁺* phenotype. The recessive *met⁺* allele from the F′ donor is maintained separately from the chromosome, so the recipient has a *met⁺/met⁻* genotype and a *met⁻* phenotype.

8. *Evaluate:* This is type 3. *Deduce:* P2 can cotransduce only genes separated by 2 minutes (2% of the *E. coli* chromosome) on the chromosome map. *trpB* and *trpC* are located at 28.4 minutes. *Solve:* The closest *cys* gene is *cysB*.

9. *Evaluate:* This is type 3. *Deduce:* Essentially all bacteria that are cotransductants for both outside genes should also be cotransductants for the middle gene. *Solve:* Selection for *ara⁺* yielded cotransductants that are also *leu⁺* and *ilv⁺* but none that are *leu⁻ ilv⁺*. This indicates that *leu* is in the middle. This is consistent with the fact that selection for *ilv⁺* yielded *leu⁺ ara⁺* cotransductants, but none are *leu⁻* and *ara⁺*. This is also consistent with the finding that selection for *leu⁺*, the putative middle gene, yields all four possible phenotypes at detectable frequencies. Therefore, the gene order is *ara-leu-ilv*.

10. *Evaluate:* This is type 1. *Deduce > Solve:* The only genotype absent from among the bacteria that received donor DNA was the *a⁺b⁻c⁺* genotype, which corresponds to integration of the donor DNA by a quadruple crossover. The only transfer procedure ruled out by that result is one in which the donor was an F′ strain, since integration of the transferred DNA into the recipient chromosome is not required in that case. Since F′ strains are not typically used to determine gene order, these results do not rule out transformation, conjugation using an Hfr strain, or cotransduction as possible methods used to do the mapping. The genotypes obtained after DNA transfer are consistent with integration of the donated DNA, which is required in all three approaches.

7

DNA Structure and Replication

7.01 Genetics Problem-Solving Toolkit

Key Terms and Concepts

Base pairing: Hydrogen bonding between the nitrogenous bases of nucleotides in DNA or RNA.

Base stacking: Hydrophobic interactions between bases in the center of a DNA double helix.

DNA ligase: An enzyme that catalyzes the formation of a phosphodiester bond between nucleotides.

Polymerase chain reaction: A procedure for amplifying the amount of a specific DNA sequence *in vitro* by performing multiple rounds of DNA synthesis.

Primosome: A complex of proteins involved in synthesizing RNA primers during DNA synthesis.

Replication bubble: A region of DNA that is currently being replicated or has recently been replicated. Every replication bubble has a least one replication fork, which is the site of new DNA synthesis.

Replisome: The complex of proteins involved in replication of DNA at replication forks. The replisome includes DNA polymerases, clamp proteins, clamp loaders, helicases, and ssDNA binding proteins.

Key Analytical Tools

Replication Bubble

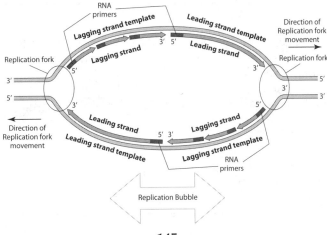

145

7.02 Types of Genetics Problems

1. Describe the structure of DNA or the evidence leading to identification of DNA as the genetic material.
2. Describe the structure, composition, or enzymology of DNA replication bubbles.
3. Describe the major strategies cells employ for genome replication.
4. Analyze or predict the results of experiments that use DNA synthesis as an analytical tool.

1. Describe the structure of DNA or the evidence leading to identification of DNA as the genetic material.

Problems of this type test your recollection of the experiments that demonstrated DNA as the genetic material. They also test your knowledge of the structure of double-stranded DNA. You may be asked about the structure of nucleotide components, the linkage between nucleotides in a strand of DNA, or the linkage between bases in antiparallel strands of DNA. These problems must be mastered in order to perform well on type 2, 3, or 4 problems in this chapter. See Problem 4 for an example.

Example Problem: The double-stranded DNA molecule shown below was synthesized in the presence of dATP that contained radioactive phosphate. The dsDNA molecule was digested with an enzyme that cleaves the phosphodiester bond connecting the phosphate group and the 5′ carbon of every nucleotide in the molecule. Which nucleosides generated by this experiment contain radioactive phosphate?

5′-TACTACTACTACTACTACTAC-3′

3′-ATGATGATGATGATGATGATG-5′

Solution Strategies	Solution Steps
Evaluate	
1. Identify the topic this problem addresses, and explain the nature of the requested answer.	1. This problem tests your understanding of nucleotide structure and the linkage between nucleotides in DNA. The answer will be a nucleoside indicated by A, T, G, or C.
2. Identify the critical information given in the problem.	2. The double-stranded DNA was synthesized in the presence of dATP that contained radioactive phosphorous. The enzyme used to digest the DNA breaks the phosphodiester bond between the phosphate group and the 5′ carbon of deoxyribose.
Deduce	
3. Redraw the DNA sequence, inserting a "p" between bases to indicate the position of phosphate in the backbone of each strand. Circle the phosphates that contain radioactive phosphorous. **TIP:** The phosphate in dATP is attached to the hydroxyl group of the 5′ carbon on deoxyribose.	3. The DNA sequence written to show phosphate groups is as follows: 5′ ⎸pTpApCpTpApCpTpApCpTpApCpTpApCpTpApCpTpApC⎸ 3′ 3′ ⎸ApTpGpApTpGpApTpGpApTpGpApTpGpApTpGpApTpGp⎸ 5′

(continued)

4. Add lines to the drawing from step 3 to show the location of the break caused by treatment with the enzyme that breaks the phosphodiester bond between each phosphate and the 5' carbon of deoxyribose.

4. The DNA sequence with lines showing the break points is as follows:

5' | pT•A•C•T•A•C•T•A•C•T•A•C•T•A•C•T•A•C•T•A•C | 3'
3' | A•T•G•A•T•G•A•T•G•A•T•G•A•T•G•A•T•G•A•T•G• | 5'

Solve

5. Identify the nucleotide that contains a circled phosphate.

5. **Answer:** The only labeled nucleotide produced from this procedure is T.

2. Describe the structure, composition, or enzymology of DNA replication bubbles.

This type of problem tests your understanding of bidirectional DNA replication with emphasis on the structure and function of replication bubbles and replication forks. The problems may ask about the biochemical properties of specific proteins or the location of those proteins within replication bubbles. The problems may ask you to draw or label replication forks and indicate the polarity of all DNA/RNA molecules present or to identify leading and lagging strand templates. See Problem 6 for an example.

Example Problem: The figure shows the template strands of a replication fork that is moving from right to left. This fork is at one end of a replication bubble. Draw the replication fork at the other end of this replication bubble, include the newly synthesized DNA strands in that fork, and label the templates for leading and lagging strand synthesis. Label the 5' and 3' ends of all DNA strands.

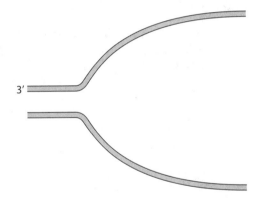

Solution Strategies	Solution Steps
Evaluate	
1. Identify the topic this problem addresses, and explain the nature of the requested answer.	**1.** This problem tests your understanding of bidirectional DNA replication. The answer will be a labeled drawing of a replication fork.
2. Identify the critical information given in the problem.	**2.** The replication fork is moving right to left. The left end of the top strand is that strand's 3' end.

(continued)

Deduce

3. Extend the lines in the center of the replication fork to complete the replication bubble.

> **TIP:** Recall that each DNA strand will have a 5′ and 3′ end, and that the two strands are antiparallel.

3.

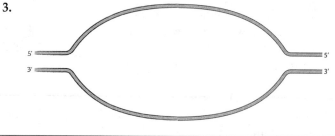

4. Draw newly synthesized DNA strands for the fork, moving left to right; label their ends; and identify which is leading and which is lagging.

> **TIP:** Both strands elongate in a 5′ to 3′ direction, and the leading strand will grow continuously in the direction that the replication fork is moving.

4.

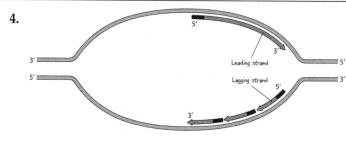

Solve

5. Answer the problem by identifying the templates for lagging and leading strand synthesis.

5. **Answer:** See figure.

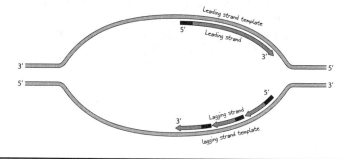

3. Describe the major strategies cells employ for genome replication.

This type of problem asks you to consider the various strategies cells employ to replicate their chromosome, including unidirectional, bidirectional, and rolling circle mechanisms. You may be asked to compare and contrast mechanisms or to recall classic experiments that distinguished among them. You may also be asked to consider strategies for replicating entire genomes, including issues having to do with the time required for genome replication and the problems posed by replication of linear chromosomes. See Problem 18 for an example.

Example Problem: The maximum doubling rate of *E. coli* is once every 20 minutes. Explain and illustrate how this is possible given that the *E. coli* chromosome is a circle containing 4×10^6 bp of DNA and a single replication origin, and that 1000 nucleotides are incorporated per second by DNA polymerase.

Solution Strategies	Solution Steps
Evaluate	
1. Identify the topic this problem addresses, and explain the nature of the requested answer.	1. This problem requires you to relate parameters of mechanism of *E. coli* chromosome replication to the rate of cell doubling. The answer will be a mechanistic explanation for replication of the *E. coli* chromosome in 20 minutes.
2. Identify the critical information given in the problem.	2. The problem states that *E. coli* cells can double once approximately every 20 minutes, that the *E. coli* chromosome is a 4×10^6 bp circle containing a single, bidirectional replication origin, and that each replication fork replicates 1000 base pairs per second.
Deduce	
3. Consider the relationship between cell division and chromosome replication.	3. At the time of cell division, each *E. coli* cell must contain at least two complete copies of its chromosome.
4. Calculate the time required to copy an *E. coli* chromosome. **TIP:** The *E. coli* replication origin initiates bidirectional replication; therefore, each replication fork replicates one-half of the chromosome.	4. The time required to replicate the *E. coli* chromosome is $$\frac{\frac{1}{2}(4 \times 10^6)}{1000} = 2000 \text{ seconds} = 33 \text{ minutes}$$
5. Relate the time required for chromosome replication to the rate of cell division.	5. Cells divide more rapidly than they replicate their chromosome; therefore, cells must already have partially replicated chromosomes at the time they are formed by cell division.
Solve	
6. Describe the process of DNA replication as it relates to the maximum rate of *E. coli* cell division.	6. **Answer:** Replication must be initiated at least once every 20 minutes. *E. coli* cells on the verge of division contain two full chromosomes, each of which is partially replicated.

(continued)

7. Illustrate the structure of the *E. coli* chromosomes found at the moment before cell division in these rapidly growing cultures.

7. See figure.

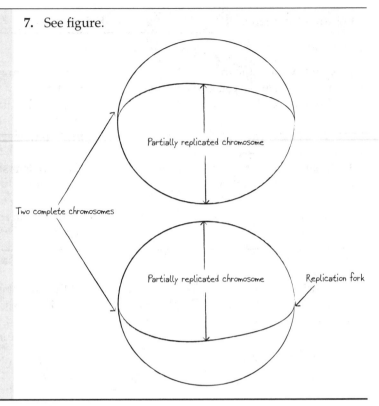

4. Analyze or predict the results of experiments that use DNA synthesis reactions as an analytical tool.

This type of problem asks you to apply your understanding of the mechanisms of DNA synthesis to the analysis of DNA by PCR or dideoxy DNA sequencing reactions. The problems may provide the results of PCR or DNA sequence analyses and ask you to interpret those results. Alternatively, you may be asked to predict the results of PCR or sequencing of a particular segment of DNA. See Problem 24 for an example.

Example Problem: Indicate the sequence that would be amplified in a PCR reaction using the DNA sequence and PCR primers shown below.

DNA sequence:

5'-ATCGGCGTACGTAAGCCCGTGGCTAGC
GTAGCGTCTCTATGTAGCGTAGACGGCCA
TTATCCGTA-3'

3'-TAGCCGCATGCATTCGGGCACCGATC
GCATCGCAGAGATACATCGCATCTGCCG
GTAATAGGCAT-5'

PCR primers:

5'-GATAATGGCCGTCTACGC-3'

5'-GTACGTAAGCCCGTGGCT-3'

Solution Strategies	Solution Steps
Evaluate	
1. Identify the topic this problem addresses, and explain the nature of the requested answer.	1. This problem tests your understanding of the polymerase chain reaction (PCR). The answer will be a dsDNA sequence.
2. Identify the critical information given in the problem.	2. The dsDNA sequence of the template and the PCR primer sequences are given.
Deduce	
3. Determine the site and strand that each PCR primer hybridizes to during the annealing step of PCR.	3. The first primer: • The primers sequence 5'-GATAATGGCCGTCTACGC-3' is identical to the bottom-strand sequence starting at the fifth nucleotide in. That primer will hybridize to the complementary strand at the sequence, 5'-GCGTAGACGGCCATTATC-3'. • The primer sequence 5'-GTACGTAAGCCCGTGGCT-3' is identical to the top-strand sequence starting at the seventh nucleotide in. That primer will hybridize to the complementary strand at the sequence 5'-AGCCACGGGCTTACGATC-3'.
Solve 5. Write the dsDNA sequence that would be amplified by PCR using these primers. **TIP:** The dsDNA sequence from the 5' end of one primer to the 5' end of the other primer will be amplified.	5. **Answer:** The amplified sequence would be 5'-GTACGTAAGCCC GTGGCTAGCGTAG CGTCTCTATGTAG CGTAGACGGCCAT TATC-3' 3'-CATGCATTCGGG CACCGATCGCATCGC AGAGATACATCGCATC TGCCGGTAATAG-5'

7.03 Solutions to End-of-Chapter Problems

1. *Evaluate:* This is type 1. *Deduce:* The result was that injection of a mixture of heat-killed SIII and living RII strains of *Pneumococcus* into mice caused the mice to die from pneumonia due to the growth of SIII bacteria. *Solve:* The living RII bacterial strain must have been "transformed" to SIII because the change from R to S and from type II to type III could not have been the result of a single mutational event. The transformed SIII bacteria continued to produce SIII bacteria after many rounds of reproduction, indicating that the transforming principle changed the genetic makeup of the RII bacteria. Since the SIII trait was heritable, the transforming factor was either genes or contained genes.

2. *Evaluate:* This is type 1. *Deduce:* The key results were those that showed that enzymes that destroyed RNA and protein did not destroy the transforming principle, whereas enzymes that destroyed DNA did. *Solve:* The transforming principle was considered to be genetic material. The most reasonable interpretation of those results was that DNA was the only essential component of the transforming principle, and therefore, the genetic material.

3.   *Evaluate:* This is type 1. *Deduce:* The structure and life cycle of phage T2 make it a good choice. First, as stated in the problem, T2 contains DNA and protein but no RNA. *Solve:* Therefore, radioactive labeling with P^{32} labels only DNA, and S^{35} labels only protein. Also, the entire T2 virus particle does not enter the bacterial cell. The protein shell remains attached to the outer cell surface, and the genetic material of the virus is injected into the bacterium, where it directs the synthesis of hundreds of new virus particles.

4.   *Evaluate:* This is type 1. *Deduce:* Hershey and Chase prepared T2 particles whose protein was labeled with the radioactive sulfur, S^{35}, and whose DNA was labeled with radioactive phosphorous, P^{32}. They used the labeled T2 to infect bacteria and then separated the infected bacteria from the empty phage shells (phage ghosts) using a blender. They found that essentially all the P^{32}-labeled T2 DNA but little to none of the S^{35}-labeled T2 protein was in the infected bacterial cells. *Solve:* Since T2 genetic material must be inside the infected cells in order to direct new virus particle synthesis, these results pointed to DNA as the genetic material of phage T2.

5a.   *Evaluate:* This is type 1. Recall that the strands of DNA in dsDNA are complementary and antiparallel. *Deduce > Solve:* 5′-GATCAGGTCGAT-3′ (same as 3′-TAGCTGGACTAG-5′)

5b.   *Evaluate > Deduce > Solve:* This is type 1. Recall that nucleotides in a strand of DNA are linked via phosphodiester bonds.

5c.   *Evaluate:* This is type 1. Recall that nucleotides in a strand of DNA are linked via phosphodiester bonds. *Deduce > Solve:* Phosphodiester bonds are covalent.

5d.   *Evaluate:* This is type 1. Recall the reaction catalyzed by DNA polymerase. *Deduce > Solve:* The phosphodiester bonds in DNA are formed when the α-phosphate of a nucleotide triphosphate reacts with the 3′ hydroxyl group of a nucleotide in the DNA chain.

5e.   *Evaluate > Deduce > Solve:* This is type 2. Recall that DNA polymerase catalyzes the formation of phosphodiester bonds.

5f.   *Evaluate:* This is type 1. Recall that the chemical interactions between strands of DNA in dsDNA. *Deduce > Solve:* Hydrogen bonds (base pairing) and hydrophobic interactions (base stacking) join the strands of DNA in a duplex.

5g.   *Evaluate:* This is type 1. Recall the chemical interactions between strands of DNA in dsDNA. *Deduce > Solve:* Noncovalent

5h.   *Evaluate:* This is type 1. Recall the terminology for DNA structure. *Deduce > Solve:* Complementary

5i.   *Evaluate:* This is type 1. Recall the terminology for DNA structure. *Deduce > Solve:* Antiparallel

6.   *Evaluate:* This is type 1. The problem concerns the chemical bonds that form base pairs in double-stranded DNA. Recall that hydrogen bonds are weak, noncovalent bonds involving two atoms sharing a hydrogen nucleus. The distance between the atoms sharing the hydrogen nucleus is critical for hydrogen bonds to form. *Deduce:* The bases in the two complementary antiparallel DNA strands are aligned such that each of the atoms that share hydrogen nuclei (N and O or N and N) in each base are positioned next to each other at a distance that allows all possible hydrogen bonds to form. *Solve:* The bases in the complementary but parallel strands are not aligned in this manner; therefore, the atoms that could form hydrogen bonds do not align and are not close enough together to allow hydrogen bonding between all possible and necessary chemical groups.

7.   *Evaluate:* This is type 1. Recall the definition of a hydrogen bond and a phosphodiester bond and the structure of dsDNA. *Deduce > Solve:* The molecule contains 29 hydrogen bonds and 24 phosphodiester bonds.

8a. *Evaluate > Deduce > Solve:* This is type 1. Recall that nucleotides in a strand of DNA are linked via phosphodiester bonds.

8b. *Evaluate > Deduce > Solve:* This is type 1. Recall that hydrogen bonds join complementary nucleotides in different strands of DNA.

8c. *Evaluate:* This is type 1. Recall the definition of a phosphodiester bond. *Deduce > Solve:* There are 12 phosphodiester bonds in the molecule.

8d. *Evaluate:* This is type 1. Recall the definition of a hydrogen bond. *Deduce > Solve:* There are 17 hydrogen bonds in the DNA molecule.

9a. *Evaluate:* This is type 1. Recall that the strands of DNA in dsDNA are complementary and antiparallel. *Deduce > Solve:*

3′-ACGCTACGTC-5′

5′-TGCGATGCAG-3′

9b. *Evaluate:* This is type 1. Recall that one phosphodiester bond connects adjoining nucleotides in a strand of DNA. *Deduce:* Each phosphodiester bond represents one covalent bond connecting nucleotides. There are nine connections between nucleotides in each strand. *Solve:* There are 18 covalent bonds connecting nucleotides in the molecule from 9a.

9c. *Evaluate:* This is type 1. Recall that AT base pairs involve two hydrogen bonds and that GC base pairs involve 3. *Deduce > Solve:* There are 26 noncovalent bonds joining complementary strands in the molecule.

10. *Evaluate:* This is type 2. Recall that biochemical functions of DNA polymerase are relevant to strand elongation. *Deduce > Solve:* DNA polymerase III determines which free nucleotide triphosphate is complementary to the base being copied. DNA polymerase III catalyzes phosphodiester bond formation between the α-phosphate of the incoming nucleotide triphosphate and the 3′ hydroxyl group of the last nucleotide added to the strand.

11a. *Evaluate:* This is type 1. *Deduce:* Compare drawing with Figure 7.7 in the text. *Solve:* Identify errors as shown in the figure. Note that errors present in both nucleotides are pointed out for only one.

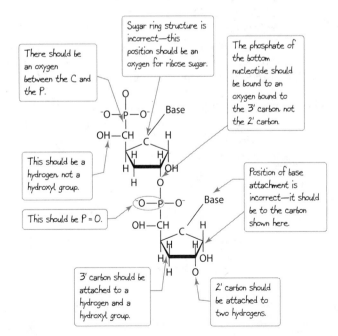

11b.   *Evaluate:* This is type 1. *Deduce:* Compare drawing with Figure 7.7 in the text. *Solve:* The 5′ carbon of the bottom nucleotide is connected to (what would normally be) the 2′ carbon position of the sugar in the top nucleotide.

11c.   *Evaluate:* This is type 1. *Deduce:* Compare drawing with Figure 7.7 in the text. *Solve:* Draw the single-stranded segment as shown.

12.   *Evaluate:* This is type 2. Recall that DNA polymerase cannot initiate DNA synthesis de novo; it can only extend an existing strand. *Deduce > Solve:* RNA is synthesized and serves as a primer for elongation by DNA polymerase.

13.   *Evaluate:* This is type 1. Recall that in dsDNA, % C = % G and % A = % T, and % C + % G + % A + % T = 1. *Deduce > Solve:* The percentages are 20% guanine, 30% adenine, 30% thymine.

14a.   *Evaluate:* This is type 2. Recall the distinct roles that DNA polymerase I and III play in DNA replication. *Deduce > Solve:* DNA polymerase I is required to remove the RNA primer and fill in the gap with DNA. DNA polymerase III is responsible for the bulk of synthesis of DNA on the leading and lagging strands.

14b.   *Evaluate:* This is type 2. Recall the role of DNA polymerase I and consider the consequences to DNA replication if everything except the function of DNA pol I was carried out. *Deduce:* RNA primers at the origin of replication and all along the lagging strands would not be removed from chromosomal DNA. *Solve:* The absence of DNA pol I will not prevent the bulk of DNA replication but will result in newly replicated DNA containing small segments of RNA and nicks at the junctions of polymerase III synthesized DNA and the 5′ end of the RNA primers.

14c.   *Evaluate:* This is type 2. Recall that DNA polymerase III is responsible for the vast majority of DNA synthesis during DNA replication. *Deduce > Solve:* An *E. coli* mutant without a functional DNA polymerase III will be unable to replicate DNA because it lacks the enzyme responsible for the bulk of DNA synthesis during replication.

15a–k. *Evaluate:* This is type 2. ***Deduce >***
Solve: Consult Figure 7.21 in the
text to inform your drawing.

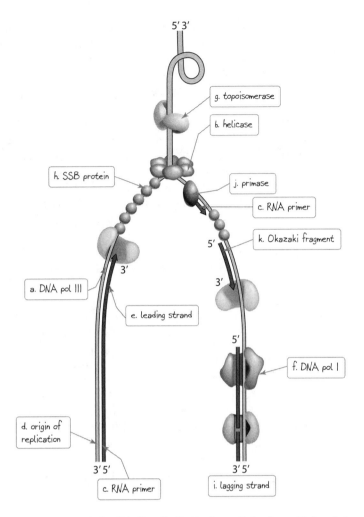

g. topoisomerase

b. helicase

h. SSB protein

j. primase

c. RNA primer

k. Okazaki fragment

a. DNA pol III

e. leading strand

f. DNA pol I

d. origin of replication

c. RNA primer

i. lagging strand

16a. *Evaluate:* This is type 1. Recall that % A = % T and that % G = % C. *Deduce:* Substitute T for A in the numerator of the equation and G for C in the denominator. *Solve:* This gives $\frac{T+G}{G+T} = 1$, which is always true.

16b. *Evaluate:* This is type 1. Recall that % A = % T and % G = % C, and % A + % T + % G + % C = 100. *Deduce:* If % A is any value except 25, then A + T is either less than or greater than 50. If % A + % T is less than 50, then % G + % C must be greater than 50, and vice versa. *Solve:* This gives $\frac{A+T}{G+C} = 1$, which is not generally a true statement.

16c. *Evaluate:* This is type 1. Recall that % A = % T and % G = % C. *Deduce > Solve:* $\frac{A}{T} = 1$ and $\frac{G}{C} = 1$, which is true.

16d. *Evaluate:* This is type 1. Recall that % A = % T and % G = % C, and % A + % T + % G + % C = 100. *Deduce:* If % A is any value except 25, then % A ≠ % C. Substitute A for T and C for G in the equation. *Solve:* This gives $\frac{A}{C} = \frac{C}{A}$, which is not generally a true statement.

16e. *Evaluate:* This is type 1. Recall that % A = % T and % G = % C. *Deduce > Solve:* $\frac{A}{T} = 1$ and $\frac{G}{C} = 1$, which is true.

17a. *Evaluate:* This is type 1. Recall that % A = % T and % G = % C. *Deduce:* Substitute A for T and G for C in the equation. *Solve:* This gives G + A = A + G, which is always true.

17b. *Evaluate:* This is type 1. Recall that % A = % T and % G = % C, and % A + % T + % G + % C = 100. *Deduce:* If % A is any value except 25, then A + T is either less than or greater than

50. If % A + % T is less than 50, then % G + % C must be greater than 50, and vice versa. *Solve:* This gives G + C = A + T, which is not generally a true statement.

17c.   *Evaluate:* This is type 1. Recall that % A = % T and % G = % C. *Deduce:* Substitute T for A and G for C in the equation. *Solve:* This gives G + T = G + T, which is always true.

18.   *Evaluate:* This is type 2. Recall the order in which the following enzymes function during DNA replication: DNA pol I, SSB, ligase, helicase, DNA pol III, and primase. *Deduce:* DNA pol I removes the RNA primer and replaces it with DNA. SSB binds to ssDNA sequences. Ligase seals the nick between the DNA synthesized by DNA pol I and DNA pol III. Helicase unwinds dsDNA. DNA pol III synthesizes DNA using the RNA primer. Primase adds the RNA primer. *Solve:* Therefore, the order in which these proteins and enzymes function is as follows: Helicase, SSB, primase, DNA pol III, DNA pol I, ligase.

19.   *Evaluate:* This is type 1. Recall that for dsDNA, % A = % T and % G = % C, whereas for ssDNA, the percentages are typically unequal. *Deduce > Solve:* Therefore, Genome 1 is ssDNA, and genome 2 is most likely dsDNA.

20.   *Evaluate:* This is type 3. Recall the Meselson-Stahl experiment and consider how the results excluded the alternatives to the semiconservative model for DNA replication. Meselson and Stahl initially cultured *E. coli* in medium containing only N15 (heavy nitrogen) until all cells contained only N15/N15 DNA. They then cultured the N15/N15 *E. coli* in normal medium (N14) and collected samples after one, two, and three rounds of DNA replication. The results showed that before transfer to N14 medium, only N15/N15 DNA was present. After one round of replication in N14 medium, all of the DNA was N15/N14; after two rounds of replication, half the DNA was N15/N14 and half was N14/N14; and after the third round of replication, one-fourth of the DNA was N15/N14 and three-fourths was N14/N14. *Deduce:* The conservative model predicted that the original N15/N15 DNA would remain throughout; therefore, the results ruled out the conservative model. The dispersive model predicted that after each round of replication, there would be only one form of DNA, which would become less and less dense. *Solve:* Although this model was not ruled out after one round of replication, the persistence of the N15/N14 DNA and the presence of two classes of DNA (N15/N14 and N14/N2) after rounds two and three ruled out the dispersive model.

21.   *Evaluate:* This is type 3. Recall the work of John Cairns and consider its impact on the current understanding of DNA replication in bacteria. **TIP:** Theta structures, which resemble the Greek letter theta, are *E. coli* chromosomes undergoing DNA replication. *Deduce > Solve:* Theta structures suggested that replication of the *E. coli* chromosome is initiated at a single site, creating a single "replication bubble." Initiation at multiple sites, as occurs in eukaryotic chromosomes, creates multiple replication bubbles.

22.   *Evaluate:* This is type 3. Cells were incubated in medium containing 3H-thymine for a short period of time (a "pulse") and then transferred to medium containing an excess of unlabeled thymine (the "chase"). The cells were then collected and their DNA was prepared for electron microscopy, which can detect replication bubbles in DNA, and for autoradiography, which reveals the location of 3H-thymine incorporation into DNA. *Deduce:* Recall that bidirectional replication from a replication origin produces a replication bubble with DNA synthesis occurring at both ends, whereas unidirectional replication results in a replication bubble with DNA synthesis occurring at one end. The results showed that DNA replication bubbles contained regions of label on both sides of the midpoint of the bubble, which correspond to the replication origin. *Solve:* If DNA synthesis was unidirectional, then the label would be present on only one side of the origin.

23.   *Evaluate:* This is type 3. Recall that eukaryotic chromosomes are significantly larger than the *E. coli* chromosome and that the rate of DNA synthesis is slower in eukaryotes as compared to *E. coli*. *Deduce:* Both types of cells must completely replicate their chromosomes before they can divide. Given the rate of DNA synthesis by *E. coli* DNA polymerase, initiation of bidirectional replication at one origin in the *E. coli* chromosome is sufficient to replicate the entire chromosome in about 30 minutes. This is consistent with known rates of *E. coli* cell growth and division; therefore, initiation at a single replication origin is sufficient to replicate the *E. coli* genome in a timely manner. *Solve:* For several reasons, eukaryotic genomes require multiple origins for timely replication. First, eukaryotic genomes are contained in multiple chromosomes and there must be at least one replication origin per chromosome. Furthermore, eukaryotic DNA polymerase functions about 10 times slower than *E. coli*, and eukaryotic chromosomes are much larger than those of *E. coli*, indicating that eukaryotic chromosome replication would take at least 10 times longer than replication of the *E. coli* chromosome if only one origin was used per chromosome. A third rationale for multiple eukaryotic origins is the presumed need for replication of different genomic regions at different times during S phase of the cell cycle, which is accomplished by initiating DNA replication at different times at different origins.

24.   *Evaluate:* This is type 2. *Deduce:* DNA helicases unwind dsDNA during DNA replication and repair. Bloom syndrome is characterized by chromosome instability and an increased rate of cancer. Chromosome instability is evident when chromosomes are lost from cells, typically because of a failure during mitosis. Mitosis fails to occur properly if chromosomes are not completely replicated. Cancer is a disease caused by accumulation of somatic mutations, which accumulate at an elevated rate if DNA repair by DNA replication is defective. *Solve:* Based on the information provided, it is reasonable to speculate that the lack of the DNA helicase encoded by the Bloom syndrome gene results in incomplete replication during S phase and during repair of DNA damage. Failure to replicate chromosomes completely could result in a failure to pass chromosomes on to progeny cells during mitosis, which would result in chromosome instability. Failure to repair DNA damage would also lead to an increased rate of somatic mutation, which would lead to cancer.

25.   *Evaluate:* This is type 3. *Deduce:* Recall the mechanisms for rolling circle replication and bidirectional replication, particularly the description of rolling circle replication of the F plasmid during conjugation. *Solve:* Rolling circle replication and bidirectional replication differ in many ways. Initiation of replication involves cleavage of one phosphodiester bond (called a nick) of one strand in rolling circle replication, whereas no strand breakage occurs in bidirectional replication. DNA helicases unwind and release the nicked strand as ssDNA in rolling circle replication; in contrast, helicases unwind dsDNA in bidirectional replication, but the unwound strand is not released as ssDNA. The 3' end of the nicked strand serves as the primer for synthesis of the complement to the un-nicked strand during rolling circle replication, whereas RNA primers are needed for synthesis of both new strands of DNA in bidirectional replication. Rolling circle replication can create multiple, head-to-tail copies of the chromosome (called a concatamer) by continued synthesis at the 3' end of the strand and continued release of ssDNA, whereas bidirectional replication cannot create concatemeric copies of the chromosome.

26a.   *Evaluate:* This is type 3. Recall that telomeres are replicated by telomerase, which uses a portion of its RNA sequence as a template to repetitively add a short, six-nucleotide sequence to the ends of chromosomes. *Deduce > Solve*: Telomeric DNA is composed of a repetitive, short DNA sequence. In many organisms, the repeated sequence is 5'-TTAGGG-3' or a variant thereof.

26b.   *Evaluate:* This is type 3. Recall that lagging strand synthesis is not complete at the ends of linear chromosomes. **TIP:** Telomerase is a reverse transcriptase that includes an RNA molecule. The RNA acts as the template for synthesis of telomeric DNA. *Deduce > Solve*: Telomerase uses a segment of its RNA as the template to add multiple copies of a simple sequence to the 3' end of each strand of DNA on a linear chromosome. This strand, which corresponds to the template for lagging strand synthesis, is copied by the normal mechanism of lagging strand synthesis after it is extended by telomerase.

26c.   *Evaluate:* This is type 3. *Deduce > Solve:* Telomeres are thought to provide two functions, one in chromosome replication and the other in chromosome protection. Telomeres provide a mechanism for replication of the ends of linear chromosomes. Without telomeres, lagging strand synthesis would fail to extend to the chromosome ends, leaving a gap at each end after each round of replication. This would shorten the chromosome and, after many rounds of replication, would result in loss of important DNA sequences (genes). Telomeres are repetitive DNA, which prevents loss of important DNA sequences if shortening occurs. Telomeres are also the binding site for telomerase, which extends the lagging strand template to compensate for sequences lost during incomplete lagging strand synthesis. Telomeres also provide a protective "cap," on the ends of linear chromosomes, that distinguishes normal chromosome ends from ends generated by double-stranded chromosome breaks (DNA damage). Without telomeric DNA and the proteins that bind telomeric DNA, the ends of chromosomes are recognized as broken chromosomes and are fused together by DNA repair enzymes. Such breakage can create chromosome end-to-end fusions, which then create dicentric chromosomes that can be broken during the next cell division, thus creating new breaks, new fusions, in an endless cycle known as the bridge-break-fusion cycle.

26d.   *Evaluate:* This is type 3. *Deduce:* Germ-line cells divide many times, whereas many somatic cells are capable of a limited number of cell divisions (some are unable to divide at all). *Solve:* Telomerase is required to ensure complete chromosome replication in germ cells, thus ensuring that every mitosis produces two daughter cells with complete chromosomes. Telomerase is not required in somatic cells, because they cannot divide enough times to result in loss of important DNA at chromosome ends. Some researchers think that the lack of telomerase in somatic cells prevents indefinite cell division because loss of DNA at chromosome ends will activate DNA damage responses that stop division and lead to cell death. This function would help protect the organism from the spread of cancerous cells.

27a.   *Evaluate:* This is type 4. This problem challenges you to integrate your knowledge of pedigrees and the inheritance of DNA markers to PCR. Recall that PCR amplifies the same segment of DNA from each individual. The size of the fragments will be proportional to the number of 10-bp repeats present at the amplified region of both homologous chromosomes in each individual. *Deduce:* The mother's genotype is 16/21, and the father's is 18/26. *Solve:* Draw the pedigree, DNA gel results, and gel labels. Finally, ensure that the pedigree and DNA gel is appropriately aligned, as shown in the figure.

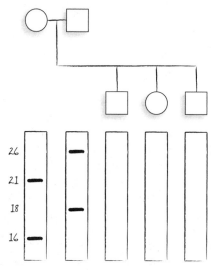

27b.   *Evaluate:* This is type 4. This problem challenges you to integrate your knowledge of pedigrees and the inheritance of DNA markers to PCR. *Deduce:* Each child will inherit an allele corresponding to 16 or 21 repeats from its mother and an allele corresponding to 18 or 26 repeats from its father. *Solve:* The possible genotypes are 16/18, 16/26, 21/18, 21/26.

27c.   *Evaluate:* This is type 4. This problem challenges you to integrate your knowledge of pedigrees and the inheritance of DNA markers to PCR. *Deduce > Solve:* The alleles are codominant because the phenotype of heterozygotes is clearly distinguishable from that of either homozygote.

28a.   *Evaluate:* This is type 4. Recall that dideoxy DNA sequencing involves the use of dideoxynucleotides that terminate elongation of the DNA strand to which they are added. **TIP:** The A lane corresponds to a reaction that produces fragments that terminate at each site of addition of a dideoxyadenosine nucleotide. *Deduce > Solve:* The reaction would have equal concentrations of deoxycytidine triphosphate, deoxythymidine triphosphate, and deoxyguanidine triphosphate. It would also have a mixture of deoxyadenosine triphosphate and dideoxyadenosine triphosphate.

28b.   *Evaluate:* This is type 4. This problem challenges you to integrate your understanding of dideoxy DNA sequencing and PCR. **TIP:** Dideoxy DNA sequencing depends on synthesis of detectable amounts of each fragment of DNA produced from a DNA sequencing reaction. *Deduce > Solve:* Dideoxy DNA sequencing uses DNA synthesis to generate labeled DNA fragments of different lengths, which are then resolved by gel electrophoresis or column chromatography. To visualize the products of DNA synthesis in traditional dideoxy sequencing, relatively high levels of template were necessary. The use of PCR allows detectable levels of DNA synthesis from much lower levels of template DNA.

28c.   *Evaluate:* This is type 4. Recall that dideoxy nucleotides lack a 3' hydroxyl group. *Deduce:* Elongation of DNA involves phosphodiester bond formation between the 5' phosphate on the incoming nucleotide triphosphate and the 3' hydroxyl group at the end of the DNA strand. *Solve:* Dideoxynucleotides contain a hydrogen group instead of a hydroxyl group on their 3' carbon. When a dideoxynucleotide is incorporated into a growing DNA strand, there is no 3' hydroxyl group present to allow phosphodiester bond formation with the next nucleotide to be added; therefore, no additional nucleotides are added to this DNA strand and its synthesis is terminated.

29.   *Evaluate:* This is type 4. Recall that each of the bands in a lane corresponds to a fragment that terminates in a dideoxynucleotide of that type (ddA, ddT, ddC, or ddG). **TIP:** The sequence reads from 5' to 3' on one strand from the bottom of the gel toward the top. *Deduce > Solve:* The double-stranded DNA sequence and polarity is 5'-TACTGATGCGATGCTAAGC-3'
3'-ATGACTACGCTACGATTCG-5'

30.   *Evaluate:* This is type 4. Recall that there will be a band on the gel corresponding to a fragment terminating in the nucleotide complementary to each nucleotide on the template strand. **TIP:** The gel should have four lanes, one for each fragment that terminates with a different dideoxynucleotide. *Deduce > Solve:* Draw an electrophoresis gel as shown in the figure.

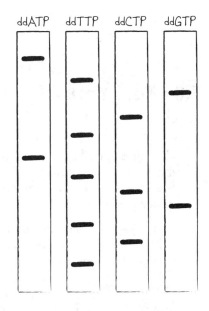

31.   *Evaluate:* This is type 4. This problem challenges you to quantify your understanding of DNA amplification during PCR. *Deduce:* The fold amplification during PCR can be calculated as 2^N, where N is the number of PCR cycles. To calculate the total number of target molecules, multiply the initial number times 2^N. *Solve:* Substitute the number of cycles for N. For $N = 10$, $1 \times 2^{10} = 1024$. After 20 cycles, there will be $1 \times 2^{20} = 1.05 \times 10^6$. After 30 cycles, there will be $1 \times 2^{30} = 1.07 \times 10^9$.

32. *Evaluate:* This is type 3. Recall that increasing the number of replication origins decreases the amount of time required to replicate large genomes. **TIP:** Replication is bidirectional; therefore, the amount of DNA replicated by initiation at one origin is two times the amount of DNA replicated by each replication fork. *Deduce:* In 5 minutes, each replication fork will cover 5 minutes × 40 nucleotides/second × 60 seconds/minute = 1.2×10^4. The DNA replicated from each origin is 2 × DNA from one fork, or 2.4 × 104 bp. *Solve:* Divide the genome size by the length of DNA replicated from initiation at one origin in 5 minutes. This gives $\frac{(1.8 \times 10^8)}{2(1.2 \times 10^4)} =$ 7500 origins of replications.

33. *Evaluate:* This is type 4. This problem challenges you to apply your understanding of the inheritance of DNA markers to PCR. *Deduce:* There are four different alleles of the A locus, which we'll designate 1, 2, 3, and 4, in order of decreasing size (1 is the largest). There are four alleles of the B locus, 1, 2, 3, and 4 (1 is the largest). There are three alleles of the C locus, 1, 2, and 3 (1 is the largest). The colt's genotype is A_1A_1, B_3B_4, C_1C_3. The mare's genotype is A_1A_2, B_1B_3, and C_3C_3. *Solve:* The colt must have received an A_1 allele, a B_4 allele, and a C_1 allele from the sire. Sire 1 does not have an A_1 allele and therefore is ruled out. Sire 3 does not have a C_1 allele and therefore is ruled out. Sire 2 has an A_1, a B_4, and a C_1 allele and therefore cannot be ruled out.

34. *Evaluate:* This is type 4. Recall that there will be a band on the gel corresponding to a fragment terminating in the nucleotide complementary to each nucleotide on the template strand. *Deduce > Solve:* The gel should have four lanes, one for fragments that terminate with each dideoxynucleotide. Draw the gel as shown in the figure.

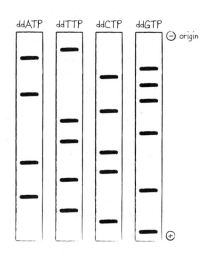

7.04 Test Yourself

Problems

1. How does the size of a eukaryotic chromosome affect the duration of S phase of the cell cycle, the number of replication origins, and the size of replication bubbles? Assume that replication is initiated at all replication origins at the same time and that replication origins are distributed uniformly along the chromosome.

2. You wish to amplify the entire DNA sequence shown below by PCR, but must first design and order PCR primers. Your supervisor wants you to design primers that are at least 20 nucleotides long and have a GC content of at least 50%. What are the sequences of the primers that you should design?

DNA sequence:

5'-ATCGGCGTACGTAAGCCCGTGGCTAGC GTAGCGTCTCTATGTAGCATAGACGGCCA TTATCCGTA-3'-

3' TAGCCGCATGCATTCGGGCACCGAT CGCATCGCAGAGATACATCGTATCTGC CGGTAATAGGCAT-5'

3. Illustrate the DNA sequencing gel that would result from dideoxy sequencing of the DNA sequence below if the primer 5'-TAGCGTAGCGTCTCTATG-3' was used as a sequencing primer.

   5'-ATCGGCGTACGTAAGCCCGTGGCTAGC GTAGCGTCTCTATGTAGCATAGACGGCCA TTATCCGTA-3'-

   3' TAGCCGCATGCATTCGGGCACCGATC GCATCGCAGAGATACATCGTATCTGCC GGTAATAGGCAT-5'

4. Describe the functions of the *E. coli* proteins DnaA, DnaB, and DnaC.

5. The consensus sequence for the 13-nucleotide repeat of oriC is 5'-GATCTATTTATTT-3'. A mutant strain of *E. coli* is isolated that contains the repeated sequence 5'-GATCTACGCATTT-3' at oriC. This strain of *E. coli* grows only at high temperatures, whereas wild-type *E. coli* can grow equally well at high and low temperatures. Provide a plausible explanation for the phenotype of this oriC mutant based on the mechanism for initiation of DNA replication.

6. *E. coli* DNA topoisomerase is required for DNA replication, even though it is not located within replication bubbles. Explain.

7. DNA pol I has two exonuclease activities, 5' to 3' and 3' to 5'. Describe the function of each of these activities, and indicate which activity is required for DNA replication.

8. You have isolated all the components of the *E. coli* replisome for a wild-type strain and a strain that is defective for DNA replication. You find that both replisome complexes are able to synthesize short fragments of DNA at about the same rate; however, the mutant replisome complex is unable to synthesize long strands of DNA. State and defend a hypothesis for which component of the replisome is defective in the mutant.

9. Telomerase is a reverse transcriptase that is composed of protein and RNA. The RNA contains a sequence that is complementary to the G-rich strand of the telomere. Design an experiment that tests the hypothesis that the telomerase RNA is used as the template for synthesis of telomeric DNA, and state the result that would support the hypothesis.

10. A list of five possible sequencing primers is shown below. Rank these primers from highest to lowest optimum annealing temperature.

   a. ATCGGCGTACGTAAGCCCGT
   b. GGCGTACGTAAGCCCGTGGC
   c. CCGTGGCTAGCGTAGCGTCT
   d. GGCTAGCGTAGCGTCTCTAT
   e. GCGTCTCTATGTAGCATAGA

Solutions

1.   *Evaluate:* This is type 3. Recall that replication proceeds in both directions away from each replication origin, creating a replication bubble. *Deduce:* The strategy for replicating large eukaryotic chromosomes in a timely manner is to use multiple replication origins, with each replication origin being responsible for initiating replication of a segment of the chromosome. The maximum size of a replication bubble is determined by the size of the chromosome segment that is replicated from initiation at one origin. *Solve:* Chromosome size and the number of replication origins are generally directly correlated, and larger chromosomes require more origins than smaller chromosomes. Chromosome size and length of S phase would be directly correlated if S-phase length was not fixed; however, S phase is typically fixed for a particular developmental stage of the organism, and the number of replication origins used varies to accommodate the need for a specific S-phase length. Chromosome size and replication bubble size are independent of each other. Instead, replication bubble size is dependent on the spacing of replication origins, and larger bubbles are created by greater distances between origins.

2.   *Evaluate:* This is type 4. Recall that PCR reactions use a pair of primers that are complementary to the 3'-to-5' sequence at opposite ends of the DNA segment to be amplified. *Deduce > Solve:* The sequence 5'-ATCGGCGTACGTAAGCCCGT-3' is 20 nucleotides long with 12/20, or 60%, GC content. The sequence 5'-TACGGATAATGGCCGTCTATGC-3' is 22 nucleotides long with 11/22, or 50%, GC content.

3.   *Evaluate:* This is type 4. Recall that the sequencing primer base-pairs with the template strand and primes DNA synthesis in a 5'-to-3' direction. *Deduce:* The primer will hybridize to the bottom strand as shown:

5'-TAGCGTAGCGTCTCTATG-3'
3'-TAGCCGCATGCATTCGGGCACCGATCGCATCGCAGAGATACATCGTATCTGCCGGTAATAGGCAT-5'

Solve: The order of nucleotides added to the primer during dideoxy sequencing will be T-A-G-C-A-T-A-G-A-C-G-G-C-C-A-T-T-A-T-C-C-G-T-A. Draw a gel similar to the figure shown.

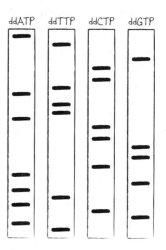

4.   *Evaluate:* This is type 2. Recall that DnaA, DnaB, and DnaC are proteins involved in initiation of DNA replication in *E. coli*. *Deduce > Solve:* DnaA proteins form a complex that binds to the 9-nucleotide-long repeats at oriC. DnaA binding promotes unwinding of the oriC region containing the 13-nucleotide-long repeat sequences. DnaC binds to DnaB and promotes binding of DnaB as a complex to the single-stranded region created by DnaA proteins. DnaB then act as a DNA helicase to unwind the dsDNA at the leading edge of replication forks.

5.   *Evaluate:* This is type 2. Recall that initiation of replication requires unwinding the region of oriC that contains the 13-nucleotide-long repeat sequence. *Deduce:* The mutant sequence contains CGC in place of TTT in the wild-type sequence. GC base pairs require more energy to break than AT base pairs. *Solve:* The mutant oriC region does not unwind at normal growth temperature, but it does unwind better at higher temperatures because the higher temperature facilitates unwinding of the mutant sequence.

6.   *Evaluate:* This is type 2. Recall that topoisomerases catalyze reactions that change the superhelical tension of DNA. *Deduce:* DNA topoisomerase unwinds supercoils that accumulate in DNA due to the action of DnaB helicase, which unwinds DNA at the leading edge of replication forks. In the absence of topoisomerase, supercoils would persist, eventually inhibiting further unwinding of DNA in front of replication forks. *Solve:* This would prevent completion (but not initiation) of DNA synthesis.

7.   *Evaluate:* This is type 2. Recall that 5'-to-3' exonuclease activity refers to removal of nucleotides from a strand of DNA or RNA one at a time, from the 5' end of the molecule, whereas 3'-to-5' exonuclease activity refers to removal of nucleotides from a strand of RNA or DNA one at a time, from the 3' end of the molecule. *Deduce > Solve:* The 5'-to-3' exonuclease activity is used to remove RNA primers during DNA replication. Pol I can then replace the RNA primer with DNA. The 3'-to-5' exonuclease activity is used for proofreading during DNA replication. Proofreading involves removal of an incorrect nucleotide from the 3' end of a growing DNA strand, which allows DNA polymerase another chance to add the correct nucleotide to the

growing chain. The 5′-to-3′ exonuclease activity is required for DNA replication, whereas the 3′-to-5′ exonuclease activity is required for high-fidelity DNA replication.

8.   *Evaluate:* This is type 2. Recall that the ability of polymerases to remain on the template over long periods of time is called processivity. **TIP:** The processivity of DNA polymerase is aided by a sliding clamp mechanism. *Deduce:* The mutant replisome appears to be defective in the processivity of DNA replication. This defect is most likely due to a defect in a protein component of the sliding clamp or the sliding clamp loader. The clamp is required to hold DNA polymerase III on the template strands at replication forks. *Solve:* In the absence of the clamp, DNA pol III will fall off the template; this event would halt DNA synthesis and lead to production of shorter DNA strands.

9.   *Evaluate:* This is type 3. Recall that the template for DNA synthesis determines the sequence of nucleotides in the DNA strand synthesized. *Deduce:* If the telomerase RNA is used as the template for telomeric DNA synthesis, then changing the template portion of the RNA sequence should result in a corresponding change in the telomeric DNA sequence. *Solve:* Mutate the template portion of the telomerase RNA gene and express this mutant telomerase RNA gene in cells. Allow the cells to divide for many generations and then sequence the ends of the chromosomes to determine if the mutant sequence is detected in telomeric DNA.

10.   *Evaluate:* This is type 1 and 4. Recall that the optimum annealing temperature for DNA sequencing and PCR primers of the same length is affected by the percentage of GC, where a higher GC content results in a higher annealing temperature. *Deduce > Solve:* The percentages of GC for these primers are (a) 60%, (b) 70%, (c) 65%, (d) 55%, and (e) 45%. Therefore, the order would be b, c, a, d, and e.

8

Molecular Biology of Transcription and RNA Processing

8.01 Genetics Problem-Solving Toolkit

Key Terms and Concepts

Coding strand: The strand of DNA that is complementary to the template strand and which is identical to the RNA transcript within the coding region, except that it contains T in place of U.

Template strand: The strand of DNA used by RNA polymerase to determine the order of nucleotides of the transcript.

Promoter: The portion of a gene that is bound by RNA polymerase plus transcription factors, and which determines the site of transcription initiation and direction of transcription.

Terminator: The portion of a gene that determines the site where transcription stops.

Band-shift experiment: An experiment based on the observation that protein plus nucleic acid complexes migrate more slowly in gels than "naked" nucleic acids, which determines whether a protein binds to a specific nucleic acid.

DNA footprint experiment: An experiment that localizes the binding site for one or more proteins within a nucleic acid by showing which regions of the nucleic acid are protected from digestion by DNase.

Posttranscriptional RNA modifications: Any change in the structure of an RNA that is not due to transcription by RNA polymerase. Examples from this chapter include 5' capping, intron splicing, polyadenylation, removal of rRNA intervening sequences, and RNA editing.

8.02 Types of Genetics Problems

1. Describe the structure of bacterial and eukaryotic genes.
2. Describe the mechanism of transcription in prokaryotes and eukaryotes.
3. Describe the mechanism of posttranscriptional RNA processing.
4. Predict or interpret the results of experiments analyzing the binding of proteins to nucleic acids.

1. Describe the structure of bacterial and eukaryotic genes.

Problems of this type require you to recall the fundamental components of genes and the functional roles of each gene component. You may be asked to describe the function of a given component or relate a given function to a specific component of a gene. You may also be asked to identify the consensus sequence of a conserved gene element and describe its function. See Problem 12 for an example.

Example Problem: The figure below represents a prokaryotic gene. Assuming that the direction of transcription is right to left, label the template strand, the coding strand, and the locations of the promoter and the terminator relative to the start of transcription (labeled as +1).

Solution Strategies	Solution Steps
Evaluate	
1. Identify the topic this problem addresses, and explain the nature of the requested answer.	1. This problem tests your understanding of prokaryotic gene structure. The answer should identify the coding and template strands and the location of the promoter and terminator of the gene sketch provided.
2. Identify the critical information given in the problem.	2. The top strand of the DNA segment is oriented 5′ to 3′ left to right, and the start of transcription is approximately in the center of the segment. Transcription is from right to left.
Deduce	
3. Consider the relationship between the direction of transcription and the structure of the mRNA.	3. Transcription from right to left means that the mRNA will have its 5′ end on the right and its 3′ end on the left.
4. Relate mRNA polarity to the strands of DNA. **TIP:** mRNAs have the same polarity as the coding strand and are antiparallel to the template strand.	4. The bottom strand of the DNA molecule has the same polarity of the mRNA, and the top strand has the opposite polarity.
5. Relate the location of the promoter and terminator to the direction of transcription and the position of the transcription start site. **TIP:** The promoter will be upstream of the transcription start site, whereas the terminator will be downstream.	5. The promoter will be to the right, and the terminator will be to the left.

(continued)

Solve

6. Label the strand identity and the promoter and terminator locations on the figure given in the problem.

6. **Answer:** The bottom strand is the coding strand, and the top strand is the template strand. The promoter is located to the right of the +1 transcription start site, whereas the terminator will be located to the left of it.

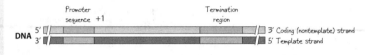

2. Describe the mechanism of transcription in prokaryotes and eukaryotes.

This type of problem tests your understanding of the molecular mechanisms of transcription. You may be asked about mechanisms of transcription initiation, elongation, or termination. You may be asked to describe the steps of each process, the identity and function of the proteins involved, or the identity and location of DNA sequences to which those proteins bind. See Problem 3 for an example.

Example Problem: The following is a list of proteins involved in transcription: RNA polymerase II, sigma factor, rho, TFIID, TFIIF, and RNA polymerase core enzyme. Indicate which proteins are involved in bacterial transcription and which are involved in eukaryotic transcription. For the eukaryotic transcription-related proteins, indicate the order in which they function during the transcription of a gene.

Solution Strategies	Solution Steps
Evaluate	
1. Identify the topic this problem addresses, and explain the nature of the requested answer.	1. This problem tests your understanding of the molecular mechanism of transcription. The answer will be a list of prokaryotic vs. eukaryotic transcription factors, and the eukaryotic factors will be ordered by function.
2. Identify the critical information given in the problem.	2. You were given this list: RNA polymerase II, sigma factor, rho, TFIID, TFIIF, and RNA polymerase core enzyme.
3. Identify the bacterial transcription factors.	3. The bacterial transcription factors are RNA polymerase core enzyme, sigma factor, and rho protein.
Deduce	
4. Recall the function of the remaining proteins, which are eukaryotic transcription factors.	4. RNA polymerase II is a protein complex that uses ribonucleotide triphosphates as substrates and a strand of DNA as the template to synthesize RNA. TFIID is the first general transcription factor to bind to the promoter. TFIIF is a general transcription factor that binds to the promoter along with RNA polymerase II.
Solve	
5. List the proteins in order of the events occurring during transcription that each protein is responsible for.	5. **Answer:** TFIID, TFIIF, and RNA polymerase II

3. Describe the mechanism of posttranscriptional RNA processing.

This type of problem tests your understanding of the molecular mechanisms of the posttranscriptional modification of RNA. You may be asked to describe the major RNA processing events or explain the events in mechanistic detail. You may be asked to describe the importance of RNA processing events in gene expression. You may also be asked to describe the structural changes that are made to RNAs during modification. See Problem 24 for an example.

Example Problem: Consider the statement, "RNA polymerase II mRNAs include sequences that are not encoded in DNA." Is this true? If so, explain.

Solution Strategies	Solution Steps
Evaluate	
1. Identify the topic this problem addresses, and explain the nature of the requested answer.	1. This problem tests your understanding of how the structure of RNA polymerase II mRNAs relates to the structure of the genes that encode them, and of the mechanism underlying the discrepancy. The answer will be yes or no; if it is yes, explain how the statement can be true.
2. Identify the critical information given in the problem.	2. The problem states that RNA polymerase II mRNAs contain sequences that are not encoded by DNA.
Deduce	
3. Recall the mechanism of transcription and pre-mRNA processing of RNA polymerase II mRNAs.	3. RNA polymerase II synthesizes primary transcripts that are complementary to the template strand of the gene. These transcripts are modified in three major ways: 5′ capping, intron splicing, and polyadenylation.
4. Consider which of the processes involving RNA polymerase II synthesis could introduce sequences that are not encoded in DNA.	4. Transcription and intron splicing do not introduce nucleotides into mRNA. 5′ capping adds a G nucleotide to the 5′ end, and polyadenylation adds 20 to 200 adenosine nucleotides to the 3′ end.
Solve	
5. Answer the problem by indicating whether the answer is true; if so, state the processes involved.	5. **Answer:** Yes, RNA polymerase II mRNAs contain a G nucleotide at their 5′ end and a stretch of 20 to 200 A nucleotides at their 3′ ends, all of which are added posttranscriptionally.

4. Predict or interpret the results of experiments analyzing the binding of proteins to nucleic acids.

This type of problem tests your understanding of DNA footprinting or band shift assays, both of which detect specific interactions between proteins and nucleic acids. You may be asked to describe the molecular basis of the experimental approach, predict the results of an experiment, or design an experiment to test for protein–nucleic acid binding. See Problem 19 for an example.

Example Problem: The figure shows RNA polymerase holoenzyme bound to a bacterial promoter on a 1000-bp fragment of DNA. It also shows the results of a DNA footprinting assay performed on this DNA fragment in the absence of RNA polymerase. Use this figure to predict the results of a DNA footprint performed on the same fragment in the presence of RNA polymerase holoenzyme.

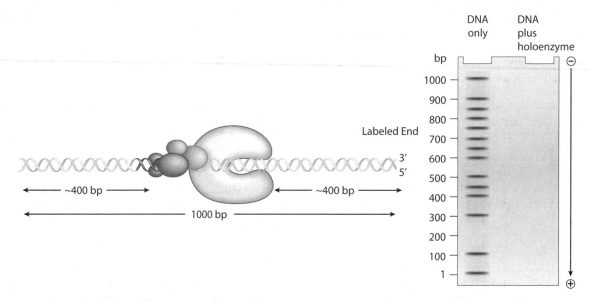

Solution Strategies	Solution Steps
Evaluate	
1. Identify the topic this problem addresses, and explain the nature of the requested answer.	1. This problem tests your understanding of DNA footprinting. The answer will involve drawing the appropriate bands on a gel corresponding to the result of DNA footprinting.
2. Identify the critical information given in the problem.	2. The location of RNA polymerase holoenzyme on a 1000-bp fragment of DNA is shown, along with the banding pattern that resulted from a DNA footprint experiment performed on the DNA fragment alone.
Deduce	
3. Determine the distance between the labeled DNA end and the beginning of the region bound by RNA polymerase holoenzyme.	3. The distance between the labeled DNA end and the beginning of the region bound by holoenzyme is 400 bp.
4. Determine the distance between the labeled DNA end and the end of the region bound by RNA polymerase holoenzyme. **TIP:** This distance corresponds to the size of the fragment minus the distance from the other end of the DNA to the holoenzyme complex.	4. The distance from the labeled DNA end to the end of the region bound by holoenzyme is 1000 − 400 = 600 bp.

(continued)

| 5. Determine which bands in the DNA-only footprint correspond to DNase cutting sites that are within the region bound by holoenzyme. | 5. Holoenzyme occupies from 400 to 600 base pairs in from the labeled end. DNase will not cut in this region; therefore, there will be no fragments between 400 and 600 base pairs in length. |

Solve

| 6. Draw all the bands into the "DNA plus holoenzyme" lane that are in the "DNA only" lane except for those deduced to be absent. | 6. **Answer:** See figure. |

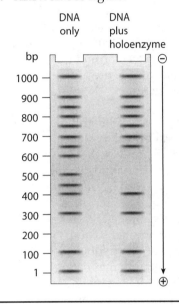

8.03 Solutions to End-of-Chapter Problems

1a. *Evaluate > Deduce > Solve:* This is type 1. With respect to the central dogma of molecular biology, a gene is a segment of DNA that contains the information for transcription of an mRNA, which codes for translation of a protein.

1b. *Evaluate > Deduce > Solve:* This is type 1. The central dogma's concept of a gene can also be stated as a segment of DNA that contains the information for the production of a gene product. Genes for rRNA and tRNA are genes in this context because the RNAs are the gene products.

2. *Evaluate > Deduce > Solve:* This is type 3. The three major modifications of mRNA are 5′ capping, intron splicing, and 3′ polyadenylation. 5′ capping refers to addition of a guanosine monophosphate by guanylyl transferase to the 5′ end of a pre-mRNA via a 5′-to-5′ triphosphate linkage and the subsequent methylation of the guanine and sometimes additional nucleotides on the pre-mRNA. Intron splicing refers to the removal of an intron from the pre-mRNA and the joining of adjacent exons by the spliceosome. 3′ polyadenylation refers to the cleavage of the pre-mRNA downstream of the polyadenylation sequence by cleavage factors and addition of 20 to 200 adenine nucleotides by polyadenylated polymerase.

3a. *Evaluate > Deduce > Solve:* This is type 2. Promoters are DNA sequences that determine the site and frequency of initiation of transcription.

3b. *Evaluate > Deduce > Solve:* This is type 2. Consensus sequences for the most common bacterial promoters often include the –35 sequence TTGACA and the –10 sequence TATAAT.

3c. *Evaluate > Deduce > Solve:* This is type 2. The consensus sequences in the β-globin promoter were the GC-rich region (GCCACACCC), the CAAT box (GGCCAAT), and the TATA box (TATAA).

3d. *Evaluate > Deduce > Solve:* This is type 2. The statement that eukaryotic promoters are more complex than prokaryotic promoters refers to the fact that there are many more different types of eukaryotic promoter sequences than are found in a prokaryote. This is true because there are multiple RNA polymerase in eukaryotes, each of which binds to a different consensus sequence. Also, there are multiple transcription factors specific to each type of polymerase, and these bind to different promoter sequences. Finally, nearby sequences in addition to the promoter consensus sequences are thought to be important in modulating the level (rate) of transcription initiation.

3e. *Evaluate > Deduce > Solve:* This is type 2. The term *alternative promoter* refers to sequences located either upstream or downstream of a "known" promoter that are used to direct transcription initiation at a different site in the gene. The consequence of this is synthesis of an mRNA that has a different 5' end than that seen when the known promoter is used. The alternative promoters may be used under different conditions in a given cell type or may be used in different cell types.

4a-c. *Evaluate:* This is type 1. Recall that the template strand is transcribed by RNA polymerase to produce an RNA that is identical in sequence to the coding strand (except that U is in place of T). *Deduce > Solve:* **(a)** The direction of transcription in the figure is from right to left. Bacterial promoter elements are located 10 and 35 nucleotides upstream (to the right of +1 in this figure) of the transcription initiation site. **(b)** The direction of transcription in the figure is from right to left. Eukaryotic promoter consensus sequences are located 90, 80, and 25 nucleotides upstream of the transcription initiation site. **(c)** The direction of transcription in the figure is from right to left. RNA polymerase III promoters are located downstream of the transcription initiation site (to the left of +1 in this figure), within the transcribed region of the gene.

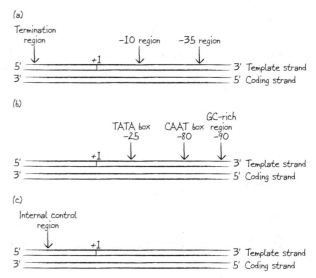

5a. *Evaluate:* This is type 1. Recall that mRNAs are transcribed in a 5'-to-3' direction. *Deduce > Solve:* The 5' end of the mRNA is on the right; therefore, the A at the left end was the last nucleotide added.

5b. *Evaluate > Deduce > Solve:* This is type 1. Recall that the template strand is complementary to the mRNA, and the coding strand has the same sequence as the mRNA except that T is in place of U. The template strand is 5'-TAGCAGTACGTCT-3', and the coding strand is 3'-ATCGTCATGCAGA-5'.

5c. *Evaluate:* This is type 1. Recall that most mRNAs in eukaryotes and all mRNAs in prokaryotes are transcribed by RNA polymerases that bind to promoter sequences located 5' to the site of transcription initiation. *Deduce > Solve:* The promoter region will be located to the right of the sequence (upstream of the 5' A).

6. *Evaluate > Deduce > Solve:* This is type 2. DNA and RNA polymerases are similar in that both (1) catalyze phosphodiester bond formation to polymerize nucleotides into nucleic acids, (2) polymerize in a 5'-to-3' direction, and (3) are dependent on a DNA sequence template. DNA and RNA polymerases differ in that (1) RNA polymerase can initiate strand synthesis, whereas DNA polymerase can only extend an existing strand; (2) most DNA polymerases can proofread using a 3'-to-5' exonuclease activity, whereas RNA polymerases cannot; and (3) DNA polymerases used deoxyribonucleotide triphosphates as substrates, whereas RNA polymerases use ribonucleotide triphosphates as substrates.

7a. *Evaluate:* This is type 1. Recall that the Pribnow box consensus sequence is TATAAT.
Deduce > Solve: Individual Pribnow box sequences are circled in the figure.

Gene 1 ...TTCCGGCTCG(TATGTT)GTGTGGA....
Gene 2 ...CGTCATTTGA(TATGAT)GCGCCCCG...
Gene 3 ...CCACTGGCGG(TGATAC)TGAGCACA...
Gene 4 ...TTTATTGCAG(TATAAT)CTGGTTACA..
Gene 5 ...TGCTTCTGAC(TATAAT)AGACAGGG...
Gene 6 ...AAGTAAACAC(TACGAT)GGGTACCACA.

7b. *Evaluate > Deduce:* This is type 1. Recall that the consensus sequence contains the most common nucleotide at each position. **TIP:** Align the sequences such that the individual Pribnow box sequences are in register *Solve:* The consensus sequence is TATGAT.

8. *Evaluate > Deduce > Solve:* This is type 3. The primary transcripts of bacterial and eukaryotic genes differ in that bacterial transcripts often contain more than one coding sequence (they are polycistronic), whereas eukaryotic transcripts do not. Polycistronic mRNAs allow for coordinate regulation of production of several proteins by controlling initiation of transcription of only one gene. Eukaryotes accomplish this by coordinate regulation of transcription of multiple genes by gene-specific transcription factors. Prokaryotic and eukaryotic primary transcripts differ in that eukaryotic transcripts are extensively modified before translation, whereas prokaryotic transcripts are not. The modification of eukaryotic transcripts includes 5' capping and 3' polyadenylation, which generate structures that are critical for regulation of the initiation of translation and for controlling the half-life of the mRNA. Mechanisms controlling translation initiation and mRNA half-life in bacteria do not involve these structures. The modification of eukaryotic transcripts also includes intron splicing, which is required for generating the complete open reading frame used in translation and allows for the generation of multiple, different (but related) mRNAs from a single primary transcript. This last mechanism increases the number of different proteins that are coded by a genome without increasing the number of genes present.

9. *Evaluate > Deduce > Solve:* This is type 2. The two types of transcription termination in bacteria are *intrinsic termination*, which is dependent only on specific DNA sequences at the 3' end of genes, and *rho-dependent termination*, which requires specific DNA sequences and the rho protein. In intrinsic termination, transcription of an inverted repeat sequence at the 3' end of genes results in an mRNA that folds into a hairpin structure. RNA polymerase pauses during transcription of the poly-A sequence that immediately follows the inverted repeat, which allows the polymerase to physically interact with the RNA hairpin. The combination of pausing and interaction with the hairpin RNA causes polymerase to release the mRNA and the gene, terminating transcription. For rho-dependent termination, transcription of inverted repeats also generates hairpin RNA structures; however, the sequences following hairpin RNA structures constitute a binding site for rho, called a rho utilization site (rut). Rho binds the rut site in the transcript and uses ATP hydrolysis to move along the RNA and melt the RNA-DNA hybrid. Disruption of base pairing between the RNA and the template DNA induces polymerase to release the transcript and the template, terminating transcription. Transcription termination in eukaryotes differs in that the 3' end of the RNA is not formed by release of the RNA from polymerase, but rather is caused by cleavage of the primary transcript (while polymerase continues to transcribe DNA downstream) and the addition of poly-A to the 3' end formed by cleavage.

10. *Evaluate > Deduce > Solve:* This is type 2. Recall that enhancers are DNA sequences that increase the level (rate) of transcription of genes in a position- and orientation-independent manner. Enhancers are binding sites for transcription factors that stimulate transcription of one or more genes. Since the expression of the transcription factors is often specific to the cell type or cell tissue, enhancers often provide for a mechanism to stimulate transcription of genes in a cell-type-specific or tissue-specific manner. Possible rationales for the lack of enhancers in bacteria include these: most bacteria lack differentiated cell types; there is little to no intergenic space on bacterial chromosomes, which makes long-range-acting enhancer sequences unnecessary; and bacterial operons make coordinate regulation of protein synthesis by enhancers unnecessary.

11.   *Evaluate > Deduce > Solve:* This is type 1. Introns are DNA sequences within genes that are transcribed and then removed from the RNA after transcription, whereas spacer sequences are non-transcribed DNA sequences that lie between genes.

12a.   *Evaluate > Deduce > Solve:* This is type 1. As shown in the figure, a bacterial promoter is a relatively short sequence with two consensus elements, both located close to the transcription initiation site. This contrasts to RNA polymerase I and II promoters in that those promoters have more complex sequences spanning 100 or more nucleotides, contain three or more consensus sequences that serve as binding sites for transcription factors, and have interspersed, non-consensus sequences that affect the efficiency of transcription.

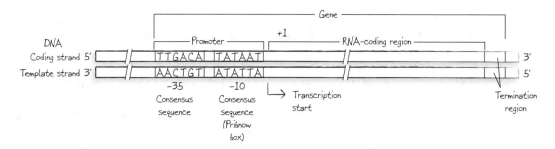

13.   *Evaluate > Deduce > Solve:* This is type 3. SR proteins bind to exonic splicing enhancers within the exons of pre-mRNAs and increase the likelihood that 5′ and 3′ splice junctions are recognized by spliceosomal components. Thus, binding of SR proteins to a potential exon in a pre-mRNA makes it more likely that the exon will be included in the mRNA. Expression of cell-type-specific SR proteins is thought to regulate cell-type-specific inclusion of exons in mRNAs for *alternatively spliced* pre-mRNAs. Alternative splicing occurs when identical pre-mRNAs in different cells produce mRNAs with different combinations of exons, ultimately producing different polypeptides. For example, if a particular exon is meant to be included in the mRNA of a gene in one cell type but not another, then the SR proteins that bind to ESEs in that exon will be expressed in the former cell type but not the latter.

14.   *Evaluate > Deduce > Solve:*
This is type 1. See figure.

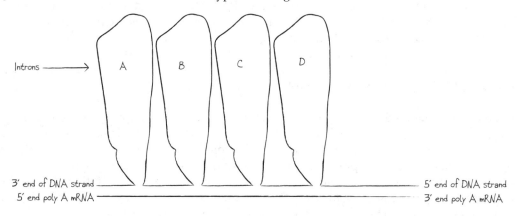

15a-b.   *Evaluate > Deduce > Solve:* This is type 1. See figure.

15c.   *Evaluate > Deduce > Solve:* This is type 1. Recall that introns are present in genes and pre-mRNAs, but not in mRNAs. The intron loops are single stranded because they are present in the DNA but not the mRNA. The only double-stranded regions are those that include exons, which are present in the DNA *and* the mRNA.

16a-b. *Evaluate > Deduce > Solve:*
This is type 2. Recall that
for intrinsic termination,
the mRNA includes an
inverted repeat in the
template strand that is
located just before a poly-A
sequence. Also recall that
this region of the mRNA
forms the stem of a stem-
loop structure. Draw figure
as shown here.

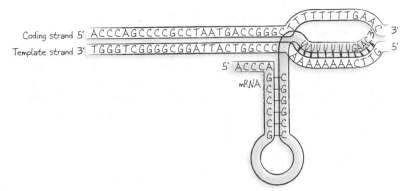

16c.   *Evaluate > Deduce > Solve:* This is type 2. RNA polymerase pauses during transcription
of the poly-A sequence that immediately follows the inverted repeat, which allows the
polymerase to physically interact with the RNA hairpin. The combination of pausing and
interaction with the hairpin RNA causes polymerase to release the mRNA and the gene,
terminating transcription.

17a.   *Evaluate:* This is type 4. This problem requires application of the band shift
assay to analyze potential DNA–protein complexes that might form under
the conditions described in the problem. Recall that DNA bound by protein
migrates more slowly than naked DNA. *Deduce > Solve:* RNA polymerase core
enzyme cannot bind to the gene. The RNA polymerase holoenzyme can bind to
and initiate transcription to form a DNA–protein complex. Rho binds to the rho
utilization sequence on mRNA and therefore cannot bind to the gene.

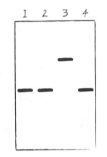

17b.   *Evaluate > Deduce > Solve:* This is type 4. Lane 1 shows the position that the naked DNA
fragment migrates to, and lane 3 shows the position of DNA migration in the presence of RNA
polymerase holoenzyme. RNA polymerase holoenzyme can bind to the DNA fragment, forming
a DNA–protein complex. RNA polymerase is large, and the DNA–RNA polymerase complex
migrates much more slowly than naked DNA.

17c.   *Evaluate > Deduce > Solve:* This is type 4. Lane 1 shows the position that the naked DNA
fragment migrates to, and lane 4 shows the position of DNA migration in the presence of rho
protein. Rho can bind to the mRNA but not to the gene; therefore, there are no DNA–protein
complexes in lanes 1 or 4, and both lanes show the migration of naked DNA.

18.   *Evaluate:* This is type 4. This problem requires application of your understanding of the band
shift assay to match each condition listed to the result shown on a gel. *Deduce:* The bands
in lanes 2 and 4 have migrated the most rapidly; therefore, they correspond to naked DNA
molecules. Lanes 1, 3, and 5 have migrated more slowly than naked DNA; therefore, they are
DNA–protein complexes. Lane 1 showed slightly higher mobility than lane 5, which showed
higher mobility than lane 3. Higher mobility indicates less protein is bound to the DNA.
Solve: Conditions 18c and 18d should result in naked DNA because 18c contains DNA only
and 18d contains DNA plus RNA pol II, which cannot bind to DNA in the absence of general
transcription factors. Therefore, 18c and 18d correspond to lanes 2 and 4 (either lane is equally
possible for either condition). Condition 18e contains the lowest number of transcription factors,
followed by condition 18a, and condition 18b has the most transcription factors. Thus, lane 1
corresponds to condition 18e, lane 5 corresponds to condition 18a, and lane 3 corresponds to
condition 18b.

19a.   *Evaluate:* This is type 4. This problem challenges you to interpret the results of a DNA footprinting experiment. Recall that DNA footprints correspond to sites where bound protein protects an end-labeled DNA fragment from digestion by the endonuclease, DNase. **TIP:** The region protected by bound protein is identified by the size of the bands that are present in the DNA-only control lane which are missing in the DNA + protein experimental lane. The size of the bands missing in the DNA + protein sample indicates the distance of the protected sites from the end of DNA that is labeled. *Deduce:* The size range of the missing bands is 500 to 800 nucleotides. *Solve:* Subtracting the lowest from the highest range, gives 300 nucleotides.

19b.   *Evaluate > Deduce > Solve:* This is type 4. Draw a figure similar to the one shown here.

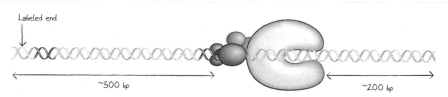

Labeled end

~500 bp   ~200 bp

19c.   *Evaluate > Deduce > Solve:* This is type 4. DNase is an endonuclease that cuts DNA randomly (it cuts in a sequence-independent manner). DNase will cut the DNA molecule wherever it can bind to it, but it cannot cut the DNA molecule in regions where other proteins are bound.

20a.   *Evaluate > Deduce > Solve:* This is type 2. The organism transcribes as a wild type.

20b.   *Evaluate > Deduce > Solve:* This is type 2. The organism transcribes slowly (i.e., is leaky).

20c.   *Evaluate > Deduce > Solve:* This is type 2. The organism does not transcribe genes.

20d.   *Evaluate > Deduce > Solve:* This is type 2. Temperature-sensitive mutant

21a.   *Evaluate:* This is type 2. Recall that rho is required for termination of transcription of only some genes. *Deduce > Solve:* At 37 degrees, all of the mRNAs normally present in *Salmonella* cells would be present in this mutant. At 40 degrees, those mRNAs that are coded by genes that use intrinsic termination would be normal; however, those that use rho-dependent termination would likely be longer than normal. Some or all of these transcripts would terminate downstream of their normal termination site.

21b.   *Evaluate > Deduce > Solve:* This is type 2. Rho function is critical only for proper termination of transcription of those mRNAs encoded by genes that use rho-dependent transcription termination.

22a.   *Evaluate:* This is type 3. Recall that AG is the consensus sequence found at the 3′ end of introns. *Deduce > Solve:* Since this mutation is in an intron but causes a defect in β-globin, it must affect splicing efficiency. The mutation replaces A with U, changing the 3′ splice site sequence AG to UG. This change is likely to affect the efficiency with which the spliceosome recognizes the end of intron 2; and it either leads to inclusion of intron 2 in the mRNA, which results in an insertion or premature termination, or causes a change in the location of the 3′ splice junction—which leads to an insertion, deletion, or frameshift mutation.

22b.   *Evaluate > Deduce > Solve:* This is type 1. This problem requires you to consider the structure of genes and identify important DNA sequences that are not part of exons. Non-exon-located mutations that could prevent gene function include mutations in the promoter or terminator sequences as well as in enhancer or silencer sequences. Mutations in the promoter would diminish or prevent transcription, which would reduce or eliminate the mRNA. Mutations in the terminator could prevent or alter termination, which would elongate the mRNA. Mutations in an enhancer would diminish transcription, which would reduce mRNA abundance. Mutations in the silencer would enhance transcription, which would increase mRNA abundance.

23a.   *Evaluate > Deduce > Solve:* This is types 1 and 3. Two features of bacterial gene expression make coupled transcription and translation possible. First, there are no membranes separating the bacterial chromosome from bacterial ribosomes; therefore, as soon as an mRNA is transcribed, ribosomal subunits have physical access to it. Second, bacterial transcripts do not contain introns; therefore, the primary transcript contains the complete open reading frame, and bacterial ribosomes can translate the mRNA without needing to wait for pre-mRNA splicing.

23b. *Evaluate > Deduce > Solve:* This is types 1 and 3. Coupled transcription and translation are not possible for single-celled eukaryotes such as yeast genes, for two reasons. First, yeast chromosomes are separated from ribosomes by the nuclear envelope. Thus, mRNAs cannot be translated until transcription is completed and the mRNA is exported to the cytoplasm. Second, some yeast genes (although they are in the minority) contain introns, which make pre-mRNA splicing necessary before useful translation can occur.

24a. *Evaluate > Deduce > Solve:* This is types 1, 2, and 3. First, the eukaryotic promoter is unlikely to be recognized by bacterial RNA polymerase holoenzyme. Second, the introns will not be removed from the pre-mRNA, which will result in production of an abnormal protein. Third, sequences required for efficient translation initiation in bacteria are not present.

24b. *Evaluate > Deduce > Solve:* This is type 1, 2, and 3. First, I would make a cDNA copy of the gene. The cDNA is a DNA copy of the mRNA sequence, which lacks introns. Second, I would place the cDNA sequence downstream of a known bacterial promoter, which will ensure that the gene is transcribed. Third, I would modify the coding sequence upstream of the ATG start codon to contain a Shine–Dalgarno sequence, which is important for proper initiation of translation. Fourth, I would place an intrinsic or rho-dependent termination sequence downstream of the cDNA to ensure efficient transcription termination.

25a. *Evaluate:* This is type 2. Recall that the template strand is the strand RNA polymerase binds and uses as template for transcription. *Deduce > Solve:* The bottom strand is the template strand. Add the RNA polymerase to the figure as shown here.

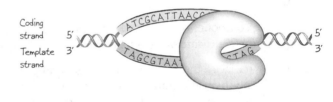

25b. *Evaluate > Deduce > Solve:* This is type 2. Recall that RNA polymerase moves along the template strand in a 3′-to-5′ direction. Therefore, it will move from left to right.

25c. *Evaluate > Deduce > Solve:* This is type 2. The RNA sequence will be the same as the DNA coding strand sequence except that U will be present in place of T. The sequence, then, is 5′-AUCGCAUUAACGAUCGAUC-3′.

25d. *Evaluate:* This is type 1. Recall that RNA polymerase moves away from the promoter during transcription. *Deduce:* RNA polymerase is moving left to right. *Solve:* Therefore, the promoter is located to the left of the figure.

26a. *Evaluate > Deduce > Solve:* This is type 4. This problem challenges you to interpret the results of a DNA footprinting experiment. Recall that DNA footprints correspond to sites where bound protein protects an end-labeled DNA fragment from digestion by the endonuclease, DNase. *Deduce:* The DNA-only and DNA + protein lanes differ because bands are missing from the DNA + protein lanes. This result indicates that the proteins are bound to the DNA. Since the proteins are transcription factors and RNA polymerase, which bind to promoters, it is reasonable to conclude that the DNA fragment contains a promoter sequence.

26b. *Evaluate:* This is type 4. This problem challenges you to interpret the results of a DNA footprinting experiment. Recall that DNA footprints correspond to sites where bound protein protects an end-labeled DNA fragment from digestion by the endonuclease, DNase. **TIP:** The region protected by bound protein is identified by the size of the bands that are present in the

DNA-only control lane, which are missing in the DNA + protein experimental lane. The size of the bands missing in the DNA + protein sample indicates the distance of the protected sites from the end of DNA that is labeled. *Deduce > Solve:* The bands from 100 to 240 nucleotides are absent. *Solve:* Therefore, 140 base pairs are protected.

26c.   *Evaluate:* This is type 4. This problem challenges you to apply your understanding of promoter structure and function to design experiments to test a DNA fragment for promoter function. **TIP:** The function of a promoter is to provide all the DNA sequences required for binding of transcription factors and RNA polymerase and for start of transcription. *Deduce > Solve:* One reasonable experiment would be to clone this DNA sequence upstream of the coding sequence for a protein whose expression is easy to assay and then introduce that chimeric construct into cells and assay for protein expression. If the result is negative, then the orientation of the fragment should be inverted to check that it was not inserted backward in the first attempt. Also, a known, control promoter should be used to confirm that the protein-coding sequence is correct and that the protein can be detected in the cells used.

8.04 Test Yourself

Problems

1. A single eukaryotic gene codes for two related 1000-amino-acid-long proteins, which differ only in a central domain of 100 amino acids. (a) How is this possible? (b) A mutation in the gene prevents production of one of the proteins but does not affect production of the other. What is the nature of the mutation?

2. Which components of the spliceosome are involved in recognition of the conserved sequences found in introns? Indicate the sequence that each spliceosomal component recognizes.

3. If you replaced the binding sites for general transcription factors for RNA polymerase III with RNA polymerase II transcription factor binding sites, would this convert an RNA polymerase III transcribed gene into an RNA polymerase II transcribed gene?

4. What is the functional significance of the ability of some enhancer-binding proteins to bend DNA?

5. Order the following consensus sequence elements from the 5'-to-3' end of a pre-mRNA: 3' UTR, coding sequence, start codon, U-rich region, 5' Cap, polyadenylation signal, cleavage site, 5' UTR.

6. What is CTD, and what is its role in gene expression?

7. Compare and contrast the processing of the 45S rRNA precursor to pre-mRNA splicing.

8. Is polyadenylation the only instance in which an mRNA contains a sequence that is not encoded by DNA? Explain.

9. A protein is capable of causing a shift in the electrophoretic mobility of a 500-bp fragment in a band shift assay for protein–DNA interactions. If you assume that the protein binds to only 100 bp of this sequence, how could you use modifications of the band shift assay to more precisely determine the binding site of this protein? What other biochemical experiment could you use to determine the binding site for this protein?

Solutions

1a.   *Evaluate > Deduce > Solve:* This is type 3. Recall that alternative pre-mRNA splicing provides a mechanism by which one gene can encode more than one protein. The pre-mRNA from this gene contains alternate exons that code for the 100-amino-acid central domain of this protein. Inclusion of one alternative exon set results in an mRNA that codes for one of the proteins, whereas inclusion of the other exon set results in an mRNA that codes for the second protein.

1b.   *Evaluate > Deduce > Solve:* This is type 3. Recall that mutation in sequences called ESEs within the exon can prevent recognition of the exon as an exon. *Deduce:* The mutation could be in the ESE for the exon for one of the proteins. This mutation would only prevent that exon from being included in the mRNA and would not affect the alternate mRNA structure.

2.   *Evaluate > Deduce > Solve:* This is type 3. The snRNPs U1 and U2 recognize the 5' splice site and the lariat branch point, respectively.

3.   *Evaluate:* This is type 2. Recall that the promoter for RNA polymerase III genes is downstream of the transcription start site, whereas the promoter for RNA polymerase II genes is upstream. *Deduce > Solve:* Replacing the TFIII binding sites with TFII binding sites would probably achieve binding of TFIIs to the former TFIII gene; however, even if these factors recruited RNA pol II to that gene, it would initiate transcription too far downstream from the proper transcription start site for the gene. Thus, the proposed strategy would not successfully convert the RNA pol III gene into a RNA pol II gene.

4.   *Evaluate > Deduce > Solve:* This is type 2. The ability of enhancer-binding proteins to bend DNA is thought to be important in bridging the distance between the enhancer and the promoter element. In this way, transcription factors bound to the enhancer can influence formation of a complete initiation complex at a distant gene's promoter.

5.   *Evaluate:* This is type 1. Refer to Figure 8.18. *Deduce > Solve:* 5' Cap, 5' UTR, start codon, coding sequence, 3' UTR, polyadenylation signal, cleavage site, U-rich region.

6.   *Evaluate > Deduce > Solve:* This is type 2 and 3. CTD stands for the C-terminal domain of RNA polymerase II. This domain is heavily phosphorylated at initiation of transcription and serves as the docking site for factors involved in pre-mRNA processing, including the capping proteins, splicing factors, and polyadenylation factors.

7.   *Evaluate > Deduce > Solve:* This is type 3. The 45S rRNA precursor and pre-mRNA both contain sequences that are not present in the final, fully processed RNAs. In both cases, internal sequences are removed (ITSs in the 45S rRNA and introns in pre-mRNA); however, the 5' end of the 45S rRNA is removed (the ETS), whereas only internal sequences are removed from pre-mRNA. These processing events also differ in that the pre-mRNA sequences that are retained are spliced together into a single mature mRNA molecule whereas the retained rRNAs, instead, form separate mature rRNAs. Another cell biological difference is that 45S rRNA is processed in the nucleolus, whereas pre-mRNA splicing occurs in many locations throughout the nucleus (these are called splicing islands).

8.   *Evaluate > Deduce > Solve:* This is type 3. No, RNA editing is a process by which nucleotides are also changed, added, or removed posttranscriptionally.

9.   *Evaluate:* This is type 4. This problem tests your ability to apply experimental approaches introduced in the chapter to determine the binding site of a protein within a DNA molecule. *Deduce > Solve:* Band shift assays determine whether a protein can bind to a given DNA molecule, but they do not indicate where the protein is bound in the molecule. Nevertheless, using 100-bp subfragments of the original 500-bp fragment could accomplish this task. If a 100-bp subfragment is shifted in the band shift assay by the protein, then it contains the binding site. A DNA footprinting assay would be an alternative method for localizing the protein-binding site within the 500-bp fragment.

9

The Molecular Biology of Translation

Section 9.01 Genetics Problem-Solving Toolkit

Section 9.02 Types of Genetics Problems

Section 9.03 Solutions to End-of-Chapter Problems

Section 9.04 Test Yourself

9.01 Genetics Problem-Solving Toolkit

Key Researchers

Shine and Dalgarno: They identified a conserved bacterial mRNA sequence that is used to identify the AUG start codon by base pairing with a segment of the 16S rRNA.

Marylin Kozak: She deduced a eukaryotic mRNA consensus sequence that helps identify the AUG start codon.

Sidney Brenner: He proposed that the genetic code would be nonoverlapping because an overlapping code would place restrictions on the number of different amino acid sequences that would be possible.

Fraenkel-Conrat: They provided experimental evidence that the code was nonoverlapping. He showed that single-nucleotide substitutions in the TMV coding sequence changed only one amino acid in the corresponding protein sequence.

Crick, Barnett, Brenner, and Watts-Tobin: They provided experimental evidence for the triplet nature of the genetic code. They isolated single-nucleotide insertion and single-nucleotide deletion mutants in a bacterial virus gene and showed that combinations of three insertions or three deletions restored gene function.

Nirenberg and Matthaei: They developed an approach for determining the identity of triplet codons. They synthesized mRNA molecules containing strings of a single nucleotide and then analyzed the polypeptides produced by in vitro translation of those mRNAs.

Har Khorana: He provided experimental evidence for the identity of triplet codons. He synthesized mRNA molecules that contained di-, tri-, and tetranucleotide repeats and then analyzed the polypeptides produced by in vitro translation of those mRNAs.

Nirenberg and Leder: They provided experimental evidence for the identity of triplet codons. They synthesized RNAs containing a single, three-nucleotide codon. They used these RNAs to purify aminoacyl-tRNAs and determined which amino acid corresponded to each three-nucleotide codon.

Francois Chapeville: He provided experimental evidence that the tRNA–mRNA interaction determined the specificity of the genetic code. He chemically converted Cys-tRNA^{Cys} to Ala-tRNA^{Cys} and showed that this modified tRNA directed the incorporation of Ala into polypeptides in place of Cys.

Gunther Blobel: He proposed the signal hypothesis to explain the mechanism by which proteins are specifically targeted to the endoplasmic reticulum (ER). In the signal hypothesis, the terminal amino acids of ER proteins serve as a signal sequence, which is required for transport into the ER and then cleaved from the protein.

Cesar Milstein: He provided experimental evidence for the signal hypothesis. He demonstrated that secretory proteins translated in vitro in the absence of the ER were about 20 amino acids longer than those secreted from cells. He showed that providing ER in vitro resulted in production of proteins of normal length.

Key Analytical Tools

Eukaryotic Gene Structure

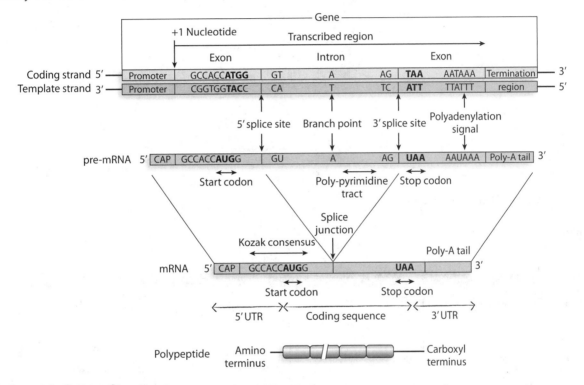

Prokaryotic Operon Structure

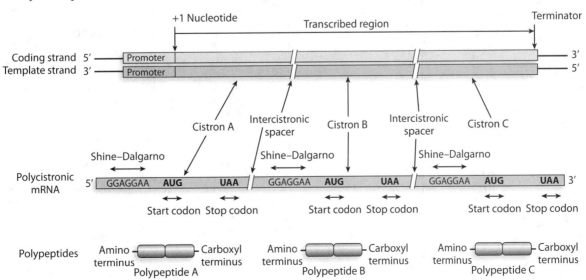

Translation Elongation

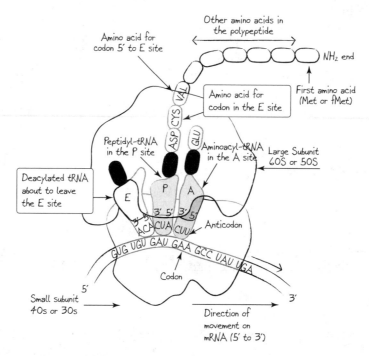

Key Genetic Relationships

Genetic Code

		Second position			
		U	**C**	**A**	**G**
First position (5' end)	**U**	UUU ⎤ Phe (F) UUC ⎦ UUA ⎤ Leu (L) UUG ⎦	UCU ⎤ UCC ⎥ Ser (S) UCA ⎥ UCG ⎦	UAU ⎤ Tyr (Y) UAC ⎦ UAA − stop UAG − stop	UGU ⎤ Cys (C) UGC ⎦ UGA − stop UGG − Trp (W)	U C A G
	C	CUU ⎤ CUC ⎥ Leu (L) CUA ⎥ CUG ⎦	CCU ⎤ CCC ⎥ Pro (P) CCA ⎥ CCG ⎦	CAU ⎤ His (H) CAC ⎦ CAA ⎤ Gln (Q) CAG ⎦	CGU ⎤ CGC ⎥ Arg (R) CGA ⎥ CGG ⎦	U C A G
	A	AUU ⎤ AUC ⎥ Ile (I) AUA ⎦ AUG − Met (M)	ACU ⎤ ACC ⎥ Thr (T) ACA ⎥ ACG ⎦	AAU ⎤ Asn (N) AAC ⎦ AAA ⎤ Lys (K) AAG ⎦	AGU ⎤ Ser (S) AGC ⎦ AGA ⎤ Arg (R) AGG ⎦	U C A G
	G	GUU ⎤ GUC ⎥ Val (V) GUA ⎥ GUG ⎦	GCU ⎤ GCC ⎥ Ala (A) GCA ⎥ GCG ⎦	GAU ⎤ Asp (D) GAC ⎦ GAA ⎤ Glu (E) GAG ⎦	GGU ⎤ GGC ⎥ Gly (G) GGA ⎥ GGG ⎦	U C A G

Third position (3' end)

Wobble Rules

3′ Nucleotide of Codon	5′ Nucleotide of Anticodon
A or G	U
U or C	G
U, C, or A	I

9.02 Types of Genetics Problems

1. Deduce the sequence of a DNA, RNA, or protein molecule.
2. Predict the results of in vitro translation of an mRNA.
3. Describe the mechanism of translation.
4. Sketch the structure of a gene or mRNA.
5. Describe the mechanism of polypeptide processing and sorting.

1. Deduce the sequence of a DNA, RNA, or protein molecule.

Problems of this type provide you with the sequence of a DNA, RNA, or protein molecule and ask you to use that sequence to deduce the sequence of the corresponding DNA, RNA, or protein molecule. The problem may ask you to move "forward" along the central dogma of gene expression (from DNA to RNA to protein) or it may ask you to move "backward," from protein to mRNA or from mRNA to DNA. Key concepts that apply to these problems include complementary base pairing, transcription, translation, and the genetic code.

Variations

(a) Determine the sequence of a polypeptide encoded by a nucleic acid sequence.

The question will indicate whether the sequence is the template or coding (nontemplate) strand of the gene or is an mRNA. **TIP:** Use the Eukaryotic and Prokaryotic Gene figures to check the relationship between the DNA sequences and the mRNA sequence. Use the genetic code to decode the mRNA sequence, starting at an AUG codon and ending at a stop (UAA, UAG, or UGA) codon. See Problem 1 for an example.

(b) Determine the sequence of an mRNA or DNA molecule corresponding to the coding sequence for a given polypeptide.

Each amino acid in the given polypeptide sequence will correspond to a codon in the mRNA sequence. **TIP:** Use the genetic code to determine how many possible mRNA codons can code for each amino acid, and use Py (pyrimidine), Pu (purine), or N (any nucleotide) to indicate the sequence at a site that is ambiguous because of the genetic code. See Problem 24 for an example.

(c) Determine a consensus sequence given multiple, specific nucleic acid sequences.

A consensus sequence is a single sequence that represents the most common nucleotide at each position of a DNA or RNA sequence element (for example, a Shine–Dalgarno sequence). First align the sequences over each other using some sequence or feature to put them in register. Then identify the most common nucleotide in each position. See Problem 26 for an example.

Example Problem: A polypeptide has the sequence Met-Thr-Ser-Val-Trp-Lys. What are the possible mRNA and dsDNA sequences that could code for this polypeptide?

Solution Strategies	Solution Steps
Evaluate	
1. Identify the topic this problem addresses, and explain the nature of the requested answer.	1. This question provides you with a polypeptide sequence and asks you to determine the corresponding mRNA and DNA sequences.
	The answers will be RNA and DNA sequences that are ambiguous at certain positions. The mRNA sequence and the DNA coding-strand sequence will be the same except for U in place of T. The template strand DNA sequence will be complementary to the coding-strand sequence.
2. Identify the critical information given in the problem.	2. The polypeptide sequence is Met-Thr- Ser-Val-Trp-Lys.
Deduce	
3. Use the genetic code to determine which nucleotide sequences correspond to each amino acid. TIP: Use Pu to indicate A or G; Py to indicate C, T, or U; and N to indicate any nucleotide.	3. The genetic code indicates that Met codons are always AUG, Thr codons are CAN, Ser codons are UCN, Val codons are GUN, Trp is UGG, and Lys is AAPu.
Solve	
4. Combine the codons to generate an mRNA sequence. Write the dsDNA sequence using T in place of U for the coding strand. The coding-strand DNA sequence should be the top strand. PITFALL: Always indicate the 5′ and 3′ ends of nucleic acid sequences.	4. **Answer:** The mRNA is 5′-AUGACNUCNGUNUGGAAPu-3′. The dsDNA sequence is 5′-ATGACNTCNGTNTGGAAPu-3′ 3′-TACTGNAGNCANACCTTPy-5′.

2. Predict the results of in vitro translation of an mRNA.

Problems of this type provide you with an mRNA sequence and ask you to indicate the polypeptide sequence(s) that would be produced by in vitro translation. The mRNAs may be natural or artificial. In these problems, translation of all possible reading frames occurs in vitro, and translation initiation is not limited to an AUG.

Variations

(a) Predict the results of in vitro translation given a nonoverlapping, three-letter code.

These are the results obtained by scientists investigating the nature of the genetic code. There are three different reading frames under these assumptions, beginning at the first, second, and third nucleotide in from the 5′ end. Use the genetic code to determine the sequence of the resulting polypeptide(s). See Problem 2 for an example.

(b) Predict the results of in vitro translation given other, nonoverlapping codes.

These questions ask you to consider the outcome given two- or four-letter codes. For a two-letter code, there are two possible reading frames, which begin at the first and second nucleotide in from the 5′ end.

For a four-letter code, there are four possible reading frames. **TIP:** Do not be concerned with the identity of the amino acid encoded by these hypothetical codons. Assume that different codons code for different (but unspecified) amino acids. See Problem 19 for an example.

(c) Predict the results of in vitro translation given overlapping codes.

These questions ask you to consider the outcome given two-, three-, or four-letter codes that overlap. Assume a maximum overlap, such that the reading frame of the sequence 5'-AUGCGA-3' assuming a two-letter code is AU UG GC CG GA, for a three-letter code is AUG UGC GCG CGA, and for a four-letter code is AUGC UGCG GCGA. For the two- and four-letter codes, assume that different codons code for different (but unspecified) amino acids. See Problem 19 for an example.

Example Problem: You are conducting an in vitro translation experiment using a synthetic mRNA made by polymerizing the four-nucleotide sequence ACGU. Predict the results that would be obtained if the genetic code were nonoverlapping and contained two-, three, or four-letter codes. Based on your predictions, could you determine whether the genetic code was a two-, three-, or four-letter code (all assumed to be nonoverlapping)?

Solution Strategies	Solution Steps
Evaluate	
1. Identify the topic this problem addresses, and explain the nature of the requested answer.	1. This question asks you to predict the results of in vitro translation of a synthetic mRNA assuming an overlapping two-, three-, or four-letter code. The question also asks you to indicate if the results would distinguish between two-, three-, and four-letter codes.
	The answers should indicate the number of different polypeptides produced and information on the amino acid sequence of each polypeptide. For two- and four-letter codes, no specific information on the amino acid sequences is required.
2. Identify the critical information given in the problem.	2. The mRNA was made by polymerizing the four-nucleotide molecule ACGU.
Deduce	
3. Determine the sequence of the mRNA by adding multiple copies of the given sequence head to tail. Then identify the reading frame under each assumption of the genetic code. **TIP:** Nucleic acid sequences written without 5' and 3' end designations are assumed to be written 5' to 3'. **TIP:** The number of possible reading frames is the same as the number of letters assumed to be in the genetic code. **TIP:** The codon sequence of mRNAs comprised of repeating sequences will repeat. There is no need to specify codons after the codon pattern has begun to repeat.	3. The mRNA sequence is 5'-ACGUACGUACGU....-3'. The two-letter reading frames are these: 5'-AC GU AC GU AC...-3' 5'-CG UA CG UA CG...-3' The three-letter reading frames are these: 5'-ACG UAC GUA CGU ACG...-3' 5'-CGU ACG UAC GUA CGU...-3' 5'-GUA CGU ACG UAC GUA...-3' The four-letter reading frames are these: 5'-ACGU ACGU ACGU...-3' 5'-CGUA CGUA CGUA...-3' 5'-GUAC GUAC GUAC...-3' 5'-UACG UACG UACG...-3'

(continued)

Solution Strategies	Solution Steps
Solve	
4. Identify reading frames that contain different codons within each assumption of the genetic code. For the two-letter and four-letter codes, assume that different codons code for different (but unspecified) amino acids. For the three-letter code, use the genetic code to determine the amino acid sequence encoded by each reading frame.	4. The predicted results are these: Assuming a two-letter code, two different polypeptides will be produced, each comprising a different, two–amino acid repeating sequence; assuming a three-letter code, one type of polypeptide with the repeating sequence Thr-Tyr-Val-Arg would be produced; and assuming a four-letter code, four different polypeptides would be produced, each containing a different, single amino acid.
5. Apply the information from steps 3 and 4 to arrive at an answer.	5. **Answer:** The predicted results differ under each assumption for the genetic code (assumptions were nonoverlapping codes containing two-, three-, or four-nucleotide codons); therefore, the results of this experiment should indicate which of these codes is correct.

3. Describe the mechanism of translation.

Problems of this type ask you to consider the mechanistic details of translation. You will be asked to consider the structure of mRNAs, tRNAs, and ribosomes as they are involved in translation initiation, elongation, and termination. Use the Translation Elongation figure to identify three sites in ribosomes where the anticodons of deacetylated-, peptidyl-, and aminoacyl-tRNAs interact with codons of mRNAs.

Variations

(a) Describe the steps of translation.

These questions ask for a description of the events that occur during initiation, elongation, and termination of translation. The identity and role of translation initiation, elongation, and termination factors is a focus of these questions. Consider each process as a sequence of ordered events. Review Figure 10.5 as well as Figures 10.7 through 10.10 from the textbook. See Problem 4 for an example.

(b) Describe ribosome structure and function.

These questions focus on ribosome structure as it relates to translation. Important key terms are *ribosomal subunits, ribosomal RNAs, ribosomal proteins*, and *peptidyl transferase activity*. The differences and similarities of eukaryotic and prokaryotic ribosomal subunits is also a focus. See Problem 10 for an example.

(c) Describe tRNA structure and function.

These questions focus on tRNA structure, and it relates to translation. Important key terms are *clover-leaf structure, L-structure, D-arm, T-arm, anticodon loop, anticodon, wobble base pairing, aminoacyl-tRNA synthase, peptidyl tRNA, aminoacyl tRNA*, and *deacetylated tRNA*. See Problem 6 for an example.

(d) Who did what?

Problems of this type ask you to recall historical information from the textbook describing the experimental strategies and results that lead to our understanding of translation. The information includes the names of one or more scientists and the specific scientific contribution each made. See Problem 2 for an example.

Example Problem: A ribosome contains a peptidyl-tRNA in the P site. The peptidyl-tRNA is base-paired to a Val codon. The sequence of the mRNA is 5'-GCAGAUGUUUCCAGAGGUAUGGCG-GUGAAGUCAU-3'. What is the sequence of the peptide attached to the peptidyl tRNA?

Solution Strategies	Solution Steps
Evaluate	
1. Identify the topic this problem addresses, and explain the nature of the requested answer.	1. This question asks for the sequence of the peptide attached to the peptidyl-tRNA in the P site of a ribosome in the act of translating an mRNA whose sequence is given. The answer will be an amino acid sequence that ends in valine.
2. Identify the critical information given in the problem.	2. The process is translation elongation. The mRNA sequence is 5′-GCAGAUGUUUCCAGAGGUAUG-GCGGUGAAGUCAU-3′. The amino acid attached directly to the tRNA is valine.
Deduce	
3. Identify the AUG start codon. Use that AUG to identify the reading frame. Write out the reading frame until reaching a valine codon. TIP: The correct AUG may or may not be the first AUG. Check to see that the selected AUG is in frame with a valine codon.	3. The AUG start codon is underlined: 5′-GCAG<u>AUG</u>UUUCCAGAGGUAUGGC GGUGAAGUCAU-3′. The reading frame is 5′-AUG UUU CCA GAG <u>GUA-3′</u>. The underlined GUA is an in-frame Val codon.
Solve	
4. Translate the open reading frame from above and write out the polypeptide sequence.	4. **Answer:** Met-Phe-Pro-Glu-Val

4. Sketch the structure of a gene or mRNA.

Problems of this type provide you with partial information on the structure of a gene or mRNA and ask you to sketch the complete gene, pre-mRNA (for eukaryotic genes), mRNA, and protein structure. Use the Eukaryotic Gene and Prokaryotic Operon figures to help you organize your sketches.

Variation

(a) Sketch the structure of a eukaryotic gene.
These questions ask for details concerning eukaryotic gene structure and function. Structures specific to eukaryotic genes include exons and introns, pre-mRNA, 5′ CAP, and 3′ poly-A tails, a polyadenylation sequence, and a transcription termination region. See Problem 17 for an example.

(b) Sketch the structure of a prokaryotic operon.
These questions ask for details concerning prokaryotic operon structure and function. Structures specific to prokaryotic operons include cistrons, polycistronic mRNA, intercistronic spacers, Shine–Dalgarno sequences, and transcription termination signals. See Problem 27 for an example.

Example Problem: The Trp operon in *E. coli* codes for five polypeptides involved in the biosynthesis of the amino acid tryptophan. For each of the following sequence elements, indicate whether the Trp operon has the element and, if so, how many copies of the element it contains: (a) promoter, (b) Shine–Dalgarno sequence, (c) cistron, (d) intercistronic spacer, (e) start codon, (f) stop codon, (g) transcription termination signal, (h) template strand, (i) coding strand, (j) intron, (k) exon, (l) Kozak consensus sequence.

Solution Strategies	Solution Steps
Evaluate	
1. Identify the topic this problem addresses, and explain the nature of the requested answer.	1. This question asks you to identify whether specific gene sequence elements are present in a bacterial operon. The answer will be either yes or no for each sequence element, and it will indicate the number of elements for those that are present in the operon.
2. Identify the critical information given in the problem.	2. The gene is a bacterial operon, and it codes for five polypeptides.
Deduce	
3. Bacterial operons contain one coding sequence, called a cistron, per polypeptide they encode. Determining the number of cistrons is important for understanding the structure of the operon.	3. The operon encodes five polypeptides; therefore, there are five cistrons.
Solve	
4. Refer to the Prokaryotic Operon Key Analytical Tool to determine which of the sequence elements are present and which are not. **TIP:** Note which sequence elements are present more than once and the relationship each element has to the protein-coding regions.	4. List the elements that are present and the ones that are not present. **Present:** • Bacterial operons contain a promoter, Shine–Dalgarno sequences, cistrons, intercistronic spacers, start codons, stop codons, a transcription termination signal, a template strand, and a coding strand. • There is one promoter, template strand, coding strand, and transcription termination signal per operon. • There is one Shine–Dalgarno sequence, start codon, and stop codon per cistron. • There is one intercistronic spacer sequence between pairs of cistrons. **Not present:** Bacterial operons do not contain introns, exons, or Kozak consensus sequences.
5. Apply the information from steps 3 and 4 to arrive at an answer.	5. **Answer:** • The Trp operon has one promoter, template strand, coding strand, and transcription termination signal. • The Trp operon has five cistrons and four intercistronic spacer sequences. • The Trp operon has five Shine–Dalgarno sequences, start codons, and stop codons.

5. Describe the mechanism of polypeptide processing and sorting.

These questions ask about events that occur after a polypeptide leaves the ribosome. The focus of this chapter is on events associated with proteins that are destined to reside in the endoplasmic reticulum (ER), the Golgi apparatus, the lysosome, the plasma membrane, or for secretion outside the cell. All such proteins contain a 20-amino-acid sequence called a signal sequence, which is required for transport into the ER and is removed after it enters the ER. Proteins with final destinations beyond the ER are transported from the ER in transport vesicles. See Problem 20 for an example.

Example Problem: The signal receptor particle (SRP) is the receptor that binds to the signal sequence during translation of a secretory protein in eukaryotic cells. Binding of SRP stops translation but leaves the ribosome intact. Binding of SRP to its receptor on the ER membrane causes SRP to release the ribosome, which then resumes translation. The polypeptide being translated passes through a channel in the ER membrane and enters the ER. Review Cesar Milstein's in vitro translation experiments and answer the following question. Besides the experiments Milstein performed, what additional result would have been obtained if Milstein did a third translation experiment that contained SRP and ER membrane but lacked the SRP receptor in the ER membrane? What does this suggest about where the SRP in Milstein's original experiment came from?

Solution Strategies	Solution Steps
Evaluate	
1. Identify the topic this problem addresses, and explain the nature of the requested answer.	1. This question asks you to consider the effect of the lack of the SRP receptor on the in vitro translation experiment of Cesar Milstein. In Milstein's experiment, he used an mRNA for a secretory protein and in vitro translation systems that contained ER membrane. His results were that the protein was 20 amino acids longer when translated in the absence of ER as compared to translation in the presence of ER. The answer will address the fate of the polypeptide encoded by the mRNA under the hypothetical conditions proposed by this question.
2. Identify the critical information given in the problem.	2. Additional information on secretory protein processing and transport into the ER was provided: the signal sequence is bound by a receptor called SRP; SRP binds to the signal sequence, but the polypeptide is only partially translated; SRP binding halts translation; the SRP-bound ribosome binds to the ER membrane, where SRP binds to its receptor; binding of SRP to its receptor causes SRP to release the signal sequence, which allows translation of the secretory protein to resume; and the secretory protein is transported through a channel into the ER as it is being translated.
Deduce	
3. Consider the events listed in step 2 under the condition where SRP is present but the SRP receptor is not. **TIP:** Follow the process up to the step that requires the function of the SRP receptor. This is where the process will halt in the absence of the SRP receptor.	3. The events that will occur under the conditions established in this question are these: a ribosome assembles on the secretory protein's mRNA and translation is initiated; the signal sequence is translated and emerges from the ribosome; the signal sequence is bound by SRP; and SRP binding halts translation.

Solution Strategies	Solution Steps
Solve	
4. Consider the fate of the secretory protein in the in vitro translation experiment considering the information in step 3. Also list the results obtained by Milstein in his original experiment.	4. The secretory protein is not fully synthesized. The portion corresponding to the signal sequence has been produced and exists as a peptidyl tRNA. The size of the protein is not much more than 20 amino acids. The protein was 20 amino acids shorter when ER was present than when ER was absent.
5. Apply the information from steps 3 and 4 to arrive at an answer. TIP: Milstein used gel electrophoresis to determine whether the secretory protein was synthesized and to determine whether it contained the signal sequence.	5. **Answer:** In the presence of SRP and ER membrane but the absence of the SRP receptor, the protein would be very small, perhaps too small to be detected. Since, in Milstein's original experiment, the lack of ER (and therefore the SRP receptor) did not inhibit translation (as it did in the new version of the experiment), the SRP must have been absent from the "no ER membrane" translation experiment. SRP must have been supplied to the in vitro translation experiment along with the ER membrane.

9.03 Solutions to End-of-Chapter Problems

1a.   *Evaluate:* This is type 1a. You were given the sequence of the template strand of a gene. *Deduce:* The mRNA sequence is complementary to the DNA template strand but contains U in place of T. The mRNA is 5′-AUGCAUCCGAUUGCCUCAUUCGAUUGA. AUG is a start codon, and UGA is an in-frame stop codon. *Solve:* Conceptual translation of the mRNA gives MHPIASFD (Met-His-Pro-Ile-Ala-Ser-Phe-Asp).

1b.   *Evaluate > Deduce > Solve:* The term is Homodimer

2a.   *Evaluate > Deduce > Solve:* This is type 3d. Nirenberg and Matthaei developed an in vitro translation system that contained everything necessary for translation except for amino acids and mRNAs. Synthetic RNA comprised of only U (poly-U) was added to 20 separate reactions, each containing a different radioactive amino acid as well as the other 19 nonradioactive amino acids. Only the reaction with radioactive phenylalanine produced radioactive protein. Since the RNA sequence poly-U contains only UUU codons, then UUU codes for phenylalanine.

2b.   *Evaluate > Deduce > Solve:* This is type 3d. Only reaction with radioactive proline produces radioactive protein. Since poly-C contains only CCC codons, CCC codes for proline.

2c.   *Evaluate > Deduce > Solve:* This is type 3d. One type of polypeptide, composed of alternating Arg and Glu amino acids, was produced.

2d.   *Evaluate:* This is type 2a. The mRNA sequence would be GCUAGCUAGCUA.... *Deduce:* There are three potential reading frames, each of them used during in vitro translation. All three reading frames contain the following codons: GCU AGC UAG CUA GCU AGC UAG CUA. GCU codes for Ala, AGC for Ser, CUA for Leu, and UAG is a stop codon. *Solve:* Since UAG appears every fourth codon in every reading frame, no polypeptides longer than three amino acids would be produced. The peptide is below the size detectable by Khorana, so he would have inferred that no polypeptides were produced.

3.   *Evaluate > Deduce > Solve:* This is type 3d. First, single- or double-nucleotide insertions or deletions in the *rII* gene resulted in a mutant phenotype, whereas three-nucleotide insertions or deletions resulted in a wild-type phenotype. Three insertions or deletions would restore the reading frame only if codons were three nucleotides long. Second, at least 20 codons are needed, one for each amino acid. There are only 4^2, or 16, different doublet nucleotide codons; whereas there are 4^3, or 64, different triplet codons. Therefore, codons were expected to be at least three nucleotides long. Third, in vitro translation of RNA composed of a dinucleotide repeat produced polypeptides with two alternating amino acids, as expected from a triplet code. Also, in vitro translation of RNA comprised of trinucleotide repeats produced three different polypeptides, each containing a single type of amino acid, as expected from a triplet code.

4.   *Evaluate > Deduce > Solve:* This is type 3a. Step 1—Preinitiation complex formation: Small ribosomal subunit and IF3 bind to the mRNA, AUG start codon is identified by 16S rRNA base-pairing with Shine–Dalgarno sequence, and AUG codon is in the ribosomal P site. Step 2—Formation of the 30S preinitiation complex: fMet-tRNA^{fMEt} bound to IF2-GTP binds to start codon in the P site, and IF1 binds. Step 3—Formation of the 70S initiation complex: 50S ribosomal subunit binds, IF2 cleaves GTP to GDP + phosphate, and IF1, IF2-GDP, and IF3 leave the complex.

5a.   *Evaluate:* This is type 1a. You are given the template-strand sequence of a gene and asked for the mRNA sequence. *Deduce:* The mRNA sequence is complementary to the template DNA sequence. *Solve:* Therefore, it will be 5'-AGCUCUAUACAUCACGCAUCGCGU-3'

5b.   *Evaluate:* This is type 1a. You are given the template-strand sequence of a gene and asked for the predicted polypeptide sequence. *Deduce:* Assuming that the first A in the mRNA (see 5a) is the first nucleotide of a codon in the correct reading frame, the codons are AGC (Ser), UCU (Ser), AUA (Ile), CAU (His), CAC (His), GCA (Ala), UCG (Ser), and CGU (Arg). *Solve:* Translation of this mRNA would yield the polypeptide N-terminus–Ser-Ser-Ile-His-His-Ala-Ser-Arg-C–terminus.

5c.   *Evaluate > Deduce > Solve:* This is type 1a. It is Isoleucine.

6.   *Evaluate > Deduce > Solve:* This is type 3c. It is tRNAs that are charged with different amino acids have unique structural features that allow them to interact with their cognate aminoacyl-tRNA synthetases. Unique features include the anticodon sequence as well as sequences and base modifications in the T-arm and D-arm.

7a–e.   *Evaluate:* This is type 3c. You are given tRNA anticodon sequences written 5' to 3' and asked to identify the amino acid attached to each tRNA. *Deduce:* The codons should correspond to the three nucleotide sequences that would base-pair with each presumed anticodon sequence. The amino acid corresponding to each codon is determined using the genetic code table, and that amino acid will be attached to the corresponding tRNA. *Solve:* **(a)** 5'-UAG-3' base-pairs with 5'-CUA-3', which is a leucine codon. **(b)** 5'-AAA-3' pairs with the codon 5'-UUU-3', which codes for phenylalanine. **(c)** 5'-CUC-3' pairs with the codon 5'-GAG-3', which codes for glutamate. **(d)** 5'-AUG-3' pairs with the codon 5'-CAU-3', which codes for histidine. **(e)** 5'-GAU-3' pairs with the codon 5'-AUC-3', which codes for isoleucine.

8a–e.   *Evaluate:* This is type 3. *Deduce:* The wobble rules state that a U in the first (5') position of an anticodon can base-pair with either A or G in the third position of a codon. The rules also state that G in the first (5') position of an anticodon can base-pair with either U or C in the third position of a codon. *Solve:* **(a)** 5'-UAG-3' can base-pair with the codons 5'-CUA-3' or 5'-CUG-3', which both code for leucine. **(b)** 5'-AAA-3' pairs only with the codon 5'-UUU-3', which codes for phenylalanine. **(c)** 5'-CUC-3' pairs only with the codon 5'-GAG-3', which codes for glutamate. **(d)** 5'-AUG-3' pairs only with the codon 5'-CAU-3', which codes for histidine. **(e)** 5'-GAU-3' can base-pair with the codons 5'-AUC-3' and 5'-AUU-3', which both code for isoleucine.

9.   *Evaluate > Deduce > Solve:* This is type 3a. UAA, UGA, and UAG are stop codons. They indicate the end of the coding sequence and trigger translation termination. When a stop codon appears in the ribosomal A site, the following occurs (based on bacterial translation termination): (1) release factor RF1 or RF2 binds to the A site; (2) the ester bond linking the polypeptide to the tRNA in the P site is hydrolyzed, releasing the polypeptide from the ribosome; (3) GTP is hydrolyzed, causing release of the RF from the A site; (4) the deacetylated tRNA in the P site is released; (5) the 70S ribosome disassembles into 50S and 30S ribosomal subunits, and the mRNA is released.

10.   *Evaluate:* This is type 3b. List the similarities and differences as shown in the table.

	Bacterial Ribosome	**Eukaryotic Ribosome**
Similarities		
Composition	RNA and protein	RNA and protein
Number of subunits	two (small and large)	two (small and large)
tRNA binding sites	three (E, P, and A)	three (E, P, and A)
Differences		
Number and size of rRNAs	three (16S, 23S, and 5S)	four (18S, 28S, 5.8S, and 5S)
Size of subunits	30S and 50S	40S and 60S
Number of proteins	21 in the small subunit and 31 in the large subunit	~35 in the small subunit and 45 to 50 in large subunit

11a. *Evaluate > Deduce > Solve:* This is type 3a. See figure.

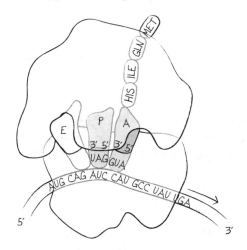

11b. *Evaluate > Deduce > Solve:* This is type 3a. The sequence is 5′-GGC-3′.

11c. *Evaluate:* This is type 3a. You are provided with an mRNA sequence. *Deduce:* Assuming translation started at the first AUG, the reading frame is AUG CAG AUC CAU GCC UAU UGA, which codes for the polypeptide Met-Gln-Ile-His-Ala-Tyr. The fourth amino acid is His, which is added to the Met-Gln-Ile polypeptide by transfer of the peptide form the peptidyl tRNA in the P site to the amino-acyl-tRNA in the A site. *Solve:* The E site is empty, the P site contains the deacetylated tRNA^Ile, and the A site contains the peptidyl tRNA^His. The next codon is GCC, which enters the ribosome A site when the ribosome translocates three nucleotides toward the 3′ end of the mRNA. Translocation, which is catalyzed by EF-G and the energy of GTP hydrolysis, moves the deacetylated tRNA in the P site to the E site and the peptidyl tRNA in the A site to the

P site. The ribosome must translocate three nucleotides toward the mRNA 3′ end. Translocation is facilitated by hydrolysis of GTP to GDP + phosphate by EF-G. Translocation moves the deacetylated tRNA^{Ile} to the E site, where it can exit the ribosome, moves the peptidyl tRNA^{His} to the P site, and leaves the A site open.

12a. *Evaluate > Deduce > Solve:* The errors in the diagram are (1) the ribosome is moving in the wrong direction along mRNA; (2) the mRNA contains Ts; (3) the terminal amino acid of the peptide is incorrect; (4) the ribosomal subunit sizes are incorrect; (5) the anticodon sequence of the tRNA in the P site is incorrect; and (6) the amino acids on tRNAs in P and A sites are incorrect.

12b. *Evaluate:* This is type 3a. Two amino acids are attached to the peptidyl tRNA in the P site, and an aminoacyl tRNA is in the A site. *Deduce:* The mRNA sequence is 5′-AUGCUGGCCACGCUUUAA-3′, which encodes the polypeptide Met-Leu-Ala-Thr-Leu. *Solve:* **TIP:** See the Translation Elongation Key Analytical Tool. The two amino acids on the peptidyl-tRNA in the P site should be attached to Leu-Met. Ala should be attached to the amino acyl tRNA in the A site. The ribosomal subunit should be 60S and 40S or 50S and 30S (but then Met should be fMET). The anticodon of the tRNA in the P site should be 5′-CAG-3′. The 5′ end of the mRNA should be closest to the E site, and the ribosome should be moving toward the 3′ end of the mRNA. Redraw the ribosome as shown here.

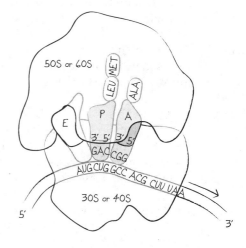

13a–d. *Evaluate:* This is type 3c. You are given four different amino acids. *Deduce > Solve:* **TIP:** See Genetic Code Key Analytical Tool. Leucine codons include 5′-CUPy-3′ (Py = C or U), 5′-CUPu-3′ (Pu = A or G), and 5′-UUPu-3′. Arginine codons include 5′-CGPy-3′, 5′-CGPu-3′, and 5′-AGPu-3′. Isoleucine codons include 5′-AUPy-3′ and 5′-AUA-3′. Lysine codons include 5′-AAPu-3′. *Solve:* **TIP:** See Wobble Rules Key Analytical Tool, and assume that all possible tRNAs with wobble nucleotides in the 5′ anticodon position exist. **(a)** There are two alternative answers for leucine. The first includes tRNAs with the anticodons 5′-GAG-3′ (can pair with 5′-CUPy-3′), 5′-UAG-3′ (can pair with 5′-CUPu-3′), and 5′-UAA-3′ (can pair with 5′-UUPu-3′). The second includes tRNAs with the anticodon 5′-IAG-3′ (can pair with 5′-CUPy-3′ and 5′-CUA-3′), 5′-GAG-3′ (can pair with 5′-CUPy-3′), and 5′-UAA-3′ (can pair with 5′-UUPu-3′). **(b)** For arginine, tRNAs with the anticodons 5′-GCG-3′ (can pair with 5′-CGPy-3′), 5′-UCG-3′ (can pair with 5′-CGPu-3′), and 5′-UCU-3′ (can pair with 5′-AGPu-3′) are needed. **(c)** For isoleucine, the tRNA with the anticodon 5′-IAU-3′ (can pair with 5′-AUPy-3′ and 5′-AUA-3′) is the only one needed. **(d)** For lysine, the tRNA with the anticodon 5′-UUU-3′ (can pair with 5′-AAPu-3′) is the only one needed.

14. *Evaluate:* This is type 3d. *Deduce:* **TIP:** See the Genetic Code Key Analytical Tool. Leucine, serine, and arginine each have six codons. Four are the same except for the third position, and two differ from each other only in which purine or pyrimidine is in the third position. Valine, proline, threonine, alanine, and glycine each have four codons that differ only in the third position. Isoleucine has three codons that differ in the third position: either U, C, or A. Phenylalanine, tyrosine, histidine, glutamine, asparagine, lysine, aspartic acid, glutamic acid, and cysteine each have two codons that differ in which purine or pyrimidine is in the third position. Methionine and tryptophan each have only one codon. *Solve:* **TIP:** See the Wobble Rules Key Analytical Tool. Amino acids with six codons require three tRNAs each; those with four codons require two tRNAs each; and those with three, two, or one codon require only one tRNA each. This is $(3 \times 3) + (5 \times 2) + (12 \times 1) = 31$.

15a.   *Evaluate:* This is type 3c. *Deduce > Solve:* mRNAs contain the information for the primary structure of a protein. The information includes the identity of the start codon, the stop codon, and the sequence of codons in between. tRNAs "decode" the information in the mRNA, relating each codon to a specific amino acid. rRNAs are important structural and functional components of the ribosome. Roles for rRNAs include identification of the Shine–Dalgarno sequence in bacteria, the peptidyl transferase catalytic activity, and binding sites for ribosomal proteins.

15b.   *Evaluate:* This is type 3c. *Deduce > Solve:* mRNAs are typically the least stable of the three RNAs. An individual tRNA or rRNA is used many times for translation of many mRNAs and is rarely targeted for destruction. mRNAs are inherently unstable to allow cells to rapidly reprogram their pattern of gene expression.

15c.   *Evaluate:* This is type 3c. *Deduce > Solve:* mRNAs are typically the least stable of the three RNAs in eukaryotic cells. Eukaryotic cells have specific mechanisms that target and destroy mRNAs in order to control the pattern of protein expression.

15d.   *Evaluate:* This is type 3c. *Deduce > Solve:* mRNAs in human cells are more stable than mRNAs in prokaryotic cells. Bacterial mRNAs do not contain specific 5′ or 3′ structures that protect them from nucleases; therefore, they begin to be degraded almost immediately after synthesis, whereas human mRNAs have structures based on the 5′ CAP and 3′ poly-A tail that protect them from nucleases for a period of time after synthesis.

16.   *Evaluate:* This is type 1a. *Deduce:* Use Gene figures in the Key Analytical Tools to review the relationship between the template strand, nontemplate strand, and mRNA sequences. Use the Genetic Code Analytical Tool to identify the corresponding amino acid and to find its three-letter and single-letter abbreviations. *Solve:* For example, the first sequence codes for N, which is asparagine, whose three-letter code is Asn. AAU and AAC in mRNA both code for Asn. The mRNA box has a C in the third position, indicating the mRNA sequence is 5′-AAC-3′. The nontemplate strand has the same sequence as the mRNA (except for T in place of U); therefore, 5′-AAC-3′ is also the nontemplate strand sequence. The template is complementary to the nontemplate and therefore is 5′-GTT-3′ (written as 3′-TTG-5′ in the table). The tRNA anticodon that pairs with 5′-AAC-3′ is 5′-GUU-3′, written as 3′-UUG-5′ in the table.

DNA	Nontemplate (5′ to 3′)	AAC	A̲TA	TGT	GAA	G̲G̲C̲	GAG	AA̲T̲	GA̲A̲	CGA
	Template (3′ to 5′)	TTG	T̲AT̲	ACA	CTT̲	CCG	CTC	TTA	CTT	GC̲T
	mRNA (5′ to 3′)	AAC̲	AUA	UGU	GAA	GGC	GAG	AA̲U	GAA	C̲GA
	tRNA (3′ to 5′)	UUG	UAU	ACA	CUU	CCG	CUC	UUA	CUU	GCU
Amino acid abbreviations	3-letter	Asn	Ile	Cys	Glu	Gly	Glu	Asn	Glu	Arg
	1-letter	N	I	C	E	G	E	N	E	R

17.   *Evaluate:* This is type 4a. *Deduce:* Use the Eukaryotic Gene Key Analytical Tool to identify which sequence elements are components of mRNA and, if so, where. *Solve:* Promoters and

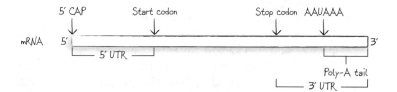

transcription termination sequences are parts of the gene but are not present in the mRNA. Introns are parts of the gene and pre-mRNA but not the mRNA. The others are present in the mRNA, and their locations are shown in the figure.

18.   *Evaluate:* This is type 4a. *Deduce:* Use the Eukaryotic Gene Key Analytical Tool to review the relationships between the dsDNA gene, the mRNA, and the polypeptide. *Solve:* All of the sequence elements listed in Problem 17 are present in either the gene or the mRNA and are shown in the figure.

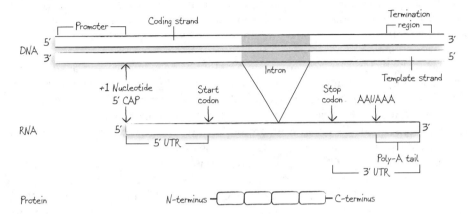

19a.   *Evaluate:* This is type 2b. Predict the results of translation of the sequence 5′-ACACACAC...-3′ assuming a nonoverlapping two-letter code. *Deduce:* There would be two reading frames, each containing a different single codon sequence: 5′-AC AC AC AC AC...-3′ and 5′-CA CA CA CA CA...-3′. *Solve:* Two different polypeptides would be produced, each one composed of a single type of amino acid.

19b.   *Evaluate:* This is type 2c. Predict the results of translation of the sequence 5′-ACACACAC...-3′ assuming an overlapping two-letter code. *Deduce:* There would be two reading frames, both containing alternating CA and AC codons. *Solve:* The resulting polypeptide would be composed of two alternating types of amino acids.

19c.   *Evaluate:* This is type 2a. Predict the results of translation of the sequence 5′-ACACACAC...-3′ assuming a nonoverlapping three-letter code. *Deduce:* There would be three reading frames, and each would contain alternating CAC and ACA codons. *Solve:* The resulting polypeptide would be composed of two alternating types of amino acids.

19d.   *Evaluate:* This is type 2c. Predict the results of translation of the sequence 5′-ACACACAC...-3′ assuming an overlapping three-letter code. *Deduce:* There would be three reading frames, each containing alternating CAC and ACA codons. *Solve:* The resulting polypeptide would be composed of two alternating types of amino acids.

19e.   *Evaluate:* This is type 2b. Predict the results of translation of the sequence 5′-ACACACAC...-3′ assuming a nonoverlapping four-letter code. *Deduce:* There would be four reading frames, two containing repeated CACA codons and two containing repeated ACAC codons. *Solve:* Two different polypeptides would be produced, each one composed of a single type of amino acid.

19f.   *Evaluate:* This is type 2b. Predict the results of translation of the sequence 5′-ACACACAC...-3′ assuming an overlapping four-letter code. *Deduce:* There would be four reading frames, each containing alternating ACAC and CACA codons. *Solve:* The resulting polypeptide would be composed of two alternating types of amino acids.

20.   *Evaluate > Deduce > Solve:* This is type 5. Soon after initiation of translation of an mRNA coding for a secretory protein, the amino terminus of the secretory protein is synthesized and is exposed on the surface of the ribosome. The amino terminus contains the signal sequence that marks this protein for cotranslational translocation into the endoplasmic reticulum (ER). The signal receptor particle (SRP) binds to the signal sequence and the ribosome and halts further translation. The SRP/ribosome/mRNA complex binds to the ER membrane, with SRP binding to its receptor and the ribosome binding to a protein translocation channel. SRP is released, translation resumes, and the growing polypeptide is extruded through the channel into the lumen of the ER. In the lumen of the ER, the signal sequence is cleaved by signal peptidase and the protein is glycosylated, folded with the help of chaperones, and packaged into a transport vesicle destined for the Golgi apparatus. The carbohydrate on the protein is modified as the protein passes through the compartments of the Golgi, and the protein is packaged into vesicles destined for transport to the plasma membrane. Fusion of the transport vesicle membrane with the plasma membrane releases the secretary protein into the extracellular fluid.

21a–b. *Evaluate:* This is type 1b. You are given a polypeptide sequence and asked to infer the corresponding mRNA and dsDNA sequences. The inferred sequence will be ambiguous at certain positions due to the degenerate nature of the genetic code. Positions of ambiguity can be written N (any nucleotide possible), Pu (a purine), or Py (a pyrimidine). *Deduce:* The polypeptide sequence is given from amino terminus to carboxyl terminus, which corresponds to an mRNA sequence from 5' to 3'. Cys codons are UGU and UGC, designated as UGPy. Pro codons are CCU, CCC, CCA, and CCG, designated as CCN. Ala codons are GCU, GCC, GCA, and GCG, designated as GCN. The Met codon is AUG. Gly codons are GGU, GGC, GGA, and GGG, designated as GGN. His codons are CAU and CAC, designated as CAPy. Lys codons are AAA and AAG, designated as AAPu. *Solve:* **(a)** Combining the codons using the ambiguous nucleotide nomenclature gives the mRNA sequence 5'-UGPyCCNGCNAUGGGNCAPyAAPu-3'. **(b)** The DNA coding-strand sequence will be the same as the mRNA sequence except that T will replace U. The coding-strand sequence is 5'-TGPyCCNGCNATGGGNCAPyAAPu-3', and the template-strand sequence is 3'-ACPuGGNCGNTACCCNGTPuTTPy-5'.

22a. *Evaluate > Deduce > Solve:* This is type 3d. Recall the work of Har Gobind Khorana and his colleagues. 5'-UGUGUGUGUGUGUGUG…-3'.

22b. *Evaluate > Deduce > Solve:* This is type 3d. Recall the work of Har Gobind Khorana and his colleagues. Cys-Val-Cys-Val-Cys-Val-Cyc-Val…

22c. *Evaluate:* This is type 2a. *Deduce:* Assuming a three-letter nonoverlapping code, there would be two possible reading frames: 5'-UGU GUG UGU GUG UGU…-3' and 5'-GUG UGU GUG UGU…-3'. *Solve:* Both possible reading frames code for a polypeptide composed of alternating Val and Cys amino acids, which is what Khorana and his colleagues found. The result was production of a polypeptide composed of alternating Val and Cys amino acids, which is what was predicted for a nonoverlapping triplet code.

22d. *Evaluate:* This is type 2b. *Deduce:* There would be two open reading frames in the mRNA, one reading 5'-UG UG UG UG…-3' and the other reading 5'-GU GU GU GU…-3'. *Solve:* Translation of the RNA using a doublet nonoverlapping code would result in two different polypeptides, each composed of single type of amino acid.

22e. *Evaluate:* This is type 2c. *Deduce:* The overlapping doublet and triplet code both result in a single reading frame (5'-UG GU UG GU…-3' and 5'-UGU GUG UGU GUG…-3'). Translation would produce a single polypeptide with two alternating types of amino acids. Thus, the result (22c) does not distinguish between a nonoverlapping triplet code and doublet and triplet codes that overlap. *Solve:* Translation of the RNA using an overlapping doublet or triplet code gives the same result; a single type of polypeptide with two alternating types of amino acids. This result does not differ from that predicted based on a nonoverlapping three-letter code, but it does differ from that predicted for a nonoverlapping two-letter code.

23a–c. *Evaluate:* This is type 2a and 2b. *Deduce:* The mRNA sequence would be 5'-UUGUUGUUGUU-GUUG…-3'. The number of reading frames depends on the number of codons assumed in the code: two reading frames for a doublet code, three reading frames for a triplet code, and four reading frames for a quadruplet code. For a doublet code, both reading frames would have repeats of the three-codon sequence 5'-UU GU UG-3'. For a triplet code, the three reading frames would differ, each composed of a single codon (5'-UUG, UGU, or GUU-3'). For a quadruplet code, all four reading frames would have repeats of the same four codons: 5'-UUGU UGUU GUUG-3'. *Solve:* **(a)** Three reading frames are possible (5'-UUG UUG UUG...-3', 5'-UGU UGU UGU...-3', and 5'-GUU GUU GUU...-3'). **(b)** Translation of 5'-UUG UUG UUG ...-3' produces polyleucine (Leu). Translation of 5'-UGU UGU UGU...-3' produces polycysteine (Cys). Translation of 5'-GUU GUU GUU...-3' produces polyvaline (Val). **(c)** The result differs from that predicted for nonoverlapping doublet or quadruplet codes, both of which predict a single type of polypeptide composed of repeats of two, or three, amino acids). Only a triplet code is predicted to produce three different single–amino acid polypeptides from the given mRNA. Doublet and quadruplet codes both predict a single type of polypeptides composed of repeats of a three–amino acid sequence.

24.   *Evaluate:* This is type 1b. You are given the length of a polypeptide. *Deduce:* There is 1 three-nucleotide codon in the mRNA for each amino acid in the polypeptide. *Solve:* $146 \times 3 = 438$ nucleotides are needed to code for the polypeptide (not including the stop codon).

25.   *Evaluate:* This is type 4a. Review the Eukaryotic Gene Figure Analytical Tool to identify sequences present in mRNA other than the coding sequence. *Deduce:* mRNAs contain a stop codon (technically not translated) and untranslated (UTR) sequences on both the 5' and 3' sides of the coding sequence. *Solve:* The names of the sequences are 5' UTR, 3' UTR, and the stop codon

26.   *Evaluate:* This is type 1c. *Deduce:* The SD consensus sequence will be a single sequence that contains the most common nucleotide found at each relative position in several specific SD examples. To compare nucleotides at each position within the SDs, align the sequences such that the SD sequences are in register. To determine the distance from the SD to the AUG, count the nucleotides from right to left, starting with the nucleotide to the left of the AUG and ending with the first nucleotide of the SD. *Solve:* There is a GG dinucleotide in each SD sequence; therefore, this GG was used to align the mRNA sequences by placing the GG of each sequence directly over each other. The bold letters indicate the nucleotides identified as SD sequences and the AUG start codons from Figure 10.6. The arrow above the araB SD sequence illustrates the distance between the AUG and the SD sequence in this mRNA.

	←−8→
E. coli araB	UUUGGAUGGAGUGAAACGAUGGCGAUUGCA 3'
E. coli lacI	CAAUUCAGGGUGGUGAAUAUGAAACCAGUA
E. coli lacZ	UUCACACAGGAAACAGCUAUGACCAUGAUU
E. coli thrA	GGUAACCAGGUAACAAGGAUGCGAGUGUUG
E. coli trpA	AGCACGAGGGGAAAUCUGAUGGAACGCUAC
E. coli trpB	AUAUGAAGGAAAGGAACAAUGACAACAUUA
λ phage *cro*	AUGUACUAAGGAGGUUGUAUGGAACAACGC
E. coli RNA polymerase B	AGCGAGCUGAGGAACCCUAUGGUUUACUCC
SD Consensus	GGAGGAA −5 to −9 from AUG

27a.   *Evaluate:* This is type 1a. *Deduce > Solve:* The genetic code is essentially universal and does not differ between human and *E. coli*; therefore, the code itself does not provide a barrier to the production of human proteins in *E. coli*.

27b.   *Evaluate:* This is type 5. *Deduce:* Human insulin is a eukaryotic gene, which differs from bacterial genes in multiple ways. **TIP:** Compare the Eukaryotic Gene and Bacterial Operon figures in the Key Analytic Tools section. Additionally, the human insulin gene encodes preproinsulin, which must be processed in the ER to produce insulin. *E. coli* does not contain ER or associated processing enzymes. *Solve:* The human insulin is unlikely to be transcribed, the pre-mRNA will not be spliced if it is produced, the mRNA is unlikely to be translated if produced, and the preproinsulin polypeptide will not be processed if it is produced. The human gene for insulin will not work, for several reasons. First, the insulin promoter will not be recognized by *E. coli* RNA polymerase and, even if it is transcribed, the insulin pre-mRNA will not be spliced. Second, the mRNA does not contain a Shine–Dalgarno sequence and therefore may not be efficiently translated in *E. coli*. Third, even if the mRNA is produced and translated, translation will produce preproinsulin, which will not be processed in *E. coli*.

27c. *Evaluate:* This is type 5. *Deduce:* **TIP:** See Figure 9.20 for details of human insulin gene structure. Insulin is a heterodimer of two polypeptides called the A and B chains. These polypeptides are part of a single polypeptide called preproinsulin and are cleaved into separate polypeptides after translation and translocation into the ER. *E. coli* does not have an ER or the enzymes for processing preproinsulin. *Solve:* To produce active insulin in *E. coli*, the A and B chains must be produced separately, by expression from two separate genes. To do this, the coding sequence for each polypeptide must be placed under control of appropriate *E. coli* transcription and translation signals.

28a. *Evaluate:* This is type 1a. The problem provides a double-stranded DNA sequence. *Deduce:* The coding sequence should begin with an ATG and contain four additional in-frame sense codons. The strand that reads "ATG" will be the coding strand, and its complement will be the template strand. There is no polarity indicated for the given sequence; therefore, consider both possible sequences polarities (top strand 5′ to 3′ and 3′ to 5′). There are three ATG sequences, only one of which is followed by only four sense codons. This sequence is on the bottom strand and reads from left to right. *Solve:* The coding-strand sequence is 5′-ATG GTA GGA GGA AGT-3′ and is the shaded portion of sequence in the figure. The coding strand is the bottom strand, and the template strand is the top strand.

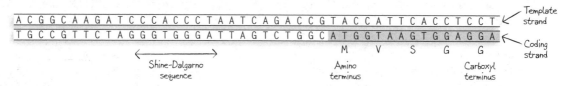

28b. *Evaluate:* This is type 1a. *Deduce:* The mRNA sequence will be the same as the DNA sequence, except that U replaces T. *Solve:* Assuming the entire sequence is transcribed, the mRNA sequence is 5′-UGCCGUUCUAGGGUGGGAUAGUCUGGCAUGGUAAGUGGAGGAA-3′.

28c. *Evaluate > Deduce > Solve:* This is type 1a. The polypeptide sequence is NH$_2$-Met-Val-Ser-Gly-Gly-COOH (NH$_2$-M-V-S-G-G-COOH).

28d. *Evaluate:* This is type 1c. *Deduce:* The Shine–Dalgarno sequence from Problem 26 contains the sequence 5′-GGAGGAA-3′ located 5 to 9 nucleotides 5′ to the ATG. *Solve:* The best match to this sequence 5′ to the ATG is 5′-GGTGGGA-3′.

28e. *Evaluate > Deduce > Solve:* This is type 1. The Shine–Dalgarno sequence is a bacterial mRNA sequence located just 5′ to the AUG start codon that is important for identifying that AUG as the start codon. This sequence functions by base-pairing with a complementary sequence on the 16S rRNA in the small ribosomal subunit. The interaction between the Shine–Dalgarno sequence and the 16S rRNA helps place the AUG start codon in the P site of the small ribosomal subunit.

29a–b. *Evaluate:* This is type 1a. You are given a 15-nucleotide sequence from the coding strand of a gene. The answer will include a DNA sequence complementary to the one given, an mRNA sequence that is the same as the sequence given (except T = U), and a polypeptide sequence based on a reading frame starting with the first three nucleotides of the mRNA sequence. *Deduce > Solve:* **(a)** 5′-AGATTCATTCTCTCC-3′ is complementary to 5′-GGAGAGAATGAATCT-3′, and is therefore the template strand. 5′-GGAGAGAAUGAAUCU-3′ is the mRNA. **(b)** The codons are GGA (Gly or G), GAG (Glu or E), AAU (Asn or N), GAA (Glu or E), and UCU (Ser or S). The polypeptide is written as GENES (single-letter code) and as Gly-Glu-Asn-Glu-Ser (three-letter code).

30a–b. *Evaluate:* This is type 1a. You are given a eukaryotic mRNA sequence. *Deduce:* The reading frame will start with the AUG closest to the 5′ end and continue in that reading frame until reaching a stop codon. *Solve:* **(a)** There are four AUGs in the mRNA. The first AUG is 12 nucleotides in from the CAP. The first in-frame stop codon is UAA, which is 16 nucleotides in from the poly-A tail. The start and stop codon are in bold print. 5'-*CAPCCAAGCGUUAC*AUGU AUGGAGAGAAUGAAACUGAGGCUUGCCACGUUUGUUAAGCACCUAUGCUACCG*AAA AAAAAAAAAAAAAAAAAAAAA*-3' **(b)** The coding sequence is **AUG** UAU GGA GAG AAU GAA ACU GAG GCU UGC CAC GUU UGU, which codes for the polypeptide Met-Tyr-Gly-Glu-Asn-Glu-Thr-Glu-Ala-Cys-His-Val-Cys (or MYGENETEACHVC in single-letter code).

31.   *Evaluate:* This is type 4a. *Deduce:* The gene and the pre-mRNA will contain introns and the associated sequences, whereas the mRNA will not. The sequences at the end of exon 1 and beginning of exon 2 in the gene and pre-mRNA will be joined in the mRNA. There will be untranslated sequence before the start codon and after the stop codon. *Solve:* Use the Eukaryotic Gene figure Key Analytical Tool to assist you in diagramming the eukaryotic gene as shown here.

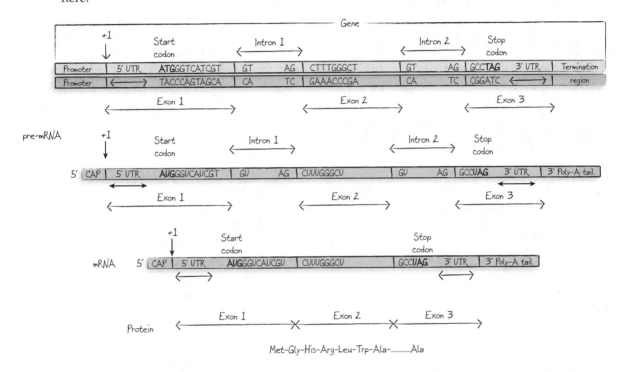

32.   *Evaluate:* This is type 1c. *Deduce:* The table indicates the percentages of sequences with each nucleotide present at each position. The most common nucleotide at each position is the one with the highest percentage score. For example, 35% of the sequences had a C at the −12 position. This is the most common nucleotide at −12; therefore, the consensus sequence has a C at −12. *Solve:* The consensus sequence is CCCGCCGCCACCAUGG. The AUG start codon is in bold.

Position	−12	−11	−10	−9	−8	−7	−6	−5	−4	−3	−2	−1	[start] +4
Percent A	23	26	25	23	19	23	17	18	25	61	27	15	[AUG] 23
Percent C	35	35	35	26	39	37	19	39	53	2	49	55	[AUG] 16
Percent G	23	21	22	33	23	20	44	23	15	36	13	21	[AUG] 46
Percent T	19	18	18	18	19	20	20	20	7	1	11	9	[AUG] 15
Consensus	C	C	C	G	C	C	G	C	C	A	C	C	**AUG** G

33a–b.   *Evaluate:* This is type 1c. *Deduce:* The table indicates the sequence of each mRNA from position −12 to + 4 with respect to the AUG start codon. The most common nucleotide at each position can be determined by calculating the percentage of times each nucleotide occupies that position. For example, −12 is a T in 3/16 sequences, a G in 2/16, an A in 5/16, and a C in 6/16. Position

−12 is a C in 6/16 or 38% of the sequences; therefore, the consensus nucleotide at −12 is C. *Solve:* **(a)** The consensus sequence is CCAGCAGCCACCAUGG. **(b)** See the table, which shows the data from Problems 32 and 33 for comparison. Percentage values were rounded to whole numbers. The consensus sequences from the two data sets agree at 11 of 13 positions. They differ at positions −10 and −7, where A is the most common nucleotide at both positions in the globin sequences but is the second most common in the data set for Problem 33.

Position	Problem No.	−12	−11	−10	−9	−8	−7	−6	−5	−4	−3	−2	−1	[start] +4
Percent A	32	23	26	25	23	19	23	17	18	25	61	27	15	[AUG] 23
	33	31	25	56	31	13	44	6	25	13	69	0	13	[AUG] 0
Percent C	32	35	35	35	26	39	37	19	39	53	2	49	55	[AUG] 16
	33	38	38	19	25	75	19	6	56	75	0	75	88	[AUG] 0
Percent G	32	23	21	22	33	23	20	44	23	15	36	13	21	[AUG] 46
	33	12	31	13	38	13	0	75	13	13	31	13	0	[AUG] 88
Percent T	32	19	18	18	18	19	20	20	20	7	1	11	9	[AUG] 15
	33	19	6	13	6	0	0	13	6	0	0	13	0	[AUG] 13
Consensus	32	C	C	C	G	C	C	G	C	C	A	C	C	AUG G
	33	C	C	A	G	C	A	G	C	C	A	C	C	AUG G

34.   *Evaluate:* This is type 1c. This problem is similar to Problems 32 and 33 but focuses on position −6 to +4 in the consensus sequence from Problem 33. The answer to Problem 33 includes a table, which lists a consensus sequence using vertebrate α-globin and β-globin sequences. *Deduce:* The consensus sequence from −6 to +4 can be taken directly from this table. *Solve:* The consensus sequence is GCCACCAUGG.

9.04 Test Yourself

Problems

1. You have isolated an enzyme from a eukaryotic cell and have cloned the gene that encodes that enzyme. When you sequence the gene, you find that it encodes a polypeptide that is 20 amino acids longer than the purified enzyme. Your colleague suggests that, based on this information, the enzyme is a component of the endoplasmic reticulum.
    a. Explain why your colleague might be correct.
    b. Explain why your colleague might be incorrect.

2. Cesar Milstein used his in vitro translation systems to test Gunter Blobel's signal hypothesis. One system contained all components necessary for translation but lacked ER. The other contained both.
    a. What were Milstein's results?
    b. What was Milstein's interpretation of those results?
    c. How would the results have differed if Milstein had used his in vitro translation systems to produce proteins that do not normally enter the ER?

3. Sketch the structure of the human insulin *gene* using the Eukaryotic Gene figure from this study guide, Figure 9.20c from the textbook, and the following information as your guide. The human preproinsulin gene contains two exons and a single intron. The intron interrupts the pro-amino acid segment. Indicate the location of the following on your drawing:
    a. The promoter
    b. The template (noncoding) strand
    c. The coding (nontemplate) strand
    d. The +1 nucleotide and the direction of transcription
    e. The position of exons and introns

  f. The location of the polyadenylation signal
  g. The location of transcription termination
  h. The pre-amino acid coding region
  i. The A chain coding region
  j. The pro-amino acid coding region
  k. The B-chain coding region

4. The dsDNA sequence below includes the coding sequence for a polypeptide with the amino acid sequence Met-Pro-Ile-Asp-Arg-Trp-Tyr-Stop. Use this information to answer the following questions.
  a. Write the sequence of the mRNA for this protein, and underline the start codon.
  b. Indicate which DNA strand is the template for transcription.
  c. Indicate the direction of transcription.

5'-ATGGTACTTAATACCATCGATCGATAGGCATGCAGTGCATACC-3'

3'-TACCATGAATTATGGTAGCTAGCTATCCGTACGTCACGTATGG-5'

5. You have isolated an aminoacyl tRNA with the anticodon sequence 5'-IAU-3'. What codon sequences can base-pair with this tRNA? Which amino acid is attached to this tRNA?

6. What is the difference between polyribosomal mRNAs and polycistronic mRNAs?

7. All tRNAs have common structures and functions. It is also true that tRNAs for different amino acids have unique structures and functions. Generate a table listing tRNA structures and functions that are common among all tRNAs in a given cell and structures and functions that differ between tRNAs for different amino acids in a cell.

8. Draw a polyribosome containing two ribosomes bound to the following mRNA sequence: 5'-CACUAGUGGAGGAAUGGCUUGCGAUGAAUUCGGACAUAUAAAACUAAACCCCCAGC-GAAGUACCGUGUGGUACUAGAAUCGC-3'

Draw the ribosomes, tRNAs, and polypeptides with the same level of detail as in your answer to Problem 11 from the textbook. One ribosome should have a peptidyl tRNA in its P site, an aminoacyl tRNA in its A site, and a codon for threonine in its E site. The other ribosome should have a peptidyl tRNA in its P site, an aminoacyl tRNA in its A site, and the codon for alanine in its E site.

9. Mutations in tRNA genes can affect tRNA function. The following are examples of alterations in tRNA function that could be caused by single-nucleotide mutations in a tRNA gene. For each example, describe a specific mutation that would cause the effect listed.
  a. A mutation in a tRNA gene causes tyrosine to be added to a polypeptide at sites that are normally sites of translation termination.
  b. A mutation in a tRNA^{Leu} gene prevents it from being charged with any amino acid.
  c. A mutation in a tRNA^{Lys} gene causes lysine to be inserted in polypeptides at sites that should contain arginine.

10. Mutations in the coding sequence of a gene do not always have an effect on the structure of the polypeptide gene product. For most codons, the effect of a mutation depends on which nucleotide position within the codon is changed. For each codon listed below, describe the effect that transition mutations (C to T, T to C, A to G, or G to A) at each position would have on the structure of the encoded polypeptide.
  a. CTG
  b. AGA
  c. TGG

11. In certain rare instances, complete translation of a polypeptide from an mRNA requires that the reading frame change by one nucleotide in the middle of the translated region. This is referred to as programmed translational frameshifting. One example is a retrotransposon in yeast. Translation of the complete polypeptide requires a change in the reading frame at a specific site in the mRNA.

The sequence of the mRNA in the region of the frameshift is

5'-AUGGUGGAUCCAGUGAGAGCACGUAACGCGAGUUCUAACCGAUCUUGAA-3'

The sequence of the polypeptide produced in cells translating this mRNA is (using single-letter code) MVDPVRARTRVLTDLE.

Write out the codons that are read to produce this polypeptide, and indicate the mRNA reading frame for each codon.

12. The sequence of the template strand of a bacterial gene is shown below. Use this sequence to answer the following questions.

3'-GTGATCACCTCCTTACCGAACGCTACTTAAGCCTGTATATTTTGATTTGGGGGGTC-GCTTCATGGCACACCATGATCTTAGCG-5'

  a. Assuming that this entire sequence is transcribed, what is the sequence of the mRNA?
  b. What is the sequence of the polypeptide encoded by this gene if translation starts at the first AUG codon? Write your answer using the single-letter amino acid code.
  c. What is the sequence of the polypeptide encoded by this gene if translation starts at the second AUG codon? Write your answer using the single-letter amino acid code.
  d. Add an additional G nucleotide after the G of the first AUG codon and then use this mutant mRNA to determine the sequence of the encoded polypeptide, using the first AUG as the start codon. Write your answer using the single-letter amino acid code.
  e. Add an additional G nucleotide after the G of the second AUG codon and then use this mutant mRNA to determine the sequence of the encoded polypeptide, using the second AUG as the start codon. Write your answer using the single-letter amino acid code.
  f. Compare the four polypeptide sequences from 12b through 12e, and indicate which two are most similar. Why are these polypeptides similar, even though one comes from a wild-type sequence and the other comes from a mutant sequence?

13. The following are tRNA anticodon triplet sequences. Identify the amino acids these tRNAs carry.
    a. 5'-AAC-3'
    b. 5'-AGG-3'
    c. 5'-GCC-3'
    d. 5'-CGG-3'
    e. 5'-UAC-3'
    f. 5'-UCC-3'
    g. Indicate which of the tRNAs 13a through 13f are isoaccepting tRNAs.

14. The figure shows a ribosome during translation elongation. There is a peptidyl tRNA in the P site, but the E and A sites are empty and not labeled. A portion of the mRNA sequence is shown, but the 5' and 3' ends of the sequence are not indicated. Use this figure to answer the following questions.
  a. Which is the 5' end of the mRNA?
  b. What direction is the ribosome moving relative to the mRNA (left or right)?
  c. Which of the sites on the ribosome is the E site and which is the A site?
  d. Which aminoacyl tRNA will be the next to bind to this ribosome's A site?

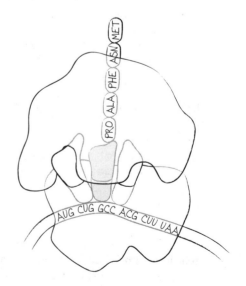

Solutions

1a–b.   *Evaluate:* This is type 5. *Deduce:* The segment of the enzyme posttranslationally removed is about the size of a signal sequence. Signal sequences direct polypeptides to the ER, where the signal sequence is removed and the protein is potentially processed and sorted further. *Solve:* **(a)** Your colleague may be correct because genes for ER proteins will encode polypeptides that include a 20–amino acid signal sequence, which is removed during transport into the ER. **(b)** Proteins that are secreted from the cell or transported to other organelles (the Golgi or the lysosome, for example) are also made with a signal sequence that is subsequently removed. Your colleague may be wrong because genes for proteins found in the Golgi and other membrane systems in eukaryotic cells also code for proteins that have signal sequences that are removed during transport to the ER.

2a–c.   *Evaluate:* This is type 5. Recall the in vitro translation experiments of Cesar Milstein. *Deduce > Solve:* **(a)** Milstein and his colleagues observed that in vitro translation of secretory proteins in the absence of ER membrane (and all components required for cotranslational transport into the ER) were about 20 amino acids longer than those translated in the presence of ER membrane. **(b)** Milstein interpreted the test results to be consistent with the signal hypothesis, which states that secretory proteins contain a signal sequence, comprising the terminal 20 amino acids, which is removed after transport through the ER membrane. **(c)** Since proteins that do not normally enter the ER do not have a signal sequence, in vitro translation of these proteins will produce the same polypeptides regardless of the presence or absence of ER membrane.

3.   *Evaluate > Deduce > Solve:* This is type 4a. You are told that the gene contains a single intron that interrupts the coding sequence within the pro-peptide coding segment. Review Figure 9.20 of the textbook to aid in drawing a figure similar to the one shown here.

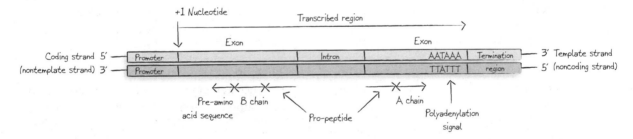

4a–c.   *Evaluate:* This is type 1b. The problem provides a dsDNA sequence and a polypeptide sequence. *Deduce:* The first amino acid, methionine, must be an ATG codon. This will then be followed by a proline codon, which will be CCN, where N is any nucleotide. ATGCCN identifies the coding DNA strand. The mRNA will have the same sequence as the coding strand except that U replaces T. There are two ATGs on the top strand, but neither is followed by CCN. There are three ATGs on the bottom strand, but only one is followed by CCN, where N = T. This ATG initiates the following reading frame: 5′-ATG CCT ATC GAT CGA TGG TAT UAA-3′. This reading frame codes for Met-Pro-Ile-Asp-Arg-Trp-Tyr-Stop. Thus, the bottom strand is the coding strand, the top strand is the template strand, transcription is right to left, and the mRNA will have the same sequence as the bottom strand, with U replacing T. *Solve:* **(a)** The mRNA sequence, which includes all of the sequence shown in the problem, is 5′-GGUAUGCACUGCAUGCCUAUCGAUCGAUGGUAUUAAGUACCAU-3′. The AUG start codon is underlined. **(b)** The top strand is the template used for transcription. **(c)** The direction of transcription is from right to left.

5.   *Evaluate:* This is type 3c. *Deduce:* U pairs with A, A pairs with U, and I (inosine) can pair with U, C, and A. *Solve:* 5′-IAU-3′ pairs with the codons 5′-AUU-3′, 5′-AUC-3′, and 5′-AUA-3′, which all code for isoleucine; therefore, the aminoacyl version of this tRNA will carry isoleucine.

6. *Evaluate > Deduce > Solve:* This is type 4b. A polyribosomal mRNA can be any mRNA that is being translated by more than one ribosome at one time. A polycistronic mRNA is encoded by bacterial operons and contains the coding sequence for more than one polypeptide. Polycistronic mRNAs that are being translated will typically be found in polyribosomes and so will also be polyribosomal mRNAs.

7. *Evaluate:* This is type 3c. *Deduce > Solve:* See table. All tRNAs have many common structures that identify them as tRNAs. They are small RNAs with some nucleotide sequence identity and that contain posttranscriptionally modified bases. All fold into a clover-leaf-like secondary structure by base pairing within the tRNA molecule and fold into an L-shaped tertiary structure. The clover-leaf structure contains three ssRNA loops (D-loop, T-loop, and anticodon loop). All tRNAs have the sequence CCA at their 3′ end and can bind to amino acids through a bond between the 3′-hydroxyl group of the A's and the carboxyl group of the amino acids. All tRNAs have common functions in translation and can be said to translate an RNA sequence into amino acid sequence. All but the initiator tRNA can bind to the same elongation factor. All except for the initiator tRNA can bind to the ribosomal A site and can exist as peptidyl-tRNAs in the ribosomal A site. All tRNAs can exist as peptidyl-tRNAs in the ribosomal P site. All tRNAs can bind to an aminoacyl tRNA synthetase enzyme and can be charged with an amino acid. Non-isoaccepting tRNAs have unique sequences that differentiate them from each other. These include specific RNA sequences and specific types of posttranscriptional modifications. Each also can be attached to a different amino acid. Non-isoaccepting tRNAs have unique functions that allow each to perform a specific task as a translator of the genetic code. Each binds specifically to different aminoacyl tRNA synthetase enzymes, which charge them with different amino acids. Each tRNA pairs with one or more codons in the mRNA that code for one specific amino acid.

Common Structures	Common Functions*	Unique Structures	Unique Function
Small RNAs	Bind to same elongation factors	Each contains sequences specific to that tRNA	Each binds to a different aminoacyl tRNA synthase
Same sequence at certain nucleotide positions	Can bind to the ribosome A site	Differences in sequence at certain nucleotide positions	Each pairs with a different codon during translation
Contain posttranscriptionally modified bases	Can form peptidyl-tRNAs in the ribosome A site	Some posttranscriptional modifications are specific	
Form cloverleaf-like secondary structure	Can exist as peptidyl-RNAs in the ribosome P site		
Form L-shaped tertiary structure	Can bind to an aminoacyl tRNA synthase enzyme		
Sequence CCA at their 3′ end	Join to amino acids through an ester linkage to their 3′ end		

* Except for initiator tRNA

8.   *Evaluate:* This is type 3a. *Deduce:* There is a single AUG start codon, which identifies the open reading frame that is being translated. In this reading frame, there is only one Thr codon and one Ala codon. *Solve:* See figure. A ribosome with the Ala codon in its E site will have translated only the AUG, GCU, and UGC codons and therefore will have a Met-Ala-Cys peptidyl tRNA in its P site. The next codon is GAU, so it will have an Asp-tRNA^Asp aminoacyl tRNA in its A site. The ribosome with a Thr codon in its E site will have translated the AUG through the GUG codons and therefore will have a Met-Ala-Cys-Asp-Glu-Phe-Gly-His-Ile-Lys-Leu-Asn-Pro-Gln-Arg-Ser-Thr-Val peptidyl tRNA in its P site. The next codon is UGG, so it will have a Trp-tRNA^Trp aminoacyl tRNA in its A site.

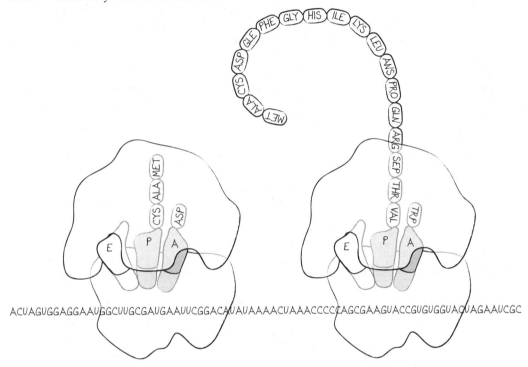

ACUAGUGGAGGAAUGGCUUGCGAUGAAUUCGGACAUAUAAAACUAAACCCCCAGCGAAGUACCGUGUGGUACUAGAAUCGC

9a–c.   *Evaluate:* This is type 3c. *Deduce:* Mutations change a single nucleotide in the tRNA gene, leading to production of a tRNA with an altered sequence. *Solve:* **(a)** Mutation causes a tRNA carrying tyrosine to base-pair with a stop codon. The mutation must be a change in a tRNA^Tyr anticodon sequence. There are two possible tRNA^Tyr anticodon sequences: 5'-GTA-3' and 5'-ATA-3'. For tRNA^Tyr with 5'-GTA-3', the mutation could have changed 5'-GTA-3' to 5'-TTA-3' or 5'-CTA-3'. For tRNA^Tyr with 5'-ATA-3', the mutation could have changed 5'-ATA-3' to 5'-TTA-3' or 5'-CTA-3'. **(b)** The mutation prevents the interaction of tRNA^Leu with its aminoacyl tRNA synthetase. The mutation could be a change in any part of the tRNA^Leu sequence that is required for binding to any aminoacyl tRNA synthetase. **(c)** The mutation causes tRNA^Lys to base-pair with an Arg codon. There are two possible tRNA^Lys anticodon sequences: 5'-TTT-3' and 5'-CTT-3'. Therefore, the mutation could have changed 5'-TTT-3' to 5'-TCT-3', or it could have changed 5'-CTT-3' to 5'-CCT-3'.

10a–c.   *Evaluate:* This is type 3c. *Deduce:* Mutations that do not change the amino acid encoded do not affect the protein. Mutations that change to an amino acid with a similar structure are likely to have little effect on protein structure. Mutations that change to an amino acid with a different structure are likely to have a significant effect on protein structure. *Solve:* **(a)** CTG codes for Leu. Mutation of the first (CTG to TTG) or third position (CTG to CTA) would have no effect, because all are Leu codons. Mutation of the second position (CTG to CCG) would alter the polypeptide structure, inserting Pro in place of Leu. This alteration is likely to be dramatic because of the difference in the structures of Leu and Pro. **(b)** AGA codes for Arg. Mutation of the third position (AGA to AGG) would have no effect, because both code for Arg. Mutation of the second position (AGA to AAA) would alter the structure of the polypeptide, inserting Lys in place of Arg. This alteration may have little effect since Lys and Arg are both positively charged amino acids. Mutation of the first position (AGA to GGA) would alter the polypeptide structure, inserting Gly in place of Arg. This alteration is likely to be dramatic because of the difference in the structures of Arg and Gly. **(c)** TGG to TAG or TGA creates a stop codon, which will truncate the polypeptide at that site. The consequence of this change depends on the location of the TGG codon in the coding sequence. In general, nonsense mutations inactivate protein function. TGG to CGG changes the codon to an Arg codon. Insertion of Arg in place of Trp is likely to have an impact on protein function because of the difference between Trp, which is a bulky, nonpolar amino acid and Arg, which is a positively charged amino acid.

11. *Evaluate:* This is type 1a. The question introduces the phenomenon of "programmed frameshifting" as a planned change in translational reading frame during translation elongation. The example of a yeast retrotransposon is described, and you are given the mRNA sequence that includes the region where translation shifts frame. You are also given the amino acid sequence of the protein in the region of programmed frameshifting. *Deduce:* The codons will come from two reading frames that differ in register by one nucleotide. There are three possible reading frames:

+1 AUG GUG GAU CCA GUG AGA GCA CGU AAC GCG AGU UCU AAC CGA UCU UGA

M V D P V R A R N A S S N R S Stop

+2 UGG UGG AUC CAG UGA GAG CAC GUA ACG CGA GUU CUA ACC GAU CUU GAA

W W I Q STOP E H V T R V L T D L E

+3 GGU GGA UCC AGU GAG AGC ACG UAA CGC GAG UUC UAA CCG AUC UUG

G G S S E S T STOP R E F STOP P I L S

Solve: Reading frame 1 starts at an AUG and codes for the polypeptide in question up until the second arginine (MVDPVRAR). The next codon and all thereafter do not match the polypeptide sequence. Reading frame 2 contains a stop codon but also has the codons for the end of the polypeptide (TRVLTDLE). Thus, the codons that are used in translation of this polypeptide are AUG GUG GAU CCA GUG AGA GCA CGU frameshift ACG CGA GUU CUA ACC GAU CUU GAA. 5'-AUG GUG GAU CCA GUG AGA GCA CGU ACG CGA GUU CUA ACC GAU CUU GAA AUG GUG GAU CCA GUG AGA GCA CGU-3' are in reading frame 1. 5'-ACG CGA GUU CUA ACC GAU CCU GAA-3' are in reading frame 2.

12a. *Evaluate:* This is type 1a. This problem provides an ssDNA template strand sequence. *Deduce > Solve:* The mRNA will be complementary and antiparallel to the template and will contain U instead of T. That sequence will be 5'-CACUAGUGGAGGAAUGGCUUGC GAUGAAUUCGGACAUAUAAAACUAAACCCCCAGCGAAGUACCGUGUGGUACU AGAAUCGC-3'.

12b. *Evaluate:* This is type 1a. *Deduce > Solve:* The reading frame starting with the first AUG is AUG GCU UGC GAU GAA UUC GGA CAU AUA AAA CUA AAC CCC CAG CGA AGU ACC GUG UGG UAC UAG, which codes for Met-Ala-Cys-Asp-Glu-Phe-Gly-His-Ile-Lys-Leu-Asn-Pro-Gln-Arg-Ser-Thr-Val-Trp-Tyr, which is MACDEFGHIKLNPQRSTVWY in the one-letter code.

12c. *Evaluate:* This is type 1a. *Deduce > Solve:* The reading frame starting with the second AUG is AUG AAU UCG GAC AUA UAA, which codes for Met-Ans-Ser-Asp-Ile, which is MNSDI in the one-letter code.

12d. *Evaluate:* This is type 1a. *Deduce > Solve:* Insertion of a G after the first AUG codon shifts the reading frame one position relative to the wild-type mRNA. The new reading frame is AUG GGC UUG CGA UGA, which codes for Met-Gly-Leu-Arg, which is MGLR in the one-letter code.

12e. *Evaluate:* This is type 1a. *Deduce > Solve:* Insertion of a G after the second AUG codon changes the reading frame to AUG GAA UUC GGA CAU AUA AAA CUA AAC CCC CAG CGA AGU ACC GUG UGG UAC UAG, which codes for Met-Glu-Phe-Gly-His-Ile-Lys-Leu-Asn-Pro-Gln-Arg-Ser-Thr-Val-Trp-Tyr, which is MEFGHIKLNPQRSTVWY in the one-letter code.

12f. *Evaluate > Deduce > Solve:* This is type 1a. The polypeptides that are answers to 12b and 12e are the most similar; they share the amino acid sequence EFGHIKLNPQRSTVWY. This is true even though the 12b polypeptide is derived from translation of a wild-type sequence and the 12e polypeptide is derived from translation of a mutant sequence. The reason for this is that the reading frame of mutant sequence in 12e is shifted 1 nucleotide by the G insertion after the second AUG codon. The reading frame from this point on is the same as the reading frame of the wild-type sequence used in 12b.

13a–g. *Evaluate:* This is type 3c. *Deduce > Solve:* **(a)** 5'-AAC-3' pairs with the codon 5'-GUU-3', which codes for valine. **(b)** 5'-AGG-3' pairs with the codon 5'-CCU-3', which codes for proline. **(c)** 5'-GCC-3' pairs with the codon 5'-GGC-3', which codes for glycine. **(d)** 5'-CGG-3' pairs with the codon 5'-CCG-3', which codes for proline. **(e)** 5'-UAC-3' pairs with the codon 5'-GUA-3', which codes for valine. **(f)** 5'-UCC-3' pairs with the codon 5'-GGA-3', which codes for glycine. **(g)** Isoaccepting tRNAs are tRNAs with different anticodon sequences that carry the same amino acid. The tRNAs in 13a and 13e are isoaccepting tRNAs for valine. The tRNAs in 13b and 13d are isoaccepting tRNAs for proline. The tRNAs in 13c and 13f are isoaccepting tRNAs for glycine.

14a–d. *Evaluate:* This is type 3a. *Deduce:* The figure indicates the sequence of the polypeptide encoded by the mRNA but not the orientation of the mRNA nor the direction the ribosome is moving. The polypeptide sequence from amino to carboxyl terminus is Met-Asn-Phe-Ala-Pro; therefore, the codons for these amino acids should be, in 3'-to-5' order, Pro (currently in the P site), Ala (in the E site), Phe, Asn, and Met (AUG). *Solve:* **(a)** The 5' end of the mRNA is to the right. **(b)** The ribosome is moving right to left. **(c)** The A site is on the left, and the E site is on the right. **(d)** Val-tRNA^Val will be the next to enter the A site.

The Integration of Genetic Approaches: Understanding Sickle Cell Disease

10.01 Genetics Problem-Solving Toolkit

Key Terms and Concepts

Human hemoglobin: The oxygen-binding molecule present in red blood cells. It is composed of two α-globin proteins and two β-globin proteins.

Human β-globin protein: A 146-amino-acid-long polypeptide. There are many known β-globin variants in which one or more amino acids differ from the wild-type form.

β^A: is the wild-type allele of the human β-globin gene and encodes a β-globin protein that contributes to formation of a normal, functional hemoglobin molecule.

β^S: is a variant allele of the human β-globin gene and encodes a β-globin protein that contributes to formation of a defective hemoglobin molecule that polymerizes within red blood cells under conditions of low oxygen pressure.

sickle cell disease (SCD): A disease caused by homozygosity for the β^S allele and characterized by severe anemia due to frequent polymerization of hemoglobin inside red blood cells, which causes them to take on a rigid sickle shape that causes the cells to lyse.

Sickle cell trait: A trait caused by heterozygosity for the β^S allele and characterized by the presence of a low level of sickle-shaped red blood cells due to the occasional polymerization of hemoglobin. The occasional sickling of red blood cells decreases their average life span, which limits the reproduction of the malaria protist and provides a level of resistance to malaria but does not lead to anemia.

Key Analytical Tools

Frequency of $\beta^S = \dfrac{(\text{number of } \beta^S \text{ alleles})}{(\text{number of } \beta^A \text{ alleles} + \text{number of } \beta^S \text{ alleles})}$, if β^S and β^A are the only alleles in the population.

Number of β^S alleles = 2 × (number of $\beta^S\beta^S$ homozygotes) + (number of $\beta^A\beta^S$ heterozygotes).

Number of β^A alleles = 2 × (number of $\beta^A\beta^A$ homozygotes) + (number of $\beta^A\beta^S$ heterozygotes).

Punnett square for the mating between carriers of sickle cell disease ($\beta^A\beta^S \times \beta^A\beta^S$):

	$\frac{1}{2}\beta^A$	$\frac{1}{2}\beta^S$
$\frac{1}{2}\beta^A$	$\frac{1}{4}\beta^A\,\beta^A$	$\frac{1}{4}\beta^S\,\beta^A$
$\frac{1}{2}\beta^S$	$\frac{1}{4}\beta^A\beta^S$	$\frac{1}{4}\beta^S\beta^S$

Eukaryotic Gene

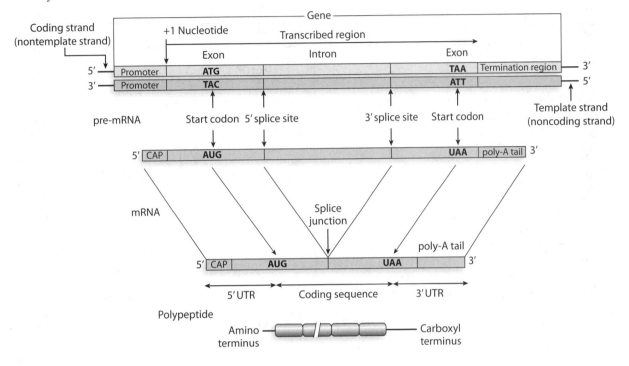

Human β-globin Gene

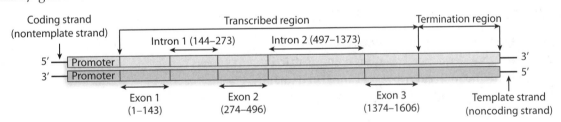

Gel Electrophoretic Phenotypes

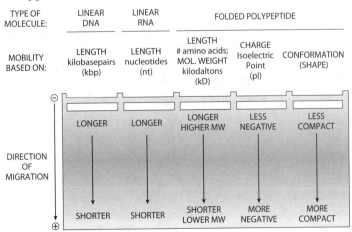

Key Genetic Relationships

The Genetic Code

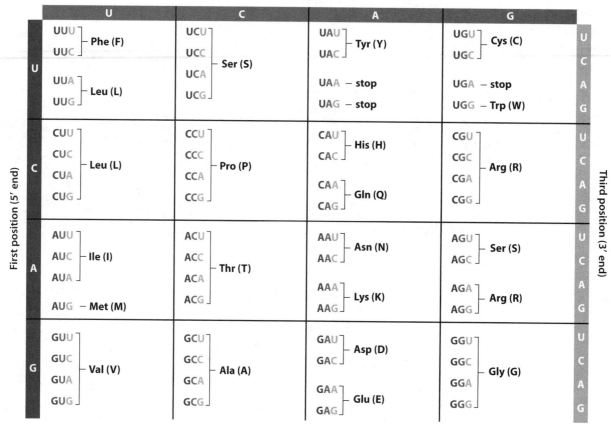

10.02 Types of Genetics Problems

1. Analyze molecules by gel electrophoresis.
2. Analyze DNA sequence variation using restriction enzymes.
3. Investigate the molecular basis of inherited anemias.
4. Explain the population genetics of sickle cell disease.

1. Analyze molecules by gel electrophoresis.

Problems of this type ask you to consider the relationship between molecular structure and the relative electrophoretic mobility of DNA, RNA, or protein molecules. You may be asked to describe the structural parameters that affect relative mobility, or you may be given information on structure and asked to predict the results of electrophoresis. For all three types of molecules, larger molecules migrate more slowly than smaller molecules of the same composition. For proteins, shape (conformation) and charge also play a role in relative mobility; more compact proteins and proteins of greater negative charge have faster rates of migration.

Variations

(a) Analyze DNA molecules by electrophoresis.

The relative mobility of linear DNA molecules is directly proportional to their length. The DNA molecules analyzed in these problems are restriction fragments generated by restriction enzyme digestion. Thus, when comparing the relative mobility of two genes (or any two DNA molecules), it is important to remember that you are comparing the relative mobility of specific restriction fragments, not necessarily the entire gene. Differences in the distance between restriction sites in the DNA molecules will determine the length of the fragments, and, therefore, their relative mobility.

(b) Analyze RNA molecules by electrophoresis.

The relative mobility of RNA molecules is directly proportional to their length. The RNA molecules analyzed in these problems are mRNAs, whose length is defined by the structure of the gene that encodes them. The relative mobility of mRNAs will be determined by the total length of all their exons. mRNAs from genes with longer total exon sequence will migrate more slowly than mRNAs from genes with shorter total exon sequence.

(c) Analyze protein molecules by electrophoresis.

The relative mobility of protein molecules depends on the conditions of gel electrophoresis.
TIP: Assume that the conditions of electrophoresis in these problems can detect changes in protein size, shape, and charge. The proteins analyzed in these problems are the products of wild-type or variant globin genes. Thus, the proteins being compared are typically the same size (same number of amino acids) but differ in their identity by one amino acid. The effect of the single amino acid difference on protein shape is complex, making the relative mobility of two globins that differ at only one amino acid difficult to predict. If the difference affects charge (substitution of a charged amino acid for an uncharged one, or vice versa, or substitution of a charged amino acid with one of opposite charge), then it is reasonable to hypothesize that the more negatively charged protein will migrate more rapidly.

Example Problem: The figure shows two alleles of a chromosomal locus that contains a gene of interest. The sites where the restriction enzymes *Eco*RI (E) and *Bam*HI (B) cut are indicated. One map shows the region that includes the wild-type gene. The other map shows the region that includes a mutant allele, which contains an additional 1-kb DNA insertion in the gene (the inserted DNA is indicated by the black box). Predict the results of gel electrophoresis of samples of these two DNA fragments under the following conditions: uncut (intact), cut with *Bam*HI alone, cut with *Eco*RI alone, and cut with *Bam*HI plus *Eco*RI together. Illustrate your answer by drawing an image of the gel and labeling the sizes of all DNA fragments.

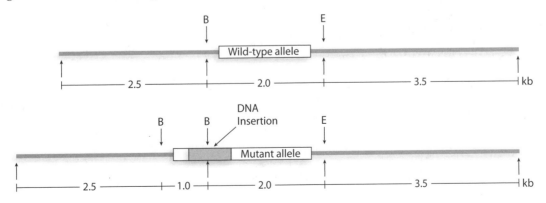

Solution Strategies	Solution Steps
Evaluate	
1. Identify the topic this problem addresses, and explain the nature of the requested answer.	1. This problem asks you to predict the results of restriction digestion and gel electrophoresis of two different versions of a chromosome region containing a gene.
	The answer will be a drawing of a gel with eight lanes, four lanes for each chromosome fragment. The lanes will show the result of electrophoresis of the uncut fragments and the fragments after digestion with *Bam*HI alone, *Eco*RI alone, or *Bam*HI and *Eco*RI together. The number and size of DNA fragments present during electrophoresis under each condition for each chromosome must be determined.
2. Identify the critical information given in the problem.	2. The two chromosomes are identical except for a 1-kb insertion into the mutant gene. Both chromosome fragments have a single *Eco*RI site. The wild-type chromosome has one *Bam*HI site. The mutant chromosome fragment has two *Bam*HI sites. The mutant chromosome fragment is 1.0 kb longer than the wild type.
Deduce	
3. Determine the number and size of DNA fragments that will be present during electrophoresis under each condition and for each chromosome. TIP: The restriction map for each chromosome fragment is linear; therefore, the relationship between numbers of fragments and number of digestion sites is 0 sites = 1 fragment; 1 site = 2 fragments; 2 sites = 3 fragments; X sites = $X + 1$ fragments. TIP: The size of each fragment is the total distance from one end of the fragment to the other, as indicated in the map.	3. The number and size of fragments for each chromosome under each condition are a. Wild-type chromosome fragment—No digestion, one 8 kb; *Eco*RI alone, one 4.5 kb and one 3.5 kb; *Bam*HI alone, one 5.5 kb and one 2.5 kb; both, one 3.5 kb, one 2.5 kb, and one 2.0 kb. b. Mutant chromosome fragment—No digestion, one 9 kb; *Eco*RI alone, one 5.5 kb and one 3.5 kb; *Bam*HI alone, one 5.5 kb, one 2.5 kb, and one 1.0 kb; both, one 3.5 kb, one 2.5 kb, one 2.0 kb, and one 1.0 kb.

(continued)

Solution Strategies	Solution Steps
Solve	
4. Draw the gel, labeling the lanes according to the DNA sample and the type of digest. Draw lines representing DNA fragments in each lane according to your answer in step 3.	**4. Answer:** 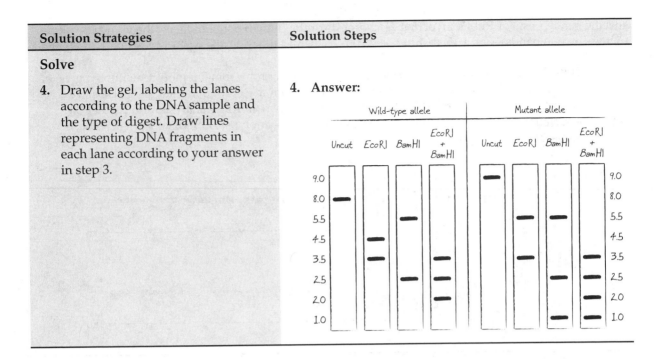

2. Analyze DNA sequence variation using restriction enzymes.

Problems of this type ask you to consider the function of restriction enzymes and their utility in the analysis of a specific form of DNA sequence variation called restriction fragment length polymorphisms, or RFLPs. The problem may focus on the restriction endonucleases and their recognition sequences. Alternatively, you may be asked to consider the structure and function of molecular probes used to identify specific restriction fragments on a Southern blot. You may also be asked to predict the results of a Southern blot analysis, given a restriction map, or to use the results of the Southern blot to assign genotypes to individuals.

Variations

(a) Describe the function of restriction enzymes.

Most common restriction endonucleases recognize palindromic sequences, so assume that this is true unless you are told otherwise. When predicting the structure of ends generated by restriction enzymes, consider whether the enzyme cuts before, at, or after the middle of the recognition sequence. Cutting before the middle produces ends with single-stranded overhangs ending with a 5′ phosphate group, cutting in the middle produces blunt ends, and cutting after the middle produces ends with single-stranded overhangs ending with a 3′ hydroxyl group.

(b) Design a molecular probe that could be used in a Southern blot analysis.

For Southern blot analysis, molecular probes are single-stranded DNA or RNA sequences that are complementary to at least a portion of the restriction fragment to be detected on the blot. To design a probe using a restriction map instead of a nucleic acid sequence, choose a region that lies within the restriction fragment you wish to identify on the Southern blot. For analysis of an RFLP, design your probe to be within a region that corresponds to restriction fragments that differ in length for each allele.

(c) Interpret an RFLP result, or predict the results of an RFLP analysis.

If you are asked to predict the Southern blot banding pattern in an RFLP analysis, you will be given restriction maps of the polymorphic region and shown the position on the map corresponding to the molecular probe. These problems are solved similarly to the DNA gel electrophoresis example problem above, except that only the restriction fragments that contain the region corresponding to the molecular probe will be detected on the Southern blot. If you are asked to interpret the RFLP Southern blot banding pattern of several individuals, you will be given information assigning specific banding patterns to specific genotypes. For RFLPs involving *Dde*I digests of the human β-globin locus, the wild-type allele (β^A) yields two bands—one 1.15 kb and one 0.2 kb—whereas the SCD allele (β^S) yields one 1.35-kb band.

Example Problem: The figure shows a restriction map of a chromosomal region containing a gene of interest. The top map is of the wild-type allele, and the bottom map is of a mutant allele. Design a molecular probe that could differentiate between these alleles in an RFLP analysis, and predict the results of such an analysis for individuals with the following genotypes: homozygous wild type, heterozygous, and homozygous mutant.

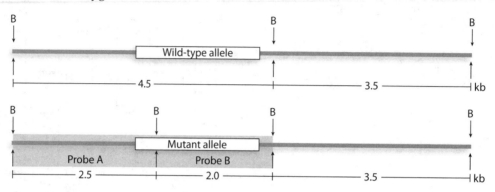

Solution Strategies	Solution Steps
Evaluate	
1. Identify the topic this problem addresses, and explain the nature of the requested answer.	1. This problem asks you to design a probe that could be useful in an RFLP analysis of the alleles shown and to predict the results of that analysis for individuals with specific genotypes. The answer will include a region of the maps to be used as a probe as well as a gel showing the banding patterns for all three genotypes.
2. Identify the critical information given in the problem.	2. The two chromosome fragments are identical except for an additional *Bam*HI site in the mutant allele. Both chromosome fragments have *Bam*HI sites at each end and a *Bam*HI site 3.5 kb in from the right end. The mutant chromosome fragment has an additional *Bam*HI site that is 2.5 kb in from the left end.
Deduce	
3. Determine the number and size of DNA fragments that will be present during electrophoresis after restriction digestion of each allele.	3. Digestion with *Bam*HI creates the following fragments: (a) wild-type allele: two fragments, one 4.5 kb and the other 3.5 kb; (b) mutant allele: three fragments, one 3.5 kb, one 2.5 kb, and one 2.0 kb.
4. Identify regions of the two alleles that are part of restriction fragments that differ in size. **TIP:** These regions will be useful as molecular probes.	4. The DNA sequence in the 4.5 kb fragment in the wild-type allele corresponds to two restriction fragments in the mutant allele: the leftmost 2.5-kb fragment and the central 2.0-kb fragment.

(continued)

Solution Strategies	Solution Steps
Solve	
5. Select a region for use as a molecular probe and then predict which fragments will be identified by that probe on the Southern blot.	5. **Answer:** See figure below for the locations of probes that could be used. 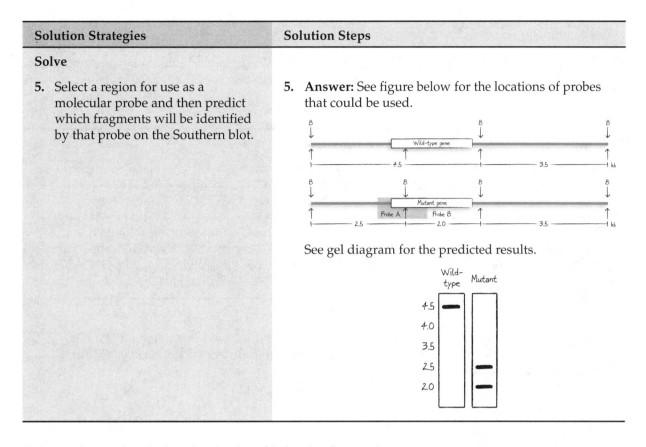 See gel diagram for the predicted results.

3. Investigate the molecular basis of inherited anemias.

These problems focus on forms of inherited anemia caused by mutations in globin genes. Some problems test your knowledge of the history of the analysis of sickle cell disease (SCD). For these problems, review the material in the textbook. Other problems provide you with partial information on globin gene mutations and ask you to complete that information. You may be given a specific amino acid substitution and asked to infer the wild-type and mutant sequences of the gene and mRNA. Alternatively, you may be given the nucleotide sequence of the mutant and asked to predict the consequence of that mutation on the structure of the globin protein.

Example Problem: You have identified a mutant β-globin allele that differs from the wild type by only one base pair. The mutant allele has one recognition site for the restriction enzyme, *Pst*I, whereas the wild-type gene has no *Pst*I sites. *Pst*I recognizes the sequence 5′-CTGCAG-3′. In the mutant allele, this sequence is from nucleotides 275 through 280. Use this information to determine the nucleotide substitution that created this mutant allele and the resulting amino acid substitution that characterizes the mutant protein.

Solution Strategies	Solution Steps
Evaluate	
1. Identify the topic this problem addresses, and explain the nature of the requested answer.	1. This problem provides partial information on a mutant β-globin allele and asks for the complete description of the allele. The answer will be the specific nature of the mutation that created the allele, and the resulting amino acid substitution caused by the mutation.

(continued)

Solution Strategies	Solution Steps
2. Identify the critical information given in the problem.	**2.** You are given the following information: the mutation is a single-nucleotide substitution; the mutation creates a *Pst*I site at position 275–280; a *Pst*I site is 5′-CTGCAG-3′.
Deduce	
3. Identify the nucleotide affected by the mutation.	**3.** The sequence from 275 to 280 is the following: wild type = 5′-CTGCTG-3′; mutant = 5′-CTGCAG-3′; nucleotide affected is 5′-CTGC<u>T</u>G-3′.
4. Identify the codon affected by the mutation, and deduce the amino acid substitution caused by the mutation.	**4.** The reading frame is the following: wild type = 5′-CTGCTG-3′; mutant = 5′-CTGCAG-3′. The codon affected is 5′-CTG-3′, which is changed to 5′-CAG-3′; 5′-CTG-3′ to 5′-CAG-3′ is Leu to Gln.
Solve	
5. Formulate the answer by indicating the nucleotide substitution and the amino acid substitution.	**5.** **Answer:** The nucleotide substitution mutation is T279A. The amino acid substitution is Leu32Gln.

TIP: Nucleotide substitution is written as (wild-type nucleotide, nucleotide position, mutant nucleotide).

TIP: Amino acid substitution is written as (wild-type amino acid, amino acid number, mutant amino acid).

4. Explain the population genetics of sickle cell disease.

These problems ask you to apply your knowledge of the molecular basis of SCD and the life cycle of the malaria protist to answer questions concerning the frequency of human β-globin alleles in different populations. The key phenotypes are sickle cell disease in $\beta^S\beta^S$ homozygotes, sickle cell trait in $\beta^S\beta^A$ heterozygotes, and normal phenotype in $\beta^A\beta^A$ homozygotes. The major evolutionary forces at work are selection against $\beta^S\beta^S$ homozygotes in all environments because of severe anemia and selection against $\beta^A\beta^A$ homozygotes in geographic regions where malaria is endemic. The problems may provide you with information on β^S allele frequencies in different populations and ask you to generate a hypothesis to explain them. Alternatively, you may be given information on the environments of two different populations and asked to predict the relative frequency of the β^S allele in the populations.

Example Problem: You have collected DNA samples from 500 individuals in a newly discovered, isolated Southeast Asian population from a region where malaria is endemic. You noticed that an inherited form of severe anemia is present at an easily detectable level in the population. You analyzed the DNA samples to determine the frequency of the β^S allele in this population by looking for the associated *Dde*I RFLP. Your probe identified two bands, 1.15 kb and 0.2 kb, from every DNA sample. No sample contained a 1.35 kb band. Assume that your results are correct (there is no mistake). How would you interpret the RFLP result, and what hypothesis would you propose to explain the result? How would you test that hypothesis?

Solution Strategies	Solution Steps
Evaluate	
1. Identify the topic this problem addresses, and explain the nature of the requested answer.	1. This problem asks you to formulate and test a hypothesis concerning hereditary anemia and β^S allele frequency in a Southeast Asian population. The answer must explain the absence of the β^S allele, the presence of hereditary anemia, and the presence of the malaria parasite.
2. Identify the critical information given in the problem.	2. The problem supplies the following information: the population included detectable numbers of individuals with hereditary anemia; Southern blot analysis using *Dde*I and a β-globin probe detects 1.15- and 0.2-kb bands but no 1.35-kb band; and malaria is endemic.
Deduce	
3. Recall information relating the results of Southern blot analysis using *Dde*I and β-globin genotype, and on hereditary anemia due to alleles other than β^S.	3. Hereditary anemia in this population must be due to an allele other than β^S. Southern blot analysis failed to detect β^S in this population. β^E is an allele associated with hereditary anemia. $\beta^E\beta^A$ heterozygotes are resistant to malaria. β^E is not detected as an RFLP using *Dde*I. β^E is due to a G-to-A mutation at position 130 in the β-globin gene, which causes a Glu-to-Lys substitution in the β-globin polypeptide.
Solve	
4. Use the information on hereditary anemia and the β^E allele to formulate a hypothesis that explains the information provided by this problem, and design a test of that hypothesis.	4. **Answer:** One hypothesis that explains the information is that the hereditary anemia in this population is due to the β^E mutation. This could be tested by assaying the DNA samples for the nucleotide substitution G to A at position 130 in the β-globin gene.

10.03 Solutions to End-of-Chapter Problems

1a. *Evaluate > Deduce > Solve: A gene cluster* is a collection of genes that are physically located close to each other on a chromosome and that have a related function, for example, the globin gene cluster in humans.

1b. *Evaluate > Deduce > Solve: Heterozygous advantage* indicates that the heterozygous genotype is more fit (reproduces more successfully, has a selective advantage) than either homozygous genotype; for example, $\beta^A\beta^S$ is more fit than $\beta^A\beta^A$ or $\beta^S\beta^S$ in an environment where malaria is prevalent.

1c. *Evaluate > Deduce > Solve: Dominant* refers to a trait (phenotype) observed in a homozygote that is also observed in a heterozygote. Heterozygotes result from crossing a dominant homozygote with a homozygote expressing a different, recessive trait. These phenotypic descriptions are commonly ascribed to genotypes (although, strictly speaking, this is not correct usage of *dominant*), so a heterozygote may be described as having both dominant and recessive alleles. The allele that corresponds to the dominant trait is referred to as dominant; for example, normal (healthy) is dominant to sickle cell anemia, so the β^A allele is said to be dominant to β^S.

1d. *Evaluate > Deduce > Solve:* An *intron* is an internal segment of the transcribed region of a gene that is removed from the pre-mRNA by a process called splicing. Introns are found in most eukaryotic protein coding genes and their pre-mRNAs; for example, the human β-globin gene and pre-mRNA have two introns.

1e. *Evaluate > Deduce > Solve:* The *hemoglobin tetramer* is a molecule composed of two α-globin polypeptides and two β-globin polypeptides, each of which contains a heme group. The hemoglobin tetramer is also simply called hemoglobin.

1f. *Evaluate > Deduce > Solve:* *Hereditary anemia* refers to all forms of anemia (abnormally low numbers of red blood cells) that are due to inheritance of a particular allele of a gene (and therefore are genetic), for example, sickle cell anemia.

1g. *Evaluate > Deduce > Solve:* An *exon* is a segment of the transcribed region of a gene that is not removed from the pre-mRNA by splicing. Every intron is flanked by exons; for example, the human β-globin gene and pre-mRNA have three exons.

1h. *Evaluate > Deduce > Solve:* *Heterozygous* refers to genotypes in which the alleles at a genetic locus differ. For example, $\beta^A\beta^S$, $\beta^A\beta^E$, and $\beta^A\beta^C$ are three examples of heterozygous genotypes.

1i. *Evaluate > Deduce > Solve:* *Recessive* refers to a trait (phenotype) that is observed in a homozygote but is not observed in a heterozygote. Heterozygotes result from crossing a dominant homozygote with a homozygote expressing a recessive trait. These phenotypic descriptions are commonly ascribed to genotypes (although, strictly speaking, this is not correct usage of *recessive*). The allele that corresponds to the recessive trait is referred to as recessive; for example, sickle cell anemia is recessive to normal (healthy), and the β^S allele is said to be recessive to β^A.

1j. *Evaluate > Deduce > Solve:* The term *molecular disease* is essentially synonymous with the term *genetic disease*, but it is used to emphasize that the genetic disease is understood at the molecular level. For example, SCD was the first genetic disease to be identified as a molecular disease when it was discovered that the sickle cell mutation changed the sequence of the β-globin polypeptide, resulting in a change in the structure and function of hemoglobin, which led hemoglobin to polymerize in red blood cells (RBC) under low oxygen conditions, RBCs to rupture, and anemia to occur.

1k. *Evaluate > Deduce > Solve:* A *restriction endonuclease* is an enzyme that catalyzes the hydrolysis of phosphate ester bonds in DNA at a specific DNA sequence. The sequence is called a restriction endonuclease recognition site, and the restriction enzyme is said to "cut" the DNA at its restriction site. For example, *Bam*HI is a restriction endonuclease that cuts DNA at the *Bam*HI site, which is 5'-GGATCC-3'.

1l. *Evaluate > Deduce > Solve:* *Homozygous* refers to genotypes in which the alleles at a genetic locus are the same. For example, $\beta^A\beta^A$, $\beta^S\beta^S$, and $\beta^C\beta^C$ are three examples of homozygous genotypes.

1m. *Evaluate > Deduce > Solve:* *Gel electrophoresis* is a technique used to separate macromolecules (DNA, RNA, or protein) based on differences in their size, charge, or shape.

1n. *Evaluate > Deduce > Solve:* A *restriction fragment length polymorphism* is literally a DNA polymorphism that is detectable by a difference in the length of restriction fragments. A DNA polymorphism is a difference in DNA sequence at the same genetic locus among individuals in a population. When the difference in DNA sequence results in a difference in the lengths of DNA fragments created by digestion with restriction enzymes, this is called a restriction fragment length polymorphism.

1o. *Evaluate > Deduce > Solve:* *SNP* is the acronym for single nucleotide polymorphism. A SNP is a DNA polymorphism in which the difference between DNA sequences at the same genetic locus is a single base pair. For example, if members of a population differ in sequence at a particular site in the genome (e.g., 5'-G<u>G</u>ATCC-3' vs. 5'-G<u>A</u>ATCC-3'), the site corresponding to the difference is called a SNP (G vs. A).

1p.   *Evaluate > Deduce > Solve: Electrophoretic mobility* refers to the rate at which a molecule migrates during electrophoresis. Often, the distance migrated after a fixed period of time is used as an indirect indicator of electrophoretic mobility.

1q.   *Evaluate > Deduce > Solve: Southern blot* is a method developed by Edwin Southern for transferring DNA from an agarose gel to a membrane. The DNA-containing membrane is referred to as a Southern blot and is typically "probed" with a DNA or RNA probe to identify a specific restriction fragment.

1r.   *Evaluate > Deduce > Solve:* A *molecular probe* is a labeled DNA, RNA, or antibody molecule that is capable of binding to a specific target molecule (DNA, RNA, or protein). For example, radioactively labeled (^{32}P) DNA molecules complementary to specific DNA fragments are used to probe Southern blots. The location of the DNA probe is detected by exposing the probed blot to X-ray film and then developing the film. A dark band will be present on the film at the site corresponding to the location of the probe.

1s.   *Evaluate > Deduce > Solve: Northern blot* is a method for transferring RNA from an agarose gel to a membrane. The RNA-containing membrane is referred to as a northern blot and is typically probed with a DNA or RNA probe to identify a specific RNA molecule.

1t.   *Evaluate > Deduce > Solve:* An *antibody probe* is a labeled antibody that binds to a specific protein. For example, radioactively labeled antibodies (^{125}I) are used to probe western blots. The location of the antibody probe is detected by exposing the probed blot to X-ray film and then developing the film. A dark band will be present on the film at the site corresponding to the location of the probe.

1u.   *Evaluate > Deduce > Solve: Western blot* is a technique for transferring protein from a polyacrylamide gel to a membrane. The protein-containing membrane is referred to as a western-blot and is probed with an antibody probe to identify a specific protein molecule.

2.   *Evaluate:* This is type 3. Recall that genetic diseases are those determined by genetic factors in the individual's own genome. On the other hand, infectious diseases are caused by foreign agents that have invaded the individual. *Deduce > Solve:* All genetic diseases are also molecular diseases since the genes involved encode molecules that are responsible for development of the disease. The only difference between the terms *genetic* and *molecular* is their point of emphasis: *genetic* emphasizes that it is not an infectious disease, whereas *molecular* emphasizes that defects in one or more molecules are the ultimate cause of the disease. Sickle cell disease is the first genetic disease for which the molecular mechanism was understood. SCD is seen in individuals homozygous for the β^S allele, which encodes an altered β-globin polypeptide that leads to formation of an altered hemoglobin protein, polymerization in red blood cells under low oxygen conditions, the cells with a sickle shape, blockage of capillaries in peripheral tissue, and ultimately pain and tissue damage.

3.   *Evaluate > Deduce > Solve:* This is type 3. Recall James Neel, Linus Pauling, and Vernon Ingram's successive contributions to the discovery of the molecular basis of sickle cell disease. Neel hypothesized that SCD was a recessive trait and that the parents of patients with SCD would, therefore, be heterozygous carriers. He confirmed this by demonstrating that both parents of multiple patients with SCD had low levels of sickled red blood cells (a phenotype called sickle cell trait). Pauling demonstrated that SCD and sickle cell trait were due to the presence of abnormal β-globin polypeptides in red blood cells. He demonstrated this by showing that the electrophoretic mobility of β-globin from patients with SCD was different from that in healthy individuals, and that heterozygous carriers contained β-globin polypeptides with normal and altered electrophoretic mobilities. Ingram revealed that the ultimate difference between β-globin polypeptides in individuals who were healthy and those with SCD is an amino acid substitution. He demonstrated this by performing a peptide fingerprint analysis on β-globin from individuals with $\beta^A\beta^A$ and $\beta^S\beta^S$, which identified the segment of β-globin that was changed by the β^S mutation. He then determined that the amino acid difference between β^S and β^A peptides was a difference in the sixth amino acid in the β-globin polypeptide (i.e., glutamic acid in β^A but valine in β^S).

4. *Evaluate:* This is type 1c. Recall that electrophoretic mobility is the rate at which a molecule migrates during electrophoresis. The electrophoretic mobility of a protein is determined by its charge, size, and shape, all of which are determined by its amino acid sequence. *Deduce > Solve:* Two proteins with different electrophoretic mobility typically have some difference in amino acid sequence. If different alleles of a protein-coding gene code for polypeptides differ in sequence, even if they differ at only one amino acid position, the charge or shape of the proteins can differ and result in differences in electrophoretic mobility.

5. *Evaluate:* This is type 1c. This problem requires you to consider the relationships between hereditary anemia, hemoglobin structure, and protein electrophoretic mobility. *Deduce > Solve:* Hereditary anemia is caused by defects in hemoglobin due either to abnormal α-globin or β-globin protein. One possibility is if the hereditary anemia is caused by variation in α-globin genes instead of β-globin genes: This person would have the genotype $\beta^A\beta^A$. A second possibility is that the person has variant forms of the β-globin gene, but these variants do not alter the electrophoretic mobility of the β-globin polypeptide. Although β-globin polypeptides with different electrophoretic mobilities almost certainly have different amino acid sequences, we cannot assume that β-globin polypeptides with different amino acid sequences necessarily have different electrophoretic mobilities.

6. *Evaluate:* This is type 3. To determine the template strand sequence using information about amino acids, we first need to determine the mRNA codon and the coding strand, respectively, since the template strand is complementary to the coding strand. Recall that an amino acid may have a couple different possible codons (see The Genetic Code in Section 10.1, The Genetics Problem-Solving Toolkit), and create a codon column in the table identifying the possibilities for each amino acid, as shown. Use Figure 10.9 in the textbook to identify the sequence of the coding and template strands for each wild-type codon. *Deduce > Solve:* Analyze the possible codons to determine what *single* nucleotide change can convert the wild-type codon into the mutant codon, and eliminate all other possibilities. Recall that the coding strand has the same sequence as the mRNA codon, except that U is replaced by T because it is a DNA sequence, and create a column identifying the coding strand for each amino acid. Finally, deduce the complementary strand and create a column with this information, called "Template Strand (DNA)," as shown.

β-Globin Form	Position	Amino Acid	Codon (mRNA)	Coding Strand (DNA)	Template Strand (DNA)
β^A (wild type)	7	Glu	5'-GAG-3' ~~GAA~~ TIP: Reference Figure 10.9 in the textbook to determine the wild type codon.	5'-GAG-3'	5'-CTC-3'

(continued)

β-Globin Form	Position	Amino Acid	Codon (mRNA)	Coding Strand (DNA)	Template Strand (DNA)
Siriraj	7	Lys	5′-AAG-3′ ~~AAA~~ TIP: Eliminate possibilities that have more than one base change.	5′-AAG-3′	5′-CTT-3′
San Jose	7	Gly	5′-GGG-3′ ~~GGU~~ ~~GGC~~ ~~GGA~~	5′-GGG-3′	5′-CCC-3′
β^A (wild type)	58	Pro	5′-CCU-3′	5′-CCT-3′	5′-AGG-3′
Ziguinchor	58	Arg	5′-CGU-3′ ~~CGC~~ ~~CGA~~ ~~CGG~~	5′-CGT-3′	5′-ACG-3′
β^A (wild type)	145	Tyr	5′-UAU-3′ ~~UAC~~	5′-TAT-3′	5′-ATA-3′
Bethesda	145	His	5′-CAU-3′ ~~CAC~~	5′-CAT-3′	5′-ATG-3′
Fort Gordon	145	Asp	5′-GAU-3′ ~~GAC~~	5′-GAT-3′	5′-ATC-3′

7. *Evaluate:* This is type 3. By comparing the sequences, we can determine that the Hb^{CS} mutation changes UAA, a stop codon, to a CAA, a Gln codon **TIP:** Reference The Genetic Code in Section 10.1, The Genetics Problem-Solving Toolkit, to determine the effect of changing U to C in the mRNA. *Deduce > Solve:* Since translation continues until a stop codon is reached, and since converting UAA to CAA removes the stop codon, translation continues past the normal stop codon until reaching the next in-frame stop codon. We can deduce that this is the cause of the difference in amino acid length, and that the next stop codon does not occur for another 31 amino acids.

8. *Evaluate:* This is type 3. Recall that mutations that increase the length of a polypeptide either change the stop codon to a sense codon or shift the reading frame before, but close to, the stop codon. By comparing the sequences, we can determine that the wild-type and mutant sequence differ immediately after codon 144. *Deduce > Solve:* The continued change in sequence after codon 144 rules out base-substitution mutants, which would affect only specific bases, and points to a shift in the reading frame caused by the insertion of AG between codon 144 and codon 145. The two-nucleotide insertion shifts the reading frame to one that includes 13 sense codons (compared with two sense codons in the wild type) before a stop codon appears, which happens to be after codon 157.

9. *Evaluate > Deduce > Solve:* This is type 3. This problem asks you to distinguish between the sickle cell disease phenotype and other phenotypes associated with heterozygosity for the β^S allele. SCD is recessive because only individuals that are homozygous for the β^S allele have the symptoms of SCD. $\beta^A\beta^S$ heterozygotes are not completely normal but do not have the spectrum of symptoms described as SCD.

10. *Evaluate:* This is type 1a, 1b, and 1c. Recall that DNA or mRNA molecules of the same length have the same electrophoretic mobility regardless of differences in nucleotide sequence and that DNA or mRNA molecules of different lengths have different electrophoretic mobilities. Also recall that proteins which differ in length or amino acid sequence typically differ in electrophoretic mobility. *Deduce > Solve:* The primary molecular parameter affecting the electrophoretic mobility of DNA and mRNA is size (typically expressed as length). For proteins, size (the number of amino acids), charge, and shape affect electrophoretic mobility.

11. *Evaluate > Deduce > Solve:* This is type 1a and 2b. Specific DNA molecules (typically restriction fragments) are detected on Southern blots by probing the blot with a "labeled" DNA or RNA molecule. If the labeled probe is radioactive, then the location of the DNA molecules detected can be revealed by exposing the blot to X-ray film long enough to allow the radioactive probe to decay and emit electrons that expose the film. The developed film is called an autoradiograph (or autoradiogram) that will contain dark bands corresponding to the position where the radioactive probe hybridized to DNA molecules on the blot.

12. *Evaluate > Deduce > Solve:* This is type 1a and 1b. Electrophoretic mobility is a measure of the size (length) of mRNA and DNA molecules; however, comparing them to each other is, in most cases, like comparing apples to oranges. The length of an mRNA is an inherent, biological property of the gene that encodes the mRNA and is not dependent on the method used to prepare the mRNA for gel electrophoresis. The size of the DNA molecule containing the gene is entirely dependent on the experimental method used to prepare it for electrophoresis. For example, the size of a DNA restriction fragment containing a gene is determined by which restriction enzyme is used. DNA fragment sizes will likely differ for different restriction enzymes and is unlikely to be the same length as the mRNA. The exception to this rule is the comparison of a cDNA molecule to an mRNA molecule. A cDNA molecule is a double-stranded DNA copy of the mRNA. If the DNA copy is perfect, then cDNA length in base pairs should be equal to the mRNA length in nucleotides.

13. *Evaluate:* This is type 2b. *Deduce > Solve:* The sequence would be complementary and antiparallel to the target sequence; therefore, it would be 5'-TATAGCGTGCCTGA-3'. The probe will be able to detect the target sequence by base-pairing with the target. The probe itself must be detectable and so must be labeled with either a radioisotope or a fluorescent molecule.

14. *Evaluate:* This is type 4. *Deduce:* The frequency of β^S will be determined by the evolutionary forces acting on the phenotypes of $\beta^S\beta^S$, $\beta^S\beta^A$, and $\beta^A\beta^A$ individuals. The β^S allele encodes an abnormal β-globin polypeptide that forms abnormal hemoglobin molecules that reduce the life span of red blood cells. This is most severe in $\beta^S\beta^S$ homozygotes and is detectable but less severe in $\beta^S\beta^A$ heterozygotes. The reduced life span of red blood cells interrupts the malaria protist's life cycle, resulting in a certain level of resistance to malaria. *Solve:* The severity of the anemia in $\beta^S\beta^S$ individuals decreases their reproductive fitness relative to $\beta^A\beta^A$, which selects against the β^S allele. The less severe effect in $\beta^S\beta^A$ heterozygotes does not affect fitness except in populations where malaria is endemic, where it increases reproductive fitness relative to $\beta^A\beta^A$. Based on this information, a reasonable hypothesis would be that malaria is more prevalent in western Africa than southern Africa.

15. *Evaluate:* This is type 3. Recall the Southern blot results for diagnosis of SCD from Figure 10.13 in the text, which shows that the 1.35-kb band corresponds to the sickle cell allele (β^S) and that the 1.15- and 0.2-kb bands correspond to the wild-type allele (β^A). *Deduce:* I-1, I-2, II-1, and II-4 have the 1.35-kb band and therefore have a β^S allele. I-1, I-2, II-2, II-3, and II-4 have the 1.15- and 0.2-kb bands and therefore have a β^A allele. *Solve:* Individuals I-1, I-2, and II-4 have both alleles and therefore have the genotype $\beta^A\beta^S$. II-1 has only the β^S allele and therefore is $\beta^S\beta^S$. II-2 and II-3 have only the β^A allele and are therefore $\beta^A\beta^A$.

16a. *Evaluate:* This is type 3. I-1 and I-2 are $\beta^A\beta^S$. *Deduce:* By using a Punnett square, we can determine that there is a (1/4) possibility that the couple's fifth child could have SCD. *Solve:* Yes, the fetus has a 1/4 chance of having sickle cell disease.

16b. *Evaluate:* This is type 3. The fetal DNA contains only the 1.35-kb band. *Deduce:* The 1.35-kb band corresponds to the β^S allele. Because there is only one band present, the genotype of the fetus is $\beta^S\beta^S$, which guarantees it will develop SCD. *Solve:* The fetus will develop SCD because it is homozygous for the β^S allele and therefore has the genotype $\beta^S\beta^S$.

17. *Evaluate > Deduce > Solve:* This is type 2a. Restriction endonucleases are enzymes that catalyze the formation of a double-strand break (they are said to "cut") at specific DNA sequences called the enzyme's recognition site. Cutting DNA with a restriction enzyme produces DNA fragments called restriction fragments. The size of each fragment indicates the distance between two restriction sites. DNA sequence variation includes differences in the relative position of an enzyme's restriction sites, which is seen as variation in the size of some restriction fragments. Thus, one can use restriction enzymes to detect DNA sequence variation by looking for variation in the size of specific restriction fragments.

18. *Evaluate > Deduce > Solve:* This is type 2a. Some restriction enzymes make a staggered double-stranded DNA cut at their recognition sequence, cutting the two DNA strands at different positions. This leaves single-stranded DNA ends, called "sticky ends" because they can form base pairs with the single-stranded ends of other restriction fragments, causing the fragments to stick together. Some restriction enzymes make a clean double-stranded DNA cut at their recognition sequence, cutting both strands at the same site. This leaves ends that are blunt in the sense that they contain no nucleotides that are not base paired. Blunt ends, therefore, are not "sticky."

19. *Evaluate:* This is type 2b. The best probe sequence will be identical to part of the sequence of one strand and will be complementary to the other strand. **PITFALL:** Do not ignore the polarity of the DNA sequences provided (for example, the sequence 3'-ACTGC-5' is not the same as 5'-ACTGC-3'). *Deduce:* The sequence 3'-ACTGCCTGATAAGAT-5' is identical to the bottom strand sequence, starting six nucleotides in from the left, and will hybridize to the top strand. None of the other three sequences are identical or complementary to a strand in the dsDNA sequence. *Solve:* Sequence (d) is the best probe. It will hybridize to nucleotides 6 through 20 (from the left) on the upper strand.

20. *Evaluate:* This is type 1. *Deduce:* Restriction enzymes break two phosphodiester bonds between the same nucleotides in the same sequence on both strands of DNA. Two sites of action suggest two active sites in the enzyme. *Solve:* Restriction enzymes typically bind as dimers; each monomer binds to the recognition sequence on opposite strands and cuts that sequence in the same location. Thus, $\begin{array}{l}\text{5'-GGTACC-3'}\\\text{3'-CCATGG-5'}\end{array}$ is bound by a *Bam*HI dimer, and each monomer binds to and positions its active site to break the phosphodiester bond between the G's.

21a. *Evaluate:* This is type 2a. *Deduce:* Lane 1 has two bands (4 kb and 10 kb); therefore, individual 1 has an R^1 and an R^3 allele. Lane 2 has one band (13 kb); therefore, individual 2 has only the R^2 allele. Lane 3 has two bands (7 kb and 13 kb); therefore, individual 3 has an R^4 and an R^2 allele.

21b. *Evaluate:* This is type 2a. *Deduce > Solve:* The R^1R^3 individual will have a 4-kb band and a 10-kb band. The R^3R^4 individual will have a 7-kb band and a 10-kb band. The R^1R^1 individual will have a 4-kb band.

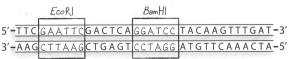

22. *Evaluate:* This is type 2a. *Deduce > Solve:* The sequence has an *Eco*RI site and a *Bam*HI site but no *Hind*III sites. Draw a box around each restriction sequence as shown in the figure.

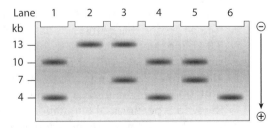

23a.   *Evaluate > Deduce > Solve:* Probe A will detect a single band of 4.0 kb. Probe B will detect two bands, 3.0 kb and 5.0 kb. Both probes together will detect all three bands (3.0, 4.0, and 5.0).

23b.   *Evaluate > Deduce > Solve:* Probe A will detect two bands, 4.0 kb and 7.0 kb. Probe B will detect two bands, 3.0 kb and 8.0 kb. Both probes together will detect all four bands (3.0, 4.0, 7.0, and 8.0).

23c.   *Evaluate > Deduce > Solve:* H^1H^2, H^1H^4, H^3H^2, and H^3H^4.

23d.   *Evaluate:* This is type 2c. *Deduce:* Probe B identifies restriction fragments containing the box on the left, whereas probe A identifies fragments containing the box on the right. Probe B detects a 3.0-kb fragment in alleles H^1 and H^2, a 5.0-kb fragment in allele H^3, and an 8.0-kb fragment in allele H^4. Probe A detects a 4.0-kb fragment in alleles H^1, H^3, and H^4 and a 7.0-kb fragment in allele H^2. The combination of probes A and B detects all fragments identified by both probes. *Solve:* For H^1H^2: probe A detects two bands, 4.0 and 7.0; probe B detects one band, 3.0. For H^1H^4: probe A detects one band, 4.0; probe B detects two bands, 3.0 and 8.0. For H^3H^2: probe A detects two bands, 4.0 and 7.0; probe B detects two bands, 3.0 and 5.0. For H^3H^4: probe A detects one band, 4.0; probe B detects two bands, 5.0 and 8.0.

24a-b.   *Evaluate:* This is type 2c. *Deduce:* The dwarf phenotype is recessive; therefore, the dwarf allele is probably a loss-of-function allele that does not code for a functional protein. The RFLP analysis identifies two bands in the heterozygous parents, a 7.5-kb band and a 4.0-kb band. The restriction map shows that the 7.5-kb band corresponds to the wild-type allele; therefore, the 4.0-kb band must correspond to the mutant allele. This agrees with the Southern blot analysis of the dwarf progeny plant, which must be homozygous mutant and shows only a 4.0-kb band. Thus, the mutational events you propose must be able to change the size of the restriction fragment identified by the probe from 7.5 kb to 4.0 kb and also must convert the wild-type stature allele to a nonfunctional allele. *Solve:* **(a)** The most obvious way to convert the 7.5-kb fragment to a 4.0-kb fragment *and* inactivate the stature gene is to delete 3.5 kb of DNA from the middle of the gene. The deletion must lie within the 7.5-kb fragment but not include the sequence corresponding to the probe. Another way is for a point mutation to occur in the coding sequence of the stature gene that inactivates the gene (a nonsense mutation, for example) and also creates a new restriction site exactly 4.0 kb in from the right end of the map. **(b)** The deletion mutation would remove 3.5 kb of transcribed DNA from the stature gene, which would result in production of an mRNA that is 3.5 kb shorter than the wild type. Whether this mRNA can be detected on a northern blot will depend on whether it is stable enough to accumulate to detectable levels. The point mutation would not alter the size the transcribed region of the stature gene, which would not alter the size of the stature mRNA. Whether this mRNA is detectable on a northern blot will depend on whether it is stable enough to accumulate to detectable levels.

25a-b.   *Evaluate:* This is type 2c. *Deduce:* Since the mutant allele results in a recessive dwarf phenotype, it is a nonfunctional allele. The mutation must therefore increase the size of the restriction fragment identified by the probe *and* inactivate or reduce the function of the gene. *Solve:* **(a)** Two straightforward explanations are a point mutation creating an RFLP or an insertion mutation. The point mutation would have to destroy the restriction site 7.5 kb in from the right end and create a missense mutation, a frameshift mutation, or a nonsense mutation. If it's a missense mutation, it must be one that reduces or destroys the gene's function. The insertion mutation must be a 3.0-kb insertion somewhere within the 7.5-kb restriction fragment, and the inserted sequence cannot have a site for that restriction enzyme. **(b)** The mRNA encoded by this mutant dwarf allele would be the same length as the wild type if the mutation is a point mutation, unless the mutation alters pre-mRNA splicing. The mRNA will most likely be present at the same levels as the wild type, unless the mutation alters mRNA stability. If the mutation is an insertion mutation, the mRNA will most likely be longer if it can be detected at all. This assumes that transcription of the 10.5-kb dwarf allele is initiated normally and then proceeds completely through the inserted sequence and ends at its normal location. It is also possible that sequences in the inserted DNA inhibit transcription through the mutant allele, resulting in shorter mRNAs or mRNAs with lengths similar to wild-type alleles or unstable mRNAs that are present at lower levels.

26.   *Evaluate:* This is type 1a and 1b. Linear DNA molecules can be considered to move through an agarose gel matrix as extended (rod-like) molecules that have the same charge-to-mass ratio regardless of length. *Deduce > Solve:* Since their shape and charge-to-mass ratio is the same, the only factor affecting the relative electrophoretic mobility of linear DNA molecules is their length. The gel will slow the longer molecule to a greater extent than the shorter molecule. The same rationale applies to mRNAs, assuming electrophoresis under conditions where mRNAs are linear.

27.   *Evaluate:* This is type 1c. Recall that size (molecular weight), shape, and charge are the three factors that affect the electrophoretic mobility of proteins. *Deduce > Solve:* A single amino acid substitution typically has a negligible effect on the molecular weight of a protein; therefore, it must affect the protein's shape, charge, or both to affect its electrophoretic mobility. For example, if aspartate is substituted for an uncharged or positively charged amino acid, then the overall charge of the protein will change (and its shape will also likely change). A reasonable hypothesis is that this change would decrease mobility by increasing charge. In practice, it is difficult to accurately predict whether an amino acid substitution will affect the electrophoretic mobility of a protein and, if so, whether the effect will be to increase or decrease its mobility.

28.   *Evaluate:* This is type 2a. Recall that restriction endonucleases are components of bacterial defense systems that protect against invasion by foreign DNA. *Deduce > Solve:* You can infer from the information provided that the restriction enzyme is able to cut foreign DNA, for example, the genome of an invading bacteriophage. DNA from a bacteriophage is digested by the restriction enzyme, which decreases the likelihood that the bacteriophage will successfully infect the bacterium.

29.   *Evaluate:* This is type 1b. Exons and introns are both present in the nuclear pre-mRNA. Only exons are present in the cytoplasmic mRNA. *Deduce:* The cytoplasmic mRNA will have five exons but no introns, totaling 1500 nucleotides. The nuclear pre-mRNA will have five exons and four introns, totaling 5000 nucleotides. *Solve:* Two types of mRNAs will be detected on the blot: a 5000-nucleotide RNA in the nuclear sample and a 1500-nucleotide RNA in the cytoplasmic sample.

30.   *Evaluate:* This is type 2c. *Deduce:* If we number the bands 1 through 4, where 1 is the largest, then ♀1 is 2,4; ♂1 is 1,3; ♂2 is 1,4; P1 is 1,2; P2 is 1,4; and P3 is 3,4. Puppies 1 and 2 received the globin corresponding to band 1 from their father, whereas puppy 3 received the globin corresponding to band 3 from its father. *Solve:* Yes, the father could be ♂1 but not ♂2, because puppy P3 has a band that is not in the mother or in ♂2 but is in ♂1.

31a-c.   *Evaluate:* This is type 2c. *Deduce:* The probe detects a 2-kb band in the D^1 allele, a 3.0-kb band in the D^2 allele, and a 3.5-kb band in the D^3 allele. *Solve:* **(a)** A single band of 3.0 kb is detected in D^2D^2. Two bands, one 3.0 kb and the other 3.5 kb, are detected in D^2D^3. One band of 3.5 kb is detected in D^3D^3. **(b)** The organism with the 2.0 and 3.5 bands is D^1D^3. The organism with the 2.0 and 3.0 bands is D^1D^2. **(c)** Organisms with the genotype D^1D^1 have only a 2.0-kb band because they are homozygous for the D^1 allele and because the probe hybridizes to DNA contained entirely within the 2.0-kb restriction fragment. The probe does not hybridize to the 1.0-kb or 1.5-kb fragments present in D^1D^1 individuals; therefore, those bands are not detected on the Southern blot.

10.04 Test Yourself

Problems

1. The figure shows a pedigree in which one individual, II-3, has sickle cell disease (genotype $\beta^S\beta^S$) and II-4 is known to be a carrier for SCD (genotype $\beta^A\beta^S$). The results of a Southern blot analysis of *Dde*I cut DNA from II-3, II-4, and II-5 is shown below the pedigree. Use this figure and answer the following questions.

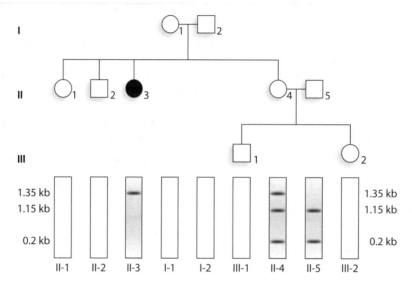

   a. Indicate the genotypes and RFLP band patterns for individuals I-1 and I-2.
   b. Indicate the genotypes and RFLP band patterns for individuals II-1 and II-2. Assume that they have different genotypes.
   c. Indicate the genotypes and RFLP band patterns for individuals III-1 and III-2. Assume that they have different genotypes.

2. Individuals with the genotype $\beta^S\beta^S$ have sickle cell disease. Individuals with the genotype $\beta^A\beta^A$ are normal. Individuals with the genotype $\beta^A\beta^S$ have sickle cell trait. Describe the types of hemoglobin molecules you expect to find in the red blood cells of individuals with each of these genotypes.

3. How does the half-life of red blood cells in individuals with the genotypes $\beta^S\beta^S$, $\beta^A\beta^S$, and $\beta^A\beta^A$ differ? Suggest a mechanism to explain the differences.

4. The website Online Mendelian Inheritance in Man (OMIM: www.ncbi.nlm.nih.gov/omim) contains information on the mutant alleles of many human genes, including the human β-globin gene (*HBB*). The names of the mutant alleles are based on the location where the mutation was discovered and the amino acid substitution that is caused by the mutation. For example, hemoglobin Alabama (*HBB*, Gln39Lys) refers to a mutation discovered in the state of Alabama in the United States that changes the glutamine codon number 39 to a lysine codon. The table lists several other known *HBB* mutant alleles and the corresponding wild-type codon sequence. For each mutation, indicate the sequence of the codons of the mutant mRNAs, and the sequence of the mutant gene coding and template strands. This information for *HBB*, Gln39Lys is provided as an example. *Hint:* Assume that only a single nucleotide is different between the wild-type and mutant sequence.

Allele Name	Mutation	Wild-Type Codon	Mutant Codon	Mutant Gene Coding Strand	Mutant Gene Noncoding Strand
Alabama	*HBB*, Gln39Lys	CAG	AAG	AAG	CTT
Alberta	*HBB*, Glu101Gly	GAG			
Baylor	*HBB*, Leu81Arg	CUC			

(*continued*)

Allele Name	Mutation	Wild-Type Codon	Mutant Codon	Mutant Gene Coding Strand	Mutant Gene Noncoding Strand
Iran	*HBB*, Glu22Gln	GAA			
Istanbul	*HBB*, His92Gln	CAC			
Tampa	*HBB*, Asp79Tyr	GAC			
Vanderbilt	*HBB*, Ser89Arg	AGU			

5. Human β-globin (*HBB*) has a methionine at amino acid position 55. Using the terminology for describing mutant *HBB* introduced in Problem 4, describe six possible mutant alleles that substitute a different amino acid for methionine at this position. Generate a table similar to that in Problem 6, and give a name to each mutant allele.

6. The coding sequence of the β-globin gene from nucleotide position 476 to 483 (1 is the transcription start site) is GTG GAT CCT and codes for the amino acids Val-Asp-Pro. This sequence includes the recognition site for the restriction enzyme *Bam*HI (5′-GGATCC-3′). Describe all possible single-nucleotide substitutions in this sequence that could create mutant alleles that contain a single amino acid substitution and could be detectable as an RFLP. Name the mutant alleles using the convention introduced in Problem 4, but also include the nucleotide substitution. For example, the Alabama allele would be called *HBB*, Gln39Lys (C299A) to indicate that nucleotide number 299 is changed from a C to an A in this allele. For each missense allele, indicate all possible single-nucleotide substitutions that could create that allele.

7. The coding sequence of the β-globin gene from nucleotide position 1395 to 1400 (1 is the transcription start site) is GAA TTC and codes for the amino acids Glu-Phe. This sequence is also the recognition site for the restriction enzyme *Eco*RI (5′-GGATCC-3′). Describe all possible single-nucleotide substitutions in this sequence that could create mutant alleles that contain a single amino acid substitution and could be detectable as an RFLP. Name the mutant alleles using the convention in Problem 6.

8. The figure shows the results of a Southern blot analysis of the human β-globin locus in 500 individuals from a particular population. The genomic DNA from each individual was digested and probed. The number of individuals with each RFLP result is shown below the gel. Use this information to assign a genotype to each banding pattern and then use that information to determine the frequency of the β^S allele in this population.

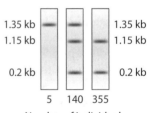

*Dde*I RFLP analysis of the beta-globin locus in 500 individuals

9. The figure shows the results of a Southern blot analysis of the human β-globin locus in 500 individuals from a particular population. The genomic DNA from each individual was digested and probed. The number of individuals with each RFLP result is shown below the gel. Use this information to assign a genotype to each banding pattern and then use that information to determine the frequency of the β^S allele in this population.

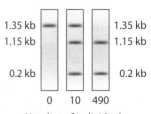

*Dde*I RFLP analysis of the beta-globin locus in 500 individuals

10. The β^S allele frequency in one human population is 0.15 whereas it is 0.01 in a different population. Speculate about the reason for this difference and suggest possible locations of each population.

Solutions

1. *Evaluate:* This is type 3. II-3 has SCD; her RFLP phenotype is one 1.35-kb band. II-4 is a carrier of SCD, and his RFLP phenotype is three bands: 1.35 kb, 1.15 kb, and 0.2 kb. The RFLP phenotype of II-5 is two bands, one 1.15 kb and one 0.2 kb. The genotypes of II-1 and II-2 must be different from each other, and the genotypes of III-1 and III-2 must be different from each other. *Deduce:* SCD is a recessive disease. Since II-3 has SCD, her genotype is $\beta^S\beta^S$ and her RFLP phenotype corresponds to the β^S allele. The genotype of II-4 is $\beta^A\beta^S$, and his RFLP phenotype corresponds to one β^A allele and one β^S allele. *Solve:* See figure. **(a)** I-1 and I-2 do not have SCD but have a child with SCD; therefore, they must be heterozygous carriers ($\beta^A\beta^S$) and will have the same RFLP phenotype as II-4. **(b)** II-1 and II-2 are unaffected siblings of II-3. Since they must differ, one must be like II-4 and the other must be $\beta^A\beta^A$. The $\beta^A\beta^A$ individual (shown as II-2 in the figure) will have an RFLP phenotype corresponding to the β^A allele only, which is 1.15-kb and 0.2-kb bands. **(c)** III-1 and III-2 are unaffected progeny of II-4 and II-5. Since they must differ, one will be like II-4 and the other will be like II-2.

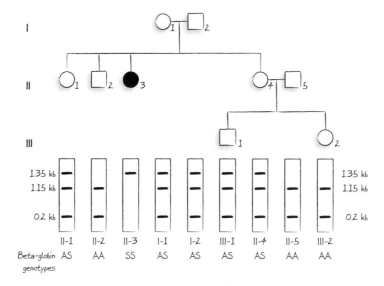

2. *Evaluate:* This is type 3. Each allele contributes equally to the pool of β-globin protein present in red blood cells. The β-globin proteins randomly associate with each other and α-globin polypeptides to form $\alpha\alpha\beta\beta$ hemoglobin (Hb) tetramers. *Deduce > Solve:* All Hb molecules in $\beta^A\beta^A$ individuals will have two normal β-globin proteins and two normal α-globin proteins. All the Hb molecules in $\beta^S\beta^S$ individuals will have two mutant β-globin proteins and two normal α-globin proteins. There will be three types of Hb molecules in $\beta^A\beta^S$ individuals. Some will have two normal β-globin proteins and two normal α-globin proteins. Some will have one normal β-globin protein, one mutant β-globin protein, and two normal α-globin proteins. The third type will have two mutant β-globin proteins and two normal α-globin proteins.

3. *Evaluate:* This is type 3. Recall the differences between hemoglobin molecules (Hb) in individuals with $\beta^A\beta^A$, $\beta^S\beta^S$, and $\beta^A\beta^S$ genotypes. *Deduce > Solve:* The red blood cells in $\beta^A\beta^A$ individuals live the longest. These cells have only normal Hb, which does not polymerize under low oxygen pressure; therefore, they will not lyse prematurely due to sickling. The red blood cells of $\beta^S\beta^S$ individuals have the shortest life span. They only have abnormal Hb, which tends to polymerize under low oxygen pressure and leads to the frequent sickling of red blood cells, which promotes their lysis. The red blood cells of $\beta^A\beta^S$ individuals will have an intermediate life span. These cells contain a mixture of normal and abnormal Hb molecules. Occasionally, red blood cells will contain sufficient levels of abnormal Hb molecules that polymerize under low oxygen pressure to cause sickling and cell lysis. This is much less frequent than in $\beta^S\beta^S$ individuals, and it results in a life span intermediate between those of $\beta^A\beta^A$ and $\beta^S\beta^S$ individuals.

4.   *Evaluate > Deduce > Solve:* This is type 3. This problem tests your ability to use the genetic code. **TIP:** Only one single-nucleotide substitution can explain the conversion of each codon in the "Wild-Type Codon" column to the codon indicated by the mutation. For example, the GAG codon for Glu can be converted to a GGG Gly codon by changing the second position from an A to a G. All other ways of changing the GAG Glu codon to a Gly codon require more than one nucleotide substitution. See table.

Allele Name	Mutation	Wild-Type Codon	Mutant Codon	Mutant Gene Coding Strand	Mutant Gene Template Strand
Alabama	*HBB*, Gln39Lys	CAG	AAG	AAG	CTT
Alberta	*HBB*, Glu101Gly	GAG	GGG	GGG	CCC
Baylor	*HBB*, Leu81Arg	CUC	CGC	CGC	GCG
Iran	*HBB*, Glu22Gln	GAA	CAA	CAA	TTG
Istanbul	*HBB*, His92Gln	CAC	CAG	CAG	CTG
Tampa	*HBB*, Asp79Tyr	GAC	UAC	TAC	GTA
Vanderbilt	*HBB*, Ser89Arg	AGU	CGU	CGT	ACG

5.   *Evaluate:* This is type 3. *Deduce:* The codon for Met is AUG. Nucleotide substitutions in the first position produce CUG, GUG, and UUG. Nucleotide substitutions in the second position produce ACG, AGG, and AAG. Nucleotide substitutions in the third position produce AUC, AUG, and AUU. *Solve:* Nucleotide substitutions in the first position change the Met codon to Leu, Val, and Leu codons, resulting in two types of amino acid substitutions: Met55Leu and Met55Val. Nucleotide substitutions in the second position change the Met codon to Thr, Arg, and Lys codons, resulting in the amino acid substitutions Met55Thr, Met55Arg, and Met55Lys. Nucleotide substitutions in the third position change the Met codon to ILe codons, resulting in the amino acid substitution Met55Ile. See table.

β-Globin Form	Position	Amino Acid	Codon (mRNA)	Coding Strand (DNA)	Template Strand (DNA)
HBB (wild type)	55	Met	5'-AUG-3'	5'-ATG-3'	5'-CAT-3'
HBB, Met55Leu *HBB*, Met55Leu	55	Leu	5'-CUG-3' 5'-UUG-3'	5'-CTG-3' 5'-TTG-3'	5'-CAG-3' 5'-CAA-3'
HBB, Met55Val	55	Val	5'-GUG-3'	5'-GTG-3'	5'-CAC-3'
HBB, Met55Ile	55	Ile	5'-AUA-3'	5'-ATA-3'	5'-TAT-3'
HBB, Met55Thr	55	Thr	5'-ACG-3'	5'-ACG-3'	5'-CGT-3'
HBB, Met55Lys	55	Lys	5'-AAG-3'	5'-AAG-3'	5'-CTT-3'
HBB, Met55Arg	55	Arg	5'-AGG-3'	5'-AGG-3'	5'-CCT-3'

6.   *Evaluate:* This is type 3. This problem requires you to consider changes in DNA sequences that alter restriction enzyme recognition sites and their potential effect on coding sequences. *Deduce:* The first G of the *Bam*HI site (G478) is the third nucleotide of the Val codon. Since the third position of Val codons can be any nucleotide, there is no substitution for this G that will create a missense mutation. The second G of the *Bam*HI site (G479) is the first nucleotide of the Asp codon. Changing this G to an A creates an AAT Asn codon, changing it to a C creates a CAT His codon, and changing it to a T creates a TAT Tyr codon. The A of the *Bam*HI site (A480) is the

second nucleotide of the Asp codon. Changing this A to a T creates a GTT Val codon, changing it to a C creates a GCT Ala codon, and changing it to a G creates a GGT Gly codon. The T of the *Bam*HI site (T481) is the third nucleotide of the Asp codon. Changing this T to an A or a G creates a Glu codon (GAA or GAG). The first C of the *Bam*HI site (C482) is the first nucleotide of the Pro codon. Changing this C to a T creates a TCT Ser codon, changing it to an A creates an ACT Thr codon, and changing it to a G creates a GCT Ala codon. The second C of the *Bam*HI site (C483) is the second nucleotide of the Pro codon. Changing this C to a T creates a CTT Leu codon, changing it to an A creates a CAT His codon, and changing it to a G creates a CGT Arg codon. *Solve:* HBB, Asp99His (G478C), *HBB*, Asp99Asn (G478A), *HBB*, Asp99Tyr (G478U), *HBB*, Asp99Ala (A479C), *HBB*, Asp99Gly (A479G), *HBB*, Asp99Val (A479T), *HBB*, Asp99Glu (T480A), *HBB*, Asp99Glu (T480G), *HBB*, Pro100Ser (C481T), *HBB*, Pro100Thr (C481A), *HBB*, Pro100Ala (C481G), *HBB*, Pro100Leu (C482T), *HBB*, Pro100His (C482A), *HBB*, Pro100Arg (C482G).

7.   *Evaluate:* This is type 3. This problem requires you to consider changes in DNA sequences that alter restriction enzyme recognition sites and their potential effect on coding sequences. *Deduce:* The first G of the *Eco*RI site (G1395) is the first nucleotide of a Glu codon. Changing this G to an A creates an AAA Lys codon, changing it to a C creates a CAA Gln codon, and changing it to a T creates a TAA stop codon. The first A of the *Eco*RI site (A1396) is the second nucleotide of a Glu codon. Changing this A to a T creates a GUA Val codon, changing it to a C creates a GCA Ala codon, and changing it to a G creates a GGA Gly codon. The second A of the *Eco*RI site (A1397) is the third nucleotide of the Glu codon. Changing it to a C or a T creates an Asp codon (GAC or GAT). The first T of the *Eco*RI site (T1398) is the first nucleotide of a Phe codon. Changing it to a C creates a CTT Leu codon, changing it to an A creates an ATC Ile codon, and changing it to a G creates a GTC Val codon. The second T of the *Eco*RI site (T1399) is the second nucleotide of the Phe codon. Changing it to a C creates a TCC Ser codon, changing it to an A creates a TAC Tyr codon, and changing it to a G creates a TGC Cys codon. The C of the *Eco*RI site (C1400) is the third nucleotide of the Phe codon. Changing that C to an A or a G creates a Leu codon (TTA or TTG). *Solve:* HBB, Glu121Lys (G1395A), *HBB*, Glu121Gln (G1395C), *HBB*, Glu121Val (A1396T), *HBB*, Glu121Ala (A1396C), *HBB*, Glu121Gly (A1396G), *HBB*, Glu121Asp (A1397T), *HBB*, Glu121Asp (A1397C), *HBB*, Phe122Leu (T1398C), *HBB*, Phe122Ile (T1398A), *HBB*, Phe122Val (T1398G), *HBB*, Phe122Ser (T1399C), *HBB*, Phe122Tyr (T1399A), *HBB*, Phe122Cys (T1399G), *HBB*, Phe122Leu (C1400A), *HBB*, Phe122Leu (C1400G).

8.   *Evaluate:* This is type 4. *Deduce:* The Southern blot shows the results of *Dde*I-digested genomic DNA. The 1.35-kb band corresponds to the β^S allele, and the 1.15-kb and 0.2-kb bands correspond to the β^A allele. The allele β^S frequency is the total number of β^S alleles divided by the total number of β^S and β^A alleles combined. Each individual has two alleles, and 500 were tested; therefore, a total of 1000 alleles were analyzed in this experiment. *Solve:* Five individuals had the genotype $\beta^S\beta^S$, which accounts for 10 β^S alleles. There were 140 individuals with the genotype $\beta^S\beta^A$, which accounts for 140 additional β^S alleles. The remaining 355 individuals had the genotype $\beta^A\beta^A$. Therefore, the frequency of β^S is $\frac{(10+140)}{1000} = \frac{150}{1000} = 0.15 = 15\%$. The first lane corresponds to the $\beta^S\beta^S$ genotype, the second corresponds to the $\beta^A\beta^S$ genotype, and the third corresponds to the $\beta^A\beta^A$ genotype. The frequency of the β^S allele in this population is 0.15.

9.   *Evaluate:* This is type 4. *Deduce:* The Southern blot shows the results of *Dde*I-digested genomic DNA. The 1.35-kb band corresponds to the β^S allele, and the 1.15-kb and 0.2-kb bands correspond to the β^A allele. The allele β^S frequency is the total number of β^S alleles divided by the total number of β^S and β^A alleles combined. Each individual has two alleles, and 500 were tested; therefore, a total of 1000 alleles were analyzed in this experiment. *Solve:* Ten individuals had the genotype $\beta^S\beta^A$, which accounts for 10 β^S alleles. The remaining 490 individuals had the genotype $\beta^A\beta^A$. Therefore, the frequency of β^S is $\frac{10}{1000} = 0.01 = 1\%$. The first lane corresponds to the $\beta^S\beta^S$ genotype, the second corresponds to the $\beta^A\beta^S$ genotype, and the third corresponds to the $\beta^A\beta^A$ genotype. The frequency of the β^S allele in this population is 0.01.

10. *Evaluate:* This is type 4. *Deduce:* The frequency of β^S will be determined by the evolutionary forces acting on the phenotypes of $\beta^S\beta^S$, $\beta^S\beta^A$, and $\beta^A\beta^A$ individuals. The β^S allele encodes an abnormal β-globin polypeptide that forms abnormal hemoglobin molecules that reduce the life span of red blood cells. This is most severe in $\beta^S\beta^S$ homozygotes and is detectable but less severe in $\beta^S\beta^A$ heterozygotes. The reduced life span of red blood cells interrupts the malaria protist's life cycle, resulting in a certain level of resistance to malaria. *Solve:* The severity of the anemia in $\beta^S\beta^S$ individuals decreases their reproductive fitness relative to $\beta^A\beta^A$, which selects against the β^S allele. The less severe effect in $\beta^S\beta^A$ heterozygotes does not affect fitness except in populations where malaria is endemic, where it increases reproductive fitness relative to $\beta^A\beta^A$. Based on this information, a reasonable hypothesis would be that malaria is more prevalent where the population in Problem 8 lives than where the population in Problem 9 lives. Based on the text, the Problem 8 population could be from western Africa, and the Problem 9 population is from somewhere else.

11

Chromosome Structure

11.01 Genetics Problem-Solving Toolkit

Key Terms and Concepts

euchromatin: A relatively loosely packed structure of a chromosome that is found in regions containing genes that are actively being expressed.

heterochromatin: A relatively densely packed structure of chromatin that is found in regions containing genes that are not being expressed and in repetitive, gene-free regions of chromosomes.

supercoiling: The twisting of the double helix on itself—including negative supercoiling, which promotes melting of dsDNA and positive supercoiling, which stabilizes dsDNA.

micrococcal nuclease: An endonuclease that is used to cleave chromatin in order to reveal the organization of nucleosomes.

G-banding: A technique using the stain, Giemsa, to identify and characterize the chromosome content of cells in karyotype analysis.

fluorescent in situ hybridization (FISH): A technique used to fluorescently label DNA molecules in cells as an alternative to G-banding in karyotype analysis.

position effect variegation (PEV): The patchwork-like expression pattern of a gene when it is relocated near heterochromatin.

Su(var): A type of *Drosophila* mutation that increases the fraction of cells expressing a gene that has been shown to be suppressed by PEV.

E(var): A type of *Drosophila* mutation that decreases the fraction of cells expressing a gene that has been shown to be suppressed by PEV.

Key Analytical Tools

Micrococcal nuclease digestion:

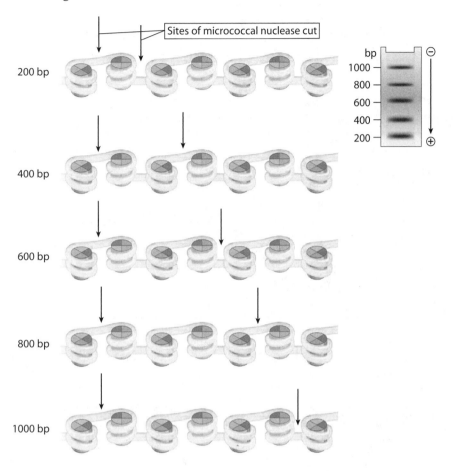

11.02 Types of Genetics Problems

1. Describe the DNA sequence content of bacterial and eukaryotic chromosomes.
2. Describe DNA supercoiling and chromatin structure.
3. Relate chromatin structure to gene expression.
4. Predict or interpret the results of experiments analyzing chromosome structure.

1. Describe the DNA sequence content of bacterial and eukaryotic chromosomes.

Problems of this type test your understanding of the structure and function of chromosomal DNA sequences. They may ask you to describe the general organization of essential chromosome sequence elements, recall the details of specific DNA sequence elements, or explain the function of each sequence element. See Problem 12 for an example.

Example Problem: In the course of evolution, two telocentric chromosomes of the same length fused at their centromeres to generate a new chromosome. Describe the new chromosome using the terminology used to describe the karyotype of a cell.

Solution Strategies	Solution Steps
Evaluate	
1. Identify the topic this problem addresses, and explain the nature of the requested answer.	1. This problem tests your understanding of terminology used to describe eukaryotic chromosomes. The answer will be a description of a new chromosome that conforms to cytogenetic nomenclature.
2. Identify the critical information given in the problem.	2. The chromosomes that fused were the same length and telocentric. They fused at their centromeres.
Deduce	
3. Recall the structure of telocentric chromosomes, and locate the centromere.	3. Telocentric chromosomes have their centromere at one end.
4. Draw two telocentric chromosomes of the same length such that their centromeres are near each other. Then draw them with their centromeres fused.	4.
Solve	
5. Describe the position of the centromere of the new chromosome relative to its length.	5. The centromere is in the middle of the new chromosome.
6. Answer the question by choosing the term for chromosome type that best fits the new chromosome.	6. **Answer:** Metacentric.

2. Describe DNA supercoiling and chromatin structure.

This type of problem tests your understanding of chromosome compaction. You may be asked to explain the strategies employed by different cells or the molecular basis of compaction in each cell type, to predict the chromatin structure of specific chromosomal regions or specific types of DNA sequences, or to relate chromosome compaction to karyotype analysis. See Problem 16 for an example.

Example Problem: You have isolated a new fluorescent chemical. You have no knowledge of its affinity for biological molecules and decide to test whether it stains mitotic cells in a karyotype analysis. The chemical causes the centromeres, and only the centromeres, of each mitotic chromosome to fluoresce. Suggest three different molecular targets (binding sites) that could explain this fluorescence pattern, and explain how to determine which of these sites are the actual binding targets.

Solution Strategies	Solution Steps
Evaluate	
1. Identify the topic this problem addresses, and explain the nature of the requested answer.	1. This problem concerns the composition of centromeres in humans. The answer will identify two components of centromeres that are not found elsewhere on mitotic chromosomes.

(continued)

Solution Strategies	Solution Steps
2. Identify the critical information given in the problem.	2. The chemical binds to centromeres of all mitotic cells but not to any other chromosomal location.
Deduce	
3. Recall the structure and composition of human centromeres in mitotic cells.	3. Human centromeres are composed of repetitive, centromere-specific DNA sequences, centromere-specific heterochromatin, and mitotic proteins that form the kinetochore.
4. Describe how you would obtain samples that could be used to test for specific binding of the chemical to each of the components identified in part 3.	4. Chromosomal DNA from interphase cells could be isolated and stripped of all proteins to provide a sample that contains centromeric DNA but not chromatin or mitotic proteins. Intact chromosomes could be isolated from interphase cells to provide a sample that contains centromeric DNA and chromatin but not mitotic proteins. The metaphase chromosomes used in karyotype analysis could be used to provide cells that contain all three components.
Solve	
5. Answer the problem by indicating the expected result of the analysis from step 4 for each potential mitotic centromere target.	5. **Answer:** The chemical would bind to all three samples if centromeric DNA was the target. The chemical would bind to the intact chromosomes from interphase and mitotic cells if centromeric heterochromatin was the target. The chemical would bind to the intact chromosomes from mitotic cells only if mitotic-specific kinetochore components were the target.

3. Relate chromatin structure to gene expression.

This type of problem requires that you consider the effect of chromatin structure on gene expression. You may be asked to predict the transcriptional state of chromatin containing certain histone modifications or to predict the types of histone modifications that are present at transcriptionally active or inactive genes. You may also be asked to relate chromosomal position to gene transcription and describe classic demonstrations of position effect variegation (PEV). See Problem 15 for an example.

Example Problem: Gene "Y" in *Drosophila* is normally repressed in all cells except for those in the germ line. In your analysis of several *Su(var)* mutants, you notice that "Y" is now expressed in somatic cells as well as in the germ line. You find that *E(var)* mutations have no effect on "Y" expression ("Y" is still on in the germ line and off in somatic cells). Explain the results.

Solution Strategies	Solution Steps
Evaluate	
1. Identify the topic this problem addresses, and explain the nature of the requested answer.	1. This problem asks you to interpret the effects of *Su(var)* and *E(var)* mutations on gene expression. The answer will be an explanation that takes what is known about the molecular function of *Su(var)* and *E(var)* into account.

(continued)

Solution Strategies	Solution Steps
2. Identify the critical information given in the problem.	**2.** Expression of "Y" is germ-line specific in wild-type flies but is expressed in somatic cells as well in *Su(var)* mutants. *E(var)* has no effect on "Y" expression.
Deduce	
3. Recall the function of *Su(var)* and *E(var)* genes and the effect of the mutations on gene function.	**3.** *Su(var)* genes encode HP-1 or the H3K9 methylase that creates the binding site for HP-1. The *Su(var)* mutations decrease the function of the genes. *E(var)* gene function is not presented in the text. The *E(var)* mutation enhances the spread of heterochromatin along chromosomes.
4. Relate the effects of *Su(var)* and *E(var)* to general issues regarding gene expression.	**4.** *Su(var)* mutations diminish heterochromatin, which should increase the expression of genes affected by heterochromatin. *E(var)* mutations cause the spread of heterochromatin, which may cause genes located nearby to be repressed.
Solve	
5. Answer the problem by relating the effect of *Su(var)* and *E(var)* mutations on gene expression to the pattern of expression of "Y" in *Su(var)* and *E(var)* mutants.	**5.** **Answer:** *Su(var)* mutations cause the expression of "Y" in somatic cells because somatic cell expression of "Y" is normally repressed by heterochromatin resulting from methylated histones in all but germ-line cells. *E(var)* mutations have no effect on "Y" expression, because "Y" is probably not located near a heterochromatic region.

4. Predict or interpret the results of experiments analyzing chromosome structure.

This type of problem asks you to apply your understanding of chromosome structure to the analysis of experimental results. You may be asked to interpret the results of an experiment investigating chromosome structure or to design an experiment that will investigate chromosome structure. The types of analyses include micrococcal nuclease digestion, G-banding, in situ hybridization, or DNase I protection assays. See Problem 24 for an example.

Example Problem: An individual with what appears to be a relatively mild version of Down syndrome has asked you to help her determine if she has the disease. You first determine her karyotype by G-banding and find that all of her chromosomes appear normal in number and structure. You take a closer look at chromosome 21 in her cells by FISH and find evidence that a small portion of chromosome 21 is duplicated and located on chromosome 14. Describe the results of FISH that would produce such evidence, and provide an explanation for why G-banding might have missed it.

Solution Strategies	Solution Steps
Evaluate	
1. Identify the topic this problem addresses, and explain the nature of the requested answer.	**1.** This problem tests your understanding of karyotype analysis using G-banding and FISH. The answer will be a description of the results from a FISH analysis and a rationale for why FISH would be more sensitive than G-banding in this case.

(continued)

Solution Strategies	Solution Steps
2. Identify the critical information given in the problem.	2. The affected individual appears to have Down syndrome, but she does not have three copies of chromosome 21. A FISH probe for chromosome 21 identifies a portion of 21 on another chromosome.
Deduce	
3. Describe how FISH could be used to detect chromosome-specific sequences.	3. FISH uses probes that label each chromosome with a different fluorescent color.
4. Describe how G-banding is used to detect chromosome-specific sequences. TIP: G-banding allows identification of chromosomes based on their centromere location, size, and banding pattern.	4. G-banding identifies chromosome 21 as two of the smallest acrocentric chromosomes in a human's karyotype.
5. Describe the result that would be obtained by FISH analysis of a person who has two normal copies of chromosome 21 and one copy of chromosome 14 containing a fragment of chromosome 21.	5. The karyotype would show that all 23 pairs of chromosomes are present; however, chromosome 14 would have a small patch of color, the same as chromosome 21.
Solve	
6. Answer the problem by relating the result from FISH analysis to that from analysis of G-banded chromosomes. TIP: The bands identified by G-banding can contain many megabase pairs and many genes.	6. **Answer:** The G-banding data requires detection of a difference between normal and abnormal based on overall chromosome structure: small differences may be difficult to detect. FISH provides a positive result by detecting chromosome-specific DNA regardless of how much its presence in another chromosome alters the structure of that chromosome.

11.03 Solutions to End-of-Chapter Problems

1. *Evaluate > Deduce > Solve:* This is type 1 and 2. A bacterial nucleoid is the location of the bacterial chromosome within a bacterium, whereas a bacterial plasmid is a circular DNA molecule that is distinct from the chromosome. They are similar in that each is a single, double-stranded, supercoiled, circular DNA molecule that contains an origin of replication. They differ in a number of ways. The nucleoid DNA molecule is much larger and is bound by proteins that are typically not present on plasmids. The nucleoid is also in a specific region of the cell, whereas the plasmid does not specify a specific subcellular location. The nucleoid contains many essential genes, whereas plasmids typically contain nonessential genes.

2. *Evaluate > Deduce > Solve:* This is type 1. Recall that the term *haploid* refers to one copy of genetic information. The terms do not conflict, because a bacterium is haploid regardless of whether its genes are contained in one or more than one chromosome.

3. *Evaluate > Deduce > Solve:* This is type 2. The mechanisms for compaction of bacterial DNA are supercoiling and folding. Supercoiling refers to the twisting of double-helical DNA such that the helix is twisted on itself. Supercoiling in the counterclockwise direction, opposite to

the right-handed turn of DNA, is negative supercoiling, whereas twisting in the same direction as the helix is positive supercoiling. Negative supercoiling promotes unwinding of dsDNA, whereas positive supercoiling is thought to promote stability of dsDNA. DNA is folded by nucleoid-associated and SMC family proteins such as HU and H-NS. Folding serves to condense the chromosome, which reduces the space within the cell that is taken up by the chromosome.

4.   *Evaluate:* This is type 2. Recall that a nucleosome includes 146 bp of DNA and that the linker sequence between nucleosomes is about 50 bp. *Deduce:* The number of nucleosomes is the length of the genome divided by the length of nucleosomal and linker DNA. *Solve:* $\frac{2.9 \times 10^9}{(146 + 50)} = 1.48 \times 10^7$. Approximately 10^7 nucleosomes are required to organize the 10-nm fiber of the human genome.

5a–i.   *Evaluate > Deduce > Solve:* **(a)** Histone proteins include H2A, H2B, H3, and H4—which bind to DNA to form the nucleosome core particles present in the 10-nm fiber—and H1, which binds to nucleosomes to form the 30-nm fiber. **(b)** A nucleosome is a complex containing two H2A/H2B dimers and two H3/H4 dimers, around which 146 bp of DNA is wrapped. **(c)** CEN sequences are DNA sequences located at centromeres. **(d)** G bands are darkly stained regions of compacted chromosomes (heterochromatin) that have been stained with Giemsa. **(e)** Euchromatin refers to the relatively less condensed regions of chromosomes, which contain DNA that is readily accessible to transcription factors and RNA polymerase. **(f)** Heterochromatin refers to more condensed regions of chromosomes, which contain DNA that is relatively inaccessible to transcription factors and RNA polymerase. **(g)** Epigenetic modification refers to covalent modification of histone proteins and DNA bases, which promotes the formation of either euchromatin or heterochromatin and, consequently, affects gene expression. **(h)** A chromosome territory refers to the region of the nucleus occupied by a particular chromosome. Each chromosome is thought to occupy a different region of the nucleus. **(i)** The nucleoid is the region of a bacterial cell that contains the bacterial chromosome. The nucleoid could be thought of as being the chromosome territory of a bacterial cell.

6.   *Evaluate:* This is type 2. Recall that each chromosome is characterized by a standard G-banding pattern that allows a cytologist to unambiguously identify each chromosome in a human karyotype. The dark bands contain heterochromatic regions, whereas the light bands contain euchromatic regions.

7a–b.   *Evaluate > Deduce > Solve:* This is type 2. **(a)** Histone H4 is located in nucleosomes, which have two histone H4 proteins per nucleosome. Histone H4 is found throughout chromosomes, whether they are organized as 10-nm fibers, 30-nm fibers, or higher-order structures. **(b)** Histone H1 is associated with nucleosomes but is not a component of the nucleosome. It secures the wrap of DNA around each histone ball, and so it is found throughout chromosomes in regions that are organized as 30-nm fibers or higher-order structures.

7c.   *Evaluate:* This is type 3. Recall that nucleosomes are spaced 200 bp apart and that there are two copies each of the nucleosomal core proteins H2A, H2B, H3, and H4 but only one copy of H1. *Deduce:* The number of each nucleosomal core protein is given by $\frac{\text{length of DNA}}{\text{nucleosome spacing}} \times 2$. The number of histone H1 proteins is given by $\frac{\text{DNA}}{\text{nucleosome spacing}} \times 1$. *Solve:* $\frac{6000}{200} \times 2 = 60$ and $\frac{6000}{200} \times 1 = 30$. Therefore, there are 60 copies of H2A, H2B, H3, and H4, and 30 copies of H1.

7d.   *Evaluate > Deduce > Solve:* Histone H1 is required for the formation of the 30-nm fiber, whereas histone H3 is required for nucleosomes. Histone H3 is also involved in higher-order folding of chromosomes, depending on whether H3 is covalently modified and on the specific nature of the modification.

8.   *Evaluate:* This is type 2. Recall that interphase chromosomes consist of both euchromatin and heterochromatin, whereas metaphase chromosomes are essentially all heterochromatin. **TIP:** Euchromatin is less condensed than heterochromatin. *Deduce > Solve:* Interphase chromosomes will be less condensed than metaphase chromosomes.

9.   *Evaluate:* This is type 2. This problem asks you to relate chromosome structure as revealed by G-banding to gene content. *Deduce:* The $\frac{\text{number of genes}}{\text{number of bands}}$ gives the gene density per band. The number of genes deleted in the given region is calculated by (gene density × number of bands deleted). There are 5 bands in the region 1q21.1 through 1q21.3. *Solve:* $\frac{22,000}{2000} = 11$ genes per band. This region contains 11 × 5 = 55 genes; therefore, 55 genes will be lost as a result of the deletion.

10.   *Evaluate > Deduce > Solve:* This is type 1. This problem asks that you consider DNA sequence elements that are essential, evolutionarily conserved components of bacterial or eukaryotic chromosomes. *Deduce:* Chromosomes must be replicated and passed on to progeny cells. Bacterial chromosomes must have all the genes essential for bacterial life: an origin of replication to initiate replication, and a site for attachment to the bacterial membrane to ensure segregation of daughter chromosomes to each cell at cell division. Eukaryotic chromosomes must contain a centromere, a telomere at each end, and multiple origins of replication. Eukaryotic chromosomes must also contain genes essential for eukaryotic life, although these genes can be dispersed among the chromosomes. Natural selection will select against chromosomes that lack sequences required for replication or segregation, and the cells that contain them, because they will be less fit than those that contain these sequences. The same argument applies to chromosomes that lack essential genes.

11.   *Evaluate > Deduce > Solve:* This is type 1. Yeast centromeres are composed of three small regions, CDEI, CDEII, and CDEIII. These sequences each contribute to formation of the kinetochore, which is the binding site for microtubules in the mitotic and meiotic spindle.

12.   *Evaluate > Deduce > Solve:* This is type 1. Bacterial chromosomes are typically circular and are attached to the bacterial cell membrane. Circular chromosomes do not require telomeres, and there would be no evolutionary advantage for a chromosome to have them; therefore, telomeres are not present on bacterial chromosomes. Bacteria do not have microtubules; therefore, bacterial chromosomes do not need a centromere to facilitate assembly of a microtubule binding site. The membrane attachment site of a bacterial chromosome serves to promote segregation of daughter bacterial chromosomes to opposite sides of a dividing cell, which is analogous to the function of eukaryotic chromosome centromeres.

13.   *Evaluate > Deduce > Solve:* This is type 1 and 2. Telomeres and centromeres are both composed of repetitive DNA sequences, although the repeated sequences in centromeres and telomeres differ. Both are packaged into heterochromatin, although the specific type and distribution of histone modifications at the two elements may differ.

14.   *Evaluate > Deduce > Solve:* This is type 1. Telomeres are composed of repetitive DNA in which the repeated sequence is a simple sequence (e.g., TAAGGC is repeated many times). Directly next to the telomere are telomere-associated sequences, which are also composed of repetitive DNA, but the repeated sequences are more complex and may include genes. Next to the telomere-associated sequences are "normal" chromosome sequences that contain genes and intergenic regions. (Note that there is variation in this sequence organization among chromosomes in the same organism and between chromosomes in different organisms.)

15.   *Evaluate:* This is type 3. Recall that position effect variation (PEV) describes cell-to-cell variation in the expression of a gene due to the location of that gene on a chromosome. PEV of *Drosophila* eye color, which appears as red and white patches of eye tissue, is due to translocation of the white gene (a functional wild-type w^+ allele produces red eyes) from its normal, euchromatic location to a location in or near heterochromatin (either the centromere or telomere). At its normal location, the w^+ allele is expressed in all cells of the eye, leading to a uniformly red pigmented eye. At a more centromeric (or telomeric) location, some cells do not express the w^+ gene because it is packaged in heterochromatin. Since some cells are red (the gene is packaged in euchromatin in these cells), it is clear that the w^+ allele is functional, so mutational inactivation of the white gene does not explain the white cells.

16a.   *Evaluate:* This is type 2. Recall that nucleosomes are spaced 200 bp apart and that after S phase, there are two copies of each chromosome. The haploid genome size of *Arabidopsis* is 10^8 bp.

Deduce: The number of nucleosomes per genome in a diploid is given by $\frac{10^8}{200} \times 2 = 5 \times 10^5$. There are twice as many nucleosomes after completion of S phase. *Solve:* $2\left(\frac{10^8}{200} \times 2\right) = 10^6$. Therefore, 10^6 nucleosomes are present after S phase.

16b.   *Evaluate > Deduce > Solve:* This is type 2. The histone proteins that were part of nucleosomes before S phase are recycled and used to form the new nucleosomes during S phase. The additional histone proteins required to double the nucleosome number are newly synthesized. Therefore, half of the histone proteins present on chromosomes after S phase is newly synthesized, and the other half were already present.

17.   *Evaluate*: This is type 4. This problem tests your understanding of the relationship between micrococcal nuclease treatment of chromatin and the relative spacing of nucleosomes. **TIP:** Both the smallest band and the space between bands represent the length of DNA that contains a single nucleosome. *Deduce > Solve:* The nucleosome spacing in species A is 200 bp, whereas the spacing in species B is 400 bp.

18 a-b. *Evaluate > Deduce > Solve:* This is type 1 and 2. **(a)** See figure. **(b)** Each chromosome is composed of a pair of sister chromatids that remain joined only at the centromeres. The length of each arm and position of the centromere in homologous chromosomes will be identical. Homologs should pair, and each pair of homologs should align on the metaphase plate during meiosis I. Chromosomes 1 and 5, chromosomes 2 and 6, and chromosomes 3 and 4 in the figure are homologs.

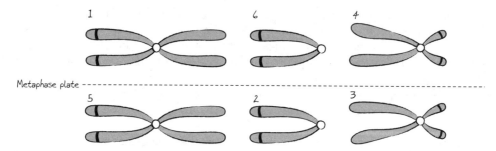

19a.   *Evaluate > Deduce > Solve:* This is type 1. The chromosome is submetacentric because the centromere is neither directly in the middle nor at an end.

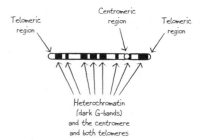

19b.   *Evaluate > Deduce > Solve:* This is type 1 and 2. No; the centromeric region is constitutive heterochromatin, as are both telomeres.

19c.   *Evaluate > Deduce > Solve:* This is type 1 and 2. Centromeric DNA is highly repetitive, composed of hundreds of copies of alpha satellite repeat sequences, whereas most other regions of the chromosome contain unique gene sequences. The nucleosomes bound to centromeric DNA contain a specialized histone H3 molecule called CENP-A, whereas nucleosomes elsewhere do not. Kinetochore proteins also bind to centromeres, but not elsewhere on the chromosome.

19d.   *Evaluate > Deduce > Solve:* This is type 3. Recall that telomeres are packaged as heterochromatin. Telomeric regions of chromosomes generally lack genes because these chromosomal regions are heterochromatic and therefore inhibit gene expression. Thus, any gene located at telomeres would be silenced unless something altered telomeric heterochromatin structure.

19e.   *Evaluate:* This is type 3. Recall that heterochromatin represses gene transcription. **TIP:** Amylase is an enzyme involved in digestion of starch. *Deduce:* Expression of the amylase gene would be repressed if it was located in a heterochromatic chromosomal region, which includes centromeric and telomeric regions. *Solve:* Therefore, the amylase gene would be expected to be found in a euchromatic region of a chromosome.

20a.   *Evaluate > Deduce > Solve:* This is type 2. Recall that the *E. coli* chromosome is 1000 times longer than an *E. coli* cell yet fits into a small region of the cell called the nucleoid. As with other bacterial chromosomes, the *Methanococcus jannaschii* chromosome is compacted by looping and supercoiling, and by the binding of proteins that fold the chromosome. The combination of supercoiling and folding (condensation) reduces the volume occupied by the chromosome, allowing it to fit into the region of the bacterial cell known as the nucleoid.

20b.   *Evaluate > Deduce > Solve:* This is type 2. Review the mechanisms of compaction of the *E. coli* chromosome. The chromosome is likely compacted in two steps. First, it is bound by nucleoid-associated proteins and SMC proteins, which loop segments of the chromosome and condense them. Second, the DNA is supercoiled by the action of topoisomerases.

20c.   *Evaluate > Deduce > Solve:* This is type 2. Recall that most bacterial chromosomes and plasmids are negatively supercoiled. Negative supercoiling folds chromosomal DNA and promotes unwinding of regions of the chromosome. This promotes access to ssDNA for enzymes such as DNA polymerase and RNA polymerase. Supercoiling of the *M. jannaschii* chromosome folds the chromosome and promotes the function of DNA and RNA polymerases.

21.   *Evaluate*: This is type 4. This problem challenges you to relate the results of micrococcal nuclease digestion of human chromosomes to nucleosomal spacing. **TIP:** The haploid size of human DNA is 3×10^9 bp, and it forms a continuous smear (not individual bands) if digested with micrococcal nuclease after all nucleosomes have been removed. *Deduce > Solve:* The fact that bands of 185 to 200 bp are detectable after micrococcal nuclease treatment of human chromosomes indicates that the vast majority of chromosomal DNA is cut into equal-sized pieces, suggesting that a regularly spaced array of proteins prevents cutting at other locations. If the spacing of nucleosomes were not regular, fragments of many different sizes would be created, appearing like a smear on a gel instead of discrete bands.

22.   *Evaluate*: This is type 2. Recall that histones interact with DNA in a sequence-independent manner to form nucleosome core particles that are conserved in structure and function, from yeast to man. Also recall that evolutionary change of a protein's sequence is under "functional constraint," which limits or prevents changes to sequences that perform essential functions. *Deduce > Solve:* Histone H4 is one of the core histones and is part of the nucleosome core particle. Most of the amino acid residues of H4 interact either with DNA or other histone proteins; therefore, most of the H4 amino acid sequence is under functional constraint (any change to the H4 amino acid sequence will likely be deleterious and therefore will be selected against by natural selection). Thus there is little change in the sequence of H4 over evolutionary periods of organisms—from pea plants to cows.

23.   *Evaluate*: This is type 3. Recall that modifications of histones correlate with chromatin structure, such that acetylated histones are found in transcriptionally active chromatin and methylated histones are found in transcriptionally inactive chromatin. *Deduce > Solve:* One hypothesis to explain the correlation is that acetylation and demethylation of the amino termini of H3 and H4 cause nucleosomes to relax into a relatively "open" structure that allows transcription factors and RNA polymerase access to DNA. The modifications serve as an epigenetic code for the genes in that region, marking them for transcription.

24.   *Evaluate*: This is type 4. Recall that all four histones are synthesized at the beginning of S phase and that, after completion of S phase, half the histone proteins present will be new and half will be left over from previous cell cycles. **TIP:** ^{35}S-containing methionine can be used in a pulse–chase experiment to radioactively mark all the histone proteins synthesized during one S phase. *Deduce > Solve:* Start with cells that contain unlabeled methionine and are in G_1 phase. Place them in medium containing ^{35}S-methionine and allow them to complete S phase. Take a sample of cells for analysis and then transfer the remaining cells to medium containing unlabeled methionine and allow them to divide and go through multiple rounds of S phase and cell division, collecting samples after each S phase. Analyze the nucleosomes of each sample by electron microscopy and autoradiography. If nucleosomes are a mixture of old and new histones, then most or all nucleosomes will be radiolabeled after the first S phase, about $\frac{1}{2}$ will be radiolabeled after the second round, and $\frac{1}{4}$ after the third round. If all nucleosomes contain either old or new histones, then about $\frac{1}{2}$ the nucleosomes will be radiolabeled after the first S phase, $\frac{1}{4}$ after the second, and $\frac{1}{8}$ after the third.

25a.   *Evaluate*: This is type 4. This problem tests your understanding of the use of micrococcal nuclease protection assays for analysis of chromosome structure. Micrococcal nuclease, like all enzymes, catalyzes reactions at a specific rate that depends on the reaction conditions. The progress of the reaction can be monitored by observing the formation of the product, which will increase over time until it reaches a maximum. The position of a band in the gel indicates its size, whereas the intensity of the band indicates its relative abundance. *Deduce > Solve:* The gel shows the same banding pattern for each sample; however, in comparing the 30-minute sample to the 1-hour sample, you will see the intensity of the upper bands decrease and that of the lowest band increase. The band intensities for the 1- and 4-hour samples are the same. The change from 30 minutes to 1 hour indicates that micrococcal nuclease continued to digest chromosomal DNA during the additional 30 minutes of incubation. The lack of change from 1 hour to 4 hours indicates that the reaction stops after 1 hour, most likely because the enzyme is active for only 1 hour under these conditions.

25b.   *Evaluate*: This is type 4. Recall that the size and spacing of the bands generated by micrococcal nuclease digestion reflect the spacing of nucleosomes. *Deduce > Solve:* If the 4-hour sample represents the pattern generated when all nuclease-sensitive sites have been cut, then the results indicate that the majority of nucleosomes are about 200 bp apart but there are regions of chromosomes where nucleosome spacing is 400, 600, and 800 bp apart. If the 4-hour sample reflects the lack of continued nuclease activity after 1 hour, then only the 1-hour sample can be interpreted, indicating nucleosome spacing of 200 bp.

26a.   *Evaluate*: This is type 4. Recall that DNase I degrades DNA that is not protected by protein. "Large amounts" of DNase I would be expected to digest DNA completely, leaving only those sequences bound by nucleosomes protected. **TIP:** Review the DNase footprinting technique introduced in Chapter 9. *Deduce > Solve:* The only DNA sequences protected from DNase I would be those in the nucleosome core particle. There would be one band, approximately 145 bp in size.

26b.   *Evaluate > Deduce > Solve:* This is type 4. Each band would represent the DNA bound by histones in the nucleosome core particle. All non-nucleosomal DNA and all linker DNA between nucleosomes would be digested.

26c.   *Evaluate*: This is type 4. Recall that the 10-nm fiber model for chromatin states that the 10-nm fiber is a linear array of nucleosomes and that the chromatin is not folded into higher-order structures. *Deduce > Solve:* The only protection against digestion by DNase I resulted from histones binding to DNA in nucleosomes, indicating that no other proteins and no higher-order packing of the chromatin at this region occurred. Higher-order structures would have generated fragments of DNA that are larger than the nucleosome core.

11.04 Test Yourself

Problems

1. Artificial chromosomes have been designed for bacteria and yeast and used to carry large chromosomal fragments. Bacterial artificial chromosomes are circular molecules about 200 kb, whereas yeast artificial chromosomes are linear molecules about 1000 kb in size.
   a. What DNA sequence elements would you expect to find on the bacterial and yeast artificial chromosomes?
   b. What would you expect to observe from a micrococcal nuclease digestion experiment of a yeast artificial chromosome that had been stripped of all proteins except histone?

2. Circular plasmid DNA isolated from bacteria was subjected to electrophoresis, and the gel was stained. A single band appeared on the gel that migrated at the same position as a linear 2-kb DNA molecular weight marker. When this plasmid was digested with an endonuclease that makes a single double-stranded DNA cut and examined by gel electrophoresis, it migrated to the same position as a linear 4-kb DNA marker. Explain these observations.

3. You have been assigned a task to develop a biochemical assay to detect DNA topoisomerase activity by your laboratory supervisor. You have circular dsDNA, buffers that provide optimum conditions for DNA topoisomerases and the reagents, and equipment to analyze DNA samples by agarose gel electrophoresis. Outline an assay that you could perform using these reagents and equipment to determine whether the extract from a plant cell contains DNA topoisomerase activity.

4. A colleague has applied a fluorescent antibody that binds specifically to the CENP-A protein to a metaphase spread of human cells. If you were able to detect fluorescence, what would you expect to see in each cell? How would that expectation differ if your colleague had used an antibody that binds to non-histone scaffolding proteins instead?

5. Order the following structures from least compact to most compact: metaphase chromosome, naked dsDNA, beads on a string, solenoid.

6. Assuming that heterochromatin protein 1 (HP-1) functions only to promote heterochromatin formation, what is the predicted relationship between the location of HP-1 on chromosomes and the location of genes that are actively being transcribed? What observation of HP-1 location and gene transcription would bring the assumption of HP-1 function into question?

7. Karyotype analysis of cells from one woman shows that the same X-chromosome is inactive in all of her cells. Provide an explanation for this observation that involves the *Xist* gene.

8. Explain the variegation that is observed in position effect variegation (PEV) of the *Drosophila* white gene expression when it is located near a centromere.

9. Relate the observation that H3/H4 dimers remain associated with sister chromatids during DNA replication to the fact that heterochromatin is inherited during mitotic cell divisions to arrive at a mechanism for epigenetic inheritance.

10. Naked bacterial DNA is packaged into nucleosomes when it is mixed with all four histones in a test tube under the appropriate conditions. How is this possible given that bacterial DNA and histone proteins are never found together in nature?

Solutions

1a. *Evaluate*: This is type 1. Recall that chromosomes must have DNA sequences that promote their replication and segregation, and they must have genes that select for their maintenance in the population. *Deduce > Solve:* Bacterial artificial chromosomes must have an origin of replication and a gene that allows for selection of cells that contain the chromosome (such as antibiotic resistance). Yeast artificial chromosomes must have an origin of replication, a centromere, and a telomere at each end and a dominant selectable marker.

1b.   *Evaluate*: This is type 4. Recall that micrococcal nuclease digestion experiments generate a ladder of DNA molecules that differ in size by the space between nucleosomes in the chromosome. ***Deduce > Solve:*** The micrococcal nuclease digestion result for the yeast artificial chromosome will be the same as that for endogenous chromosomes. The spacing between bands will be ~200 bp.

2.   *Evaluate*: This is type 4. Recall that plasmids isolated from bacteria are expected to be supercoiled, which reduces their size. ***Deduce > Solve:*** The digested plasmid is a linear molecule devoid of supercoils and migrates at a rate equivalent to a 4-kb DNA fragment, which is an accurate estimation of its size. Compacted DNA migrates more rapidly than would be predicted based on size alone. The apparent 2-kb size of the plasmid isolated from bacteria is consistent with it being highly supercoiled.

3.   *Evaluate*: This is type 4. Recall that supercoiled DNA migrates more rapidly during agarose gel electrophoresis than relaxed DNA. ***Deduce > Solve:*** One possibility is to incubate the dsDNA molecule in topoisomerase buffer with and without the plant cell extract sample. If there is topoisomerase activity in the plant extract, that sample will contain supercoiled DNA that migrates more rapidly than the dsDNA alone or the dsDNA in the sample that did not receive plant cell extract.

4.   *Evaluate*: This is type 2. Recall that CENP-A is a centromere-specific variant of histone H3 and that scaffolding proteins bind all along chromosomes and facilitate chromosome condensation. ***Deduce > Solve:*** The fluorescent anti-CENP-A antibodies will be detectable at each centromere, giving a result of 46 dots per cell (one per chromosome). The anti-scaffolding protein antibodies will be detectable all along the long-axis metaphase chromosomes, resulting in fluorescence of each chromosome in each cell.

5.   ***Evaluate > Deduce > Solve:*** This is type 2. The order is naked dsDNA (3 nm), beads on a string (10 nm), solenoid (34 nm), metaphase chromosome (1400 nm).

6.   *Evaluate*: This is type 3. Recall that transcriptionally active genes are found in euchromatic regions. ***Deduce > Solve:*** Assuming that HP-1 is present only in heterochromatin, it is expected that HP-1 will not be located at sites of active gene transcription. If HP-1 was observed to bind to regions that were being actively transcribed, then the assumption that HP-1 promotes only heterochromatin formation would come into question.

7.   *Evaluate*: This is type 2 and 3. Recall that females are typically mosaic for X-chromosome inactivation; their maternal X-chromosome is inactive in some cells, and their paternal X-chromosome inactive in others. Also recall that only the *Xist* gene on the inactive X-chromosome is transcribed and that *Xist* RNA binds to that chromosome and promotes or maintains its compaction. ***Deduce > Solve:*** One explanation is that the *Xist* gene on one X-chromosome is mutant (defective), which prevents that X-chromosome from being inactivated. Since one X-chromosome is inactivated in every cell, the other X-chromosome will always be the one that is inactivated.

8.   *Evaluate*: This is type 3. Recall that expression of the w^+ allele of the white gene is required for red eye color. ***Deduce > Solve:*** The variegated pattern of white expression is a reflection of the white gene being expressed in some cells and their descendants but not expressed in other cell lineages. PEV typically is observed when the normal position of the gene has changed due to a chromosomal rearrangement. For example, if the gene is nearer the centromere, expression can be affected by the extent to which centromeric heterochromatin spreads out from the centromere. In one cell during early eye development, heterochromatin may spread to cover the nearby white gene, repressing its expression. This pattern of chromatin structure is heritable at the cellular level such that the descendants of this cell will also package the white gene as heterochromatin, leading to a patch of white cells representing a lineage of cells in which white gene expression is repressed. In a different cell during early eye development, centromeric heterochromatin may not spread to cover the white gene, which will therefore be expressed in that cell and its descendants. This leads to a patch of red cells representing a lineage of cells in which the white gene is expressed.

9.   *Evaluate*: This is type 2. This problem challenges you to consider the mechanism by which chromatin structure could be inherited. *Deduce > Solve:* The nucleosomal proteins, H3 and H4, undergo post-translational modifications that are proposed to determine the type of chromatin that is assembled at the location of those nucleosomes. One mechanism by which chromatin specified by H3 and H4 modifications could be inherited during chromosome replication would be to ensure that H3 and H4 are partitioned to both daughter DNA molecules. This would "replicate" the location of H3 and H4 modifications of the parent chromosome in each daughter chromosome. Since H3 and H4 modification specify the type of chromatin that will be assembled, both daughter chromosomes would have the same pattern of euchromatin and heterochromatin as the parental molecule.

10.   *Evaluate*: This is type 2. Recall that histones bind to DNA in a sequence-independent fashion. *Deduce > Solve:* Since nucleosome formation is sequence independent, any DNA molecule can be packaged into nucleosomes, regardless of its evolutionary history. Thus bacterial DNA should be packaged into nucleosomes as easily as any other DNA, including eukaryotic DNA.

12

Gene Mutation, DNA Repair, and Homologous Recombination

Section 12.01 Genetics Problem-Solving Toolkit

Section 12.02 Types of Genetics Problems

Section 12.03 Solutions to End-of-Chapter Problems

Section 12.04 Test Yourself

12.01 Genetics Problem-Solving Toolkit

Key Terms and Concepts

Base-pair substitution: A change of a single base pair in a nucleotide sequence. The mutation may be a transition (purine to purine and pyrimidine to pyrimidine) or a transversion (purine to pyrimidine and pyrimidine to purine). It is also known as a *point mutation*.

Mutation rate: A measure of the number of mutations that occur per unit of DNA, per unit of time (e.g., mutations per genome per generation).

Mutation frequency: A measure of the occurrence of new mutations in a population of individuals or DNA molecules (e.g., new mutants per total number of individuals). Often used as an estimate of mutation rate.

Recombination repair: A DNA damage repair mechanism that uses homologous recombination between sister chromatids or homologous chromosomes to repair and replace damaged sites.

Heteroduplex DNA: DNA that is comprised of strands originally part of different DNA duplexes. According to the Holliday and double-stranded break models for homologous recombination, heteroduplex DNA forms during every homologous recombination event.

Gene conversion: The nonreciprocal exchange of genetic information between homologous chromosomes. Gene conversion is thought to be the result of the repair of mismatched base pairs in heteroduplex DNA formed during meiotic homologous recombination.

Key Analytical Tools

Genetic Code (see Chapter 1, page 3)

12.02 Types of Genetics Problems

1. Describe or analyze the molecular mechanisms of mutations.
2. Describe, analyze, or predict the effect of mutations on DNA, RNA, or protein.
3. Describe, analyze, or predict the effect of mutations on organismal phenotype.

4. Describe mechanisms that ensure the accuracy of DNA replication or repair DNA damage.

5. Describe, analyze, or predict the effect of homologous recombination during meiosis.

1. Describe or analyze the molecular mechanisms of mutations.

Problems of this type test your knowledge of the molecular mechanisms that explain spontaneous and induced mutations. You may be asked to recall the mechanism by which errors in DNA replication or base-tautomeric shifts result in spontaneous mutations. You may be asked to recall the mechanism of action of chemical mutagens or radiation. You may also be asked to calculate the spontaneous or induced rate of mutations. See Problem 6 for an example.

Example Problem: The Ames test determines whether chemical agents are mutagenic by testing whether they increase the rate of reversion of *his⁻* auxotrophic bacteria to *his⁺* prototrophs. The test is performed by culturing bacteria in suspension in liquid medium containing histidine, and plating the samples on medium lacking histidine but containing a filter disk that contains either the chemical in question or a buffer (no chemical) control. If 10^7 *his⁻* bacterial cells are plated and 10 colonies grow on the control plate and 110 colonies grow on the chemical plates, what are the spontaneous and induced mutation rates for this bacterial strain expressed as number of mutation/chromosome/round of DNA replication?

Solution Strategies	Solution Steps
Evaluate	
1. Identify the topic this problem addresses, and explain the nature of the requested answer.	1. This problem requires being able to distinguish between spontaneous and induced mutation, and being able to calculate mutation rate given a mutation frequency. The answers will be two mutation rates, each expressed as number of mutations per chromosome, per round of DNA replication.
2. Identify the critical information given in the problem.	2. 10^7 *his⁻* bacterial cells were plated on each plate. Ten colonies grew on the control plate, and 110 colonies grew on the chemically treated plate.
Deduce	
3. Relate the number of colonies on each plate to the number of *his⁺* reversion mutants.	3. Each bacterial colony represents a single *his⁻*-to-*his⁺* reversion mutant.
4. Relate the number of *his⁺* reversion mutants to the number of *his⁺* reversion mutations that occurred during each bacterial cell division.	4. Each *his⁺* reversion mutant is considered to represent a single, independent *his⁺* reversion mutation during one bacterial division.
5. Determine the number of spontaneous and induced mutations detected on each plate. **TIP:** The spontaneous mutation rate is considered constant, regardless of whether there is a mutagen present.	5. The 10 *his⁺* reversion mutants on the control plate and 10 of the 110 reversion mutants on the chemically treated plate are caused by spontaneous mutations. One hundred of the mutants on the chemically treated plate are caused by induced mutations, whereas none are on the control plate.

(continued)

Solution Strategies	Solution Steps
Solve	
6. Calculate the number of spontaneous and induced *his+* mutations that occurred per *his–* cell plated.	6. The numbers of mutations per cell plated are: spontaneous $\frac{10 \text{ colonies}}{10^7 \text{ cells plated}} = 10^{-6}$ induced $\frac{100 \text{ colonies}}{10^7 \text{ cells plated}} = 10^{-5}$.
7. Express the value from step 6 in terms of mutation/chromosome/round of DNA replication. TIP: Each mutation is considered to have occurred on one chromosome during one round of DNA replication.	7. **Answer:** The spontaneous mutation rate is 10^{-6} mutations per chromosome per round of DNA replication. The induced mutation rate is 10^{-5} mutations per chromosome per round of DNA replication.

2. Describe, analyze, or predict the effect of mutations on DNA, RNA, or protein.

Problems of this type test your understanding of how mutations affect genes, mRNAs, and proteins. You may be given a DNA sequence and specific mutations to that sequence, and asked to determine the consequences for the mRNA or protein products. You may be asked to classify the nature of mutations based on how they change DNA, mRNA, or protein. Your knowledge may be challenged by problems that ask you to work backward from wild-type and mutant protein sequences to deduce the sequences of wild-type and mutant mRNA or DNA. See Problem 8 for an example.

Example Problem: A portion of a protein has the amino acid sequence N-terminus…-Thr-Arg-Leu-Ile-…C-terminus. Four independent mutations are isolated, and the amino acid sequence of each mutant is determined. The sequences are as follows:

    mutant 1 N-terminus…-Met-Arg-Leu-Ile-…C-terminus
    mutant 2 N-terminus…-Thr-C-terminus
    mutant 3 N-terminus…-Thr-Arg-Trp-Ile…-C-terminus
    mutant 4 N-terminus…-Thr-Arg-Leu-Lys…-C-terminus

Use this information to determine the sequence of wild-type mRNA that codes for the wild-type protein sequence.

Solution Strategies	Solution Steps
Evaluate	
1. Identify the topic this problem addresses, and explain the nature of the requested answer.	1. This problem tests your understanding of the genetic code and the effect of mutations on coding sequences. The answer will be a sequence of four codons corresponding to the wild-type coding sequence.

(*continued*)

Solution Strategies	Solution Steps

2. Identify the critical information given in the problem.	2. The protein sequences are as follows:
	Wild type: N-terminus...-Thr-Arg-Leu-Ile-...
	   C-terminus
	Mutant 1: N-terminus...-Met-Arg-Leu-Ile-...
	   C-terminus
	Mutant 2: N-terminus...-Thr-C-terminus
	Mutant 3: N-terminus...-Thr-Arg-Trp-Ile...
	   -C-terminus
	Mutant 4: N-terminus...-Thr-Arg-Leu-Lys...
	   -C-terminus

Deduce	
3. Write the wild-type coding sequence using only the wild-type protein sequence. Indicate the identity of ambiguous nucleotides using a slash or N. For example, A/G indicates a purine, U/C indicates a pyrimidine, and N indicates any base.	3. The codons for each wild-type amino acid are as follows: Thr = ACN; Arg = AGA/G and CGN; Leu = UUA/G or CUN; Ile = AUU/C/A. The wild-type mRNA sequence in this region is either ATN AGA/G (or CGN) UUA/G AUU/C/A or ATN AGA/G (or CGN) CUN AUU/C/A.
4. Identify the nature of the mutation in each mutant referring to the effect of the mutation on the protein sequence. Write the sequence of the mRNA codon corresponding to the wild-type and mutant codon that was changed in each mutant.	4. Mutant 1 is a missense mutation, Thr (ACN) to Met (AUG); mutant 2 is a nonsense mutation, Arg (AGA/G or CGA) to Stop (UGA), which limits the Arg codon possibilities to AGA or CGA; mutant 3 is a missense mutation, Leu (UUA/G or CUN) to Trp (UGG); mutant 4 is a missense mutation, Ile (AUU/C/A) to Lys (AAA/G).
5. Identify the wild-type codon(s) that could be converted to the mutant codon by a single-nucleotide substation (point mutation) for each mutant.	5. To create mutant 1, ACN could be converted to AUG by a C-to-U change if N is G. The wild-type Thr codon is ACG. To create mutant 2, AGA or CGA could be converted to UGA by an A-to-U or C-to-U change. The wild-type Arg codon is AGA or CGA. To create mutant 3, UUA/G or CUN could be converted to UGG by a U-to-G change if the codon was UUA/G and the A/G is a G. The wild-type Leu codon is UUG. To create mutant 4, AUU/C/A could be converted to AAA/G by a U-to-A change if U/C/A is A. The wild-type Ile codon is AUA.
Solve	
6. Combine the wild-type codons deduced in step 5 to create a wild-type mRNA sequence.	6. **Answer:** The wild-type mRNA sequence is 5'-...ACGA/CGAUUGAUA...-3'.

3. Describe, analyze, or predict the effect of mutations on organismal phenotype.

This type of problem tests your understanding of how mutations affect phenotype. You may be asked to consider forward mutations, reverse mutations, or suppressor mutations and their potential impact on organismal phenotype. You may be asked to analyze or predict the results of phenotypic assays on yeast or bacterial mutants, including complementation tests and the Ames test. In all of these problems, you will be required to consider the relationship between genotype and phenotype and the effect of different types of mutations on gene function. See Problem 28 for an example.

Example Problem: A total of 10^7 cells of the yeast mutant ade^-, which is defective for adenine biosynthesis, was mistakenly plated on minimal medium lacking adenine. Unexpectedly, two ade^+ colonies grew, and they could be grown on medium lacking adenine for multiple generations. To understand the origin of the ade^+ strains, R_1 and R_2, they are crossed with a wild-type ade^+ strain, and 100 random ascospores from each cross are isolated. The results are as follows: R_1 × wild type produced 100 ascospores that could grow on medium lacking adenine, and R_2 × wild type produced 75 ascospores that could grow on medium lacking adenine and 25 that could not. Use this information to deduce the most likely origin of both ade^+ yeast strains.

Solution Strategies	Solution Steps
Evaluate	
1. Identify the topic this problem addresses, and explain the nature of the requested answer.	1. This problem requires an understanding of compensatory mutations and the ability to apply it to the results of crosses involving yeast revertants. The answer will be a description of the mutations that converted the adenine auxotroph to adenine prototrophs for both revertants.
2. Identify the critical information given in the problem.	2. Two independent, spontaneous ade^+ revertants were isolated and crossed to wild-type yeast. Revertant 1 × wild type produced only ade^+ progeny. Revertant 2 × wild type produced 75% ade^+ and 25% ade^- progeny.
Deduce	
3. Recall the possible types of mutations that could convert a mutant (adenine auxotroph) to wild-type phenotype (adenine prototroph). Write a proposed genotype that corresponds to the revertant colony for each mutation type. **TIP:** Review forward mutation and reversion in Section 12.2 of the text.	3. The types of mutations that could convert an adenine auxotroph to an adenine prototroph include a reversion mutation that converts the ade^- gene to ade^+; and a second-site compensatory mutation, called a suppressor mutation (sup^-), that converts a different wild-type gene to a mutant gene, which compensates for the ade^- mutation. The reversion mutation would create a revertant that is ade^+. The suppressor mutation would create a revertant that is a double mutant, ade^- sup^-.
4. Predict the results that would be obtained from crossing each of the mutants in step 3 to wild type. **TIP:** Assume different genes show independent assortment.	4. The cross of ade^+ × wild type (ade^+) will produce only ade^+ progeny. The cross ade^- sup^- × wild type (ade^+ sup^+) will produce 1/4 ade^- sup^-, 1/4 ade^- sup^+, 1/4 ade^+ sup^-, and 1/4 ade^+ sup^+.
5. Assign phenotypes (ability to grow on medium lacking adenine) to the progeny genotypes predicted in step 4. **TIP:** Suppressor mutations (sup^-) can be assumed to have no phenotype other than suppression; ade^+ sup^- strains are wild type.	5. All progeny from the first cross are ade^+ and will be able to grow on medium lacking adenine. Progeny from the second cross include ade^- sup^-, 1/4 ade^+ sup^-, and 1/4 ade^+ sup^+ that will be able to grow on medium lacking adenine, and 1/4 ade^- sup^+ that will not be able to grow on medium lacking adenine.

(continued)

Solution Strategies	Solution Steps
Solve	
6. Calculate the predicted frequency of adenine prototrophs and auxotrophs from the crosses in step 5, and match those to the results described in the problem.	6. Cross 1 produces 100% adenine prototrophs, which matches the results from the cross of R_1. Cross 2 produces 75% adenine prototrophs, which matches the results from the cross of R_2.
7. Describe the nature of the spontaneous compensatory mutation that created each revertant colony.	7. **Answer:** R_1 was created by a reversion mutation that converted the mutant ade^- allele to a wild-type ade^+ allele. R_2 was created by a suppressor mutation in a gene unlinked to ade^- that suppresses the ade^- mutant phenotype.

4. Describe mechanisms that ensure the accuracy of DNA replication or repair DNA damage.

This type of problem tests your understanding of mechanisms that act to prevent errors in DNA replication and to repair damaged DNA. You may be asked to identify which DNA repair pathway is most appropriate to counter a particular error in DNA replication or type of DNA damage. You may be asked to describe how a specific repair mechanism operates. You may also be asked to match specific DNA repair proteins to DNA repair pathways or DNA replication errors/DNA damage events. See Problem 18 for an example.

Example Problem: A mutant yeast strain that has a spontaneous mutation rate 10 times higher than the wild-type control is isolated. The strain is suspected to be deficient for a process that protects the genome from mutations, so it is subjected to a number of different mutagens in an attempt to identify the defective pathway. The strain responds to UV light, ionizing radiation, and intercalating agents in the same way as a wild-type strain, but is significantly more sensitive to nucleotide analogs than is the wild-type control. Generate a hypothesis for the nature of the genetic defect of the yeast strain, and suggest how study of this yeast mutant could help researchers better understand a specific heritable human disease.

Solution Strategies	Solution Steps
Evaluate	
1. Identify the topic this problem addresses, and explain the nature of the requested answer.	1. This problem tests your knowledge of mutagenesis and DNA repair. The answer will be the identification of a specific DNA repair pathway and a heritable human disease that results from defects in that pathway.
2. Identify the critical information given in the problem.	2. The yeast strain in question has a high spontaneous mutation rate and is sensitive to base analogs. The strain shows normal sensitivity to UV light, ionizing radiation, and intercalating agents.

(continued)

Solution Strategies	Solution Steps
Deduce	
3. Describe how nucleotide analogs act, and identify the DNA repair pathway that is responsible for countering the effect of nucleotide analogs but not UV irradiation, ionizing irradiation, or intercalating agents.	3. Nucleotide analogs can be incorporated into DNA during DNA replication in one of two tautomeric forms. One tautomer base-pairs in the same manner as the normal nucleotide, whereas the other pairs improperly, leading to mismatched base pairs. The DNA repair pathway responsible for repairing mismatched bases is the mismatch repair pathway. Mismatch repair is not responsible for repair of damage due to UV or ionizing radiation or intercalating agents.
4. Explain how a defect in the repair pathway identified in step 3 could lead to a high spontaneous mutation rate.	4. Mistakes in DNA replication lead to the formation of mismatched base pairs, which are normally corrected by the mismatch repair pathway. The lack of a mismatch repair pathway would allow more mismatches to remain and therefore more mutations to accumulate.
Solve	
5. State the nature of the defect in the yeast mutant strain, and identify a heritable human disease that could be studied by further investigation of this yeast mutant.	5. **Answer:** The yeast strain is defective for mismatch repair. This strain would be useful for learning about nonpolyposis colon cancer (NPCC), which is a human heritable disease due to defects in mismatch repair.

5. Describe, analyze, or predict the effect of homologous recombination during meiosis.

This type of problem tests your understanding of homologous recombination during meiosis and the relationship between the proposed mechanism for homologous recombination and gene conversion. These problems ask you to analyze the results of crosses using yeast or other fungi, to indicate if gene conversion is evident, and to explain how gene conversion occurs. You may also be asked to explain how gene conversion in fungi contributes to our understanding of the mechanism of homologous recombination. See Problem 34 for an example.

Example Problem: Two yeast strains that differ at three linked genetic loci are crossed. One strain is a^+ $b^+ c^+$ and the other is $a^- b^- c^-$. You are told that you may see evidence of gene conversion at the b locus in one or more asci obtained that show evidence of recombination between a and c. Generate a hypothesis to describe two different ascus types that would provide evidence of gene conversion at the b locus.

Solution Strategies	Solution Steps
Evaluate	
1. Identify the topic this problem addresses, and explain the nature of the requested answer.	1. This problem tests your understanding of gene conversion and your ability to apply it to the analysis of a yeast cross. The answer will be two different yeast ascus types that show evidence of gene conversion at the b locus.

(continued)

Solution Strategies	Solution Steps
2. Identify the critical information given in the problem.	2. The parent yeast strains have the genotype $a^+\,b^+\,c^+$ and $a^-\,b^-\,c^-$.
Deduce	
3. Describe gene conversion in general terms.	3. Gene conversion is the nonreciprocal exchange of genetic information between homologous chromosomes. When gene conversion occurs during meiosis, aberrant ratios of allele segregation are detected.
4. Give an example of normal and aberrant allele segregation during meiosis of a yeast diploid heterozygous at the b locus ($b^+\,b^-$).	4. A $b^+\,b^-$ yeast diploid normally produces 2 b^+ and 2 b^- spores. Aberrant segregation due to gene conversion would be 3 b^+ and 1 b^- or 3 b^- and 1 b^+.
5. Describe the genotype of the linked a and c loci in an ascus resulting from the cross described in this problem that shows a single crossover between a and c.	5. The parental genotypes for a and c are $a^+\,c^+$ and $a^-\,c^-$; therefore, an ascus showing a crossover between a and c would have spores of the genotypes $a^+\,c^+$, $a^+\,c^-$, $a^-\,c^+$, and $a^-\,c^-$.
Solve	
6. Describe two ascus types, each of which shows a crossover between a and c, that provide evidence of gene conversion at the b locus.	6. **Answer:** The two ascus types would be as follows: Type 1: $a^+\,b^+\,c^+$, $a^+\,b^+\,c^-$, $a^-\,b^+\,c^+$, $a^-\,b^-\,c^-$ Type 2: $a^+\,b^+\,c^+$, $a^+\,b^-\,c^-$, $a^-\,b^-\,c^+$, $a^-\,b^-\,c^-$

12.03 Solutions to End-of-Chapter Problems

1. *Evaluate:* This is type 1. Recall that chemical mutagens include those that chemically modify bases, those that intercalate between nucleotides, and those that act as nucleotide analogs. *Deduce > Solve:* Chemical mutagens (1) can act as base analogs and (2) can chemically modify bases. Analogs are recognized by DNA polymerase and incorporated into DNA in place of nucleotides and then cause mutations by base-pairing in a manner that differs from the analogous nucleotide. For example, 5-BrdU can be incorporated opposite an A during replication and then pair as a C during the next round of replication, causing a TA-to-CG transition. Chemical modification of bases alters their base-pairing properties such that a modified purine will base-pair with the wrong pyrimidine, and vice versa. For example, EMS is an alkylating agent that converts guanine to O^6-ethylguanine, which base-pairs with T to create a GC-to-AT transition.

2. *Evaluate > Deduce > Solve:* This is type 1. Recall that 5-BrdU is a base analog and that nitrous acid is a deaminating agent. 5-BrdU is a thymidine analog that can base-pair like thymine or like cytosine. If 5-BrdU is incorporated in place of thymidine and then base-pairs like cytosine in the subsequent round of replication, it causes an AT-to-GC transition. If 5-BrdU is incorporated into DNA in place of cytidine and then base-pairs like thymine in the following round of replication, it causes a GC-to-AT transition. Nitrous acid converts cytosine to uracil, which will base-pair with adenine in the next round of replication and cause a CG-to-TA transition. Nitrous acid can

also convert adenine to hypoxanthine, which will base-pair with cytosine in the next round of replication and cause an AT-to-GC transition.

3a. *Evaluate:* This is type 2. Recall that a transition mutation involving A is a change of A to G. *Deduce > Solve:*

> ...GCCC...
> ...CGGG...

3b. *Evaluate:* This is type 2. Recall that a transversion mutation involving A is a change of A to T or A to C. *Deduce > Solve:*

> ...GCAC...
> ...CGTG...

> or

> ...GCGC...
> ...CGCG...

4a. *Evaluate*: This is type 2. Identify the most amino-proximal amino acid difference between the wild-type and mutant protein sequence. Then, determine if this is the only difference or if the sequence after that site has also changed. *Deduce:* The first difference is a Leu in place of Tyr, and the remainder of the mutant sequence differs from the wild type. Recall that substitution mutations change a single codon, whereas frameshift mutations alter the remainder of the sequence from the site of the mutation to the carboxyl terminus. *Solve:* The mutation is a frameshift mutation (insertion or deletion).

4b. *Evaluate:* This is type 2. This problem requires knowledge of the genetic code to "reverse translate" a protein sequence into a coding sequence. Refer to the genetic code (page 3) to identify all the possible codons corresponding to each amino acid in the wild-type sequence, and use the mutant coding sequence to deduce the correct wild-type codon at each position. Since the mutation is a frameshift, the mutant coding sequence will differ from wild type by an insertion or a deletion. Once the precise insertion or deletion is identified, the mutant reading frame can be deduced and the mutant amino acid sequence used to determine the wild-type codons. *Deduce:* The wild-type codons are Arg, AGA/G or CGN; Met, ATG; Tyr, TAT/C; Thr, ACN; Leu, CTN or TTA/G; Cys, TGT/C; Ser, AGT/C or TCN. (1) The third mutant amino acid is Leu (CTN or TTA/G). An insertion of T after ATG and before TAT/C creates a Leu codon. (2) The next mutant codon is T/CAC. For this to code for Tyr, T/C must be a T, indicating that the third wild-type codon, Tyr, is TAT. (3) The next mutant codon is either NCT or NTT and should code for Ala. Ala is GCN, so N must be G, resulting in the mutant sequence GCT, which indicates that the fourth wild-type codon, Thr, is ACG. (4) The next mutant codon is N/T/ATG and should code for Leu. Leu is either CTN or TTA/G; therefore, N/T/A must be a C and the mutant sequence is CTG, which indicates that the fifth wild type, Leu, is CTC. (5) The next mutant codon is T/CAG or T/CTC and should code for Phe. Phe is TTT/C; therefore, T/C must be a T and the mutant sequence must be TTC, indicating that the sixth wild-type codon, Cys, is TGT and the last wild-type codon, Ser, is TCN. (6) *Solve:* The wild-type sequence for the coding strand can be written as AGA/G or CGN-ATG-TAT-ACG-CTC-TGT-TCN, and finally the template strand is TCT/G-TAC-ATA-TGC-GAG-ACA-AGN.

5. *Evaluate > Deduce > Solve:* This is type 1. Recall how spontaneous mutations are caused by tautomeric shifts of the bases in nucleotides. Draw a diagram similar to the figure shown.

Thymine (enol)   Guanine (keto)

6a.   *Evaluate > Deduce > Solve:* This is type 1. The solution is pyrimidine dimers.

6b.   *Evaluate:* This is type 1. Recall that thymidine dimers do not base-pair normally. *Deduce > Solve:* If thymidine dimers created by UV irradiation persist until DNA replication, DNA polymerase is likely to make a mistake reading the template sequence at TT and insert something besides AA on the complementary strand. The altered sequence on the newly synthesized strand will be copied during the next round, or replication, thus creating a mutant sequence.

6c.   *Evaluate:* This is type 4. *Deduce > Solve:* Recall that T/T dimers can be repaired by photoreactivation and by nucleotide excision via UV repair. Photoreactivation is a light-dependent mechanism involving photolyase, an enzyme present in many microbes that removes the cross-links created by UV irradiation and restores the normal DNA structure. Nucleotide excision repair involves recognition of crosslinks created by UV irradiation, excision of a segment of the strand containing the crosslinks, DNA synthesis using the complementary strand to fill in the gap, and ligation to seal the nick between the newly synthesized DNA and the original DNA strand.

7a.   *Evaluate:* This is type 1. *Deduce > Solve:* Chemical mutagens can induce transitions, transversion, insertions, or deletions. EMS and hydroxylamine are two examples of chemical mutagens that cause transition mutations. EMS is an alkylating agent that can add an ethyl group to guanine to create O^6-ethylguanine, which base-pairs with thymine, causing a GC-to-AT transition. Hydroxylamine is a hydroxylating agent that can add a hydroxyl group to cytosine to create hydroxylaminocytosine, which base-pairs with adenine, causing a CG-to-TA transition.

7b.   *Evaluate:* This is type 1. *Deduce > Solve:* Radiation can induce transitions, transversions, insertions, or deletions. UV irradiation and γ irradiation are two examples of radiant energy that cause mutations. UV irradiation induces pyrimidine dimers in DNA. Pyrimidine dimers in the template strand of DNA during DNA replication can cause mistakes in DNA synthesis that lead to mutations. The γ irradiation causes double-strand breaks in DNA, and if these breaks are repaired improperly, they lead to mutations.

8.   *Evaluate:* This is type 2. This problem concerns the relationship between the coding sequence of a gene and the function of the encoded protein. Recall that the effect of a single-nucleotide substitution depends on whether the substitution changes the meaning of the codon and, if so, whether that change has a significant impact on the structure of the protein. *Deduce > Solve:* Nucleotide substitutions can result in silent, missense, or nonsense mutations. Silent mutations do not change the amino acid sequence of the protein and therefore have no effect on protein function. Missense mutations change one amino acid in a protein; therefore, the effect on the function of the protein depends on the importance of the amino acid that was replaced and the functional similarity (or lack thereof) of the amino acid side chain (R group) on the substituted amino acid to that of the wild-type amino acid. Nonsense mutations end the protein prematurely, almost always leading to lack of function.

9.   *Evaluate:* This is type 2. This problem tests your understanding of the relationship between the coding sequence of a gene and the function of the encoded protein. Recall that recessive alleles often have little or no function, whereas dominant alleles are functional. *Deduce > Solve:* The difference between the two proteins is that A_1 has a His at position 2 whereas A_2 has a Gln. The alleles could possibly code for proteins with different functional capability, since His typically carries a slight positive charge and Gln is polar but uncharged. But it cannot be determined which version is functional and which is nonfunctional based on sequence information alone. Neither His nor Gln is inherently a "better" amino acid; which one contributes to a functional protein depends on which of them occupied this position when the protein evolved.

10.   *Evaluate:* This is type 2. This problem tests your understanding of the effect of spontaneous mutations on gene function. Recall that spontaneous mutations are typically nucleotide substitutions and, if detected, are not silent. *Deduce > Solve:* (1) Recessive mutations are typically loss-of-function mutations. Wild-type gene sequences have been selected during evolution for optimum function; therefore, any change (mutation) to that sequence is likely to replace a nucleotide maintained by natural selection with one that reduces the function of the gene. (2) Forward mutations include all mutations in a gene that convert it from wild type to mutant, whereas reverse mutations are only those that precisely reverse a specific mutation to wild type. Thus the number of possible nucleotide changes corresponding to a forward mutation is much greater than those that reverse a given mutation, making forward mutations far more frequent than reversion.

11a–b. *Evaluate:* This is type 3. This problem concerns the distinction between forward, reverse, and suppressor mutations. *Deduce > Solve:* **(a)** The mutation creating a nonsense codon in the *ade-1* gene is a forward mutation because it converts a wild-type sequence into a mutant sequence. **(b)** The mutation creating an altered tRNA sequence is a forward mutation because it converts a wild-type sequence into a mutant sequence.

11c. *Evaluate:* This is type 3. This problem concerns genetic suppressors. *Deduce > Solve:* The original protein sequence contained a Trp at the position of the nonsense mutation. The suppressor tRNA inserts Trp at UGA stop codons at some frequency that depends on the relative abundance of the suppressor tRNA in the cell. Whenever it binds to the mutant nonsense codon of the *ade-1* gene, Trp will be inserted into the position that normally contains Trp. This would create a wild-type protein; therefore, the double mutant should synthesize some wild-type Ade-1 protein. The rate at which the double mutant will be able to grow on medium lacking adenine will depend on the frequency of insertion of the Trp at the UGA nonsense codon and how much of the *ade-1* gene product is required for growth.

12a. *Evaluate:* This is type 3. Consider the close evolutionary relationship between mice and humans and the experimental utility of the mouse as a model research organism. *Deduce > Solve:* The mouse (*Mus musculus*) is a widely used model organism for genetic analysis of mammalian development and physiology, specifically in relation to human disease because many of these processes in mice and humans are evolutionarily conserved. The other advantage is that with mice, researchers can perform experimental manipulations that are possible when studying humans or even nonhuman primates.

12b. *Evaluate:* This is type 3. Consider the differences between mice and humans. *Deduce > Solve:* Although the mouse (*Mus musculus*) is a mammal, there are many developmental, behavioral, and physiological differences between mice and humans. In addition, not every human gene has a homolog in the mouse genome. Therefore, in cases where the physiology or genetics of mice and humans differ, mutations in a mouse homolog to a human disease gene may not provide useful information on the human disease process.

13a. *Evaluate:* This is type 4. Recall that the accuracy of DNA replication depends on the function of DNA polymerase and the action of the DNA repair pathway called mismatch repair. *Deduce > Solve:* DNA polymerase contributes to the accuracy of replication in two ways. First, it ensures that the correct nucleotide is added to the growing DNA strand. Second, it removes incorrect nucleotides immediately after they are added.

13b. *Evaluate:* This is type 4. Recall that DNA polymerase has a 3′-to-5′ exonuclease activity that it uses for proofreading. *Deduce > Solve:* DNA polymerase proofreads each nucleotide after it is added to the growing DNA strand. If it detects a mismatched base pair, it removes the last nucleotide added using its 3′-to-5′ exonuclease activity and then selectively adds the correct nucleotide in its place.

13c. *Evaluate:* This is type 4. *Deduce:* Mistakes in DNA replication lead to the formation of incorrect, or "mismatched," base pairs. These mismatched base pairs are predominantly A - C and G - T (purine–pyrimidine) base pairs. *Solve:* The most likely abnormality is mismatched base pairs.

13d. *Evaluate:* This is type 4. See part (b) for more information on the mechanism of proofreading. *Deduce:* Mismatched base pairs are recognized by the DNA repair pathway called mismatch repair. The mismatch DNA repair pathway recognizes the mismatched base pair, determines which strand is the newly synthesized DNA strand, removes nucleotides on that strand, and then resynthesizes that portion of the DNA strand. *Solve:* Proofreading by DNA polymerase and mismatch repair can correct this kind of abnormality.

13e.   *Evaluate:* This is type 1. *Deduce:* Mistakes in replication typically result from incorporation of the incorrect pyrimidine (C instead of T or T instead of C) or purine (G instead of A or A instead of G), which leads to A-C or T-G mismatches. For example, if a C was improperly added to the growing strand opposite an A in the template, the A-C mismatched base pair created will be converted to A-T and G-C base pairs in the next round of replication. The strand with the G-C base pair has a transition mutation. *Solve:* Nucleotide substitutions (point mutations) that are typically transition mutations will occur.

13f.   *Evaluate:* This is type 4. *Deduce:* DNA is methylated sometime after replication, such that daughter DNA molecules produced by DNA replication initially contain a methylated strand and a non-methylated strand. Mismatch repair recognizes the methylated strand as the template strand and preferentially repairs the non-methylated strand at the site of mismatches. *Solve:* Methylation is the distinguishing characteristic.

14.   *Evaluate:* This is type 1. This problem tests your ability to calculate dominant mutation rates. Recall that each birth represents two gametes. *Deduce:* The 322,182 births provide information on the genotype of 644,364 gametes. Apert syndrome is autosomal dominant; therefore, each gamete containing a dominant Apert syndrome mutation would be detected by the birth of a child with Apert syndrome. *Solve:* The rate of new Apert syndrome mutations would

be $\frac{\text{number of Apert syndrome births}}{(\text{number of births} \times 2)}$. Substituting the known quantities into the equation results in $\frac{2}{(322,182 \times 2)} = \frac{1}{322,182}$. Therefore, the mutation rate is 1 per 322,182 gametes.

15.   *Evaluate:* This is type 1. This problem tests your ability to calculate dominant mutation rates. Recall that each birth represents two gametes and that penetrance refers to the percentage of individuals with the mutant genotype that show the mutant phenotype. *Deduce:* The 40,000 births provide information on the genotype of 80,000 gametes. Polydactyly is autosomal dominant but only 70% penetrant; therefore, only 7 out of every 10 gametes containing a polydactyly allele will be detected by the birth of a child with polydactyly. *Solve:* The rate of new polydactyly mutations would be $\frac{\text{number of polydactyly births}}{(\text{number of births} \times 2)} \times \frac{10}{7}$. Substituting the known quantities into the equation results in $\frac{1}{(40,000 \times 2)} \times \frac{10}{7} = \frac{1}{56,000}$. Therefore, the mutation rate is 1 in 56,000 births.

16a.   *Evaluate:* This is type 1. This problem tests your ability to calculate the rate of dominant mutations. Recall that each birth provides information on two gametes. *Deduce:* The mutation rates are $\frac{20}{10^6} = 2 \times 10^{-5}$ mutations/gamete for retinoblastoma, $\frac{80}{10^6} = 8 \times 10^{-5}$ mutations/gamete for achondroplasia, and $\frac{220}{10^6} = 2.2 \times 10^{-4}$ mutations/gamete for neurofibromatosis. *Solve:* The number of live births expected for each dominant condition is equal to $\frac{\text{mutation rate} \times \text{number of births} \times 2 \text{ gametes}}{\text{birth}}$. For retinoblastoma, this is $\frac{2 \times 10^{-5} \text{ mutations/gametes} \times 50,000 \text{ births} \times 2 \text{ gametes}}{\text{birth}} = 2$ births. For achondroplasia, this is $\frac{8 \times 10^{-5} \text{ mutations/gamete} \times 50,000 \text{ births} \times 2 \text{ gametes}}{\text{birth}} = 8$ births. For neurofibromatosis, this is $\frac{2.2 \times 10^{-4} \text{ mutations/gamete} \times 50,000 \text{ births} \times 2 \text{ gametes}}{\text{birth}} = 22$ births. Therefore, 2 births with retinoblastoma, 8 births with achondroplasia, and 22 births with neurofibromatosis are expected.

16b.   *Evaluate:* This is type 2. This problem concerns the mechanism underlying differences in the observed mutation rates of different genes. *Deduce:* The relative observed mutation rates for different genes depend on the rate that mutations occur within in each gene and the effect of each mutation on the function of the genes. Factors affecting the rate of mutations in each gene include (1) the relative size of the genes and (2) whether the genes contain hotspots (mutation rates that are higher than normal). The factors that comprise the effect of mutations on gene function include (1) the fraction of each gene that is noncoding or otherwise dispensable for function, (2) the fraction of mutations that result in a dominant lethal allele that cannot be detected, and (3) the relative penetrance of mutations that do have a phenotypic consequence. *Solve:* Since the mutation rate for neurofibromatosis (NF1) is tenfold higher than that of retinoblastoma (RB1), it is reasonable to hypothesize that NF1 is larger than RB1, contains hotspots for mutations that are not present in RB1, has a higher percentage of the mutations that have a phenotypic consequence, or has a higher penetrance than RB1 mutations. Or, another hypothesis is that a higher percentage of mutations in RB1 are dominant lethal (and thus cannot be detected). Two factors that could explain why the neurofibromatosis (NF1) mutation rate is higher than retinoblastoma (RB1) mutation rate are (1) that the *NF1* gene is larger than the *RB1* gene, and (2) that a higher percentage of mutations within *NF1* affect NF1 function as compared to the percentage of mutations within *RB1* that affect function.

17. *Evaluate:* This is type 4. This problem tests your understanding of repair of DNA damage caused by UV light. Recall that bacteria and fungi have a pathway for repair of UV-induced DNA damage called photoreactive repair, which requires visible light to be activated. *Deduce:* Both culture samples contained the same number of UV-damaged *E. coli* bacteria; therefore, the difference in the number of colonies on plates 1 and 2 is due to the number of *E. coli* cells plated that could grow (were viable). Since plate 2 was not protected from visible light, UV-damaged cells on that plate activated their photoreactive repair DNA repair pathway. Plate 1 was protected from visible light; therefore, UV-damaged cells on that plate did not activate photoreactive repair. *Solve:* More colonies appeared on plate 2 than plate 1 because the level of DNA damage inflicted by UV light in this experiment was high enough to require the photoreactive repair pathway to repair the DNA damage sufficiently to prevent cell death.

18. *Evaluate:* This is type 4. This problem tests your knowledge of the function of the *E. coli RecA* gene. *Deduce > Solve:* The *RecA* gene is required for both recombination and repair of DNA damage. *E. coli* with a null mutation in *RecA* would lack *RecA* function and would be deficient in recombination repair. Recombination repair is used to fill in a single-stranded DNA gap created by the lack of replication of a region due to DNA damage (for example, a UV photoproduct). Several steps in recombination repair are catalyzed by *RecA*, including two-strand invasion events and a single-stranded DNA cleavage event. The strain also would not be able to undergo homologous recombination.

19. *Evaluate > Deduce > Solve:* This is type 5. Gene conversion is the nonreciprocal exchange of genetic information between homologous chromosomes, such that the allele on one homolog is converted to the allele on the other homolog. In mitotic cells, gene conversion can convert cells with the genotype *Aa* to *AA* (*a* converted to *A*) or *aa* (*A* converted to *a*). In meiotic cells, gene conversion can result in a cell with the genotype *Aa* producing three *A*-containing gametes and one *a*-containing gamete (*a* converted to *A*), or one *A*-containing gamete and three *a*-containing gametes (*A* converted to *a*). Gene conversion often occurs at a rate much higher than can be accounted for by mutation. Gene mutation is a change in DNA sequence, which creates a different allele that can be any allele and is not restricted to alleles that are already present in the cell. For example, in a cell with the genotype *AA*, one of the *A* alleles can mutate to become *a*, and in cells with the genotype *aa*, *a* can mutate to become *A*.

20. *Evaluate:* This is type 1 and 4. This problem tests your understanding of mutation and homologous recombination. *Deduce > Solve:* Mutation is defined as a change in DNA sequence. Gene conversion resulting from recombination changes the DNA sequence of a chromosome; therefore, it fits within the definition of mutation. Recombination also combines chromosome sequences in new ways, creating new stretches of DNA sequence; therefore, recombination also is arguably a form of mutation. However, mutation is typically reserved to describe DNA sequence changes that are due to processes other than homologous recombination and gene conversion.

21.   *Evaluate:* This is type 5. This problem tests your understanding of heteroduplex DNA and gene conversion. *Deduce > Solve:* Heteroduplex DNA is double-stranded DNA composed of strands originating from different duplex DNA molecules. Heteroduplex DNA forms during recombination between homologous chromosomes. If the heteroduplex region includes portions of the homologs that are heterozygous, then it will contain one or more mismatched base pairs. These mismatches are repaired in a random manner, resulting in the heteroduplex containing one allele or the other. If the manner of mismatch repair during homologous recombination changes the genotype of the cell for this chromosome (for example, from *Aa* to *AA* or *aa*), then gene conversion is said to have occurred.

22.   *Evaluate:* This is type 5. This problem tests your understanding of homologous recombination. Review the mechanism of homologous recombination. *Deduce:* The current hypothesis for the mechanism of homologous recombination includes strand invasion and the formation of double Holliday structures. Based on this prevailing hypothesis, double-stranded DNA duplexes containing DNA strands originally from different double-stranded DNA duplexes will form during every homologous recombination event. It is important to note that the heteroduplex will contain mismatched base pairs only if the region involves portions of different sequences on the homologs that differ in base sequence. *Solve:* Yes; heteroduplex DNA is always created during homologous recombination.

23a.   *Evaluate:* This is type 5. This problem tests your understanding of recombination during meiosis in yeast and the phenomenon of gene conversion. *Deduce:* Before recombination, the meiotic yeast *Ala-b/ala-b* heterozygote contained four *Ala-B* genes: two were *Ala-B* alleles, and two were *ala-b* alleles. The resulting ascus contains three spores with *Ala-B* alleles and one spore with an *ala-b* allele, indicating conversion of one *ala-b* allele to an *Ala-B* allele. This indicates that a heteroduplex formed on the chromatid originally containing the *ala-b* allele, creating a mismatch involving the *Ala-B* A‑T base pair and the *ala-b* G‑C base pair. *Solve:* For gene conversion of *ala-b* to *Ala-B* to have occurred, the mismatch was repaired to create an A‑T base pair. This converted the *ala-b* allele originally on that chromatid to an *Ala-B* allele.

23b.   *Evaluate:* This is type 5. This problem tests your understanding of recombination during meiosis in yeast and gene conversion. *Deduce:* Before recombination, the meiotic yeast *Ala-b/ala-b* heterozygote contained four *Ala-B* genes: two were *Ala-B* alleles, and two were *ala-b* alleles. The resulting ascus contains three spores with *ala-b* alleles and one spore with an *Ala-B* allele, indicating conversion of one *Ala-B* allele to an *ala-b* allele. From these results, it can be assumed that a heteroduplex formed on the chromatid originally containing the *Ala-B* allele that created a mismatch involving the *Ala-B* A‑T base pair and the *ala-b* G‑C base pair. *Solve:* For gene conversion of *Ala-B* to *ala-b* to have occurred, the mismatch was repaired to create a G‑C base pair. The result is that the *Ala-B* allele originally on that chromatid was converted to an *ala-b* allele.

23c.   *Evaluate:* This is type 5. This problem tests your understanding of meiosis in yeast and the principle of segregation as revealed by yeast ascus analysis. *Deduce:* The meiotic yeast *Ala-b/ala-b* heterozygote contained four *Ala-B* genes: two were *Ala-B* alleles, and two were *ala-b* alleles. *Solve:* In the absence of gene conversion, meiotic chromosome segregation and ascospore formation result in two spores with the *Ala-B* genotype and two with the *ala-b* genotype, exactly as predicted by the principle of segregation.

24.   *Evaluate:* This is type 5. Recall that gene conversion is rare and results in the conversion of the genotype of one gamete out of four during meiosis. *Deduce > Solve:* The four products of meiosis in multicellular eukaryotes are not identifiable as such; instead, they are pooled with those of hundreds if not thousands of other meioses. Furthermore, these gametes are detected only by mating individuals and observing the phenotype of the resulting progeny. The high numbers of gametes produced and the random sampling of gametes during zygote formation make statistically significant identification of aberrant 3:1 segregation impossible.

25.   *Evaluate:* This is type 1 and 3. This problem tests your understanding of the mechanisms of mutagenesis in the context of forward and reverse mutations. Recall that base analogs induce transition mutations. *Deduce > Solve:* The increased rate of mutation among bacteria suggests that the chemicals spilled into the pond include mutagens. The mutant bacteria can be reverted to wild type by base analogs, which induce transition mutations. Since transition mutations can revert only transition mutations, the chemical mutagens spilled into the pond were transition-inducing mutagens.

26a.   *Evaluate:* This is type 2. This problem concerns the effect of mutations on DNA sequence. *Deduce > Solve:* *Sma*I and *Pvu*II both recognize 6-bp palindromic sequences. Two common features of these sites that would be useful to a researcher searching for mutations that disrupt restrictions sites are as follows: (1) neither recognition sequence includes a stop codon and therefore they could be located in the coding sequence of genes, which could facilitate the search by allowing the researchers to select mutants first by looking for mutant phenotypes before analyzing the mutants by restriction digestion; (2) 6-bp recognition sequences occur once every 4096 bp on average, which provides many possible mutation sites for the researchers to screen as they look for mutants.

26b.   *Evaluate:* This is type 1. This problem concerns the mechanisms of mutagenesis as they pertain to changes that could disrupt restriction enzyme recognition sequences. *Deduce > Solve:* The search for mutations that disrupt or create new *Sma*I and *Pvu*II recognition sites will include spontaneous and induced mutations, including transition mutations; transversion mutations; insertions and deletions.

27a.   *Evaluate:* This is type 3. This problem tests your ability to analyze yeast mutant phenotypes. Recall that prototrophic yeast can synthesize all organic metabolites except for glucose. *Deduce:* Colonies 1 and 4 are able to grow on minimal medium containing no additional supplements, whereas colonies 2, 3, and 5 cannot. *Solve:* Therefore, colonies 1 and 4 are prototrophic strains.

27b.   *Evaluate:* This is type 3. This problem tests your ability to analyze yeast mutant phenotypes. Recall that auxotrophic mutant yeast cannot synthesize all organic metabolites. *Deduce:* Colonies 1 and 4 are able to grow on minimal medium containing no additional supplements, whereas colonies 2, 3, 5, and 6 cannot. *Solve:* Colonies 2, 3, 5, and 6 are auxotrophic mutant strains.

27c.   *Evaluate:* This is type 3. This problem tests your ability to analyze yeast mutant phenotypes. *Deduce:* Yeast in colonies 2, 3, and 4 cannot grow on minimal medium. Yeast in colony 2 can grow on minimal medium supplemented with arginine, yeast in colony 3 can grow on minimal medium supplemented with histidine, and yeast in colony 5 can grow on minimal medium supplemented with leucine. *Solve:* Colony 2 is an Arg⁻ mutant, colony 3 is a His⁻ mutant, and colony 5 is a Leu⁻ mutant.

27d.   *Evaluate:* This is type 3. This problem tests your ability to analyze yeast mutant phenotypes. *Deduce:* Yeast that can grow on minimal medium can synthesize arginine, histidine, and leucine. Yeast that can grow only on minimal medium plus histidine can synthesize arginine and leucine but not histidine. Yeast that can grow only on minimal medium plus arginine can synthesize histidine and leucine but not arginine. Yeast that can grow only on minimal medium plus leucine can synthesize histidine and arginine but not leucine. *Solve:* Colony 1 is Arg⁺ His⁺ Leu⁺. Colony 3 is Arg⁺ His⁻ Leu⁺. Colony 5 is Arg⁺ His⁺ Leu⁻.

27e.   *Evaluate:* This is type 3. This problem tests your ability to analyze yeast mutant phenotypes. *Deduce:* Colony 6 contains yeast that cannot grow on minimal medium or any of the supplemented minimal media. Other media would have to be used to determine the genotype of the yeast in colony 6. Therefore, colony 6 represents an auxotrophic mutant for which no genotypic information was obtained by the analysis shown. *Solve:* Yes; colony 6.

28a.   *Evaluate:* This is type 2. *Deduce > Solve:* To "characterize" a mutant, the DNA sequences must be determined. Consult the genetic code (see page 3) to find the codons that can potentially define each amino acid. The sequences of each mutant is shown in the table.

Wild type		Thr	His	Ser	Gly	Leu	Lys	Ala
Mutant 1 protein		Thr	His	Ser	Val	Leu	Lys	Ala
Mutant 1 codons	Missense	ACN	CAU/C	UCN or AGU/C	GUN	UUA/G or CUN	AAA/G	GCN
Mutant 2 protein		Thr	His	Ser	Stop			
Mutant 2 codons	Transversion	ACN	CAU/C	UCN or AGU/C	UGA			
Mutant 3 protein		Thr	Thr	Leu	Asp	Stop		
Mutant 3 codons	Frameshift	ACN	ACU	CUG	GAU/C	UAG/A		
Mutant 4 protein		Thr	Gln	Leu	Trp	Ile	Glu	Gly
Mutant 4 codons	Frameshift	ACN	CAA/G	CUC	UGG	AUU	GAA	GGC

28b.   *Evaluate:* This is type 2. This problem tests your understanding of how mutations affect a protein-coding sequence and your ability to analyze the mutant protein sequences to determine the wild-type coding sequence. *Deduce:* Based on the wild-type protein sequence, the wild-type mRNA sequence can be deduced but will contain many ambiguities due to the degenerate nature of the genetic code. The ambiguities in the wild-type sequence can be resolved by systematically analyzing each mutant. (1) The mutation in 2 is a G-to-T transversion in the DNA and also identifies the third nucleotide of the wild-type Gly codon as an A (mutation 2 converts GGA to UGA). Thus, the wild-type Gly codon is GGA. (2) Mutant 3 is a frameshift mutant (either an insertion or a deletion) that changes codons 2 through 4 and introduces a stop codon in the fifth position. The wild-type His codon (CAU/C) can be converted to a Thr (ACU/A) by deletion of the C in the first position of the His codon. This shifts the reading frame and requires that the third position of the wild-type His codon is a C. The next amino acid in the mutant 3 protein is a Leu; its codon must begin with the second nucleotide of the wild-type codon 3 (Ser), which is either C or G. Since Leu codons are UUA/G or CUN, this indicates that the wild-type Ser codon must be UCU. The next amino acid in the mutant 3 protein is Asp (GAU/C), which fits all possibilities of the wild-type sequence. The next codon in the mutant 3 sequence is a stop codon, which requires that the third position of the wild-type codon 5 (Leu) is A/G, which indicates that the wild-type Leu codon at position 5 is U/CUA/G. (3) Mutant 4 is a frameshift mutant (either an insertion of deletion) that alters all the amino acids after the first Thr but does not introduce a nonsense codon. The mutant 4 protein sequence is shown in row 11. The wild-type His codon (CAU/C) can be converted to a Gln codon (CAA/G) by insertion of a G or A after the CA of the His codon. This shifts the reading frame, making the next codon in mutant 4 CUC (add G/A after CA in the updated wild-type sequence shown in row 10), which codes for Leu, the next amino acid in mutant 4. The next mutant 4 codon would be UGG, which codes for Trp, the next amino acid in mutant 4. The next mutant 4 codon would be either AUU or ACU. Since the next mutant 4 amino acid is Ile, this mutant 4 codon must be AUU. This indicates that the wild-type codon for Leu at position 5 must be UUA/G. The next mutant 4 codon would be A/GAA. Since the next mutant 4 amino acid is Glu, this mutant 4 codon must be GAA. This indicates that the wild-type codon for Leu at position 5 must be UUG. The next mutant 4 codon would be A/GGC. Since the next mutant 4 amino acid is Gly, this mutant 4 codon must be GGC. This indicates that the wild-type codon for Lys at position 6 is AAG. The wild-type mRNA sequence is given in the last row of the table.

Wild-type protein	Thr	His	Ser	Gly	Leu	Lys	Ala
Wild-type codons	ACN	CAU/C	UCN or AGU/C	GGN	UUA/G or CUN	AAA/G	GCN
Wild-type codons + mutant 2	ACN	CAU/C	UCN or AGU/C	GGA	UUA/G or CUN	AAA/G	GCN
Wild-type codons + mutant 3	ACN	CAC	UCU	GGA	UUA/G or CUA/G	AAA/G	GCN
Wild-type codons + mutant 4	ACN	CAC	UCU	GGA	UUG	AAG	GCN

28c.   *Evaluate > Deduce > Solve:* Mutant 1 is a missense mutant created by a G-to-T nucleotide substitution. Mutant 2 has a nonsense mutation created by a G-to-T substitution. Mutant 3 has a frameshift mutation caused by deletion of the fourth nucleotide (C) of the sequence. Mutant 4 has a frameshift mutation caused by insertion of a G or an A after the fifth nucleotide (A) of the wild-type sequence.

29.   *Evaluate:* This is type 2. This problem tests your understanding of how mutations affect a protein coding sequence. *Deduce > Solve:* The possible codons for Gly at position 211 are GGN. The possible codons for Arg are AGA/G, and those for Glu are GAA/G. For single-nucleotide substitutions to change the same Gly codon to Arg and Glu, the Gly codon must have been GGA/G. The Gly-to-Arg mutation would be G to A (GGA/G to AGA/G), and the Gly-to-Glu mutation would be G to A (GGA/G to GAA/G). The possible codons for Ser 235 are TCN or AGT/C. The possible codons for Leu are CTN or TTA/G. For a single-nucleotide substitution to change Ser to Leu, the Ser codon must have been TCA/G. The Ser-to-Leu mutation would be T to C (TCA/G to CTA/G). The possible codons for Gln 243 are CAA/G. The possible stop codons are TAA/G and TGA. For a single-nucleotide substitution to change Gln to Stop, the Gln codons could have been CAA/G and the stop codons could have been TAA/G. The Gln-to-Stop mutation would be C to T (CAA/G to TAA/G).

30a.   *Evaluate:* This is type 3. This problem tests your ability to analyze yeast mutant phenotypes. *Deduce > Solve:* Prototrophic yeast are able to grow on minimal medium, whereas auxotrophic yeast cannot. Yeast in colonies 4 and 5 grew on complete medium at 25°C but not on minimal medium at either temperature; therefore, colonies 4 and 5 correspond to auxotrophic yeast mutants. The remainder can grow on minimal medium at 25°C and thus are prototrophic yeast.

30b.   *Evaluate:* This is type 3. This problem tests your ability to analyze yeast mutant phenotypes. *Deduce > Solve:* The yeast in colonies 1 and 2 can grow on all media at 25°C but not on any of the media at 37°C. These yeast mutants are temperature sensitive for growth. The yeast in colony 5 cannot grow on minimal medium at either temperature but can grow on minimal plus adenine at both temperatures. This yeast mutant is an adenine auxotroph.

30c.   *Evaluate:* This is type 3. This problem tests your ability to analyze yeast mutant phenotypes. *Deduce > Solve:* The yeast in colony 4 have two separate mutant phenotypes. This mutant cannot grow on complete medium at 37°C and therefore has a temperature-sensitive growth phenotype. This mutant also cannot grow on minimal medium at 25°C but can grow on minimal plus adenine at 25°C; therefore, it is also an adenine auxotroph. The yeast mutant corresponding to colony 4 has two different mutant phenotypes, which indicates that this mutant carries two separate mutations: one affects growth independently of adenine metabolism, and a second affects only adenine metabolism.

31a.   *Evaluate:* This is type 2. This problem tests your ability to analyze wild-type and mutant DNA sequences. *Deduce > Solve:* The wild-type DNA sequence read directly off the gel is 5'-TGGCGTAAAGTCTGGCATCC-3'. The double-stranded DNA sequence would therefore be

> 5'-TGGCGTAAAGTCTGGCATCC-3'
> 3'-ACCGCATTTCAGACCGTAGG-5'

The mutant sequence is identical to the wild type except that only two of the three A's in the middle of the gel are present. The double-stranded mutant DNA sequence is

> 5'-TGGCGTAAGTCTGGCATCC-3'
> 3'-ACCGCATTCAGACCGTAGG-5'

31b.   *Evaluate:* This is type 2. This problem tests your ability to analyze wild-type and mutant DNA sequences and identify coding and template strands. *Deduce > Solve:* The coding strand contains an ATG start codon, which is located in the bottom strand; therefore, the bottom strand is the coding strand.

> 5'-TGGCGTAAAGTCTGGCATCC-3' template strand
> 3'-ACCGCATTTCAGACCGTAGG-5' coding strand

The mutant sequence is identical to the wild type except that only two of the three A's in the middle of the gel are present. The double-stranded mutant DNA sequence is

> 5'-TGGCGTAAGTCTGGCATCC-3' template strand
> 3'-ACCGCATTCAGACCGTAGG-5' coding strand

31c.   *Evaluate:* This is type 2. This problem tests your ability to analyze wild-type and mutant mRNA sequences. *Deduce > Solve:* The mRNA sequence is the same as the coding strand sequence except that U replaces T. The ATG start codon identifies the reading frame, and each triplet after ATG is a codon.

The wild-type mRNA sequence is 5'-GG AUG CCA GAC UUU ACG CCA-3', and the mutant mRNA sequence is 5'-GG AUG CCA GAC UUA CGC CA-3'.

31d.   *Evaluate:* This is type 2. This problem tests your understanding of how mutations affect the coding sequence of a gene. *Deduce > Solve:* The wild-type amino acid sequence is Met-Pro-Asp-Phe-Thr-Pro. The mutant amino acid sequence is Met-Pro-Asp-Leu-Arg.

31e.   *Evaluate:* This is type 1. This problem tests your understanding of the mechanism by which a deletion mutation occurred. *Deduce > Solve:* The information provided indicates that the mutation is a deletion of one T nucleotide of the TTT wild-type codon.

32a.   *Evaluate:* This is type 2. This problem tests your ability to analyze Southern blot results to determine the genotype of individuals in a family. *Deduce:* BamHI digestion of the wild-type allele results in production of 2.5-kb, 3.0-kb, and 4.0-kb fragments. BamHI digestion of the mutant allele results in production of 3.0-kb and 6.5-kb fragments. All of these fragments are detected by the combined use of probes A and B because probe A is internal to the 3.0-kb fragment and probe B straddles BamHI site 3, which indicates that it can detect the 3.5-kb and 4.0-kb wild-type fragments and the 6.5-kb mutant DNA fragment. The Southern blot pattern of the mother shows wild-type–specific fragments (2.5 and 4.0) and the mutant-specific fragment (6.5); therefore, she is heterozygous (*Aa*). The father has the same pattern, so he is also *Aa*. Child C1 has wild-type–specific bands but no mutant-specific band; therefore, C1 is homozygous dominant (*AA*). Child C2 lacks the wild-type–specific bands but has mutant-specific bands; therefore, C2 is homozygous recessive (*aa*). **Solve:** C2 has alkaptonuria.

32b.   **Evaluate:** This is type 2. This problem tests your ability to relate genotypes and the restriction pattern of wild-type and mutant gene alleles to the expected results of a Southern blot analysis. **Deduce > Solve:** The possible genotypes of their children are *AA*, like C1 in part (a); *Aa*, like the mother in part (a); and *aa*, like C2 in part (a). See figure.

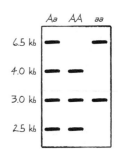

32c.   **Evaluate:** This is type 2. This problem tests your understanding of the effect of a mutation on a restriction enzyme recognition site and the resulting effect on a gene's coding sequence. **Deduce > Solve:** The *Bam*HI site is 5'-GGATCC-3'. The missense mutation changes a Ser codon (TCN or AGT/C) to a Thr codon (ACN). For the mutation to inactivate the *Bam*HI site and alter a Ser codon, at least part of the Ser codon must be present in the *Bam*HI site. The only possible Ser codon in the *Bam*HI site is 5'-TCC-3'. To change the Ser to a Thr codon requires a T-to-A transversion (TCC to ACC). This inactivates the *Bam*HI recognition sequence to 5'-GGAACC-3', which is not recognized or cut by *Bam*HI.

33a.   **Evaluate:** This is type 5. This problem tests your understanding of the mechanism of recombination during meiosis in fungi. **Deduce:** The ascus identified contains the mutant *a* allele in the four meiotic products in one-half of the asci and the wild-type *A* alleles (indicated by a "+") in the four products in the other half. Since the alleles at the *A* locus—*ade1* and *his2*—are linked, one would expect the parental genotypes to be the most common among the meiotic products. **Solve:** The parental genotypes were + *ade1 his2* and *a* + +; therefore, the ascus genotype most common for this type of ascus would contain *a* + +, *a* + +, *a* + +, *a* + +, + *ade1 his2*, + *ade1 his2*, + *ade1 his 2*, and + *ade1 his2*.

33b.   **Evaluate:** This is type 5. This problem tests your understanding of the mechanism of recombination and the production of asci that show aberrant allele segregation ratios. Recall that the mechanism of recombination involves the formation of heteroduplex DNA, which can contain mismatched base pairs and therefore result in aberrant allele segregation ratios. **Deduce > Solve:** The ascus given contains five products with the mutant *ade1* genotype and three with wild-type *ade1* alleles (+). This is a 5:3 segregation ratio, which differs from the 4:4 ratio expected from meiotic segregation of alleles. The aberrant ratio is indicative of a gene conversion event. For fungi with ordered 8-spored asci, there are two mitotically derived products for each meiotic product. A 5:3 segregation ratio indicates that one meiotic product divided to produce two different types of spores (one + and one *ade1*). One explanation for this is that one meiotic product contained an unrepaired heteroduplex, which included both + and *ade1* strands. Replication of this heteroduplex would give one + mitotic product and one *ade1* mitotic product, resulting in the 5:3 segregation ratio observed.

33c.   **Evaluate:** This is type 5. This problem tests your understanding of the mechanism of recombination and the production of asci that show aberrant allele segregation ratios. Recall that the mechanism of recombination involves the formation of heteroduplex DNA, which can contain mismatched base pairs and therefore result in aberrant allele segregation ratios. **Deduce > Solve:** The ascus given contains six products with the mutant *his2* genotype and two with wild-type *ade1* alleles (+). This is a 6:2 segregation ratio, which differs from the 4:4 ratio expected from meiotic segregation of alleles. The aberrant ratio is indicative of a gene conversion event. For fungi with ordered 8-spored asci, there are two mitotically derived products for each meiotic product. A 6:2 segregation ratio indicates that a wild-type *his2* allele (+) was converted to a mutant allele (*his2*), resulting in three *his2* products and one + product. Normal mitotic division resulted in an ascus containing six *his2* spores and two + spores.

34a. *Evaluate:* This is type 5. This problem tests your understanding of the composition of 8-spored asci produced by fungi. *Deduce:* Among the 8 spores, there are 4 *ala*+ and 4 *ala*−, 4 *cty*+ and 4 *cty*−, and 6 *brp*+ and 2 *brp*−. *Solve:* The ascus shows 6:2 segregation of *brp*+: *brp*−.

34b. *Evaluate:* This is type 5. This problem tests your understanding of the process of gene conversion and its relationship to recombination during meiosis. *Deduce > Solve:* The aberrant ratio of 6 *brp*+: 2 *brp*− indicates that gene conversion in the *brp* locus occurred during the meiosis producing this ascus. Gene conversion is associated with recombination, which indicates that a recombination event was initiated in the region between *ala* and *cty* during this meiosis.

34c. *Evaluate:* This is type 5. This problem tests your understanding of the genotype of asci that show evidence of gene conversion. *Deduce > Solve:* Recombination is accompanied by formation of heteroduplex DNA between the two Holliday junctions that form. A region of the heteroduplex can contain mismatched base pairs if that region in the homolog is not identical. In this case, the heteroduplex included the region containing the sequence difference that distinguishes *bry*− from *bry*+; therefore, it contained mismatched bases. These mismatches are repaired by mismatch repair, but the direction of the repair is not controlled, such that a homolog that should have *bry*− could be repaired to contain *bry*+ information. The heteroduplex occupies only a portion of the region undergoing recombination, which in this case did not include the *ala* or *cty* loci. Therefore, there was no gene conversion at *ala* or *cty*, and those alleles segregated as normal, 4:4.

12.04 Test Yourself

Problems

1. A chemical mutagen induces single-nucleotide insertions at a rate of 1 per 10 fruit-fly genomes per generation. If the coding sequence of genes makes up 10% of the fruit-fly genome, how many frameshift mutations in genes would be induced by this chemical if it was used to treat male flies that then fathered 100,000 progeny?

2. Your colleague has just learned that base analogs, such as 2-amino purine, can be powerful mutagens. He decides to give it a try using yeast, collects some yeast cells, and places them on ice and then adds 10 times more 2-amino purine (2AP) than is necessary to induce mutations. He leaves the cells in the presence of 2AP on ice for 30 minutes, collects the yeast, washes away all the excess 2AP, and then plates them on a medium that selects for drug-resistant mutants. He finds that there is no increase in the mutation rate over that due to spontaneous mutations. If the 2AP he used was active, and it is known from previous experiments that transition mutagens can induce drug-resistant mutants, why didn't he get an increase in mutation rate?

3. The wild-type amino acid in the fourth position of a metabolic enzyme is not known. A number of single-base-pair substitutions in the codon that codes for this amino acid have been isolated. The amino acids at position four in these mutants are histidine, isoleucine, arginine, and glutamic acid. Based on this information, what is the wild-type amino acid at this position of that enzyme?

4. A mutation in a protein-coding gene in yeast is shown to be a complete loss-of-function mutation. When you examine yeast cells of this mutant, you find that no protein or mRNA of this gene is present in the cell. You sequence the entire transcribed region of the gene and do not find any difference in the mutant gene and wild-type gene sequence. Give one explanation for how this mutation could cause the phenotype described.

5. You have isolated a revertant of a *his*− yeast mutant by selecting spontaneous mutation on medium lacking histidine. You cross this *his*+ revertant to a wild-type strain and find that 95% of the progeny are *his*+ and 5% are *his*−. Based on this information, what was the nature of the mutation that reverted the *his*− strain to *his*+?

6. You have lost your records for three *his*− *Salmonella* strains that you normally use in the Ames test. To determine which strain corresponds to which type of mutation, you treat each of them with the

following mutagens: base analogs, intercalating agents, and oxidizing agents. You find that base analogs and oxidizing agents, but not intercalating agents, increase the rate of reversion to *his+* with strain 1. You also find that oxidizing agents, but not base analogs or intercalating agents, increase the rate of reversion to *his+* with strain 2. You also find that only intercalating agents increase the rate of reversion to *his+* with strain 3. Assuming that all *his+* revertants are true revertants, use these results to deduce the nature of the *his⁻* mutations in strains 1, 2, and 3.

7. You have isolated a yeast mutant that has a high spontaneous mutation rate. You notice that if you alter the intracellular pools of deoxynucleotide triphosphates such that they are unbalanced (e.g., very high dATP vs. dCTP, dGTP, and dTTP), the mutation rate increases further. You have determined that the mutation that is in your yeast mutants is not in the same complementation group as any other known mismatch repair mutants. Based on this information, what non-mismatch repair gene mutation could explain the phenotype of your strain?

8. You have isolated a yeast mutant that is sensitive to ionizing radiation. Propose two different DNA damage repair pathways that could be defective in your mutant. Name one gene that is required for each of these pathways.

Solutions

1.   *Evaluate:* This is type 1. This problem tests your understanding of use of mutation rates to determine the number of mutations created by a specified mutagen. *Deduce:* Single-nucleotide insertions in the coding sequence of genes result in frameshift mutations. Since 10% of the fruit-fly genome is coding sequence, 25% of the insertion mutations will be frameshift mutations. *Solve:* The number of frameshift mutations in 100,000 fly progeny of treated males will be $\left(\frac{100,000}{10}\right)(0.10) = 1000$.

2.   *Evaluate:* This is type 1. This problem tests your understanding of the mechanism by which nucleotide analogs function. *Deduce > Solve:* The 2AP, as well as all other nucleotide analogs, must be taken up into the cell and then incorporated into DNA during DNA replication in order to act as mutagens. Since the researcher kept the yeast cells on ice while they were exposed to the drug, it is likely that they took up very little drug and were not undergoing DNA replication while in the presence of the drug. Thus, 2AP would not be expected to act as a powerful mutagen under these conditions, and the results obtained by this researcher are not surprising.

3.   *Evaluate:* This is type 2. This problem tests your understanding of the effect of mutations on the coding sequence of genes (see genetic code on page 3). *Deduce:* Each mutant amino acid must correspond to a codon that can be changed into the wild-type codon by a single-base-pair substitution. If using the genetic code table, this corresponds to moving sideways along a row or up or down within a column. The wild-type amino acid should be located at the intersection of the mutant amino acids. The intersection of isoleucine, arginine, histidine, and glutamic acid is the lysine codon, AAA. *Solve:* The wild-type amino acid at this position is lysine.

4.   *Evaluate:* This is type 2. This problem tests your understanding of the mechanism by which mutations can inactivate the function of genes. *Deduce:* The mutation prevents the accumulation of mRNA and protein and is not located in the transcribed region of the gene, which rules out mutations that prevent translation or cause mRNA instability (turnover). The mutation must be located in a portion of the gene that is not transcribed but is nevertheless essential for gene expression. *Solve:* The mutation is most likely in the promoter of the gene. The mutation is in the promoter or other regulatory sequence required for transcription.

5.   *Evaluate:* This is type 3. This problem tests your understanding of the nature of reverse mutations and the use of crosses to analyze revertants. *Deduce:* The *his⁺* revertant is either a *his⁺* genetically because the *his⁻* gene was converted to *his⁺* by the reverse mutation, or it is still *his⁻* but also carries a suppressor mutation. When crossed to wild type, the suppressor mutation will

segregate away from the *his⁻* mutation at a frequency that depends on whether the suppressor is linked to *his⁻* and, if so, the degree of genetic linkage. Since the cross to wild type produced *his⁻* progeny, the revertant carried a *his⁻* allele and a suppressor mutation. If the suppressor mutation was unlinked to *his⁻*, then 75% of the progeny would be *his⁺*. Since 95% of the progeny were *his⁺*, the suppressor and the *his⁻* gene are linked (they are separated by 10% recombination). *Solve:* The revertant carries a suppressor mutation that is linked to the *his⁻* mutation.

6.   *Evaluate:* This is type 3. This problem tests your understanding of the Ames test and knowledge of the mechanism of mutagenesis by base analogs, oxidizing agents, and intercalating agents. *Deduce:* Base analogs induce transition mutations and therefore can revert only transition mutants. Oxidizing agents induce both transitions and transversion mutations and therefore can revert both types of mutation. Intercalating induces insertions or deletions and therefore can revert only frameshift mutations. Reversion of strain 1 is induced by base analogs; therefore, strain 1 is a transition mutant. Reversion of strain 2 is induced by oxidizing agents but not base analogs; therefore, strain 2 is a transversion mutant. Reversion of strain 3 is induced by intercalating agents; therefore, strain 3 is a frameshift mutant. *Solve:* Strain 1 is a transition mutant, strain 2 is a transversion mutant, and strain 3 is a frameshift mutant.

7.   *Evaluate:* This is type 4. This problem tests your understanding of mechanisms that prevent errors in DNA replication. *Deduce:* One mechanism by which unequal concentration of deoxynucleotides could contribute to elevated mutation rates is if they caused an increase in the rate of mistakes in DNA replication. A mutation that could exacerbate this effect would be a mutation that alters the ability of DNA polymerase to select the correct nucleotide for incorporation or to proofread and remove incorrect nucleotides. A defect in either function of DNA polymerase could explain the phenotype of the yeast mutant in question. *Solve:* A mutation in DNA polymerase explains the phenotype of this strain.

8.   *Evaluate:* This is type 4. This problem tests your understanding of DNA damage repair pathways that repair damage caused by ionizing radiation. *Deduce:* Ionizing radiation induces double-stranded breaks in DNA. *Solve:* Two pathways for repair of double-stranded breaks exist in eukaryotes: nonhomologous end-joining (NHEJ) and synthesis-dependent strand annealing. One gene that is required for NHEJ is *Ku70*. One gene that is required for synthesis-dependent strand annealing is the *RecA* homolog, *Rad51*.

Chromosome Aberrations and Transposition

Section 13.01 Genetics Problem-Solving Toolkit

Section 13.02 Types of Genetics Problems

Section 13.03 Solutions to End-of-Chapter Problems

Section 13.04 Test Yourself

13.01 Genetics Problem-Solving Toolkit

Key Terms and Concepts

Euploid: A cell or individual that has a chromosome number that is a whole-number multiple of the haploid chromosome number (e.g., n, $2n$, $3n$, $4n$).

Aneuploid: A cell or individual that is not euploid. Typically, aneuploids have one additional or one fewer chromosome than the euploid (e.g., $2n - 1$, $2n + 1$).

Polyploid: A cell or individual that contains three or more sets of chromosomes (e.g., $3n$, $4n$, $5n$).

Deletion: A chromosomal aberration in which the aberrant chromosome is missing DNA sequence as compared to wild type.

Duplication: A chromosomal aberration in which the aberrant chromosome contains added DNA sequence as compared to the wild type, and the added sequence is normally present elsewhere in the genome.

Inversion: A chromosomal aberration in which a segment of a chromosome is inverted relative to wild type. The inversion may include (pericentric) or exclude (paracentric) the centromere.

Translocation: A chromosomal aberration in which a segment of a chromosome is found on another chromosome.

Reciprocal translocation: A chromosomal aberration in which the terminal segments of two nonhomologous chromosomes have exchanged places.

Robertsonian translocation: A chromosomal aberration in which the short arms of two acrocentric chromosomes have fused together.

Trivalent: A meiosis I synaptic structure in which three chromosomes pair.

Monovalent: A meiosis I structure in which a chromosome is unpaired.

Nondisjunction: Aberrant chromosome segregation that results in the formation of unbalanced gametes.

Key Analytical Tools

Predicting the consequences of recombination in inversion heterozygotes (see Figure 13.17 in main text chapter)

Chromosome segregation in individuals heterozygous for a reciprocal translocation (see Figure 13.20 in main text chapter)

13.02 Types of Genetics Problems

1. Describe or predict the effects of chromosome aberrations on chromosome alignment, segregation, and recombination during meiosis.
2. Describe or predict the effects of chromosome aberrations on organismal phenotype.
3. Describe the mechanisms of transposition or predict the consequences of transposition on chromosome structure and organismal phenotype.
4. Describe the mechanisms that lead to changes in ploidy or the effects of ploidy changes on chromosome segregation and organismal phenotype.

1. Describe or predict the effects of chromosome aberrations on chromosome alignment, segregation, and recombination during meiosis.

This type of problem requires understanding of common aberrations in chromosome structure and of how those aberrations affect meiosis. You may be asked to describe aberrant chromosomes using accurate terminology or draw aberrant chromosomes based on written descriptions. You may be asked to describe or draw the alignment of aberrant chromosomes during synapsis and relate that alignment to the formation of viable or nonviable gametes. You may be asked to consider the consequences of meiotic recombination in individuals that are heterozygous for inversions or translocations. See Problem 14 for an example.

Example Problem: You have two pure-breeding lines of fruit flies that are wild type in appearance. Each line has a different chromosomal abnormality: one line is homozygous for a reciprocal translocation involving chromosome 2 and 3, and the other is homozygous for a paracentric inversion involving a large segment of chromosome 2. You crossed each line to a fly line that lacks chromosome aberrations but is homozygous recessive for genes on chromosome 2. The labels on the vials containing the F_1 flies from these crosses have come off, and you don't know which is which because all the F_1 fly lines are wild type in appearance. Describe a cross, and the results of that cross, that could help you distinguish between the two types of F_1 flies.

Solution Strategies	Solution Steps
Evaluate	
1. Identify the topic this problem addresses, and explain the nature of the requested answer.	1. This problem concerns the specific effects of different chromosomal aberrations on gamete formation. The answer will be a description of a cross that distinguishes between flies that are heterozygous for a paracentric inversion versus those that are heterozygous for a reciprocal translocation.

(continued)

Solution Strategies	Solution Steps
2. Identify the critical information given in the problem.	2. There are two pure-breeding lines of fruit flies, one homozygous for a reciprocal translocation and the other homozygous for a paracentric inversion. Each was crossed to a fly line that is homozygous recessive for genes on chromosome 2 but that has no chromosomal aberrations.
Deduce	
3. Describe the F$_1$ flies that are in question.	3. Both sets of F$_1$ flies are heterozygous for genes on chromosome 2. One set of F$_1$ flies are heterozygous for a paracentric inversion on chromosome 2. The other set of F$_1$ flies are heterozygous for a reciprocal translocation involving chromosomes 2 and 3.
4. Describe the effect of heterozygosity for each of the chromosomal aberrations on gamete formation.	4. Strains heterozygous for a paracentric inversion will show a lower than predicted recombination frequency for genes in the interval that is inverted. Strains heterozygous for a reciprocal translocation will be semisterile, producing a lower number of viable gametes and therefore fewer progeny than wild-type strains.
5. Describe a cross that could be performed on both F$_1$ flies that would detect heterozygosity for the chromosomal aberrations. TIP: Any cross should reveal the semisterility, but only a cross that measures recombination frequency will identify the inversion heterozygote.	5. A test cross of the F$_1$ using a fly strain homozygous recessive for genes on chromosome 2 will detect a decreased recombination frequency along chromosome 2 in the inversion heterozygote. The same test cross will detect semisterility in the flies heterozygous for the reciprocal translocation.
Solve	
6. Answer the problem by describing the cross and the expected results for each type of chromosomal aberration.	6. **Answer:** Perform a test cross for both sets of F$_1$ flies using a fly line that is homozygous recessive for genes on chromosome 2. The cross that produces fewer than normal numbers of progeny involves the line heterozygous for the reciprocal translocation. The cross that shows reduced recombination frequencies along a large segment of chromosome 2 involves the line that is heterozygous for the paracentric inversion.

2. Describe or predict the effects of chromosome aberrations on organismal phenotype.

Problems of this type require understanding of the structure and phenotypic consequences of common chromosomal aberrations. You may be asked to explain the basis of abnormal phenotypes associated with deletions, duplications, inversions, or translocations. You may be asked to infer what type of chromosomal aberration is present in a parent based on the phenotype of the children. You may also be asked to use deletions to map the location of point mutations on a chromosome. See Problem 24 for an example.

Example Problem: A normal, healthy couple has a child with Williams-Beuren syndrome, a congenital disorder characterized by supravalvular aortic stenosis (SVAS), mental retardation, and distinctive facial features. They are told that the disorder is caused by heterozygosity for a segmental deletion on 7q11.23 and are concerned that one of them may be a carrier for the deletion. Describe what features of a karyotype would indicate that one of the parents is a carrier for Williams-Beuren syndrome.

Solution Strategies	Solution Steps
Evaluate	
1. Identify the topic this problem addresses, and explain the nature of the requested answer.	1. This problem concerns chromosomal aberrations called segmental deletions that cause abnormal phenotypes when heterozygous. The answer will be a description of a karyotype indicative of someone being a carrier for a disorder caused by deletion of the human chromosome region 7q11.23.
2. Identify the critical information given in the problem.	2. The parents are normal but have a child with Williams-Beuren syndrome, which is due to heterozygosity for deletion of 7q11.23. The concern is that one of the parents is a carrier for the disorder.
Deduce	
3. Describe the general features necessary for an individual to be a carrier for a disorder caused by heterozygosity for a segmental deletion.	3. In general, the individual must be heterozygous for the chromosome with the deletion but also have a copy of the deleted DNA elsewhere in their genome.
4. Specify one way in which one of the parents could specifically meet the criteria described in step 3.	4. One parent will have a normal chromosome 7 and a mutant chromosome 7 that is deleted for the region 7q11.23. This parent would also have an additional copy of 7q11.23 on another chromosome.
Solve	
5. Answer the problem by describing one potential result of a karyotype analysis that would be consistent with the individual being a carrier for Williams-Beuren syndrome.	5. **Answer:** One parent would be heterozygous for a deletion of 7q11.23. That individual would have another chromosome that is longer and has a different banding pattern than normal, which is consistent with it containing the region corresponding to 7q11.23.

3. Describe the mechanisms of transposition or predict the consequences of transposition on chromosome structure and organismal phenotype.

This type of problem requires understanding of mobile genetic elements, including bacterial insertion sequences and transposons and eukaryotic transposons. You may be asked to describe the structure of transposons and how their structure relates to their mechanism of transposition. You may be asked about the effect of transposition on cellular or organismal phenotype. See Problem 28 for an example.

Example Problem: You are conducting crosses with two fly lines. Line 1 is wild type for all traits, whereas line 2 is homozygous for insertion mutations in multiple genes. When you cross females from line 2 with males from line 1, the F_1 are fertile. When you cross males from line 2 with females from line 1, the F_1 are infertile. You have a hypothesis to explain this phenomenon from your reading of Chapter 13 but wish to test it. What is the basis of the infertility of the F_1 from the second cross, and what test could you perform to test it?

Solution Strategies	Solution Steps
Evaluate	
1. Identify the topic this problem addresses, and explain the nature of the requested answer.	1. This problem concerns the effect of transposition on phenotype. The answer will be a hypothesis for the results described and a test of that hypothesis.
2. Identify the critical information given in the problem.	2. Two fly lines are used to perform reciprocal crosses. One fly line is wild type, and the other is homozygous for multiple insertion mutations. A cross of a male from the insertion mutant line to a wild-type female yields only sterile progeny, whereas the reciprocal cross produces progeny with normal fertility.
Deduce	
3. Consider a molecular basis for the insertion mutations present in fly line 2.	3. Insertion mutations are caused by the insertion of DNA into genes. The insertion could be a result of random breakage and insertion of chromosome fragments, or transposon insertions.
4. Consider which of the proposed mechanisms from step 3 would give the reciprocal cross results described.	4. The results are consistent with hybrid dysgenesis. The insertion mutations in line 2 would be a result of P element insertions. The sterility caused by crossing a line 2 male with line 1 females would be caused by massive transposition of the P elements during embryonic development. The lack of sterility when crossing a line 2 female with a line 1 male is due to the transposition repressor in the cytoplasm of line 2 eggs.
5. Consider predictions about the genotype or phenotype of the F_1 in both crosses.	5. A key prediction of the hybrid dysgenesis hypothesis is that the sterile F_1 flies will have P elements in different locations than in the line 2 flies. The sterile flies may also have more copies of the P element than the line 2 flies.
Solve	
6. Describe an experiment that would test a prediction from step 5. **TIP:** The test should detect the location or copy number (or both) of P elements.	6. Perform a Southern blot of restriction-digested DNA from line 2 flies (one fly per lane on the gel) and both types (sterile and fertile) F_1 flies. Use a restriction enzyme that does not cut within the P element, and use P element DNA as the probe.
7. **Answer** the problem by describing results of the test proposed in step 6 that are consistent with the hypothesis.	7. **Answer:** If the hybrid dysgenesis hypothesis is correct, then the bands detected in line 2 and the fertile F_1 flies should be the same, and the size (and potentially the number) of bands in the sterile F_1 flies should differ.

4. Describe the mechanisms that lead to changes in ploidy or the effects of ploidy changes on chromosome segregation and organismal phenotype.

This type of problem tests your understanding of changes in ploidy. You may be asked to describe or explain mechanisms that lead to changes in ploidy. You may be asked to predict the consequences of altered ploidy on chromosome synapsis during meiosis I, gamete formation and fertility, or organismal phenotype. See Problem 16 for an example.

Example Problem: Two distantly related plants of unknown karyotype are crossed to produce hybrids. Over the course of many crosses, three different types of hybrids were identified based on cytological analysis of meiotic cells. The meiotic cells from type 1 contain 27 monovalents, and type 1 hybrids are sterile. The meiotic cells from type 2 contain 22 bivalents and 16 monovalents, and type 2 hybrids are sterile. The meiotic cells from type 3 contain 54 bivalents, and type 3 hybrids are fertile. How many chromosomes were present in the two plants used to create the hybrids?

Solution Strategies	Solution Steps
Evaluate	
1. Identify the topic this problem addresses, and explain the nature of the requested answer.	1. This problem concerns the mechanism of allopolyploid formation. The answer will be the number of chromosomes that are present in the two plant lines used to create the hybrids.
2. Identify the critical information given in the problem.	2. Three types of hybrids were created. Type 1 and 2 are sterile, but type 3 is fertile. The meiotic cells of the type 1 hybrid contain 27 monovalents, type 2 contains 22 bivalents and 16 monovalents, and type 3 contains 54 bivalents.
Deduce	
3. Explain how hybrids are formed, and describe the implication of seeing only monovalents in meiotic cells of the hybrid.	3. Hybrids can be formed by mating related plant strains. If normal gametes fuse to create a hybrid, there will be no homologous chromosomes, and only monovalents will be seen during meiosis. The number of monovalents equals the total number of chromosomes in both parental gametes.
4. Describe the implication of seeing both bivalents and monovalents in meiotic cells of the hybrid.	4. If one gamete is diploid and the other haploid, then there will be both bivalents and monovalents in meiotic cells of the hybrid. The number of bivalents will equal the number of homologous pairs in the diploid gamete, and the number of monovalents will be equal to the number of chromosomes in the haploid gamete of the other parent.
5. Describe the implication of seeing only bivalents in meiotic cells of the hybrid.	5. If both gametes are diploid, then there will be only bivalents in meiotic cells of the hybrid. The number of bivalents will equal the total number of homologous pairs in both types of gametes.

(continued)

Solution Strategies	Solution Steps
Solve	
6. Calculate the number of chromosomes in the somatic cells of both parent plants.	6. Normal gametes from each parent are haploid. Since there were 27 monovalents in hybrid 1, and these represent the total number of chromosomes in both parents' haploid gametes, the total number of chromosomes in the somatic (diploid) cells of both parents is 54. Hybrid 2 contains 22 bivalents from a diploid gamete from one parent; therefore, the diploid number for that parent is 22. There are 16 monovalents in hybrid 2, which must correspond to the haploid number of chromosomes from the other parent. Therefore, the other parent's diploid chromosome number is 32. This is consistent with hybrid 3, proposed to be formed by fusion of diploid gametes from both parents. This should have 22 + 32, or 54, bivalents in meiosis, which it does based on the data in the problem.
7. Answer the problem by referring to the calculation in step 6 and stating the diploid number of chromosomes in the parent plants.	7. **Answer:** One parent has a diploid number of 32, and the other has a diploid number of 22.

13.03 Solutions to End-of-Chapter Problems

1. *Evaluate:* This is type 4. This problem concerns the effect of polyploidy on chromosome synapsis. Recall that bivalents contain two homologs, trivalents contain three homologs, and univalent contain one chromosome. *Deduce > Solve:* The diploid will have 18 bivalents whereas the triploid will have a combination of trivalents, bivalents, and univalents.

2. *Evaluate:* This is type 4. This problem concerns the effect of triploidy on chromosome synapsis and segregation. Recall that bivalents are pairs of homologs attached to opposite meiotic spindle poles and that monovalents are individual chromosomes attached to one spindle pole 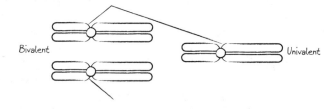 or the other. *Deduce > Solve:* The C1 and C2 chromosomes of the bivalent will segregate to opposite poles. The C3 chromosome of the monovalent will segregate to the same pole as C1 half the time and to the same pole as C2 the other half of the time. $\frac{1}{4}$ of the gametes will contain C1C3, $\frac{1}{4}$ will contain C2, $\frac{1}{4}$ will contain C2C3 and $\frac{1}{4}$ will contain C1. Draw a figure similar to the one shown here.

3a–h.   *Evaluate:* This is type 4. Recall that haploid chromosome number refers to the number of different chromosomes in a species. *Deduce > Solve:* If the haploid number is 4, then
    (a) a *diploid* species would have two copies of each chromosome, for a total of 8 chromosomes;
    (b) a *pentaploid* species would have five copies of each chromosome, for a total of 20 chromosomes;
    (c) an *octaploid* species would have eight copies of each chromosome, for a total of 32 chromosomes; **(d)** a *trisomic* species would have two copies of three chromosomes and one copy of the fourth, for a total of 9 chromosomes; **(e)** a *triploid* species would have three copies of each chromosome, for a total of 12 chromosomes; **(f)** a *monosomic* species would have two copies of three chromosomes and one copy of the fourth, for a total of 7 chromosomes; **(g)** a *tetraploid* species would have four copies of each chromosome, for a total of 16 chromosomes; **(h)** a *hexaploid* species would have six copies of each chromosome, for a total of 24 chromosomes. Species with an odd number of chromosome sets (triploid and pentaploid) would be infertile. Aneuploid species (monosomic and trisomic) would have reduced fertility.

4a–i. *Evaluate:* This is type 2. Recall that abnormalities in chromosome number or structure can result in abnormal cell phenotype by affecting gene structure, gene expression, or gene dosage. *Deduce > Solve:* **(a)** a *pericentric inversion* is not expected to affect phenotype unless the breakpoint of the inversion lies within a gene or the inversion makes expression of a gene subject to position effect variegation; **(b)** an *interstitial deletion* is expected to affect phenotype by removing or disrupting one or more genes in the deleted segment or by making the expression of a gene subject to position effect variegation; **(c)** a *duplication* is expected to affect phenotype by altering the dosage of genes in the duplicated segment; **(d)** a *terminal deletion* is expected to affect phenotype by removing or disrupting one or more genes in the deleted segment or by making the expression of a gene subject to position effect variegation; **(e)** *trisomy* is expected to affect phenotype by increasing the dosage of every gene on the trisomic chromosome; **(f)** a *reciprocal balanced translocation* is not expected to affect phenotype unless the breakpoint of the translocation lies within a gene or changes the location of a gene such that its expression is subject to position effect variegation; **(g)** a *paracentric inversion* is not expected to affect phenotype unless the breakpoint of the inversion lies within a gene or the inversion makes expression of a gene subject to position effect variegation; **(h)** *monosomy* is expected to affect phenotype by decreasing the dosage of every gene on the monosomic chromosome; **(i)** *polyploidy* is expected to affect phenotype by increasing the dosage of every gene in the organism's genome.

5a. *Evaluate:* This is type 4. This problem concerns the consequences of hybrids formed by the fusion of gametes containing different numbers of chromosomes on gamete formation by the hybrids. *Deduce > Solve:* Horses have a diploid number of 64 and produce gametes carrying 32 chromosomes. Donkeys have a diploid number of 62 and produce gametes carrying 31 chromosomes. The union of horse and donkey gametes produces a mule that has a total number of 63 chromosomes. Meiotic mule germ cells contain 30 bivalents and one trivalent or 31 bivalents and one univalent, which results in the production of unbalanced gametes that do not form viable zygotes.

5b. *Evaluate:* This is type 4. This problem concerns the chromosome number of viable zygotes formed by fusion of horse and mule gametes. *Deduce:* Horse gametes contain 32 chromosomes. For a horse–mule mating to result in a viable zygote, the zygote must have the chromosome complement of a horse or that of a mule. *Solve:* The horse chromosome number is 64, and the mule chromosome number is 63.

5c. *Evaluate:* This is type 2. This problem concerns the effect of chromosome composition on phenotype. *Deduce:* The horse–mule offspring will have a complete horse chromosome set plus whichever chromosomes assort together during gametogenesis in the mule. The mule gamete is unlikely to carry the 32-chromosome complement of the horse, since the 32 horse chromosomes in the mule will assort independently of each other. *Solve:* Assuming that recombination does not occur, the chance that a gamete will contain all 32 horse chromosomes will be $\left(\frac{1}{2}\right)^{32}$.

6. *Evaluate:* This is type 3. This problem concerns the effect of transposition on phenotype. *Deduce:* *P* element transposition results in the insertion of *P* element transposons into a new site on a *Drosophila* chromosome. If the site of insertion is within a gene or gene-regulatory element, then a mutation can result. *Solve:* Limiting the transposition of *P* elements is advantageous because it reduces the likelihood that transposition will lead to gene mutations.

7. *Evaluate:* This is type 3. This problem concerns the relationship between retrotransposons and retroviruses. *Deduce > Solve:* The *copia* and yeast *Ty* elements are retrotransposons. Retrotransposons are related to retroviruses in structure and function. For *copia* and *Ty* elements, the structural similarity includes the presence of long terminal repeats (LTRs) at their ends, and *gag* and *pol* genes. An essential feature of both retrotransposition and insertion of RNA viral genomes into host chromosomes is the reverse transcription of RNA into DNA, which requires the *pol*-encoded reverse transcriptase.

8. *Evaluate:* This is type 3. This problem concerns the *IS* transposable elements found in bacteria. *Deduce:* *IS* elements in bacteria are composed of a transposase gene flanked by short inverted repeat sequences. The transposase enzyme recognizes any DNA flanked by the appropriate inverted repeats and catalyzes transposition of the entire DNA segment from one end of the inverted repeat to the other and everything in between. If an *IS1* mutant is unable to transpose, then it is likely mutant for an essential component of the inverted repeat. The mutation could also be in the transposon gene if the *IS1* element is the only copy of *IS1* in the genome; however, if an additional *IS1* element is present and encodes a functional *IS1* transposase, then all *IS1* elements with intact inverted repeat sequences will be able to transpose regardless of whether their own transposase gene is functional. *Solve:* It can be concluded that the mutation is located in one of the inverted repeat sequences at the ends of the element.

9.   *Evaluate:* This is type 4. This problem concerns the meaning of the term *mosaic* as it applies to chromosome content. *Deduce > Solve:* An individual is mosaic for chromosome content when it contains somatic cells that differ in chromosome number, structure, or both. Somatic individuals are formed by nondisjunction during embryonic mitotic divisions, resulting in the formation of tissues populated by cells with different chromosome numbers. For example, some individuals with Turner syndrome have mixtures of cells with 45X and 46XX karyotypes.

10.   *Evaluate:* This is type 4. This problem concerns the formation of unusual gametes due to nondisjunction. *Deduce > Solve:* Yellow-bodied females are homozygous for the y^- allele on X, and wild-type males contain y^+, a wild-type allele, on X. For a progeny of this cross to be both yellow and female, it must be homozygous y^- and have a ratio of two X chromosomes to one autosome. Yellow-bodied females can be produced from this cross if nondisjunction in the female parent produces an egg with two X chromosomes, and the egg is fertilized by a sperm containing the Y chromosome. The XXY zygote will develop as female and will be homozygous for the recessive yellow-body allele.

11a.   *Evaluate:* This is type 4. This problem concerns gamete formation by tetraploids and the prediction of progeny types in tetraploid crosses. *Deduce:* The cross is $R_1 R_1 R_1 R_2 \times R_1 R_1 R_1 R_2$. Both parents produce the diploid gametes $R_1 R_1$ and $R_1 R_2$ in equal proportions ($\frac{1}{2}$ each). The progeny genotypes are $\frac{1}{4} R_1 R_1 R_1 R_1$; $\frac{1}{2} R_1 R_1 R_1 R_2$; and $\frac{1}{4} R_1 R_2 R_1 R_2$. *Solve:* Therefore, the frequencies and phenotypes are $\frac{1}{4}$ dark red, $\frac{1}{2}$ light red, and $\frac{1}{4}$ pink.

11b.   *Evaluate:* This is type 1. This problem tests your understanding of gamete formation by tetraploids and your ability to predict progeny types in tetraploid crosses. *Deduce:* The cross is $R_1 R_1 R_1 R_2 \times R_1 R_2 R_2 R_2$. The light red parent produced $\frac{1}{2} R_1 R_1$ and $\frac{1}{2} R_1 R_2$ gametes. The light pink parent produces $\frac{1}{2} R_1 R_2$ and $\frac{1}{2} R_2 R_2$ gametes. Thus the progeny genotypes are $\frac{1}{4} R_1 R_1 R_1 R_2$, $\frac{1}{2} R_1 R_1 R_2 R_2$, and $\frac{1}{4} R_1 R_2 R_2 R_2$. *Solve:* Therefore, the frequencies and phenotypes are $\frac{1}{4}$ light red, $\frac{1}{2}$ pink, and $\frac{1}{4}$ light pink.

12a.   *Evaluate:* This is type 1. This problem concerns inversion loops that form during meiosis in organisms heterozygous for a paracentric inversion. *Deduce:* The cell described is heterozygous for a paracentric inversion. All regions of the homologs pair, and the inverted regions form an inversion loop. *Solve:* Draw a diagram similar to the figure shown here.

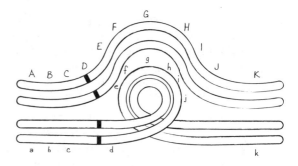

12b.   *Evaluate:* This is type 1. This problem concerns the consequences of meiotic crossing over within an inverted segment in an inversion heterozygote. *Deduce:* The cell described is heterozygous for a paracentric inversion. All regions of the homologs pair, and the inverted regions form an inversion loop. A crossover in the interval between *F* and *G* is in the inverted segment and will result in the formation of two normal gametes (one identical to each of the parental chromosomes), a dicentric chromosome duplicated for *ABCD* and deficient for *K*, and an acentric chromosome duplicated for *K* and deficient for *ABCD*. *Solve:* The resulting gametes and viability status are as follows: *ABC•DEFGHIJK*, viable; *ABC•DEFghijd.cba*, not viable; *kefGHIJK*, not viable; and *abc•djihgfek*, viable.

12c.   *Evaluate:* This is type 1. This problem concerns the consequences of meiotic crossing over outside the inverted segment in an inversion heterozygote. *Deduce:* The cell described is heterozygous for a paracentric inversion. All regions of the homologs pair, and the inverted regions form an inversion loop. A crossover in the interval between *A* and *B* is outside the inverted segment and leads to the formation of four viable gametes, two parental and two recombinant. *Solve:* The resulting gametes and viability status are as follows: *ABC•DEFGHIJK*, viable; *Abc•djihgfek*, viable; *aBC•DEFGHIJK*, viable; and *abc•djihgfek*, viable.

13a.   *Evaluate:* This is type 1. This problem concerns chromosomal aberrations. *Deduce:* Chromosome 1 contains one copy of each gene, whereas chromosome 2 contains an additional copy of the segment including genes *D, H,* and *B*. Chromosome 2 appears to carry a duplication of *D, H,* and *B* inserted between *B* and *G* of chromosome 1. *Solve:* The term used to describe this situation is partial duplication.

13b.   *Evaluate:* This is type 1. This problem concerns the duplication loops formed during meiosis in organisms heterozygous for a partial chromosomal duplication. *Deduce:* The cell described is heterozygous for a partial chromosomal duplication. All regions of the homologs pair, and the duplicated segment forms an unpaired loop. *Solve:* Draw a diagram similar to the figure shown here.

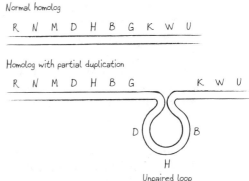

13c.   *Evaluate:* This is type 1. This problem concerns the structures that form during meiosis in an organism heterozygous for a partial chromosome duplication. *Deduce > Solve:* The term used to describe this structure is an unpaired loop.

13d.   *Evaluate:* This is type 1. This problem concerns the structures that form during meiosis in an organism heterozygous for a partial chromosome duplication. *Deduce > Solve:* In an inversion heterozygote, both homologs must loop out with one loop twisted around to accommodate synapsis of all homologous regions. During meiosis in a heterozygote for a partial duplication, all regions of the normal homolog can pair without looping out; therefore, only the chromosome containing the partial duplication needs to loop out.

14a.   *Evaluate:* This is type 1. This problem concerns the cruciform structures formed during meiosis in an individual heterozygous for a reciprocal translocation. *Deduce:* The animal indicated is heterozygous for a balanced reciprocal translocation. All homologous regions of all four chromosomes will synapse during meiosis, resulting in formation of a tetravalent, cruciform-shaped structure. *Solve:* Draw a diagram similar to the figure shown here.

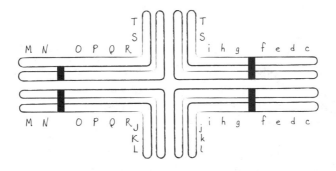

14b.   *Evaluate:* This is type 1. This problem concerns gametes formed by individuals heterozygous for a balanced, reciprocal translocation. *Deduce:* Tetravalent, cruciform-shaped structures that form during meiosis can segregate by alternate, adjacent-1, or adjacent-2 segregation mechanisms. The alternate pattern segregates homologous centromeres, which creates four viable gametes. Two gametes contain the balanced, reciprocal translocation chromosomes and two contain normal chromosomes. *Solve:* Alternate segregation produces four viable gametes. Two contain the chromosomes MN•OPQRST + cdef•ghijkl and two contain the chromosomes MN•OPQRjkl + cdef•ghiST.

14c.   *Evaluate:* This is type 1. This problem concerns gametes formed by individuals heterozygous for a balanced, reciprocal translocation. *Deduce:* Tetravalent, cruciform-shaped structures that form during meiosis can segregate by alternate, adjacent-1, or adjacent-2 segregation mechanisms. The adjacent-1 pattern segregates homologous centromeres but creates four nonviable gametes. Two gametes contain the normal version of one chromosome and the translocation version of the other, and two have the reciprocal chromosome arrangement. *Solve:* Adjacent-1 segregation produces four nonviable gametes that contain duplications and deletions. Two contain the chromosomes MN•OPQRST + cdef•ghiST and two contain the chromosomes MN•OPQRjkl + cdef•ghijkl.

14d.   *Evaluate:* This is type 1. This problem concerns gametes formed by individuals heterozygous for a balanced, reciprocal translocation. *Deduce:* Tetravalent, cruciform-shaped structures that form during meiosis can segregate by alternate, adjacent-1, or adjacent-2 segregation mechanisms. The adjacent-2 pattern segregates nonhomologous centromeres and creates four nonviable gametes. Two gametes contain both copies of one chromosome, one normal and one with the translocation. The other two gametes carry two copies of the other chromosome, one normal and one with the translocation. *Solve:* Adjacent-2 segregation produces four nonviable gametes that contain duplications and deletions. Two contain the chromosomes MN•OPQRST + MN•OPQRjkl and two contain the chromosomes cdef•ghijkl + cdef•ghiST.

14e.   *Evaluate:* This is type 1. This problem concerns gametes formed by individuals heterozygous for a balanced, reciprocal translocation. *Deduce:* Tetravalent, cruciform-shaped structures that form during meiosis can segregate by alternate, adjacent-1, or adjacent-2 segregation mechanisms. The adjacent-2 pattern is least frequent because it requires two nonhomologous centromeres to segregate to opposite poles during meiosis I. *Solve:* Adjacent-2 is the least likely to occur because it requires that nonhomologous centromeres segregate during meiosis I.

15a.   *Evaluate:* This is type 4. This problem concerns the formation of allopolyploids. *Deduce:* Tomato gametes will have $\frac{1}{2}$ of 48 chromosomes, which is 24. Potato gametes will have $\frac{1}{2}$ of 12 chromosomes, which is 6. *Solve:* The hybrid will have 24 + 6 = 30 chromosomes.

15b.   *Evaluate:* This is type 1. This problem concerns gamete formation by allopolyploids. *Deduce:* The hybrid will have 30 chromosomes, but none will exist as homologous pairs; therefore, chromosomes will not pair during meiosis and chromosome segregation will be random, resulting in the formation of unbalanced, nonviable gametes. *Solve:* The hybrid will be infertile.

15c.   *Evaluate:* This is type 1. This problem concerns gamete formation by allopolyploids. *Deduce > Solve:* The hybrid has 30 chromosomes; therefore, it will have 60 chromosomes after a chromosome doubling event.

15d.   *Evaluate:* This is type 3. This problem concerns the effect of abnormal chromosome composition on phenotype. *Deduce > Solve:* The phenotype of an allopolyploid cannot be predicted. The combination of genes from the two genomes results in tens of thousands of genetic interactions that have never been seen before. To predict the phenotype of the allopolyploid, it would be necessary to predict the outcome of all of these genetic interactions, and this is not yet possible.

16a.   *Evaluate:* This is type 4. This problem concerns the DNA marker phenotype of individuals with abnormal chromosome numbers. *Deduce:* Nondisjunction occurred during maternal meiosis I; therefore, the gamete contributing two copies of chromosome 21 to the child will contain both maternal alleles and one of the two paternal alleles. *Solve:* Two PCR marker combinations are possible: 290, 310, 380 and 310, 340, 380.

16b. *Evaluate:* This is type 4. This problem concerns the DNA marker phenotype of individuals with abnormal chromosome numbers. *Deduce:* Nondisjunction occurred during maternal meiosis II; therefore, the gamete contributing two copies of chromosome 21 to the child will contain two copies of one maternal allele and one copy of the two paternal alleles. The two maternal alleles will be detected as one PCR product (either 310 bp or 380 bp). *Solve:* Four PCR marker combinations are possible: (1) 290, 310; (2) 310, 340; (3) 290, 380; and (4) 340, 380.

16c. *Evaluate:* This is type 4. This problem concerns the DNA marker phenotype of individuals with abnormal chromosome numbers. *Deduce:* Nondisjunction occurred during paternal meiosis I; therefore, the gamete contributing two copies of chromosome 21 to the child will contain both paternal alleles and one of the two maternal alleles. *Solve:* Two PCR marker combinations are possible: 290, 310, 340 and 290, 340, 380.

16d. *Evaluate:* This is type 4. This problem concerns the DNA marker phenotype of individuals with abnormal chromosome numbers. *Deduce:* Nondisjunction occurred during paternal meiosis II; therefore, the gamete contributing two copies of chromosome 21 to the child will contain two copies of one paternal allele and one copy of the two maternal alleles. The two paternal alleles will be detected as one PCR product (either 290 bp or 340 bp). *Solve:* Four PCR marker combinations are possible: (1) 290, 310; (2) 290, 380; (3) 310, 340; and (4) 340, 380.

17a. *Evaluate:* This is type 4. This problem concerns chromosome alignment and segregation during meiosis in trisomic individuals. *Deduce:* The three copies of chromosome IV in the male will form either a trivalent or a bivalent and a monovalent. Regardless, meiosis I will result in two copies segregating to one pole and one to the other. The three chromosomes can segregate six possible ways: ++ and *ey*, +*ey* and +, + and + *ey*. *Solve:* The proportions will be $\frac{1}{3}$ + *ey*, $\frac{1}{3}$ +, $\frac{1}{6}$ ++, and $\frac{1}{6}$ *ey*.

17b. *Evaluate:* This is type 4. This problem concerns the results of matings involving trisomic individuals. *Deduce:* The paternal gametes are $\frac{1}{3}$ + *ey*, $\frac{1}{3}$ +, $\frac{1}{6}$ ++, and $\frac{1}{6}$ *ey*. Random fusion of these gametes with female gametes, all of which are *ey*, results in $\frac{1}{3}$ + *ey ey*, $\frac{1}{3}$ + *ey*, $\frac{1}{6}$ ++ *ey*, and $\frac{1}{6}$ *ey ey*. *Solve:* $\frac{1}{6}$ will be eyeless and $\frac{5}{6}$ will be normal, resulting in a ratio of 1:11.

18a. *Evaluate:* This is type 2. This problem concerns the consequence of chromosomal abnormalities on phenotype. *Deduce:* Cri-du-chat is due to heterozygosity for a terminal deletion on chromosome 5. The child's karyotype is consistent with that genotype. *Solve:* Yes; because cri-du-chat is due to heterozygosity for a terminal deletion of chromosome 5, which is evident in the child's chromosomes.

18b. *Evaluate:* This is type 1. This problem concerns the nature of chromosomal abnormalities. *Deduce:* Chromosomal abnormalities visible by karyotype analysis include alterations in chromosome structure, which are evident by differences in chromosome size, banding pattern, or both. Chromosomes 5 and 12 in the father appear different from the chromosomes in the mother and from the expected pattern for normal chromosome 5. The difference appears to be a reciprocal translocation of the terminal segments of these chromosomes. The father appears to have a reciprocal balanced translocation involving the end of the short arm of chromosome 5 and the short arm of chromosome 12. Based on altered banding patterns, the break points appear to be at 5p 15.1 and 12p 13.1.

18c. *Evaluate:* This is type 2. This problem concerns the consequence of chromosomal aberrations on phenotype. *Deduce:* Cri-du-chat is due to heterozygosity for a terminal deletion on chromosome 5. The father contains two copies of this region of chromosome 5: one on the end of his normal chromosome 5 and one on the end of his abnormal chromosome 12. Since he has two copies of this region, he does not have cri-du-chat syndrome. He also contains two complete copies of the information on the end of chromosome 12. *Solve:* The father is not affected, since he has not lost any genetic material in the translocation.

18d.   *Evaluate:* This is type 1. This problem concerns the pairing of homologous chromosomes in individuals that are heterozygous for a reciprocal translocation. *Deduce:* All homologous regions of the four chromosomes will pair, resulting in a tetravalent that has a cruciform-like structure. *Solve:* Draw a diagram similar to the figure shown here.

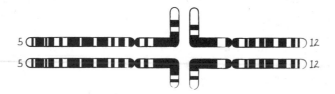

18e.   *Evaluate:* This is type 1. This problem concerns the pattern of chromosome segregation during meiosis in individuals that are heterozygous for a reciprocal translocation. *Deduce:* Adjacent-1 segregation involves segregation of homologous centromeres but includes one translocation chromosome and one normal chromosome. The child received the father's translocation chromosome 5 and his normal chromosome 12. *Solve:* Therefore, the gamete that contributed to this child was formed by adjacent-1 segregation.

18f.   *Evaluate:* This is type 1. This problem concerns the formation of gametes by individuals that are heterozygous for a reciprocal translocation. *Deduce:* The main patterns of chromosome segregation in heterozygotes for a reciprocal translocation are alternate and adjacent-1. *Solve:* Since one of the two types of gametes formed by adjacent-1 segregation results in the cri-du-chat phenotype and neither of the gametes from alternate give cri-du-chat, one of four, or 25%, of the children fathered by this man will have cri-du-chat.

18g.   *Evaluate:* This is type 1. This problem concerns gamete formation by individuals that are heterozygous for a reciprocal translocation. *Deduce:* Approximately half of the gametes produced by individuals that are heterozygous for a reciprocal translocation are unbalanced and could be expected to be nonviable. *Solve:* Yes; the karyotype does help explain the abortions because reciprocal translocation heterozygosity is consistent with semisterility due to the high probability of the father producing sperm with chromosome abnormalities.

19a.   *Evaluate:* This is type 2. This problem concerns the phenotype of individuals with chromosomal aberrations. *Deduce > Solve:* The boy with Down syndrome must have three copies of at least part of chromosome 21. Since he has only 46 chromosomes, one of his chromosomes must contain two copies of chromosome 21 information. Since his sisters are phenotypically normal but have 45 chromosomes, they must have a chromosome that contains all the information from two chromosomes. This is reminiscent of a Robertsonian translocation involving chromosomes 14 and 21. This could explain why the boy has Down syndrome; he would have one normal chromosome 14, two normal copies of 21, and one 14:21 Robertsonian translocation.

19b.   *Evaluate:* This is type 2. This problem concerns gamete formation by individuals with chromosomal aberrations. *Deduce:* If the boy with Down syndrome has one copy of 14, two copies of 21, and a 14:21 Robertsonian translocation, then one of his parents must carry the 14:21 translocation. *Solve:* Since both parents are phenotypically normal, the parent carrying the 14:21 chromosome must have only one copy of chromosomes 14 and 21 and, therefore, only 45 total chromosomes. One parent is expected to have a normal karyotype and the other is expected to have 45 chromosomes; one 14, one 21, and one 14:21.

19c.   *Evaluate:* This is type 4. This problem concerns the terminology for structural chromosome abnormalities. *Deduce > Solve:* A fusion of two acrocentric chromosomes is called a Robertsonian translocation.

19d.   *Evaluate:* This is type 1. This problem concerns gamete formation by individuals with abnormal chromosome number or structure. *Deduce:* The parent with a normal karyotype will produce normal gametes; therefore, the frequency of normal versus abnormal gametes produced by the parent carrying the 14:21 chromosome will determine the frequency of their having normal children versus children with Down syndrome. The parent carrying the 14:21 chromosome will make three types of viable gametes: (1) 14:21, (2) 14 + 21, and (3) 14:21 + 21. Only one of these gamete types will form phenotypically normal individuals with 46 chromosomes when fused with a normal gamete. *Solve:* Therefore, this couple's next child has a $\frac{1}{3}$ chance of being phenotypically normal with 46 chromosomes.

20. *Evaluate:* This is type 1. This problem concerns the identification of differences in chromosome structure based on karyotype analysis. *Deduce > Solve:* The human and orangutan chromosome have identical banding patterns along their entire lengths, and all four species have the same chromosome 5 banding pattern from band 5q14.1 to the q-arm telomere. In comparison to human chromosome 5, the chimpanzee chromosome has undergone a pericentric inversion with breakpoints at approximately 5p13.2 and 5q13.3. The gorilla chromosome differs from the human chromosome from 5q13.3 to the telomere of 5p. It may have undergone a balanced translocation with another chromosome.

21a. *Evaluate:* This is type 1. This problem concerns the identification of differences in chromosome structure based on karyotype analysis. *Deduce:* The chromosome banding patterns appear identical from the telomere of the q arm through band q3.1 on the mainland chromosome. The patterns also appear identical for the last two bands at the end of the p arm. The region between q3.1 and p2.1 on the mainland chromosome appears to be present but inverted on the island chromosome. *Solve:* Two island deer chromosomes appear to have undergone a pericentric inversion that includes the bands q2 through p2.1 on the mainland chromosome.

21b. *Evaluate:* This is type 1. This problem concerns the alignment of homologs during meiosis in individuals that are heterozygous for a pericentric inversion. *Deduce:* All homologous sequences pair during synapsis. For inversion heterozygotes, an inversion loop must form to promote synapsis. In this problem, the loop includes the region between q3.1 and p2.1 on the mainland chromosome and the region from q2.1 to p2 on the island chromosome. *Solve:* Draw a diagram similar to the figure shown here.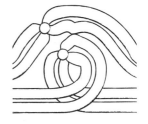

21c. *Evaluate:* This is type 1. This problem concerns the consequences of crossing over during meiosis within the inverted region in an individual who is heterozygous for a pericentric inversion. *Deduce:* The crossover will occur between the q1 region of the mainland chromosome and the homologous p1 region of the island chromosome. The resulting gametes include two non-crossover gametes that are identical to the mainland and island chromosomes, respectively, and two recombinant gametes that contain duplications and deletions. *Solve:* Draw a diagram similar to the figure shown here.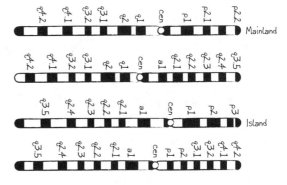

21d. *Evaluate:* This is type 1. This problem concerns the consequences of crossing over during meiosis within the inverted region in an individual who is heterozygous for a pericentric inversion. *Deduce:* In 60% of meioses, there are no crossovers within the inverted region. These meioses produce all viable gametes. In 40% of meioses, there are crossovers within the inverted region. These meioses produce $\frac{1}{2}$ viable and $\frac{1}{2}$ nonviable gametes. The nonviable gametes are due to duplications and deletions that are created by the crossover within the inverted segment. *Solve:* The fraction of viable gametes is $(0.4 \times 0.5) + (0.6 \times 1) = 0.2 + 0.6 = 0.8$ or 80%. The decrease in viability is due to the formation of duplications and deletions by crossing over within the inverted chromosome segment.

22.   *Evaluate:* This is type 4. This problem concerns the consequence of chromosomal abnormalities on phenotype. *Deduce > Solve:* Mosaicism refers to the condition in which the individual has cells with more than one karyotype. The range of phenotypic effects observed in these sex chromosome mosaics depends on the relative percentages of cells with each karyotype at different stages of development, which can vary greatly among $\frac{XX}{XO}$ and $\frac{XY}{XO}$ mosaics. Greater percentages of XO cells during critical stages of development correspond to phenotypes that are more similar to those with Turner syndrome.

23.   *Evaluate:* This is type 4. This problem concerns the consequences of abnormal chromosome number on gamete formation. *Deduce > Solve:* Tetraploids produce diploid gametes, and diploids produce haploid gametes; therefore, a hybrid of the tetraploid and diploid will be a triploid. The triploid will have either bivalents plus monovalents or all trivalents during synapsis in meiosis I. It will produce unbalanced, nonviable gametes and thus will be seedless.

24.   *Evaluate:* This is type 2. This problem concerns the use of strains carrying interstitial deletions to map the location of point mutations. *Deduce:* The progeny of crosses between strains homozygous for an interstitial deletion and a point mutation are heterozygous for the deletion and point mutation. A mutant phenotype represents noncomplementation, indicating that the mutation maps to the region of the deletion; a wild-type phenotype represents complementation, indicating that 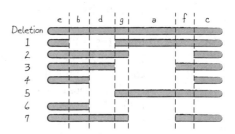 the mutation does not lie within the deleted region. Mutations that do not complement several deletions must lie in the region that all the deletions have in common. Mutation *a* lies in a region common to deletions 2, 3, 4, 6, and 7. Mutation *b* lies in a region common to deletions 1, 5, and 6. Mutation *c* lies in a region unique to deletion 6. Mutation *d* lies in a region common to deletions 1, 4, 5, and 6. Mutation *e* lies in a region unique to deletion 5. Mutation *f* lies in a region common to deletions 2, 4, and 6. Mutation *g* lies in a region common to deletions 3, 4, and 6. *Solve:* Draw a diagram similar to the figure shown here.

25a.   *Evaluate:* This is type 4. This problem concerns the consequences of allopolyploidy on gamete production. *Deduce:* The hexaploid line produces triploid gametes that contain 24 chromosomes. The tetraploid line produces diploid gametes that contain 16 chromosomes. Experimental variety 1 is the result of fusion of normal gametes from both lines: 24 + 16 = 40. Experimental variety 2 is the result of fusion of a normal gamete from the hexaploid line with a tetraploid gamete from the tetraploid line: 24 + (2 × 16) = 56. *Solve:* Neither line is expected to be fertile, because the hexaploid line contributes 3 of each of its 8 different homologous chromosomes to the experimental varieties. These chromosomes will form trivalents or bivalents plus a monovalent during synapsis in meiosis I, thus preventing the formation of balanced gametes.

25b.   *Evaluate:* This is type 4. This problem concerns the mechanism of allopolyploid formation. *Deduce:* The hexaploid line produces gametes containing 24 chromosomes, and the tetraploid line produces gametes containing 16 chromosomes. Experimental variety 1 is the result of fusion of normal gametes from both lines: 24 + 16 = 40. Experimental variety 2 is the result of fusion of a normal gamete from the hexaploid line with a tetraploid gamete from the tetraploid line: 24 + (2 × 16) = 56. *Solve:* The hexaploid line contributed 24 chromosomes to both experimental varieties.

25c.   *Evaluate:* This is type 4. This problem concerns the mechanism of allopolyploid formation. *Deduce:* The hexaploid line produces gametes containing 24 chromosomes, and the tetraploid line produces gametes containing 16 chromosomes. Experimental variety 1 is the result of fusion of normal gametes from both lines: 24 + 16 = 40. Experimental variety 2 is the result of fusion of a normal gamete from the hexaploid line with a tetraploid gamete from the tetraploid line: 24 + (2 × 16) = 56. *Solve:* The tetraploid line contributed 16 chromosomes to experimental variety 1 and 32 chromosomes to experimental variety 2.

26. *Evaluate:* This is type 1. This problem concerns the effect of chromosomal abnormalities on recombination frequency. *Deduce:* The genetic map distance predicts 12% recombination between *dwarf* and *peach* and 17% recombination between *peach* and *oblate*. The cross is $\frac{DPO}{dpo} \times \frac{dpo}{dpo}$. The progeny showing crossovers between dwarf and peach were dwarf, smooth, round (13 progeny, 1 crossover each); tall, peach, oblate (17 progeny, one crossover each); dwarf, smooth, oblate (1 progeny, two crossovers); tall, peach, round (0 progeny). These results indicate a recombination frequency of $\frac{31}{1000}$ or 3.1% recombination. The progeny showing crossovers between peach and oblate were dwarf, peach, round (8 progeny); tall, smooth, oblate (12 progeny); dwarf, smooth, oblate (1 progeny); tall, peach, round (0 progeny). These results indicate a recombination frequency of $\frac{22}{1000}$ or 2.2%. *Solve:* The results show that recombination frequency in the region between dwarf and oblate is much lower than expected. This result is consistent with inversion of the chromosome region containing *peach* and the presence of heterozygosity for the inversion in the trihybrid line. Inversion heterozygosity is suppressing the appearance of most of the crossover chromosomes in this cross.

27a. *Evaluate:* This is type 3. This problem concerns the effect of transposition on genotype and phenotype. *Deduce > Solve:* The wild-type *white* allele is detected as a 5-kb band on a Southern blot. If a 3-kb *P* element transposon inserts into the *white* gene to create a *white* mutant, then this would be detected as an 8-kb band on the Southern blot as long as the restriction enzyme used to digest the DNA did not cut within the *P* element. Draw a diagram similar to the figure shown here.

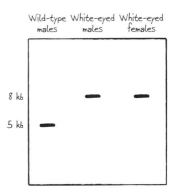

27b. *Evaluate:* This is type 3. This problem concerns the effect of transposition on phenotype. *Deduce > Solve:* The mutant *white* allele was due to insertional inactivation of *white* by the *P* element. The males inherit this allele from their female parent and therefore have that allele in all of their somatic and germ-line cells. The *P* element can precisely excise from the *white* locus in somatic cells that contribute eye development. Those cells and all their mitotic progeny will have a wild-type *white* allele and will lead to the formation of red eye tissue. The excision event is random (stochastic), occurring at different times in different cells during eye development within a fly eye, between the two eyes of the same fly, and between eyes of different flies. This leads to the generation of patches of different sizes, where larger patches correspond to excision events early in development and small patches correspond to excision later. This process also leads to varying numbers of red patches in different eyes, where more patches correspond to more frequent excisions and fewer patches to less frequent excisions.

27c. *Evaluate:* This is type 3. This problem concerns the effects of transposition on phenotype and genotype. *Deduce > Solve:* Yes; the white sector will have a *white* locus in which the *P* element remains, which will be seen as an 8-kb band on the Southern blot. The red sector will have a *white* locus in which the *P* element has excised, which will be seen as a 5-kb band on the Southern blot.

28a. *Evaluate:* This is type 3. This problem concerns the mechanism of transposition by DNA transposons and retrotransposons. *Deduce:* The *P* element is 2.5 kb in length but contains an additional 1-kb intron. Because *P* elements transpose by excision of the DNA element and reinsertion of the entire element in the genome, the entire 3.5-kb *P* element will transpose. *Solve:* Therefore, the length of the new *P* element will be 3.5 kb.

28b. *Evaluate:* This is type 3. This problem concerns the mechanism of transposition by DNA transposons and retrotransposons. *Deduce:* The *copia* element is 6 kb in length but contains an additional 1-kb intron. Because *copia* is a retrotransposon and transposes via an RNA intermediate, the intron will be removed and the new insertion site will contain only the unmodified 6-kb *copia* element. *Solve:* Therefore, the length of the new *copia* element will be 7.0 kb.

28c. *Evaluate:* This is type 3. This problem concerns the mechanism of transposition by DNA transposons and retrotransposons. *Deduce:* The *P* element transposes by excision of the DNA element and reinsertion of the entire element in the genome; therefore, the entire 3.5-kb modified *P* element will transpose. The *copia* element is a retrotransposon and transposes via an RNA intermediate; therefore, the intron will be removed from the RNA before reverse transcription to DNA, and the new insertion site will contain only the unmodified 6-kb *copia* element.

29. *Evaluate:* This is type 3. This problem concerns the mechanism of transposition and the use of transposons as delivery vehicles for introduction of foreign DNA into organisms. *Deduce:* The key components of this experiment include (1) design of the *P* element carrying the mutant flapper allele, (2) expression of *P* element transposase to catalyze integration of the modified *P* element into a fly genome, (3) a method for introduction of the modified *P* element into flies, and (4) a screen to identify transgenic flies that contain the modified *P* element in germ cells (germ-line transformants). *Solve:* (1) The 31-bp inverted repeats at ends of *P* elements must be at the ends of the modified *P* element, and the flapper allele should be between them. The *P* element should also have an additional dominant gene that will allow transgenic flies to be visually identified (typically, a wild-type *white* allele). (2) The *P* element transposase gene must be supplied by a separate DNA molecule that is co-introduced along with the flapper-containing modified *P* element. This ensures that the flapper allele flanked by 31-bp inverted repeats is recognized by the transposase and inserted somewhere into the genome. (3) Microinjection can be used to introduce DNA into an early *Drosophila* embryo. The DNA will transpose into chromosomes of some but not all nuclei in the fly embryo; therefore, a screen is necessary to identify which embryo incorporated the modified *P* element into a cell that is part of its germ line. (4) A cross of potentially transgenic flies to a white-eyed mutant will identify germ-line transformants because their progeny will have red eyes.

30a. *Evaluate:* This is type 3. This problem concerns the mechanism of transposition. *Deduce:* The target site for insertion is cut on the top strand before the T nucleotide, which is 8 bp in from the left end of the sequence, and on the bottom strand, after the C, which is 6 bp in from the right end of the sequence. This makes a staggered cut in the insertion site, which is filled in by DNA replication to create an 8-bp direct repeat that will flank the inserted transposon. The 868-bp transposon sequence will be located between the 8-bp direct repeats. *Solve:* Draw a diagram similar to the figure shown here.

30b. *Evaluate:* This is type 3. This problem concerns the mechanism of transposition. *Deduce:* The target site for insertion is cut by transposase on the top strand before the T nucleotide, which is 8 bp in from the left end of the sequence, and on the bottom strand, after the C, which is 6 bp in from the right end of the sequence. This makes a staggered cut in the insertion site, which is filled in by DNA replication to create an 8-bp direct repeat that will flank the inserted transposon. The 868-bp transposon sequence will be located between the 8-bp direct repeats.

31a. *Evaluate:* This is type 3. This problem concerns the effect of transposition on genotype and phenotype. *Deduce:* For the *Ty1* transposon to inactivate the *Db* gene, it must insert into the gene, which is contained on an 8.5-kb *Not*I fragment. *Solve:* If *Ty1* does not contain a *Not*I recognition site, then a *Db* allele with *Ty1* inserted will be 8.5 + 5.6 = 14.1 kb in length.

31b. *Evaluate:* This is type 3. This problem concerns the effect of transposition on genotype and phenotype. *Deduce:* The *Not*I fragment containing the mutant *Db* gene, which contains a *Ty1* insertion, is 14.1 kb in length. The *Not*I fragment containing the wild-type *Db* gene is 8.5 kb in length. A diploid heterozygous for the *Db* mutation will have both *Not*I fragments. *Solve:* The DNA fragments detected are 8.5 kb and 14.1 kb.

31c. *Evaluate:* This is type 3. This problem concerns the effect of transposition on genotype and phenotype. *Deduce:* Loss-of-function mutations are mutations that inactivate the function of a gene. Insertion of a *Ty1* element into the yeast *Db* gene would be expected to cause loss of *Db* function because the insertion is highly likely to disrupt the *Db* coding sequence or promoter and prevent production of the *Db* protein. Loss of *Db* protein production would fit the definition of a loss-of-function mutation in *Db*.

32a. *Evaluate:* This is type 1. This problem concerns the mechanism underlying the formation of abnormal gametes. *Deduce:* The child with Klinefelter syndrome and hemophilia must have the genotype X^hX^hY because XXY is the karyotype for Klinefelter syndrome and X^hX^h is the genotype for hemophilia. Since neither parent is a hemophiliac, the man must be X^HY; therefore, the woman must be X^hX^H. For the son to receive X^hX^h, nondisjunction must have occurred at meiosis II during oogenesis in the woman. *Solve:* The man is X^HY, the woman is X^hX^H, the child is X^hX^hY, and nondisjunction occurred in the woman during meiosis II.

32b. *Evaluate:* This is type 1. This problem concerns the mechanism underlying the formation of abnormal gametes. *Deduce:* For the child to receive two Y chromosomes, nondisjunction must have occurred at meiosis II during spermatogenesis in the man. For the child to have hemophilia, the woman must have been a carrier for hemophilia. *Solve:* The man is X^cY, the woman is X^hX^H, the child is X^hYY, and nondisjunction must have occurred in the man during meiosis II.

32c. *Evaluate:* This is type 1. This problem concerns the mechanism underlying the formation of abnormal gametes. *Deduce:* The man is color blind, and so he is X^cY. The child with Turner syndrome has normal vision and therefore inherited an X^C from the woman. Since the child is X^CO, nondisjunction must have occurred in the man and could have been at either meiosis I or meiosis II. *Solve:* The man is X^cY, the woman has at least one X chromosome with the wild-type C allele ($X^CX^?$), and the child is X^CO. Nondisjunction must have occurred in the man at meiosis I or II.

32d. *Evaluate:* This is type 1. This problem concerns the mechanism underlying the formation of abnormal gametes. *Deduce:* The man who is color-blind and hemophiliac must be $X^{ch}Y$. For the child to have three X chromosomes and have hemophilia but not color blindness, she must carry three mutant *h* alleles and have at least one wild-type *C* allele. The child cannot have inherited all her X chromosomes from the man and therefore must have at least one X carrying *h* and one X carrying a *C* from the woman. The woman is not a hemophiliac, and so she must be X^hX^H, and at least one of these chromosomes is carrying *C*. The abnormal gamete contributing to the child could have come from nondisjunction during meiosis II in the man to produce a sperm with $X^{ch}X^{ch}$, which fused with an egg carrying X^{Ch}. Alternatively, nondisjunction could have occurred during meiosis II in the woman to produce an egg with $X^{Ch}X^{Ch}$, which fused with a sperm with X^{ch}. *Solve:* The man is $X^{ch}Y$, and the woman is $X^{hC}X^{H?}$. The child could be $X^{ch}X^{hC}X^{hC}$, where nondisjunction has occurred in the woman during meiosis II; or the child could be $X^{ch}X^{ch}X^{Ch}$, where nondisjunction has occurred during meiosis II in the man.

13.04 Test Yourself

Problems

1. The figure shows chromosomes in a cell that is heterozygous for a reciprocal translocation. The translocation places the recessive *a* and *b* alleles on different chromosomes.

    a. Draw these chromosomes as they would appear during meiosis at metaphase I.

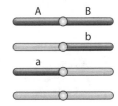

b. Determine the genotypes of gametes produced by a meiosis in which there was a crossover in the region between the *B* locus and the translocation breakpoint.

c. What is the genotype of the gametes that will result in formation of semisterile progeny if they were to fuse with chromosomally normal gametes?

2. The five chromosomes illustrated below (a through e) are evolutionarily related. The most recent version of this chromosome is chromosome (a). What is the evolutionary relationship between (a) and the other chromosomes? Indicate this by stating which chromosome is most closely related to (a) and what change occurred to derive (a) from that chromosome. Then indicate which chromosome is most closely connected to that chromosome and what change occurred to convert one into the other. Continue until all five chromosomes are positioned in the lineage and you have described the change that occurred at each step during evolution of this chromosome.

  a. *AD•CGFGFEH*

  b. *AD•CEFGH*

  c. *ABC•DEFGH*

  d. *AD•CBEFGH*

  e. *AD•CGFEH*

3. The figure shows the structure of a series of fruit-fly deletion mutants. The gap in each line represents the segment of the chromosome that is deleted in each mutant. Strains that were heterozygous for each deletion mutation and a point mutation (a through f) were created and analyzed for wild-type (+) or mutant (−) phenotype. The results of the analyses are shown in the table. Use this information to draw a map that shows the location of each point mutation on the chromosome map.

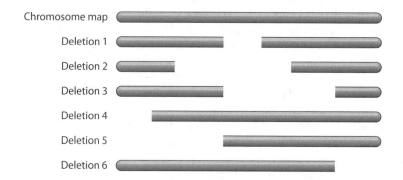

	Mutant a	Mutant b	Mutant c	Mutant d	Mutant e	Mutant f	Mutant g
Deletion 1	+	+	+	−	+	+	+
Deletion 2	+	+	+	−	−	+	−
Deletion 3	+	−	+	−	+	+	−
Deletion 4	+	+	−	+	+	+	+
Deletion 5	+	+	−	+	−	−	+
Deletion 6	−	+	+	+	+	+	+

4. Segmental deletions are common DNA polymorphisms in humans. Many cause no known phenotype, whereas others cause congenital birth defects. A child is born with a congenital form of neuropathy. A karyotype reveals that she is heterozygous for a deletion of the region 17q9 through 17q13. A screen of individuals in the general population reveals that this form of congenital

neuropathy is seen in individuals with the following deletions on chromosome 17: 17q11–17q15, 17q9–17q12, and 17q11–17q17. Use this information to determine the cytological location of the gene or genes involved in the congenital neuropathy.

5. RNA copies of the yeast *Ty1* element are found in virus-like particles in the cytoplasm of yeast cells that contain this retrotransposon. The *Ty1*-containing particles are not infectious, however. How does this observation relate to the hypothesis that retrotransposons and retroviruses are evolutionarily related?

6. *Drosophila P* elements can excise from their location by either precise or imprecise excision events. Precise excision events leave the genomic location exactly as it was before transposon insertion. Imprecise excision results in removal of all or part of the *P* element plus adjacent chromosomal sequences. Relate this information to the ability of insertion mutations caused by *P* element transposition to revert and to the potential utility of *P* element excision to induce deletion mutation in the *Drosophila* genome.

7. Colchicine is a plant chemical that destabilizes microtubules in fungal cells. Treatment of fungi with colchicine causes cells to arrest at metaphase of mitosis. Prolonged treatment of cells causes some cells to "forget" that they have not divided. These cells continue into the next mitotic cell cycle without dividing. Based on this information, explain how colchicine could be used to generate tetraploid fungi starting with haploid fungal cells.

8. Two closely related diploid plants are crossed to produce a hybrid. The hybrid is semisterile. Examination of the hybrid's meiotic cells reveals 22 bivalents and two monovalents. In situ hybridization experiments using DNA from the parent species indicates that the monovalents are homologous to one chromosome from each parent and that each bivalent contains a chromosome from both parents. Use this information to deduce the chromosome number of the parent plants. Also explain why the hybrid is only semisterile and not completely sterile.

Solutions

1a–c. *Evaluate:* This is type 1. This problem concerns the consequences of chromosomal aberrations on meiosis and gamete formation. *Deduce > Solve:* All homologous regions of the chromosomes will pair, resulting in formation of a tetravalent with a cruciform shape. The crossover will occur at the X. Alternate segregation will result in formation of viable gametes with the genotype shown. Gametes number 3 and 4 will form semisterile individuals if they fuse with normal gametes.

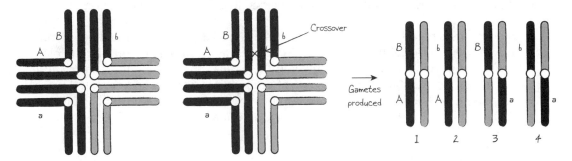

2. *Evaluate:* This is type 1. This problem concerns chromosomal aberrations and how they contribute to chromosome evolution. *Deduce > Solve:* The most recent chromosome is (a), which has the structure *AD•CGFGFEH*. This is most closely related to (e), which is *AD•CGFEH*. A duplication of *GF* changes (e) to (a). The next most closely related chromosome is (b), which is *AD•CEFGH*. An inversion of *EFG* changes (b) to (e). The next most closely related chromosome is (d), which is *AD•CBEFGH*. A deletion of B changes (d) to (b). The next most closely related chromosome is (c), which is *ABC•DEFGH*. An inversion of *BC•D* changes (c) to (d).

3.   *Evaluate:* This is type 2. This problem concerns the effect of chromosomal aberrations on phenotype and the use of deletions to map the location of point mutations. *Deduce:* Point mutation *a* is located within the region deleted only in deletion 6. Point mutation *b* is located in the region deleted only in deletion 3. Point mutation *c* is located within the region deleted in deletion mutants 4 and 5. Point

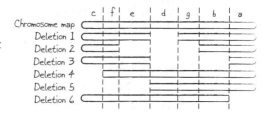

mutation *d* is located within the region deleted in deletion mutants 1, 2, and 3. Point mutation *e* is located within the region deleted in 2 and 5. Point mutation *f* is located in the region deleted only in deletion 5. Point mutation *g* is located in the region deleted in deletion mutants 2 and 3. *Solve:* Draw a diagram similar to the figure shown here.

4.   *Evaluate:* This is type 2. This problem concerns the effects of chromosomal aberrations on phenotype. *Deduce:* The location of the gene or genes involved in the congenital neuropathy will be a deleted region that is common to all the segmental deletions identified in individuals with the disorder. *Solve:* The gene or genes that are responsible for the neuropathy are located in band 17q11 or 12.

5.   *Evaluate:* This is type 3. This problem concerns the structure and function of retrotransposons and their relationship to retroviruses. *Deduce > Solve:* Retroviruses are viruses whose genome is RNA. They use viral-encoded reverse transcriptase to copy their RNA genome into DNA, which is then inserted into a host-cell chromosome. The chromosomal copy of the viral genome is then transcribed to produce RNA, which codes for viral proteins and serves as the genome (or the template for replication of the genome). *Ty1*, like other retrotransposons, shares the essential features of this replication cycle, which contributes to the hypothesis that retroviruses and retrotransposons are related. In addition, like retroviruses, *Ty1* RNA is packaged into virus-like particles, although the particles are not infectious. Thus *Ty1* appears to be a defective yeast retrovirus that has lost the ability to escape the host cell and infect other host cells.

6.   *Evaluate:* This is type 3. This problem concerns the mechanism of transposition by *P* elements. *Deduce > Solve:* The problem indicates that *P* elements can insert into genes, causing insertion mutations, and then excise from those genes either precisely or imprecisely. This explains the reversion of *P* element insertion mutations to wild type because the precise excision restores the wild-type sequence of the gene. This also provides a means to create deletion mutations in regions that contain *P* element insertions by screening for imprecise excisions. In fact, imprecise excision of existing *P* elements is a common laboratory method to induce deletion mutations in nearby genes.

7.   *Evaluate:* This is type 4. This problem concerns the mechanism of polyploidy formation. *Deduce > Solve:* Since colchicine causes cells to arrest at metaphase of mitosis, haploid fungal cells will arrest growth of 2*N* DNA content. If these cells continue into the next cell cycle without dividing, they will double their DNA content again during S phase and become 4*N*. If colchicine is still present or is added back to the cells in this second cell cycle, they will arrest at metaphase of mitosis with 4*N* DNA content. Any cells that continue into the next cell cycle without dividing will be tetraploid (4*N*).

8.   *Evaluate:* This is type 4. This problem concerns allopolyploids and the effect of ploidy on gamete formation. *Deduce > Solve:* Since each bivalent and one of the monovalents are from each parent and the hybrid has 24 chromosomes, each parent contributed 12 chromosomes to the hybrid. Since the parents were diploids, each has a chromosome number of 24. Eleven of the chromosomes from the parents synapse to form bivalents, thus supporting the statement that the two parents are closely related. These 22 bivalents will segregate normally during meiosis I, contributing to gametes that have 11 of the 12 chromosomes needed for viability. The remaining monovalents have two options: both can migrate to the same pole, resulting in nonviable gametes; or they can migrate to opposite poles, forming viable gametes. These options predict that about $\frac{1}{3}$ of meioses will result in viable gamete formation, a condition consistent with semisterility.

14

Regulation of Gene Expression in Bacteria and Bacteriophage

Section 14.01 Genetics Problem-Solving Toolkit

Section 14.02 Types of Genetics Problems

Section 14.03 Solutions to End-of-Chapter Problems

Section 14.04 Test Yourself

14.01 Genetics Problem-Solving Toolkit

Key Terms and Concepts

Allostery: A change in the structure and function of a protein caused by binding to an allosteric regulator.

Allosteric regulator: A metabolite that binds to the allosteric site of a protein and causes a conformational change that alters function of the protein.

Allolactose: A metabolite of lactose that is the allosteric regulator of the *lac* repressor.

Attenuation: The mechanism by which cotranscriptional translation of a leader mRNA peptide regulates premature termination of transcription of an operon.

Cis-dominant: A mutation in a cis-acting regulatory sequence of an operon.

Inducible operon: An operon transcribed only in the presence of an environmental inducing substance.

Merodiploid (also called partial diploid): A strain that is diploid for only a portion of its genome.

Repressible operon: An operon transcribed only in the absence of an environmental repressing substance.

Trans-acting: A DNA sequence element that codes for a protein.

*trp*L: The *trp* operon mRNA leader sequence.

Key Analytical Tools

Structure and regulation of the *lac* operon (see Figure 14.7, page 465)

Structure and regulation of *trp* operon by attenuation (see Figures 14.18 and 14.19, pages 477–478)

14.02 Types of Genetics Problems

1. Describe the biological significance and logic of bacterial gene regulation mechanisms.
2. Describe the mechanisms regulating operon transcription, and interpret the results of mutational and molecular analyses of transcriptional regulation.
3. Describe attenuation or antisense regulation of translation, and interpret the results of mutational analysis of these regulatory mechanisms.

1. Describe the biological significance and logic of bacterial gene regulation mechanisms.

This type of problem tests your understanding of the general mechanisms and strategies bacteria and bacteriophage use to control gene expression. These problems ask you to consider how these mechanisms accommodate and facilitate the life cycle of bacteria and the life cycle of bacteriophage. You may be asked to explain the difference between regulation of transcription initiation and attenuation and how these differences are exploited to fine-tune control of gene expression. You may be asked to compare and contrast inducible versus repressible transcriptional regulation and explain how each fits the needs of bacteria to respond to their environment. You may be asked to describe the alternative life cycles of bacteriophage lambda and explain how the decision between lysis and lysogeny is made at the molecular level. See Problem 4 for an example.

Example Problem: The *lac* and *trp* operons described in the textbook are well-studied examples of how transcription initiation of bacterial operons is regulated. Although both use negative regulatory mechanisms involving a repressor protein that binds to an operator and prevents initiation of transcription, *lac* is said to be inducible, whereas *trp* is said to be repressible. Explain why the terms *inducible* and *repressible* are used correctly, even though both mechanisms involve a repressor protein.

Solution Strategies	Solution Steps
Evaluate	
1. Identify the topic this problem addresses, and explain the nature of the requested answer.	1. This problem tests your understanding of *inducible* and *repressible* as they relate to the mechanisms regulating bacterial operon transcription. The answer will require an explanation of inducible and repressible operons and of the role of repressor proteins in each.
2. Identify the critical information given in the problem.	2. The mechanisms regulating transcription initiation of the *lac* and *trp* operons are presented in the textbook chapter.
Deduce	
3. Describe the general strategy regulating the *lac* operon, focusing on the role of the *lac* repressor in making the operon inducible.	3. In the absence of lactose, transcription of the *lac* operon is repressed when the *lac* repressor binds to the operator. The presence of lactose in the environment induces transcription of the *lac* operon because lactose is imported into the cell and converted to allolactose, which binds to the repressor and prevents the repressor from binding to the operator. Thus, lactose induces *lac* operon transcription.

(continued)

Solution Strategies	Solution Steps
4. Describe the general strategy regulating the *trp* operon, focusing on the role of the *trp* repressor in making the operon repressible.	4. Transcription of the *trp* operon occurs in the absence of tryptophan because the *trp* repressor cannot bind to the operator on its own. The presence of tryptophan in the environment represses *trp* operon transcription because tryptophan is imported into the cell and binds to the *trp* repressor, and the *trp* repressor–tryptophan complex then binds to the operator to repress transcription. Thus, tryptophan represses *trp* operon transcription.
Solve	
5. Answer the problem by pointing out the contrasting ways that the *lac* and *trp* repressors link environmental signals to operon transcription.	5. **Answer:** The *lac* and *trp* repressors link environmental signals to transcription in opposite ways. For *lac*, the environmental signal (lactose) prevents repression and therefore induces transcription. For *trp*, the environmental signal (tryptophan) activates repression and therefore represses transcription.

2. Describe the mechanisms regulating operon transcription, and interpret the results of mutational and molecular analysis of transcriptional regulation.

This type of problem tests your understanding of mechanisms regulating transcription initiation of the *lac*, *trp*, and *ara* operons, and bacteriophage lambda genes. You may be asked to describe the mechanisms in detail; analyze mutant phenotypes to deduce the function of regulatory elements and structural genes, including the analysis of partial diploids; predict the phenotype of specific regulatory mutants and partial diploids; or interpret the results of DNase I footprinting experiments. See Problem 20 for an example.

Example Problem: Two new *E. coli* mutants have been isolated. Like wild type, they do not transcribe the *lac* operon in medium containing glucose and lacking lactose, and they do transcribe the *lac* operon in medium containing lactose and lacking glucose. Their mutant phenotype is revealed when grown in medium containing glucose and lactose: the *lac* operon is transcribed. One mutant is cis-dominant and the other is trans-dominant. What is the molecular nature of the mutations in these strains?

Solution Strategies	Solution Steps
Evaluate	
1. Identify the topic this problem addresses, and explain the nature of the requested answer.	1. This problem tests your understanding of the mechanisms regulating *lac* operon transcription. The answer will identify possible mutations that result in the mutant phenotype described in the problem.
2. Identify the critical information given in the problem.	2. Both mutants transcribe the *lac* operon in medium containing lactose and glucose but not in medium containing glucose but lacking lactose. One mutant is cis-dominant and the other is trans-dominant.

(continued)

Solution Strategies	Solution Steps
Deduce	
3. Identify the regulatory mechanisms that are normal and those that are abnormal in the mutants.	3. The mutants require lactose for *lac* operon transcription, which indicates that the operator and the *lac* repressor are functioning normally. Transcription of the *lac* operon is not repressed by glucose in the mutants, which indicates that the link between glucose metabolism and *lac* operon transcription is faulty.
4. Describe the details of the regulatory mechanism identified as faulty in the mutants.	4. Transcription of the *lac* operon normally requires binding of CAP–cAMP to its binding site in the *lac* promoter. In the presence of glucose, cAMP levels are low; therefore, CAP–cAMP complexes cannot form, and *lac* transcription does not occur.
5. Indicate the elements of *lac* operon regulation that could mutate to create the cis-dominant and trans-dominant phenotypes described for the mutants. **TIP:** The cis-dominant mutations are in protein-binding sites, whereas trans-dominant mutations are in protein-coding genes.	5. The *lac* promoter would be a locus for a cis-dominant mutation affecting glucose regulation of *lac* transcription. The CAP gene would be a locus for a trans-dominant mutation affecting glucose regulation of *lac* transcription.
Solve	
6. Answer the problem by describing the nature of the mutations in the loci identified in step 4.	6. **Answer:** The cis-dominant mutant could contain a mutation in the *lac* promoter that makes RNA polymerase binding independent of CAP–cAMP. The trans-dominant mutation could be in the CAP gene that either makes its binding to the *lac* promoter cAMP-independent or increases the affinity of CAP for cAMP such that the complex forms at low cAMP levels. These mutations would allow transcription of *lac* in the presence of glucose as long as lactose was also present.

3. Describe attenuation or antisense regulation of translation, and interpret the results of mutational analysis of these regulatory mechanisms.

This type of problem tests your understanding of attenuation and translational control by antisense RNA. You may be asked to describe the mechanism of attenuation control of the *trp* operon, predict the phenotype of *trp* attenuator mutants, or interpret the phenotype of *trp* attenuator mutants. You may also be asked to describe the mechanism of antisense regulation of *IS10* translation or predict the phenotype of mutations affecting *IS10* sense-strand or antisense-strand synthesis. See Problem 14 for an example.

Example Problem: You have analyzed the control of *trp* operon gene expression in a mutant strain of *E. coli*. Like a wild-type strain, the mutant repressed transcription initiation of the operon in the presence of tryptophan and de-repressed (activated) transcription initiation in the absence of tryptophan. However, the mutant failed to produce full-length *trp* operon mRNA when tryptophan levels were low. You determined that low levels of the amino acid cysteine were required in addition to low levels of tryptophan to carry out full-length *trp* operon mRNA production and that the mutation was not linked to the *trp* operon. Provide a reasonable explanation of the mutation.

Solution Strategies	Solution Steps
Evaluate	
1. Identify the topic this problem addresses, and explain the nature of the requested answer.	1. This problem tests your understanding of attenuation regulation of the *trp* operon. The answer will be an explanation of a mutation that makes *trp* operon attenuation responsive to cysteine levels.
2. Identify the critical information given in the problem.	2. Transcription initiation of *trp* operon is normal in the mutant. Initiation is repressed by the presence of tryptophan and de-repressed (activated) by the absence of tryptophan. Attenuation regulation is not normal. Transcription termination occurs even though tryptophan levels are low or absent. Transcription termination is prevented only when cysteine levels are low.
Deduce	
3. Recall the mechanistic relationship between low tryptophan levels and attenuation of *trp* operon transcription.	3. Transcription of *trp* operon mRNA and translation of a peptide in the leader sequence of the mRNA (*trpL*) is coupled. Low levels of tryptophan reduce the levels of charged $tRNA_{trp}$, which causes ribosomes translating the *trpL* mRNA to pause at one of the two tryptophan codons. The paused ribosome promotes formation of the antitermination stem–loop structure, which allows RNA polymerase to continue past the leader and transcribe the structural genes of the *trp* operon.
4. Relate the mutant phenotype to the mechanism of antitermination of *trp* operon transcription.	4. The mutant appears to regulate antitermination of transcription by cysteine levels, suggesting that antitermination loop structure is affected by charged $tRNA_{cys}$ instead of charged $tRNA_{trp}$.
5. Consider the relationship between $tRNA_{trp}$ and $tRNA_{cys}$. TIP: What are the codons for tryptophan and cysteine?	5. The *trp* codons in *trpL* are UGG. Cysteine codons are UGU and UGC.
6. Consider how a mutation not linked to the *trp* operon could cause substitution of cysteine for tryptophan in attenuation control. TIP: What type of mutation causes insertion of cysteine in place of tryptophan during translation?	6. A mutation in a $tRNA_{cys}$ that converts the anticodon from GCA, which base-pairs with cysteine codons UGU and UGC, to UCA or CCA, which base-pair with the tryptophan codon UGG.

(continued)

Solution Strategies	Solution Steps
Solve	
7. Answer the problem by relating the tRNA$_{cys}$ mutation to the effect of cysteine levels on attenuation control of the *trp* operon.	7. **Answer:** The mutation is in a tRNA$_{cys}$ anticodon, causing tRNA$_{cys}$ to base-pair with the tryptophan codon in *trpL* and insert cysteine into the leader peptide when cysteine levels and charged mutant tRNA$_{cys}$ levels are high. This prevents ribosome stalling in *trpL* when tryptophan levels are low but cysteine levels are high. Only when tryptophan levels are low (transcription initiation occurs) and cysteine levels are low (ribosome pauses in *trpL*) does the antitermination loop form, which prevents transcription termination and allows transcription of the *trp* operon structural genes.

14.03 Solutions to End-of-Chapter Problems

1. *Evaluate:* This is type 1. This problem concerns the biological significance of bacterial operons. Recall that operons contain multiple, functionally regulated protein-coding sequences under a common regulatory mechanism. *Deduce > Solve:* Operons have the advantage of affording the organism the opportunity to simultaneously regulate transcription of multiple genes whose products are active in the same process. By sharing a single promoter, a single operator, and perhaps other components such as an attenuator sequence, one regulatory region can control the transcription of a polycistronic mRNA that encodes the polypeptides expressed by the operon genes.

2a–i. *Evaluate:* This is type 2. This problem concerns the function of the regulatory components of operons and how they interact. *Deduce > Solve:* **(a)** A DNA sequence that binds a regulatory protein, such as the *lac* operator sequence. **(b)** A regulatory protein that binds DNA, such as the *lac* repressor protein. **(c)** A compound that induces or activates transcription, such as lactose. **(d)** A compound that interacts with another protein or compound to form an active repressor, such as the *trp* corepressor. **(e)** A DNA sequence that binds RNA polymerase and regulates transcription, such as the *lac* promoter. **(f)** A process of transcription regulation through which the binding of regulatory proteins to DNA activates transcription, such as the CAP binding site of the *lac* promoter. **(g)** A process by which the stereochemistry of a protein is altered to change its interaction capabilities, such as the *lac* repressor protein. **(h)** A process of transcriptional regulation through which binding of regulatory proteins to DNA blocks transcription, such as the *lac* repressor protein binding to *lac O*. **(i)** A mechanism of transcriptional regulation in which transcription level is modified (attenuated) to meet environmental requirements, such as *trp* operon attenuation.

3. *Evaluate:* This is type 1. This problem concerns the biological significance of regulation of gene expression by bacteria. *Deduce > Solve:* Like all organisms, bacteria have genes whose expression is always required (constitutive gene expression) and others whose expression is required only under certain conditions (regulated gene expression). Examples of constitutively expressed genes are those required for protein synthesis. Examples of regulated genes are those required for bacteria to adapt their metabolism to match changes in the environment, such as nutrient availability and elevated temperature. If all genes were constitutively expressed, a great deal of unnecessary transcription and translation would occur in order for bacteria to be ready for all environments; and the many proteins produced might be in competition, to the detriment of the organism.

4a–c. *Evaluate:* This is type 1. This problem concerns the relationship between the regulatory sequences of inducible and repressible operons. *Deduce > Solve:* **(a)** Inducible and repressible operons are similar in that both have promoter and operator regulatory sequences. Inducible operons are bound by a repressor protein to block transcription and may also be bound by a positive regulator to help activate transcription. Repressible operons are bound by a corepressor plus the pathway end product to repress transcription. Repressible operons often use attenuation. **(b)** Both types of regulatory systems use allostery in regulating transcription. The mechanism and consequences of allostery in inducible and repressible operon control differ. In *lac* operon regulation, the repressor protein binds the operator, but allosteric change caused by the repressor binding to allolactose prevents repressor binding to the operator. In *trp* operon regulation, the corepressor protein cannot bind the operator until its shape is changed by binding to tryptophan. **(c)** Both types of operons contain multiple genes that share a single promoter and a single operator sequence. Repressible operons often use attenuation and contain a transcribed leader sequence that participates in determining structural gene transcription, whereas inducible operons do not.

5. *Evaluate:* This is type 2. This problem concerns the mechanisms of induction of *lac* operon transcription by lactose. Recall that lactose permease, which is encoded by the *lac* operon, is required for lactose to move across the cell membrane. *Deduce > Solve:* The reversible nature of the repressor protein–operator interaction means that occasionally and spontaneously, the repressor will drop off the operator sequence. This process allows occasional and sporadic transcription of the polycistronic mRNA and translation of the *lac* operon polypeptides. The few molecules of permease usually present in *lac+* bacteria in the absence of lactose are sufficient to "prime the pump" when it comes to facilitating the movement of lactose across the cell membrane when the sugar is available and needed for growth.

6. *Evaluate:* This is type 3. This problem concerns the mechanism of attenuation and the concept of allostery. *Deduce > Solve:* Attenuation does not involve allosteric changes in protein structure. Attenuation is the result of transcription of a leader sequence that undergoes translation. Coupling of transcription and translation dictates whether transcription continues past the leader sequence and into the structural genes.

7. *Evaluate:* This is type 3. This problem concerns the mechanism by which stem-loop secondary structures in the *trp* operon leader sequence regulate transcription of the operon structural genes. *Deduce > Solve:* The *trp* operon stem-loop structures are regions of the mRNA transcript that contain complementary base pairs (termed regions 1, 2, 3, and 4) that form short, double-stranded segments. Each double-stranded stem is topped with a single-stranded loop. Formation of the 2–3 stem loop is favored by inadequate tryptophan availability. Known as the antitermination stem loop, it permits transcription to continue into the structural genes of the operon, leading to production of the enzymes needed to synthesize tryptophan. Formation of the 3–4 stem loop is favored when tryptophan availability is adequate. Known as the termination stem loop, it contains repeat region 4 and is immediately followed by a poly-U string. Formation of the 3–4 stem loop, and the presence of the poly-U string, are the structural basis for intrinsic termination of transcription. In this case, transcription stops before RNA polymerase reaches the structural genes of the operon.

8. *Evaluate:* This is type 2. This problem concerns the mechanism of positive regulation of *lac* operon transcription by the CAP–cAMP complex. *Deduce > Solve:* The CAP binding site is part of the *lac* promoter and is located at approximately –60. The CAP–cAMP complex binds to this region and alters the structure of the promoter to allow efficient RNA polymerase binding at the *lac* promoter.

9. *Evaluate:* This is type 1. This problem concerns the biological significance of regulation of *lac* operon transcription by the CAP–cAMP complex. *Deduce > Solve:* Cyclic AMP is produced when glycolysis is not active, that is, when glucose is low or absent. CAP is a constitutively produced protein that binds with cAMP to form a dimeric molecule that binds the CAP binding site of the *lac* promoter to carry out positive control of operon gene transcription.

10.   *Evaluate:* This is type 2. This problem concerns the effect of loss-of-function mutations in the *cap* gene on transcription of the *lac* operon. *Deduce > Solve:* A *cap⁻* mutation would prevent active CAP protein synthesis. The required positive regulation of transcription would not occur, and *lac* operon transcription would be minimal. The strain would be *lac⁻*.

11.   *Evaluate:* This is type 1. This problem concerns the biological significance of attenuation in bacterial operon regulation and its potential utility in regulation of gene expression in eukaryotes. *Deduce > Solve:* Attenuation refers to the regulation of transcription elongation by premature termination of transcription, which is controlled by the rate of cotranscriptional translation of leader mRNA–encoded peptides. This mechanism allows for the fine-tuning operon transcription rates to sensitively respond to small changes in amino acid availability. Since transcription and translation occur in distinct subcellular compartments in eukaryotes, attenuation is not expected to be found in eukaryotes.

12.   *Evaluate:* This is type 2. This problem concerns the mechanism of *ara* operon regulation by araC. *Deduce > Solve:* In the absence of arabinose, araC protein negatively regulates transcription by binding at $araO_1$, *araI*, and $araO_2$. The araC proteins at *araI* and $araO_2$ dimerize, inducing the formation of a DNA loop that prevents RNA polymerase from binding to P_{ara}, thus blocking transcription of the *ara* operon genes. When arabinose is present, it binds to araC proteins at the *ara* operator site, which prevents dimerization of araC proteins at *araI* and $araO_2$. This prevents DNA loop formation and makes a CAP binding site available for binding. CAP–cAMP activates *ara* operon gene transcription.

13.   *Evaluate:* This is type 1 and 2. This problem concerns the biology of bacteriophage λ and the roles of λ repressor protein and cro in regulating the phage life cycle. *Deduce > Solve:* Bacteriophage λ infect and kill host bacterial cells by lysis. The lytic cycle of phage involves expression of phage genes, assembly of phage proteins and particles, and packaging of phage DNA into progeny phage that are released by host cell lysis. Under certain circumstances, λ phage can insert itself into the bacterial host chromosome by a site-specific recombination event. This lysogenic event allows the phage DNA to be replicated along with host DNA. At a later time, the lysogen is excised and completes the lysis of the host. The decision between lytic and lysogenic life cycles is determined by which of two phage gene promoters, P_R or P_{RM}, is used immediately after infection of the bacterial cell. The λ repressor protein and cro protein bind competitively at three operator sequences (O_{R1}, O_{R2}, and O_{R3}) that are part of P_R and P_{RM}. If cro succeeds in binding O_{R3} within P_{RM}, the lytic cycle is established. If, on the other hand, λ repressor successfully binds O_{R1} and O_{R2} within P_R, the lysogenic cycle is established.

14.   *Evaluate:* This is type 1 and 3. This problem concerns the biological significance and mechanism of antisense RNA regulation of translation. *Deduce > Solve:* Antisense RNAs are single-stranded RNAs that are complementary to a portion of specific mRNA transcripts. Blocking translation prevents the production of proteins that might initiate unnecessary or harmful actions.

15a–h. *Evaluate:* This is type 3. This problem concerns the mechanism of *trp* operon attenuation and the analysis of *trp* attenuator mutations. The mutations listed prevent the involvement of the corresponding region in stem-loop formation. **TIP:** Tryptophan regulates operon transcription at two levels: transcription initiation and attenuation. *Deduce:* The 2–3 stem loop prevents termination but forms only when tryptophan levels are low, whereas the 3–4 stem loop induces termination but forms only when tryptophan levels are high. If no stem loops form, termination cannot occur. *Solve:* **(a)** There is no significant effect on attenuation, and operon transcription occurs. **(b)** There is no significant effect on attenuation, and operon transcription does not occur. **(c)** Antitermination is prevented, and operon transcription occurs. **(d)** Antitermination is prevented, and operon transcription does not occur. **(e)** Termination and antitermination are prevented, and operon transcription occurs. **(f)** Termination and antitermination are prevented, and operon transcription does not occur. **(g)** Termination is prevented, and operon transcription occurs. **(h)** Termination is prevented, and operon transcription occurs.

16a–e. *Evaluate:* This is type 2. This problem concerns the mechanism of *lac* operon regulation and requires analysis of *lac* operon mutations. *Deduce:* The promoter is required for initiation of transcription under any condition; therefore, promoter mutants cannot transcribe the *lac* operon. The operator is required to prevent transcription in the absence of lactose; therefore, operator mutants transcribe the *lac* operon constitutively. *lacI* encodes the lac repressor, which is required to prevent transcription in the absence of lactose; therefore, *lacI* mutants transcribe the *lac* operon constitutively. CAP–cAMP binds to the CAP binding site to promote transcription of the *lac* operon in the absence of glucose therefore, CAP binding-site mutants will not transcribe the *lac* operon. *Solve:* **(a)** Transcription is blocked; **(b)** transcription is constitutive; **(c)** transcription is blocked; **(d)** transcription is constitutive; **(e)** transcription is blocked.

17a–g. *Evaluate:* This is type 2. This problem concerns the mechanism of *lac* operon regulation and requires the analysis of *lac* mutations. *Deduce: lacI* encodes the *lac* repressor, *lacI$^+$* is wild type, and *lacI$^-$* mutants do not make repressor. *lacP* is the promoter, *lacP$^+$* is wild type, and *lacP$^-$* mutants prevent RNA polymerase binding. *lacO* is the operator, *lacO$^+$* is wild type, and *lacO^C* mutants cannot be bound by repressor. *lacZ* encodes β-galactosidase, *lacZ$^+$* is wild type, and *lacZ$^-$* mutants do not make β-galactosidase. *lacY* encodes lactose permease, *lacY$^+$* is wild type, and *lacY$^-$* does not make permease. Expression of lactose permease and β-galactosidase is required for the strain to be *lac$^+$* since permease is needed to bring the lactose into the cell, and β-galactosidase is needed to break down lactose into glucose and galactose and to form allolactose. *Solve:* **(a)** $I^+ P^+ O^+ Z^+ Y^-$ is *lac$^-$* and the genes are not transcribed. **(b)** $I^+ P^+ O^C Z^- Y^+$ is *lac$^-$* and the genes are constitutively transcribed. **(c)** $I^- P^+ O^+ Z^+ Y^+$ is *lac$^+$* and genes are constitutively transcribed. **(d)** $I^+ P^- O^+ Z^+ Y^+$ is *lac$^-$* and the genes are not transcribed. **(e)** $I^+ P^+ O^+ Z^- Y^+$ is *lac$^-$* and the genes are inducibly transcribed. **(f)** $I^+ P^+ O^C Z^+ Y^-$ is *lac$^-$* and the genes are constitutively transcribed. **(g)** $I^+ P^+ O^C Z^+ Y^+$ is *lac$^+$* and the genes are constitutively transcribed.

18. *Evaluate:* This is type 2. This problem concerns the mechanism of *lac* operon regulation and the analysis of *lac* operon partial diploids. *Deduce: lacI* encodes the *lac* repressor; *lacI$^+$* is wild type, *lacI$^-$* mutants do not make repressor, and *lacS* mutants make super repressor. *lacI^S* is trans-dominant to *lacI$^+$* and *lacI$^-$*, and *lacI$^+$* is trans-dominant to *lacI$^-$*. *lacP* is the promoter, *lacP$^+$* is wild type, and *lacP$^-$* mutants prevent RNA polymerase binding. *lacP$^+$* and *lacP$^-$* are cis-dominant. *lacO* is the operator, *lacO$^+$* is wild type, and *lacO^C* mutants cannot be bound by repressor. *lacO$^+$* and *lacO^C* are cis-dominant, and *lacO^C* cannot be bound by *lacI^S*. *lacZ* encodes β-galactosidase, *lacZ$^+$* is wild type, and *lacZ$^-$* mutants do not make β-galactosidase. *lacZ$^+$* is trans-dominant to *lacZ$^-$*. *lacY* encodes lactose permease, *lacY$^+$* is wild type, and *lacY$^-$* does not make permease. *lacY$^+$* is trans-dominant to *lacY$^-$*. Expression of lactose permease and β-galactosidase is required for the strain to be *lac$^+$*. *Solve:* See table.

Genotype	β-Galactosidase		Permease		Phenotype
	Lactose	**No Lactose**	**Lactose**	**No Lactose**	
Example: $I^+ P^+ O^+ Z^+ Y^+$	+	–	+	–	*lac$^+$*
a. $\dfrac{I^S P^+ O^+ Z^+ Y^+}{I^- P^+ O^+ Z^+ Y^+}$	–	–	–	–	*lac$^-$*
b. $\dfrac{I^- P^+ O^+ Z^- Y^+}{I^+ P^+ O^C Z^+ Y^-}$	+	+	+	–	*lac$^+$*
c. $\dfrac{I^+ P^+ O^+ Z^- Y^+}{I^+ P^- O^+ Z^+ Y^-}$	–	–	+	–	*lac$^-$*
d. $\dfrac{I^- P^+ O^C Z^+ Y^+}{I^+ P^- O^+ Z^+ Y^+}$	+	+	+	+	*lac$^+$*
e. $\dfrac{I^+ P^+ O^C Z^+ Y^-}{I^+ P^+ O^+ Z^+ Y^-}$	+	+	–	–	*lac$^-$*
f. $\dfrac{I^+ P^+ O^+ Z^- Y^+}{I^S P^+ O^+ Z^+ Y^-}$	–	–	–	–	*lac$^-$*
g. $\dfrac{I^S P^+ O^+ Z^- Y^+}{I^+ P^+ O^C Z^+ Y^-}$	+	+	–	–	*lac$^-$*

19a–d. *Evaluate:* This is type 2. This problem tests your understanding of the mechanism of *lac* operon regulation and ability to determine the *lac* operon genotype based on a strain's *lac* phenotype. *Deduce > Solve:* **(a)** Constitutive *lac* operon transcription could result from lack of a repressor (*lacI⁻*) or a defective operator (*lacO^C*). The *lac⁻* phenotype could result from lack of β-galactosidase (*lacZ⁻*) or lactose permease (*lacY⁻*)—*I⁻ P⁺ O⁺ Z⁻ Y⁺* and *I⁺ P⁺ O^C Z⁻ Y⁺* (for either genotype, *Y⁻* is also possible). **(b)** Basal-level (low level) transcription could be due to a promoter mutation (*lacP⁻*) or to the presence of super repressor (*lacI^S*)—*I^S P⁺ O⁺ Z⁺ Y⁺* and *I⁺ P⁻ O⁺ Z⁺ Y⁺*. **(c)** Inducible transcription indicates that regulation is normal. The *lac⁻* phenotype could be due to lack of β-galactosidase (*lacZ⁻*)—*I⁺ P⁺ O⁺ Z⁻ Y⁺*. **(d)** Constitutive *lac* operon transcription could be due either to lack of repressor (*lacI⁻*) or a defective operator (*lacO^C*)—*I⁺ P⁺ O^C Z⁺ Y⁺* and *I⁻ P⁺ O⁺ Z⁺ Y⁺*.

20a–b. *Evaluate:* This is type 2. This problem concerns the mechanism of *lac* operon regulation and the analysis of *lac* operon partial diploids. *Deduce > Solve:* **(a)** One genotype from Problem 19a was *I⁻ P⁺ O⁺ Z⁻ Y⁺*, thus the partial diploid is $\frac{I^- P^+ O^+ Z^- Y^+}{I^+ P^+ O^+ Z^+ Y^+}$. *lacI⁺* is dominant to *lacI⁻* and *lacZ⁺* is dominant to *lacZ⁻*; therefore, this diploid will have inducible transcription of *lacZ* and *lacY*. The second genotype from Problem 19a was *I⁺ P⁺ O^C Z⁻ Y⁺*, thus the partial diploid is $\frac{I^+ P^+ O^C Z^- Y^+}{I^+ P^+ O^+ Z^+ Y^+}$. *lacO^C* is cis-dominant and *lacZ⁺* is dominant to *lacZ⁻*; therefore, the partial diploid will have constitutive transcription of *lacY* and inducible transcription of *lacZ*. One genotype from Problem 19b was *I⁺ P⁻ O⁺ Z⁺ Y⁺*, thus the partial diploid is $\frac{I^+ P^- O^+ Z^+ Y^+}{I^+ P^+ O^+ Z^+ Y^+}$. *lacP⁻* and *lacP⁺* are cis-dominant and therefore will have inducible transcription of both genes. The other genotype from Problem 19b was *I^S P⁺ O⁺ Z⁺ Y⁺*, thus the partial diploid is $\frac{I^S P^+ O^+ Z^+ Y^+}{I^+ P^+ O^+ Z^+ Y^+}$. *lacI^S* is trans-dominant to *lacI⁺*, therefore, both genes will be transcribed at a low, basal level. The genotype for Problem 19c was *I⁺P⁺O⁺Z⁻Y⁺*, thus the partial diploid is $\frac{I^+ P^+ O^+ Z^- Y^+}{I^+ P^+ O^+ Z^+ Y^+}$. *lacZ⁺* is dominant to *lacZ⁻*, therefore, transcription of both genes will be inducible. One genotype for Problem 19d was *I⁺P⁺O^C Z⁺Y⁺*, thus the partial diploid will be $\frac{I^+ P^+ O^C Z^+ Y^+}{I^+ P^+ O^+ Z^+ Y^+}$. *lacO^C* and *lacO⁺* are cis-dominant, therefore, transcription of genes from the first operon will be constitutive and transcription of genes from the second operon will be inducible. The other genotype for Problem 19d was *I⁻P⁺O⁺Z⁺Y⁺*, thus the partial diploid is $\frac{I^- P^+ O^+ Z^+ Y^+}{I^+ P^+ O^+ Z^+ Y^+}$. *lacI⁺* is trans-dominant to *lacI⁻*, therefore, transcription of both genes will be inducible. **(b)** *lac⁺* strains produce β-galactosidase and lactose permease whereas *lac⁻* strains fail to produce one protein or the other. All partial diploids except for $\frac{I^S P^+ O^+ Z^+ Y^+}{I^+ P^+ O^+ Z^+ Y^+}$ are able to produce β-galactosidase and lactose permease. All partial diploids will be *lac⁺* except for $\frac{I^S P^+ O^+ Z^+ Y^+}{I^+ P^+ O^+ Z^+ Y^+}$, which will be *lac⁻*.

21. *Evaluate:* This is type 2. This problem concerns the mechanism of *lac* operon regulation and the analysis of *lac* operon partial diploids. *Deduce:* Mutants A and B are *lac⁻* and therefore do not express *lacZ* or *lacY*. Since Mutant A is inducible for *lac* transcription, it does not contain a promoter or *lacI^S* mutation; therefore, it is *lac⁻* because it is *lacZ⁻* or *lacY⁻*. Since Mutant B is uninducible for *lac* transcription, it is either *lacP⁻* or *lacI^S*. Since the partial diploid A × B remains inducible for *lac* transcription, mutation B does not act in trans and therefore is not *lacI^S*. Mutants C and D are constitutive for *lac* operon transcription and therefore are either *lacI⁻* or *lacO^C*. Since the partial diploid C × D is inducible for *lac* transcription, one mutant must be *lacI⁻* and the other *lacO^C*. Since the partial diploids C × B and C × A are inducible, Mutant C must have been *lacI⁻*. This predicts that mutant D is *lacO^C* and that the partial diploids D × A and D × B should be constitutive, and this prediction is confirmed by the results. *Solve:* Mutant A has a mutation in *lacZ* and is *lacZ⁻*. Mutant B has a mutation of the promoter and is *lacP⁻*. Mutant C has a repressor gene mutation and is *lacI⁻*. Mutant D has a mutation of the operator and is *lacO^C*.

22a. *Evaluate:* This is type 2. This problem concerns the mechanism of *lac* operon regulation and the analysis of *lac* operon partial diploids. *Deduce:* *cap⁻* indicates that the CAP–cAMP binding site is mutant and is not bound by CAP–cAMP. The partial diploid $\frac{cap^+ I^+ P^+ O^+ Z^- Y^+}{cap^+ I^- P^+ O^+ Z^+ Y^-}$ contains an uninducible operon as well as an inducible operon that does not contain a functional lactose permease. The absence of lactose permease expression makes the partial diploid *lac⁻* because it cannot import lactose and will not produce the inducer, allolactose. *Solve:* No, because lactose permease is not fully expressed.

22b.   *Evaluate:* This is type 2. This problem concerns the mechanism of *lac* operon regulation and the analysis of *lac* operon partial diploids. *Deduce:* The partial diploid $\frac{cap^-\ I^+\ P^+\ O^+\ Z^-\ Y^+}{cap^+\ I^-\ P^+\ O^+\ Z^+\ Y^-}$ contains only one functional promoter, which is part of the inducible *lac* operon that is *lacZ*$^+$ and *lacY*$^-$. *Solve:* Neither β-galactosidase nor permease expression will be fully inducible, because permease is not fully expressed.

22c.   *Evaluate:* This is type 2. This problem concerns the mechanism of *lac* operon regulation and the analysis of *lac* operon partial diploids. *Deduce > Solve:* The *lacI* gene has its own promoter and is not affected by *lac* operon regulatory of gene mutations. The *cap*$^-$ mutation minimizes transcription, but repressor protein produced from this chromosome is trans-active and binds *lacO*$^+$ on the other chromosome to render *lacZ* expression inducible. *lacY*$^+$ on the uninducible operon cannot complement *lacY*$^-$ on the inducible operon, thus *lacY* expression remains uninducible.

23.   *Evaluate:* This is type 2. This problem concerns the mechanism of regulation of a hypothetical operon similar to a *lac* operon and the analysis of partial diploids. *Deduce:* The operator and regulatory protein mutants will be constitutive for enzyme synthesis, whereas the enzyme mutant will be unable to produce enzyme. *S* and *T* mutants are constitutive, and *R* mutants cannot produce enzymes; therefore, *R* is the gene for the enzyme. The operator mutant will be cis-dominant, whereas the repressor gene mutant will be recessive. Since $\frac{R^+\ S^-\ T^+}{R^-\ S^+\ T^+}$ is constitutive, and *R*$^+$ is on the same molecule as *S*$^-$, *S*$^-$ is cis-dominant and *S* is the operator. That leaves *T* as the repressor gene. The phenotype of the other partial diploids confirms this answer. *Solve:* Gene *T* produces the repressor protein, *S* is the operator, and gene *R* produces the enzyme.

24.   *Evaluate:* This is type 2. This problem concerns the mechanism of regulation of a hypothetical, *trp* operon-like operon and the analysis of partial diploids. *Deduce:* The operator and regulatory protein mutants will be constitutive for enzyme synthesis. *G* and *W* are constitutive mutants, whereas *Z* mutant cannot produce enzyme. Therefore, *G* and *W* are regulatory mutants and *Z* is the enzyme gene. If *G* is the operator, then *G*$^-$ *Z*$^+$ will be constitutive for enzyme synthesis regardless of the *W* genotype. If *W* is the operator, then *Z*$^+$ *W*$^-$ will be constitutive for enzyme synthesis regardless of the *G* genotype. $\frac{G^-\ Z^+\ W^+}{G^+\ Z^-\ W^-}$ is constitutive, whereas $\frac{G^+\ Z^+\ W^-}{G^-\ Z^-\ W^+}$ is not; therefore, *G* is the operator. This leaves *W* as the repressor. *Solve:* Gene *Z* is the enzyme, gene *W* is the repressor, and gene *G* is the operator.

25a–j.   *Evaluate:* This is type 3. This problem concerns the mechanism of *trp* operon attenuation and the analysis of *trp* attenuator mutations. The mutations listed prevent the involvement of the corresponding region in stem-loop formation. **TIP:** Tryptophan regulates operon transcription at two levels: transcription initiation and attenuation. *Deduce:* The 2–3 stem loop prevents termination but forms only when tryptophan levels are low, whereas the 3–4 stem loop induces termination but forms only when tryptophan levels are high. If no stem loops form, termination cannot occur. *Solve:* **(a)** Attenuation cannot occur because neither the antitermination (2–3) nor the termination (3–4) stem loop can form. **(b)** Attenuation cannot occur because the termination (3–4) stem loop cannot form. **(c)** Attenuation cannot occur because the attenuator is missing. **(d)** Attenuation cannot occur. The antitermination (2–3) stem loop will form exclusively since no ribosome will attach to the transcript. **(e)** An effect on attenuation is unlikely because an attached ribosome can cover region 1 and region 2. **(f)** Attenuation is likely to be affected. The antitermination (2–3) stem loop will form exclusively since an attached ribosome cannot span both region 1 and region 2. **(g)** There is no effect on attenuation since the mutation causes only the loop portion of the antitermination (2–3) stem loop to get larger. **(h)** A two-base insertion will shift the reading frame and will likely alter the two *trp* codons so translation will no longer pause in the absence of *trp*. Termination should occur frequently. **(i)** Since the ribosome is never binding to the RNA, loop 1 will always pair with 2 and 3, and loop 4 will lead to termination almost every time. **(j)** Attenuation will no longer occur. Without the poly-U region, the pairing of 3–4 will no longer form a rho-independent termination signal.

26a–d. *Evaluate:* This is type 2. This problem concerns the mechanism of *ara* operon regulation. Recall that araC protein plays a dual role in *ara* operon transcription: it acts as a repressor by binding at $araO_2$ and *araI* in the absence of arabinose, and it acts as a positive regulator by binding at *araI* in the presence of arabinose. *Deduce > Solve:* Binding of araC protein to *araI* and $araO_2$ is most critical to *ara* operon regulation. **(a)** *araI* mutants do not regulate *ara* operon expression normally and cannot transcribe the operon under any condition because araC protein binding at *araI* is required for transcription. **(b)** The $araO_1$ mutation may cause a defect in *ara* operon repression because transcription of araC will not be prevented. Increased production of araC protein could lead to binding of araC protein dimers at *araI* and the promotion of *ara* operon transcription. **(c)** *ara* operon repression will not occur because the DNA loop, which requires araC binding at both *araI* and $araO_2$, cannot form. **(d)** *ara* operon repression cannot occur because it requires araC protein binding at $araO_2$.

27.   *Evaluate:* This is type 2. This problem concerns the mechanism of bacteriophage λ gene regulation and the effect of P_{RE} mutations on the determination of lysis versus lysogeny. *Deduce > Solve:* Transcription from the λ phage promoter P_{RE} is activated to establish lysogeny. A mutation that decreases the promoter's transcription efficiency will inhibit lysogeny establishment. A mutation that increases transcription from the promoter will make it difficult to reverse lysogeny.

28a–f. *Evaluate:* This is type 2 and 3. This problem concerns the mechanism bacteriophage λ transcription and antitermination and the effect of mutations on the determination between lysis and lysogeny. *Deduce > Solve:* **(a)** *cI* codes the lambda repressor, which inhibits transcription from P_R and therefore promotes lysogeny. A *cI* mutation will, therefore, be incapable of establishing lysogeny. **(b)** *cII* is required, along with *cIII*, for activating transcription from P_L, which is required for lysogeny. *cII* mutants will, therefore, be incapable of establishing lysogeny. **(c)** *cro* is a repressor of P_{RM} transcription and a promoter of P_R transcription, which promotes the lytic life cycle. *cro* mutants will be unable to initiate transcription from P_R or enter the lytic life cycle. **(d)** *int* is required to catalyze the integration of the λ chromosome into the bacterial chromosome, which is required for lysogeny. Therefore, *int* mutants will be unable to establish lysogeny. **(e)** *cII* is required for transcription from P_L and lysogeny, whereas *cro* is required for transcription from P_R and lysis. The *cII cro* double mutant will be unable to carry out lysis due to the *cro* mutation, and it will be unable to establish lysogeny due to the *cII* mutation. **(f)** *N* is required for antitermination at terminators t_1, t_2, and t_3. *N* mutants cannot transcribe genes beyond these terminators and therefore will not be able to undergo lysogeny.

29a–b. *Evaluate:* This is type 3. This problem concerns the mechanism of *IS10* translation regulation by antisense RNA. *Deduce > Solve:* **(a)** Transcription from P_{OUT} produces antisense RNA, which inhibits translation of the transposase gene. Decreased synthesis of antisense RNA will result in increased expression of transposase and increased *IS10* transposition rates. The mutant will freely produce transposase and will very actively transpose. **(b)** Transcription from P_{IN} produces transposase mRNA, which is required for *IS10* transposition. Elimination of transcription from P_{IN} will prevent transposase expression and eliminate *IS10* transposition.

30a–g. *Evaluate:* This is type 2. This problem concerns the mechanism of *lac* operon regulation and northern blot analysis. *Deduce > Solve:* **(a)** *lac* transcription in a $I^+ P^+ O^C Z^+ Y^+$ strain is independent of lactose but is still under carbon catabolite repression; therefore, low levels of *lac* mRNA are present in glucose medium. **(b)** *lac* transcription is independent of *lacZ*; therefore, mRNA accumulation in a $I^+ P^+ O^+ Z^- Y^+$ strain is the same as for wild type. **(c)** *lac* transcription in a $I^+ P^- O^C Z^+ Y^+$ strain cannot occur regardless of conditions. **(d)** *lac* transcription in a $I^- P^+ O^C Z^+ Y^+$ strain is independent of lactose but is still under carbon catabolite repression; therefore, low levels of *lac* mRNA will be present in glucose medium. **(e)** *lac* transcription in a $I^+ P^+ O^C Z^- Y^-$ strain is independent of lactose but is still under carbon catabolite repression; therefore, low levels of *lac* mRNA are present in glucose medium. **(f)** No band in lane 1, and a

full-intensity band in lane 2, since promoter is fully functional with no reduction from O^C.
(g) *lac* transcription in a strain where CAP–cAMP cannot bind to the *lac* promoter is dependent on lactose but will be at low levels similar to those seen with wild type in medium where both lactose and glucose are present.

31a–b. *Evaluate:* This is type 2. This problem concerns the mechanism repression of transcription and DNA footprint analysis. *Deduce > Solve:* **(a)** The DNA sequence is determined by reading the DNA sequencing gel from the bottom to the top.

5′-TGATCGAAATACAGTTACACTTACAATCAAGATTC-3′

3′-ACTAGCTTTATGTCAATGTGAATGTTAGTTCTAAG-5′

(b) In DNase I footprint analysis, the region of DNA protected by a protein corresponds to the region on the gel where DNA bands that are seen in the naked DNA control are absent. When the repressor is bound to DNA, bands from region 20 to 31 are not present. When RNA polymerase is bound to DNA, bands from 5 through 23 are absent. Nucleotide positions 20 through 31 are protected by the repressor protein, and positions 5 through 23 are protected by RNA polymerase.

32a–e. *Evaluate:* This is type 2. This problem concerns the mechanism of *lac* operon regulation and requires the deduction of a partial diploid phenotype from its genotype. *Deduce > Solve:* **(a)** *lacI⁻* is recessive to *lacI⁺* (*lacI⁺* is trans-dominant); therefore, this partial diploid has a wild-type phenotype, which is inducible *lacZ* mRNA synthesis and a *lac⁺* phenotype. **(b)** *lacOᶜ* is cis-dominant and *lacZ⁺* is dominant to *lacZ⁻*; therefore, the partial diploid will be constitutive for *lacZ* mRNA synthesis and will be *lac⁺*. **(c)** *lacIˢ* is trans-dominant; therefore, this partial diploid will be uninducible for *lacZ* mRNA synthesis and will be *lac⁻*. **(d)** *lacP⁻* is cis-dominant (as is *lacP⁺*), and *lacZ⁺* is in cis to *lacP⁻*; therefore, this partial diploid will be inducible for *lacZ* mRNA synthesis but *lac⁻*. **(e)** Permease is required for import of lactose, which is required for synthesis of the inducer, allolactose; therefore, this partial diploid will be uninducible and *lac⁻*.

Genotype	*LacZ* mRNA Synthesis	*Lac* Phenotype
a. $\dfrac{I^- P^+ O^+ Z^+ Y^+}{I^+ P^+ O^+ Z^+ Y^+}$	inducible	*lac⁺*
b. $\dfrac{I^+ P^+ O^C Z^+ Y^+}{I^+ P^+ O^+ Z^- Y^+}$	constitutive	*lac⁺*
c. $\dfrac{I^S P^+ O^+ Z^+ Y^+}{I^+ P^+ O^+ Z^+ Y^+}$	uninducible	*lac⁻*
d. $\dfrac{I^+ P^+ O^+ Z^- Y^+}{I^+ P^- O^+ Z^+ Y^+}$	uninducible	*lac⁻*
e. $\dfrac{I^+ P^+ O^+ Z^+ Y^-}{I^+ P^+ O^+ Z^+ Y^-}$	uninducible	*lac⁻*

33.   *Evaluate:* This is type 1. This problem concerns the mechanism of *lac* operon regulation and the analysis of *lac* operon mutants. ***Deduce > Solve:*** *lacI⁻* and *lacO^C* lead to constitutive expression of β-galactosidase. The A^- and C^- mutants are constitutive for β-galactosidase; therefore, one is *lacI⁻* and the other is *lacO^C*. *lacO^C* is cis-dominant and *lacI⁻* is recessive. A^- is cis-dominant and C^- is recessive; therefore, A is *lacO* and C is *lacI*. This leaves B to be *lacZ*.

14.04 Test Yourself

Problems

1.   You have been asked to design the cis- and trans-acting regulatory regions of a bacterial operon that codes for enzymes required for the utilization of the amino acid proline as an energy source. The operon should be expressed only when needed, and the mechanism of operon regulation should be based on a model presented in Chapter 14 of the textbook. Provide one possible design.

2.   You have been asked to design the cis- and trans-acting regulatory regions of a bacterial operon that codes for enzymes required for the synthesis of vitamin B_6. The operon should be expressed only when needed, and the mechanism of operon regulation should be based on a model presented in Chapter 14 of the textbook. Provide one possible design.

3.   Explain how the term *allostery* could be applied to regulation of araC protein function.

4.   How does cooperative binding of cro protein and cI protein to O_{R1}, O_{R2}, and O_{R3} determine whether transcription is initiated predominantly from the lambda P_{RE} or P_{RM} promoters?

5.   Explain why one could consider a stalled ribosome at a *trpL* tryptophan codon and the lambda phage N protein analogous with respect to regulation of gene expression.

Solutions

1.   *Evaluate:* This is type 1. This problem concerns bacterial operon regulation and the biological relationship between the regulatory mechanism and the needs of a bacterial cell. **TIP:** An energy source is a molecule that can be imported into the cell and metabolized to yield ATP. ***Deduce > Solve:*** The operon should be transcribed only when proline is present in the environment. The cis-acting regulatory region should contain a promoter, an operator, and a CAP–cAMP binding site. The cell should express a trans-acting repressor that is allosterically regulated by proline: the repressor will bind to the operator unless it is bound to proline. The organization of these elements should be like the *lac* operon.

2.   *Evaluate:* This is type 1. This problem concerns bacterial operon regulation and the biological relationship between the regulatory mechanism and the needs of a bacterial cell. ***Deduce > Solve:*** The operon should be transcribed only when vitamin B_6 is not present in the environment. The cis-acting regulatory region should include a promoter and an operator. The cells should express a repressor that is allosterically regulated by vitamin B_6: the repressor will bind to the operator only when it is also bound to vitamin B_6. The organization of these elements should be like the *trp* operon.

3.   *Evaluate:* This is type 2. This problem concerns the regulation of araC protein function by binding to arabinose. ***Deduce > Solve:*** araC protein is capable of binding to three different molecules. araC binds to its operator. Operator binding is not under allosteric control. araC can also bind another araC protein to form a homodimer. araC binding to araC is allosterically regulated by arabinose. Binding to arabinose prevents araC binding to araC, which is an example of allosteric control.

4.   *Evaluate:* This is type 2. This problem concerns the mechanism of regulation of λ phage gene expression and the concept of cooperative binding. ***Deduce > Solve:*** The cro and cI proteins are both capable of binding to all three operator sequences; however, cro protein binds most tightly to O_{R3}, and its binding to O_{R3} promotes binding of a second cro protein to O_{R2} and then another to O_{R1}. Thus, binding of cro protein to O_{R3}, O_{R2}, and O_{R1} is cooperative. cro protein binding favors transcription from P_{RE} and expression of genes required for a lytic infection. cI protein binds most tightly to O_{R1}, and its binding to O_{R1} promotes binding of a second cI protein to O_{R2} and then another to O_{R3}. Thus, binding of cI protein to O_{R1}, O_{R2}, and O_{R3} is cooperative. This favors transcription from P_{RM} and expression of genes required for lysogeny.

5.   *Evaluate:* This is type 3. This problem concerns the regulation of gene expression by attenuation/antitermination and requires a comparison of the role of translation of the *trpL* peptide to the antitermination by the λ phage N protein. ***Deduce > Solve:*** Ribosome stalling at tryptophan codons in *trpL* drives formation of the antiterminator stem-loop structure, which prevents termination of transcription of the *trp* operon by RNA polymerase. Bacteriophage λ N protein binds to terminator structures that prevent expression of lytic and lysogenic genes. Both events prevent transcription termination and promote expression of a gene downstream of terminators, and therefore the events (i.e., proteins) involved are analogous.

15

Regulation of Gene Expression in Eukaryotes

Section 15.01 Genetics Problem-Solving Toolkit

Section 15.02 Types of Genetics Problems

Section 15.03 Solutions to End-of-Chapter Problems

Section 15.04 Test Yourself

15.01 Genetics Problem-Solving Toolkit

Key Terms and Concepts

Cis-acting regulatory sequence: A DNA sequence that affects transcription initiation by binding to proteins.

Trans-acting regulatory protein: A protein that affects transcription by binding to cis-acting regulatory sequences.

Enhancer: A cis-acting regulatory sequence that promotes transcription initiation at promoters in a distance and orientation independent manner and includes upstream activator sequences (UAS) in yeast.

Insulator: A cis-acting regulatory sequence that influences which promoters are subject to regulation by which enhancers.

Silencer: A cis-acting regulatory sequence that inhibits transcription initiation at promoters in a distance and orientation independent manner.

Transcription factor: A trans-acting regulator of transcription initiation that promotes or inhibits transcription initiation at promoters by one of many possible mechanisms.

RNA interference (RNAi): A general term describing multiple mechanisms that regulate gene expression through the action of small, noncoding RNAs.

Chromatin remodeling: Reorganization or restructuring of nucleosomes, usually for the purpose of activating or repressing transcription initiation.

Histone acetylation or deacetylation: Regulation of chromatin structure by the addition or removal of acetyl groups to histone amino–terminal amino acids within nucleosomes on chromatin.

CpG islands: Clusters of CpG dinucleotides that are often found in the promoter regions of genes and are subject to cytosine methylation.

DNA methylation: Methylation of cytosine bases and correlates with the formation of heterochromatin.

Key Analytical Tools

Levels of regulation of gene expression (see Figure 15.1, page 496)

15.02 Types of Genetics Problems

1. Describe the structure and function of cis-acting eukaryotic transcriptional regulatory elements.
2. Interpret the results from molecular or genetic analyses of cis-acting gene regulatory sequences.
3. Describe the structure and function of trans-acting eukaryotic regulators, and predict the effect of mutations in those regulators on the levels of gene expression.
4. Describe the role of RNAi in regulation of gene expression.

1. Describe the structure and function of cis-acting eukaryotic regulatory elements.

This type of problem requires understanding of the cis-acting regulators of gene expression. You may be asked to describe the structure and function of cis-acting elements, which include promoters, enhancers, silencers, insulators, and CpG islands. Answers could include descriptions of sequences or the location and orientation relative to the start of transcription. See Problem 2 for an example.

Example Problem: Compare and contrast the structure and function of silencers and insulator elements presented in Chapter 15 of the textbook.

Solution Strategies	Solution Steps
Evaluate	
1. Identify the topic this problem addresses, and explain the nature of the requested answer.	1. This problem tests your understanding of the structure and function of silencers and insulators. The answer will describe similarities and differences between silencers and insulators, with respect to the regulation of gene expression.
2. Identify the critical information given in the problem.	2. The problem refers you to the discussion of silencer and insulator sequences in Chapter 15 of the textbook.
Deduce	
3. Describe the structure and function of silencers.	3. Silencers are DNA sequences that bind proteins that negatively regulate gene transcription. Like enhancers, they can be located at a great distance from the genes they regulate, and they function in an orientation-independent manner. One possible mechanism they use is to interfere with the function of enhancers.
4. Describe the structure and function of insulator sequences.	4. Insulators are DNA sequences that bind proteins that influence expression of genes. Insulators may act to prevent the action of an enhancer on one gene while directing that enhancer's function to a different gene. They also may block the spread of heterochromatin, which would otherwise inhibit the transcription of nearby genes. The mechanism of insulator function is not clear, but it may include the formation of DNA loops that group together enhancers and the genes they regulate while isolating enhancers from genes that they should not regulate.

(continued)

Solution Strategies	Solution Steps
Solve	
5. Answer the problem by pointing out common and contrasting features of silencers and insulators.	5. **Answer:** Silencers and insulators are both DNA sequences that function by binding to proteins that affect gene transcription. Whereas silencer function is negative with respect to transcription of genes under its control, insulators may have positive or negative effects on gene expression. The role of insulators depends on the context of the insulator, the genes under its control, and the physiology of the cell.

2. Interpret the results from molecular or genetic analyses of cis-acting gene regulatory sequences.

This type of problem requires the interpretation of results from molecular or genetic analyses of cis-acting regulatory sequences. You may be asked to interpret the results from promoter deletion analyses or promoter–*lacZ* fusions to identify positive and negative cis-acting DNA sequences; to interpret the results of Southern blot analysis to infer the presence or absence of DNA methylation; or to interpret the results of northern or western analysis of gene expression to deduce the level at which gene expression is regulated. See Problem 18 for an example.

Example Problem: The figure shows a mammalian muscle protein gene that contains six exons (labeled 1 through 6) and two different promoters (P_H and P_S). P_H is used in heart muscle cells, and P_S is used in skeletal muscle cells. The figure also shows the locations of four different deletion mutations (labeled *A* through *D*) and the results of a northern blot analysis of heart and skeletal muscle in strains harboring wild type or each mutant allele. The probe for the northern blot analysis corresponds to exon 3. Use this figure to determine the function of regions A, B, C, and D.

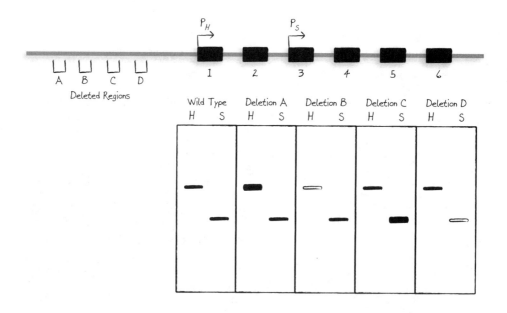

Solution Strategies	Solution Steps
Evaluate	
1. Identify the topic this problem addresses, and explain the nature of the requested answer.	1. This problem tests your understanding of deletion analysis of cis-acting regulatory sequences and ability to interpret the results of a northern blot. The answer will identify the deleted regions and neutral, positive-, or negative-acting regulatory sequences.
2. Identify the critical information given in the problem.	2. The muscle protein gene is expressed in heart and skeletal muscle. The heart muscle protein gene contains six exons, and the skeletal muscle protein gene contains four of the six exons. Transcript accumulation has been analyzed on northern blots for wild type and four mutants, each carrying a different deletion 5′ to the promoters. The results of the analysis are shown in the figure.
Deduce	
3. Compare the level of the heart muscle transcript in wild type to that in each mutant. TIP: The transcript level is indicated by the intensity of the signal (thickness of the band) on the northern blot.	3. The heart muscle transcript accumulates at wild-type levels in mutants C and D, at levels higher than wild type in mutant A, and at levels lower than wild type in mutant B.
4. Compare the level of the skeletal muscle transcript in wild type to that in each mutant.	4. The skeletal muscle transcript accumulates at wild-type levels in mutants A and B, at levels higher than wild type in mutant C, and at levels lower than wild type in mutant D.
5. Generalize the relationship between the effects of a deletion mutation on transcript levels to the function of deleted region.	5. Deletions that do not affect transcript levels are considered to have no role in transcription. Deletions that increase transcript levels are considered to contain negative-acting regulatory elements. Deletions that decrease transcript levels are considered to contain positive-acting regulatory elements.
6. Relate the position and function of each regulatory element identified in step 5 to the types of cis-acting elements described in the textbook.	6. The positive and negative regulatory elements act over a distance. Such positive-acting elements are called enhancers. Such negative-acting elements are called silencers.
Solve	
7. Answer the problem by naming the cis-acting regulatory elements controlling transcription of the muscle protein gene.	7. **Answer:** Region A contains a silencer of P_H transcription, B an enhancer of P_H transcription, C a silencer of P_S transcription, and D an enhancer of P_S transcription.

3. Describe the structure and function of trans-acting eukaryotic regulators, and predict the effect of mutations in those regulators on the levels of gene expression.

This type of problem concerns the function of regulatory proteins that enhance or repress transcription, including transcription factors, histones, chromatin-remodeling proteins, histone-modifying enzymes, and DNA-modifying enzymes. You may be asked to describe the regulatory elements or explain how they function to affect transcription, or to predict the effect of mutations in one or more of these regulators on gene expression. See Problem 4 for an example.

Example Problem: You wish to test the model presented in the textbook for regulation of the yeast galactose metabolism genes. You make loss-of-function mutations in the following regulatory elements: UAS_G, GAL4, GAL80, and GAL3. Predict the phenotype of each mutant by indicating whether the mutant can grow on galactose as the sole carbon source and whether gal mRNAs will accumulate in the presence or absence of galactose.

Solution Strategies	Solution Steps
Evaluate	
1. Identify the topic this problem addresses, and explain the nature of the requested answer.	1. This problem tests your understanding of the mechanism of yeast gal gene regulation and the analysis of gal regulatory mutants. The answer will be a description of each mutant's phenotype.
2. Identify the critical information given in the problem.	2. Four different yeast mutants are described: inactivation of UAS_G, GAL4, GAL80, and GAL3.
Deduce	
3. Recall the model for gal gene regulation presented in the textbook. TIP: See Figure 15.6.	3. gal gene transcription is activated by binding of Gal4 protein to UAS_G. Gal80 protein inhibits gal gene transcription by binding to Gal4, preventing Gal4 from binding to UAS_G. Gal3 protein activates gal gene transcription in the presence of galactose by binding to Gal80 protein, preventing Gal80 from binding to and inhibiting Gal4 function.
4. Consider the effect of each mutation on gal gene transcription.	4. Loss of UAS_G prevents Gal4 protein binding and activation of gal gene transcription. Loss of GAL4 function also prevents activation of gal gene transcription. Loss of GAL80 function causes Gal4 protein to bind constitutively to UAS_G and activate gal gene transcription regardless of galactose levels. Loss of GAL3 function causes Gal80 to bind constitutively to Gal4, which prevents gal gene transcription regardless of galactose levels.

(continued)

Solution Strategies	Solution Steps
Solve	
5. Answer the problem by describing the gal phenotype of each mutant and by drawing a table showing the effect of each mutant on *gal* mRNA accumulation in the presence or absence of galactose.	5. **Answer:** UAS_G, *GAL4*, and *GAL3* mutants will not be able to use galactose. *GAL80* mutants will be able to use galactose.

Strains	*Gal* gene mRNA Accumulation	
	Galactose Present	Galactose Absent
Wild type	+	–
UAS_G mutant	–	–
Gal4 mutant	–	–
Gal80 mutant	+	+
Gal3 mutant	–	–

4. Describe the role of RNAi in regulation of gene expression.

This type of problem concerns regulation of gene expression by mechanisms involving the production of small interfering RNAs (siRNAs) or microRNA (miRNAs). You may be asked to describe mechanisms that produce these RNAs or that use the RNAs to inhibit gene expression via destruction of mRNAs, inhibition of mRNA translation, or inhibition of transcription via heterochromatin. See Problem 2 for an example.

Example Problem: Compare and contrast the mechanisms presented in the textbook for the production and function of siRNAs and miRNAs in preventing mRNA translation.

Solution Strategies	Solution Steps
Evaluate	
1. Identify the topic this problem addresses, and explain the nature of the requested answer.	1. This problem concerns the posttranscriptional regulation of gene expression by RNAi. The answer will compare and contrast the mechanisms of siRNA and miRNA production and function.
2. Identify the critical information given in the problem.	2. The problem refers you to the discussion of RNAi in the textbook.
Deduce	
3. Describe the mechanism of siRNA production.	3. siRNAs are derived from double-stranded RNA, which originates from double-stranded RNA virus infection or from transcription of repetitive DNA in the nucleus. The double-stranded RNA is recognized and bound by Dicer, which cleaves the RNA into segments between 21 and 25 bp long. These siRNAs are bound by RISC, which denatures them and delivers one strand to the complementary region of target mRNAs. The RISC–siRNA–mRNA complex is recognized by nucleases that destroy the mRNA.

(*continued*)

Solution Strategies	Solution Steps
4. Describe the mechanism of miRNA production.	4. miRNAs are derived from transcription of nuclear genes and are processed to pre-miRNAs. pre-miRNAs contain complementary regions that can base-pair to form a hairpin RNA structure. The pre-miRNA is recognized and bound by Dicer, which cleaves the RNA into segments between 21 and 25 bp long. These miRNAs are bound by RISC, which denatures them and delivers one strand to partially complementary regions of target mRNAs. The RISC–miRNA–mRNA complex is unable to be translated and is sometimes destroyed by nucleases.
Solve 5. Answer the problem by pointing out similar and contrasting features of siRNA and miRNA production and function.	5. **Answer:** siRNAs and miRNAs are small, single-stranded RNAs that are produced through the function of Dicer and RISC complexes. siRNAs typically induce the destruction of complementary target mRNAs, whereas miRNAs typically inhibit translation of partially complementary mRNAs.

15.03 Solutions to End-of-Chapter Problems

1a–e. *Evaluate:* This is type 1. This problem concerns the structure and function of transcriptional regulatory elements and components of the RNAi pathway. *Deduce > Solve:* **(a)** A promoter is a DNA region whose sequences are recognized and bound by transcriptionally active proteins such as transcription factors and RNA polymerase. Promoters regulate the initiation of transcription of genes and contribute to controlling the timing, level, and cells in which transcription of particular genes occurs. **(b)** Enhancers are DNA sequences that bind regulatory proteins that interact with promoter-bound proteins to activate the transcription of specific genes. **(c)** Silencers are transcription-regulating DNA sequences that bind proteins, which inhibit the expression of genes, often by interfering with the function of enhancers. **(d)** RISC is the RNA-induced silencing complex. It is a protein complex that is part of the RNA interference (RNAi) mechanism. RISC denatures short double-stranded RNAs to single strands that carry out RNAi. **(e)** Dicer is an enzyme complex that is active in RNAi, where it cuts double-stranded regulatory RNAs into 21-bp to 26-bp segments that are subsequently denatured by RISC.

2a–e. *Evaluate:* This is type 1. This problem concerns the function of regulatory elements or processes that regulate gene expression. *Deduce > Solve:* **(a)** UAS elements are found in the yeast genome, where they operate as enhancer-like regulatory sequences. Gal4 protein binds yeast UAS elements to activate transcription of galactose utilization genes. **(b)** Insulator sequences shield genes from enhancer effects. The mechanism of action may be through the formation of specific DNA loops that protect particular genes from enhancers. **(c)** Silencer sequences prevent transcription of particular genes. The mechanism of action may be through competitive protein binding at silencer sequences that overlap with enhancer sequences. The yeast Mig1 and Tup1 proteins bind a silencer sequence during glycolysis to prevent transcription of galactose utilization genes. **(d)** The protein complexes that assemble at enhancers to facilitate transcription are known as enhanceosomes. The enhanceosome complex known as Mediator assembles at yeast enhancers. It contacts promoter-bound proteins to activate transcription. **(e)** RNA interference describes the posttranscriptional regulation of mRNAs by regulatory RNA molecules. Prevention of translation of the Dnmt3 DNA methyltransferase transcript by RNA interference leads to the development of fertile queens in honeybees.

3. *Evaluate:* This is type 3. This problem concerns chromatin remodeling and its role in regulation of transcription. Recall that transcription in eukaryotes is controlled by nucleosomes, whose positioning or structure can prevent access to DNA by RNA polymerase and transcription factors. *Deduce > Solve:* Chromatin remodeling describes several processes that alter chromatin structure by altering the positioning of nucleosomes with respect to specific DNA sequences or the composition of histones within nucleosomes. These alterations can be used to activate transcription of a gene by opening its enhancer or promoter sequences or to repress transcription of a gene by hiding its enhancer or promoter sequences.

4. *Evaluate:* This is type 3. This problem concerns the role of histone acetylation in regulation of gene expression. Recall that acetylation occurs when acetyl groups are added to amino acids of histone proteins by acetylase enzymes. *Deduce > Solve:* Histone acetylation events are most often associated with transcription activation, although there are many exceptions.

5. *Evaluate:* This is type 3. This problem concerns histone and DNA methylation and its role in transcription regulation. Recall that methylation occurs when methyl groups are added to amino acids of histone proteins or to nucleotides, most often cytosines that are part of CpG dinucleotides. *Deduce > Solve:* Histone amino acid methylation is most often associated with repression of transcription, although there are many exceptions. This is a general eukaryotic gene regulatory mechanism that operates in eukaryotes as diverse as yeast and mouse. Cytosine (CpG) methylation is associated with the inactivation of mammalian promoter activity, since CpG islands are common near these promoters. This mechanism of gene regulation is less universal than histone methylation. For example, it occurs in mammals, including mouse, but not in *C. elegans*.

6. *Evaluate:* This is type 3. This problem concerns the structure and function of sequence-specific DNA-binding transcription factors. *Deduce > Solve:* The retention of structural similarities among DNA-binding regulatory proteins is an evolutionary consequence of positive natural selection favoring the capability of these proteins to bind to DNA. The conserved structures, termed motifs, allow the side chains of amino acids to make contact with exposed chemical groups within the major and minor grooves of double-stranded DNA. The specificity of these interactions is determined by the chemical contacts between the amino acids in a particular motif and the nucleotides in a specific DNA sequence.

7. *Evaluate:* This is type 1. This problem concerns the role of CpG islands in regulation of transcription. Recall that CpG islands are clusters of repeated CG dinucleotides, which are often found near the 5′ promoters of genes. *Deduce > Solve:* The cytosines in CpG dinucleotides in mammalian genomes are frequent targets for cytosine methylation. Cytosine methylation correlates with closed chromatin structure and repression of transcription. A role of CpG islands would be to provide a DNA marker for identifying promoters that should be incorporated into repressive, closed chromatin.

8. *Evaluate:* This is type 1, 2, and 3. This problem concerns the multiple levels and molecular mechanisms employed by eukaryotic cells to regulate gene expression. *Deduce > Solve:* Several factors can be cited: (1) the presence of a nucleus in eukaryotic cells, which separate transcription from translation and therefore allow RNA processing and transport to be used in regulation; (2) the chromatin structure of eukaryotic genomes, which allows the differential packaging of genes to be used in regulation; (3) multicellularity and cell-type specificity that is frequent in eukaryotes and requires organisms to regulate large suites of genes in a cell-type-specific manner; and (4) posttranslational processing of proteins, which allows cells to regulate the function of proteins after their synthesis by covalent modification, transport, or both.

9. *Evaluate:* This is type 1, 2, and 3. This problem requires understanding of bacterial *lac* operon regulation and yeast *GAL* gene regulation. *Deduce > Solve:* Both galactose and lactose are alternative sugars that must be metabolized to yield glucose that can then enter the glycolytic pathway to provide energy for the organism. The enzymes catalyzing galactose and lactose metabolism are produced from genes whose transcription is activated only when these sugars are present and glucose has been depleted. Gene transcription from the bacterial *lac* operon is induced by the binding of the lactose metabolite, allolactose, to the repressor protein. This process opens the *lac* operator and promoter region for RNA polymerase and activates transcription. Induction of transcription of *GAL* genes by galactose is more indirect. The yeast *GAL* genes have their transcription activated by binding of gal4 protein to UAS elements affiliated with each gene. Gal4 binding is induced when it is released from Gal80, which is caused by the binding of Gal3 to Gal80, which occurs only when Gal3 is bound to galactose and cellular glucose has been depleted.

10. *Evaluate:* This is type 3. This problem concerns the effect of chromatin structure on access to DNA by non-chromatin proteins. *Deduce:* Heterochromatin is the most compact form of chromosome packaging and is thought to prevent access of all enzymes to DNA. For chromosomal regions that do not contain genes, the only enzymes that require access to DNA are replication and DNA repair enzymes. Replication enzymes require access during S phase, whereas DNA repair enzymes require access whenever DNA damage is detected. *Solve:* Heterochromatin regions will decondense for DNA replication during the S phase of the cell cycle to allow replisome access, and whenever DNA damage is detected to allow DNA repair enzymes access.

11. *Evaluate:* This is type 1. This problem concerns the structure and function of cis-acting elements that control eukaryotic gene transcription. *Deduce > Solve:* Promoters recognized by RNA polymerase II are located upstream and within a few dozen nucleotides of the genes they control, whereas enhancers can be located nearby or at great distances from the gene they regulate. Promoters have a specific orientation that controls initiation of transcription in a specific direction; therefore, each promoter's function is orientation dependent and the promoter must be either upstream or downstream of its gene. In contrast, enhancer function is orientation independent; therefore, enhancers can be upstream or downstream of genes they regulate.

12a–c. *Evaluate:* This is type 2. This problem concerns the analysis of deletion mutations in the cis-acting sequences regulating gene transcription. *Deduce > Solve:* **(a)** Enhancers are sequences located at a distance from the promoter, and they increase transcription of a gene above its basal level. Deletion of an enhancer will reduce transcription levels but typically will not eliminate expression. Mutants A and C both reduce expression levels; however, mutant C removes sequences close to the transcription start site. Thus, mutant A best fits the expectation for deletion of an enhancer. Mutant A has an enhancer mutation. This deletion is located well upstream of the start of transcription and substantially reduces transcription. **(b)** Silencers are sequences located at a distance from the promoter, and they decrease transcription of a gene. Deletion of a silencer will increase transcription over that observed in wild type. Mutant B is the only mutation that increases expression over wild-type levels. Mutant B affects a silencer sequence. This deletion results in a substantial increase in the level of transcription. **(c)** Promoters are sequences located (typically) just upstream of the transcription start site and are required for full-level transcription. Deletion of a promoter or part of a promoter will prevent or reduce transcription levels. Mutants C and D affect sequences proximal to the transcription start site, and both reduce expression levels. Both mutants C and D are promoter mutations. Their location immediately upstream of the transcription start and the reduced levels of transcription from these mutants are consistent with promoter mutations.

13a–c. *Evaluate:* This is type 2. This problem concerns the analysis of deletion mutations in the cis-acting sequences regulating gene transcription. Recall that not all sequences within a promoter are important for function and that different regions of a promoter may be important in different cell types. *Deduce > Solve:* **(a)** The deletion in mutant F has no effect on *UG4* expression in either tissue, indicating that this region contains no sequences required for promoter action in those cells. Mutant B only mildly affects *UG4* transcription in leaves, but knocks it out almost entirely in stems. This indicates the use of different promoter sequences in the transcription of *UG4* in these tissues. **(b)** Mutation D results in greater than wild-type level expression in leaves but not stems. Mutations that increase expression typically remove silencers; therefore, mutant D lacks a silencer sequence that regulates the level of transcription of *UG4* in leaf tissue but not in stem tissue. **(c)** Mutation E eliminates a region distal to the transcription start site and prevents expression in leaves but has no effect in stems. Mutations in regions at a distance from the transcription start site that prevent expression typically affect an enhancer; therefore, mutation E deletes a required enhancer sequence for *UG4* transcription in leaf tissue but not in stem tissue.

14a–c. *Evaluate:* This is type 2. This problem concerns the analysis of promoter–*lacZ* fusion experiments. Recall that the data is analyzed by comparing the results from truncated or deleted promoters to the full-length promoter–lacZ fusion control. *Deduce:* The fusion DNA constructs divide the regulatory sequences into eight regions, as shown in the figure. All constructs that contain region 1 are expressed at some level, whereas both constructs that lack this region are not expressed; therefore, this region likely contains promoter sequences required

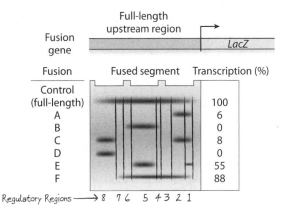

for transcription. Constructs A and C produce only low-level expression as compared with control and constructs E and F; therefore, A and C lack one or more positive-acting elements, probably enhancers. This finding locates enhancers to regions 3, 4, 5, 6, or 7 and indicates that region 2 contains no enhancer. Constructs E and F produce expression above basal level (although still less than the control); therefore, they contain one or more enhancers. This finding locates at least one enhancer to region 5, which is present in both E and F. The difference in expression levels between constructs E and F may indicate that F has additional enhancers missing in A, C, and E (regions 3, 6, or 7) or that the structure of the promoter in F better facilitates the action of the enhancer in region 5. *Solve:* **(a)** An enhancer is in region 5, and additional enhancers may be in regions 3, 6, or 7. **(b)** Region 1 contains the promoter. **(c)** E and F differ either because F contains enhancers not present in E or because the location of the enhancer that is present in both E and F functions better in F, which is likely due to its more distant location from the promoter.

15a–b. *Evaluate:* This is type 1. This problem concerns structure and function of genes and their regulatory sequences. *Deduce > Solve:* **(a)** The parents must be heterozygous for the mutant allele because they are carriers for the mutation. The wild-type mRNA is 1250 nt long; therefore, the mutant allele encodes a 1020-nucleotide-long mRNA. Thus the genotype of I-1, I-2, and II-2 is 1250/1020; the genotype of II-1 and II-3 is 1020/1020; and the genotype of II-4 is 1250/1250. I-1, I-2, and II-2 are 1250/1020; II-1 and II-3 are 1020/1020; II-4 is 1250/1250. **(b)** The mutant allele encodes a 1020-nt mRNA, which is exactly 230 nt shorter than wild type. Exon 3 is 230 nt long. The absence of exon 3 sequences from the mutant mRNA is the most likely explanation of the results, indicating that the mutation affects pre-mRNA splicing such that exon 2 is spliced to exon 4, skipping exon 3. The mutant allele is most likely a deletion of exon 3.

16a–b. *Evaluate:* This is type 1. This problem concerns the structure and function of cis-acting regulatory sequences. Recall that CpG dinucleotides are often methylated on cytosines. *Deduce > Solve:* **(a)** The cytosines in each of the CpG dinucleotides of the restriction sequences are the most likely methylation targets. **(b)** Recall that the restriction enzyme *Msp*I cuts the sequence CCGG regardless of whether the middle C is methylated, and that *Hpa*II cuts the sequence CCGG only if the middle C is not methylated.

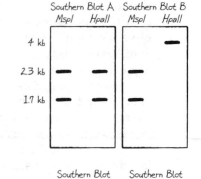

17a. *Evaluate:* This is type 2. This problem concerns the structure and function of genes and their regulatory elements as investigated by Southern blot analysis. *Deduce: Hind*III will produce two bands from this muscle gene region. Probe *a* will hybridize to the shorter band, and probe *b* will hybridize to the longer band. Heart and skeletal muscle are identical with respect to the genes; therefore, the Southern blots will be identical. *Solve:* See figure.

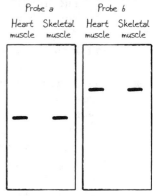

17b. *Evaluate:* This is type 2. This problem concerns the structure and function of genes and their regulatory elements as investigated by northern blot analysis. *Deduce:* Heart muscle cells will use P_H and produce an mRNA that contains all six exons. Skeletal muscle cells will use P_S and produce an mRNA that contains exons 3–6. Transcription from P_H will produce a longer mRNA than transcription from P_S. Probe *a* will hybridize only to transcripts from P_H, whereas probe *b* will hybridize to transcripts from either P_H or P_S. *Solve:* See figure.

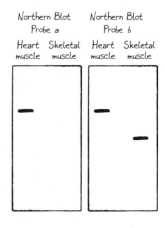

17c. *Evaluate:* This is type 2. This problem concerns the structure and function of genes and their regulatory elements as investigated by western blot analysis. *Deduce:* Heart muscle cells will use P_H and produce mRNAs that contain all six exon coding sequences, whereas skeletal muscle cells will use P_S and produce mRNAs that contain exons 3–6. Heart muscle cells will synthesize a larger protein than skeletal muscle cells. *Solve:* See figure.

18a–c. *Evaluate:* This is type 2. This problem concerns the analysis of promoter–*lacZ* fusion experiments. Recall that the data is analyzed by comparing the results from truncated or deleted promoters to the full-length promoter–*lacZ* fusion control. *Deduce:* The fusion DNA constructs divide the regulatory sequences into nine regions as shown in the figure. All constructs that contain region 1 are expressed at some level; therefore, region 1 contains the *ME1* promoter. Constructs C and F are expressed at basal levels; therefore, they are missing an enhancer element located in region 6, which is the only region deleted in C and F but not in A, B, or E. Construct E is expressed at higher-than-control levels, therefore, region 9, which is deleted in E but not F, contains a silencer. *Solve:* **(a)** The enhancer is in region 6, and the silencer is in region 9. **(b)** The deletion in construct D removes the promoter; therefore, there is no *lacZ* expression by this construct. **(c)** Expression of *ME1* is regulated by an enhancer and a silencer, which can be regulated such that *ME1* expression is activated at some developmental stages and then repressed at others.

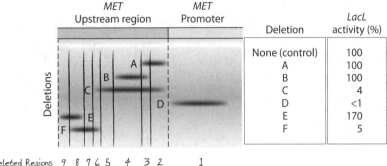

15.04 Test Yourself

Problems

1. The promoters of Gene *X* and Gene *Y* are unlinked genes containing CpG islands that are required for repression of their expression of lung tissue. In a sample of lung tissue from a cancer patient, Gene *X* and Gene *Y* are both expressed, even though both contain wild-type CpG islands. Explain how a mutation in a trans-acting regulator of gene expression could account for the expression of Gene *X* and Gene *Y* in these cells.

2. It has been proposed that Gene *Z*, which is expressed at high levels in lung cancer but not in normal lung tissue, is normally regulated by a histone acetylation and deacetylation. To test this hypothesis, you isolate fragments of chromosome containing Gene *Z* from cancer cells and normal cells and analyze whether histone H4 is acetylated in the DNA fragments. Describe the expected results if the hypothesis is correct.

3. You are studying the promoter of a eukaryotic gene. Cells that express this gene are thought to lack nucleosomes in this region, whereas cells that do not express the gene are thought to have a single nucleosome that covers the middle portion of the region. The gene is expressed in adult kidney cells and embryonic heart cells but not in adult heart, adult or embryonic thymus, or embryonic kidney cells. Draw a gel representing the results of a DNase I footprint analysis of the promoter from cells that express the gene and those that do not. Include a naked DNA control lane.

4. Gene *K*, which is expressed only in adult kidney cells, is controlled by an enhancer element called EE1 located 50 kb 5′ to the gene's promoter. A transcription factor called TF1 binds to this enhancer and activates the gene. A mutant that does not express TF1 has recently been isolated and has been found not to express gene *K*. Further analysis determines that the TF1 mutant also fails to express a number of additional adult kidney cell genes, all of which are located within a few hundred kilobases of EE1. Use this information to generate a hypothesis for how TF1 regulates Gene *K* and the other kidney cell genes that are dependent on TF1 for expression.

5.   You are studying the regulation of gene *S* in yeast. Analysis of the chromatin at the gene *S* promoter indicates that nucleosomes remain positioned at the promoter when it is in its open conformation. Generate a hypothesis for regulation of gene *S* by chromatin remodeling that is consistent with this result, and describe one way to test the hypothesis.

6.   You have transformed mouse with two promoter–*lacZ* fusion constructs to study the cis-acting regulatory sequences of a gene expressed only in adult kidney cells. Construct A contains 20 kb 5´ to the transcription start site, whereas construct B contains only 5 kb. Six mouse transformants were created with construct A and six with construct B. Each transformant contains one construct integrated into a different mouse chromosome. All six mice transformed with construct A expressed *lacZ* in adult kidney cells. Each mouse transformed with construct B expressed *lacZ* in different tissue at different developmental stages, and only one expressed *lacZ* in adult kidney cells. The other transformants expressed *lacZ* in embryonic heart cells, adult heart cells, embryonic thymus cells, adult thymus cells, or embryonic kidney cells. **(a)** Explain why construct A is expressed in adult kidney cells in all six transformants. **(b)** Explain why construct B is not expressed in adult kidney cells in all six transformants, and propose a hypothesis to explain why each transformant expresses *lacZ* in a specific cell type and developmental stage.

Solutions

1.   *Evaluate:* This is type 1 and 3. This problem concerns the regulation of gene transcription by CpG islands and the role of trans-acting factors in CpG island–dependent control. *Deduce:* CpG islands are clusters of CpG dinucleotides that are often near promoters of genes and can be methylated on cytosines by DNA methylase. Since DNA methylation correlates with the formation of transcriptionally repressive heterochromatin, loss of DNA methylase activity could have led to less de-repression of Gene *X* and Gene *Y* transcription. *Solve:* The lung cancer cells do not express a cytosine DNA methylase.

2.   *Evaluate:* This is type 3. This problem concerns regulation of gene transcription by histone acetylation and deacetylation. *Deduce > Solve:* If histone Gene *Z* is regulated by histone acetylation and deacetylation, then one would expect the expressed version of the gene to contain higher levels of acetylated histone H4 than the repressed version. In the proposed experiment, normal cells should contain lower levels of acetylated histone H4 levels than cancer cells.

3.   *Evaluate:* This is type 2. This problem concerns the analysis of promoters using DNase I footprinting. *Deduce > Solve:* DNase I cleaves DNA at sites that are not protected by proteins such as histones. The naked DNA will be cut at every site, resulting in a ladder of bands covering the entire region. The promoter region in cells that express the gene will have the same bands corresponding to DNase I cutting at all positions, whereas the promoters in cells that do not express the gene will lack bands from the region in the middle protected by the single nucleosome.

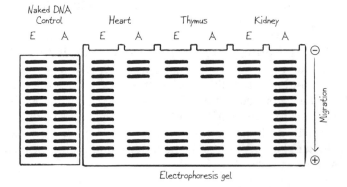

Electrophoresis gel

4.   *Evaluate:* This is type 2. This problem concerns the function of enhancer sequences. *Deduce > Solve:* TF1 is required for transcription of gene *K*, which is regulated by the enhancer EE1. Since transcription factor binding to enhancers is a known mechanism for regulation of gene transcription, it is reasonable to hypothesize that TF1 regulates gene *K* by binding to EE1. Since a number of other genes in the vicinity of EE1 also require TF1 for expression, it is reasonable to suggest that TF1 binding to EE1 turns on expression of the entire set of kidney cell genes identified in the problem.

5.   *Evaluate:* This is type 3. This problem concerns the regulation of transcription by chromatin remodeling complexes. *Deduce > Solve:* Chromatin remodeling complexes act by altering nucleosome structure at promoters such that RNA polymerase and associated transcription factors can access the promoter. They do this by causing the loss or repositioning of nucleosomes at promoters or by catalyzing the exchange of normal histone proteins with variant histones. Since the open (accessible) conformation of the gene *S* promoter contains nucleosomes, a reasonable hypothesis is that the chromatin remodeling complex regulates gene *S* by replacing a normal histone with a histone variant (for example, H2A with H2AZ). One test of this hypothesis would be to isolate fragments of chromatin containing the promoter and test for the presence of H2AZ.

6a–b.   *Evaluate:* This is type 2. This problem concerns the regulation of transcription by cis-acting elements and the analysis of results of promoter–*lacZ* fusion constructs. *Deduce > Solve:* The expression of each promoter–*lacZ* construct will be determined in part by the cis elements present in the construct and in part by the chromosomal elements near the site of integration of the construct. Construct A contains all the elements required for expression in its proper tissue and at the proper developmental stage. This likely includes an enhancer that directs transcription in adult kidney cells and silencers or insulators that prevent nearby chromosomal enhancers from promoting its transcription in other tissue or at other developmental stages. **(b)** Construct B does not contain all the necessary sequences to determine its proper tissue-specific and developmental-specific expression. One hypothesis that explains the pattern of gene expression in each construct B mouse transformant is that *lacZ* expression is determined by chromosomal regulatory elements located near the site of integration of the construct. For example, the transformant that expressed *lacZ* in embryonic heart cells would have construct B inserted near cis-acting elements that direct embryonic heart-specific gene expression.

16

Forward Genetics and Recombinant DNA Technology

16.01 Genetics Problem-Solving Toolkit

Key Terms and Concepts

Cloning by complementation: The identification of DNA clones carrying specific genes by determining which clones can complement the phenotype of a specific mutant.

Positional cloning: The identification of DNA clones carrying specific genes by combining genetic and physical mapping data to correlate the position of a mutation with a specific DNA clone.

DNA library: A collection of recombinant DNA clones that represent all or part of the genome of an organism. The DNA library can be constructed with genomic DNA (genomic library) or with cDNA (cDNA library).

Sanger sequencing (dideoxy sequencing): Technique for determining the DNA sequence of cloned DNA that depends on termination of DNA synthesis by incorporation of dideoxynucleotides.

Massively parallel (high throughput) sequencing: "Next generation" sequencing technology that does not require DNA cloning and that determines DNA sequences by monitoring synthesis.

Genetic screen: A procedure designed to identify a specific class of mutants from among a large collection of random mutants.

Key Analytical Tools

The probability that a specific sequence will occur at any site in a DNA molecule is $\left(\frac{1}{4}\right)^N$, where N is the number of bases in the sequence.

The size of a PCR primer that would hybridize to only one site in a genome is given by N using the equation $N = \dfrac{\log\left(\frac{1}{\text{genome size in bp}}\right)}{\log\left(\frac{1}{4}\right)}$, assuming 50% GC content.

16.02 Types of Genetics Problems

1. Design strategies for creating recombinant DNA clones.
2. Analyze DNA using restriction enzymes and PCR.
3. Describe how to sequence DNA or analyze DNA sequences.
4. Describe how to conduct genetic screens for specific mutant phenotypes.

1. Design strategies for creating recombinant DNA clones.

This type of problem requires understanding of recombinant DNA techniques, including the use of restriction enzymes, DNA ligase, plasmid cloning vectors, and PCR. You may be asked to design a strategy to clone a genomic or cDNA molecule into a plasmid vector, either by simply inserting the DNA molecule into the plasmid or by cloning the DNA molecule in a specific orientation using two different restriction enzymes. You may also be asked to design strategies to identify specific DNA clones from among libraries of recombinant DNA molecules using either homology (molecular hybridization) or functional complementation of a mutant phenotype. See Problem 12 for an example.

Example Problem: You are using pUC18 to construct a genomic DNA library using yeast DNA. You have cut the yeast DNA and pUC18 with *Bam*HI, purified the digested DNA to remove all restriction enzymes, mixed the DNAs together in a ligation reaction, and then transformed a *lacZ*⁻ strain of *E. coli*. You plated the same number of transformants on three different petri plates containing medium made at different times by different technicians during the previous week. Plate one contains 1000 colonies, 50% white and 50% blue. Plate two contains about a million colonies, most of them white. Plate 3 contains 1000 colonies, all of them white. You suspect that plate 1 contains medium that was properly prepared and that a different component of the medium was left out in plates 2 and 3. What media components were not added to the improperly prepared media?

Solution Strategies	Solution Steps
Evaluate	
1. Identify the topic this problem addresses, and explain the nature of the requested answer.	1. This problem concerns the methods used to select for and identify recombinant DNA clones made using the plasmid pUC18. The answer will describe the media components that are needed for proper selection and identification of recombinant DNA clones.
2. Identify the critical information given in the problem.	2. A yeast genomic DNA library was constructed using pUC18 and plated on three different media, one that gives expected results (1000 colonies, half white and half blue), and the other two that give unexpected results (1000 colonies that are all white, and a million colonies that are mostly white).

(continued)

Solution Strategies	Solution Steps
Deduce	
3. List the components in bacterial media that are needed for selection and identification of recombinant clones, and state the function of each component.	3. Ampicillin and X-gal are two critical components needed for selection and identification of recombinant DNA clones. Ampicillin selects for growth of bacteria that contain pUC18, which carries the ampicillin resistance gene. X-gal is converted to a blue pigment by a reaction catalyzed by β-galactosidase. β-galactosidase is encoded by the *lacZ* gene, which contains the *Bam*HI site that was used to insert genomic DNA fragments. Since the *E. coli* strain used lacks a functional *lacZ* gene, insertional inactivation of *lacZ* allows for identification of recombinant DNA clones because they cannot produce β-galactosidase, and therefore are white. Transformants that carry pUC18 without an insert in the *lacZ* gene are blue.
4. Describe the consequence of plating transformants on medium that lacks one of the critical components described above.	4. Medium that lacks X-gal will have normal numbers of transformants on it; but all transformants will be white, making identification of the recombinant clones more difficult. Medium that lacks ampicillin will allow every bacterium placed on the plate to grow into a colony, whether the bacterium carries pUC18 DNA or not. This makes identification of transformants difficult, if not impossible.
Solve	
5. Answer the problem by indicating which medium lacked which component.	5. **Answer:** Plate 2 results are as expected if the medium lacked ampicillin. Plate 3 results are as expected if the medium lacked X-gal.

2. Analyze DNA using restriction enzymes and PCR.

This type of problem requires knowledge and understanding of the use of restriction enzymes and PCR in the analysis of DNA. You may be asked to estimate the frequency of occurrence of restriction enzyme sites or to use restriction enzyme digests to map the location of restriction enzymes sites in a chromosome or recombinant DNA molecule. You may also be asked to design PCR primers that would be specific to one genetic locus within an entire genome. See Problem 14 for an example.

Example Problem: *Sau*3A recognizes the 4-bp sequence 5'-GATC-3' and cuts before the first G. *Bam*HI recognizes the sequence 5'-GGATCC-3' and cuts between the G's. Genomic DNA is cut with *Sau*3A under conditions where only one in every 20 *Sau*3A sites is cut. The fragments are mixed with *Bam*HI and cut pUC18 and ligated. There are 128 recombinant plasmids recovered and analyzed. What is the expected average size of the genomic DNA inserts in these clones, and how many of the clones are expected to originate from genomic DNA inserts with *Bam*HI sites at both ends?

Solution Strategies	Solution Steps
Evaluate	
1. Identify the topic this problem addresses, and explain the nature of the requested answer.	1. This problem concerns the probability calculations of the frequency of *Sau*3A restriction enzyme recognition sites and the relationship between *Sau*3A sites and *Bam*HI sites. The answers will include the expected insert size of the specified recombinant clones and the number *Sau*3A inserts that are also *Bam*HI fragments.
2. Identify the critical information given in the problem.	2. Genomic DNA was cut at one out of every 20 *Sau*3A sites, and the fragments were cloned into the *Bam*HI site of pUC18. There were 128 recombinant clones analyzed.
Deduce	
3. Derive an equation that calculates the expected frequency of *Sau*3A site digestion under the conditions described in the problem. Use this equation to calculate the average size of *Sau*3A fragments. TIP: The expected frequency of any sequence is $\left(\frac{1}{4}\right)^N$, where N is the number of base pairs in the target sequence. The average size of restriction fragment is $\frac{1}{\text{frequency of cutting}}$.	3. The expected frequency of *Sau*3A sites in random DNA is $\left(\frac{1}{4}\right)^4$. If one out of every 20 sites is cut, this frequency is $\left(\frac{1}{4}\right)^4 \times \frac{1}{20}$. The average size of *Sau*3A fragment is $\frac{1}{\left(\frac{1}{4}\right)^4 \times \frac{1}{20}}$.
4. Derive an equation that calculates the fraction of *Sau*3A sites that are also *Bam*HI sites in the target DNA.	4. *Sau*3A cuts at GATC. For a *Sau*3A site to also be a *Bam*HI site, the nucleotide before the G must be a G (probability is $\frac{1}{4}$), and the nucleotide after the C must be a C (probability is $\frac{1}{4}$).
Solve	
5. Calculate the average size of genomic DNA if *Sau*3A cut one in every 20 sites. TIP: The average size of a restriction fragment is given by $\frac{1}{\text{frequency of the digestion}}$.	5. The frequency of digestion is $\left(\frac{1}{4}\right)^4 \times \frac{1}{20} = \frac{1}{5120 \text{ bp}}$. The average-size fragment is 5120 bp.
6. Calculate how many of the 128 clones have *Bam*HI sites at both ends.	6. The number of clones with inserts that came from *Bam*HI sites is $\left(\frac{1}{4}\right)^4 \times 128 = 8$.
7. Answer the problem, presenting the information from your calculations in the context of the requested answer.	7. **Answer:** The average size insert among the clones is expected to be 5120 bp. The number of clones that originated from sequences with *Bam*HI sites at both is expected to be 8.

3. Describe how to sequence DNA or analyze DNA sequences.

This type of problem requires understanding of Sanger sequencing and massively parallel sequencing technologies. The problems may ask you to describe the sequencing strategies or to interpret the results from sequence analysis. You may also be asked to predict the DNA banding pattern from DNA sequencing reactions that are run on a gel. See Problem 20 for an example.

Example Problem: You have cloned a cDNA of a human gene into the multiple cloning site (MCS) of pUC18 and analyzed both ends of the clone by Sanger sequencing. The results from the sequence analysis using the M13 forward and reverse primers are shown in the figure. Which sequence represents the 5′ end of the cDNA, and which one represents the 3′ end?

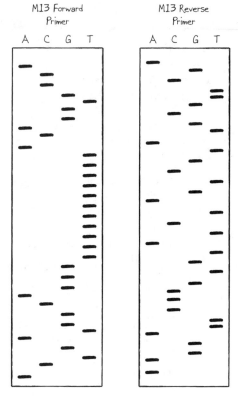

Solution Strategies	Solution Steps
Evaluate	
1. Identify the topic this problem addresses, and explain the nature of the requested answer.	1. This problem concerns the analysis of DNA sequencing data and identification of the sequence corresponding to the 3′ end of the operon.
2. Identify the critical information given in the problem.	2. The results of Sanger sequencing of a cDNA clone using M13 forward and reverse primers is shown.
Deduce	
3. Deduce the sequence of each end of the cDNA clone. TIP: The sequence reads 5′ to 3′ starting at the bottom of the gel.	3. The sequence derived from the M13 forward primer from 5′ to 3′ is ACTGATGGCAGGGTTTTTTTTTTTACAGCTGCCA. The sequence derived from the M13 reverse primer from 5′ to 3′ is AAGGATTCCCGTGTATCTAGTCGTATGCGTTCGA.
4. Determine which end contains the sequence corresponding to the poly-A sequence at the 3′ end of the mRNA.	4. The sequence derived from the M13 forward primer contains the sequence corresponding to the mRNA poly-A tail.

(continued)

Solution Strategies	Solution Steps
Solve	
5. Answer the problem by stating which sequence corresponds to the cDNA 5′ end and which corresponds to the cDNA 3′ end.	5. **Answer:** The T7-primer-derived sequence corresponds to the cDNA 3′ end, and the T3-primer-derived sequence corresponds to the cDNA 5′ end.

4. Describe how to conduct genetic screens for specific mutant phenotypes.

This type of problem tests your understanding of the genotype–phenotype relationship and how it can be used to identify mutants that affect specific biological processes. You should be familiar with methodology associated with the model organisms, *Saccharomyces cerevisiae* (yeast), *Drosophila*, and *Arabidopsis*. See Problem 22 for an example.

Example Problem: Exposure to UV irradiation correlates with increased rates of skin cancer, which is caused by UV-induced mutations. To understand how cells combat UV-induced DNA damage, you wish to identify yeast genes required for UV-induced DNA damage repair. Describe a genetic screen using yeast that would identify UV-induced DNA damage repair mutants, which is based on detecting an increase in the frequency of UV-induced cell death.

Solution Strategies	Solution Steps
Evaluate	
1. Identify the topic this problem addresses, and explain the nature of the requested answer.	1. This problem tests your understanding of how to use a genetic screen to identify yeast mutants that are defective in repair of UV-induced DNA damage. The answer will be a procedure that could be followed to identify such mutants.
2. Identify the critical information given in the problem.	2. DNA damage caused by UV irradiation causes an increase in mutation rate and cell death. Mutants defective in repair of UV-induced DNA damage will die at exposure levels that do not kill wild type.
Deduce	
3. Describe how you could measure UV-induced cell death. TIP: Yeast cell viability is measured by the ability of yeast cells to form colonies on agar medium.	3. Yeast cells would be plated on agar medium, which would then be exposed to UV irradiation. The ability of yeast to form colonies after irradiation would be a measure of viability.
4. Describe a method to expose yeast cells to UV light under conditions that should detect mutants with increased sensitivity to UV light. TIP: Determine the amount of UV exposure that results in a slight decrease in the viability of wild-type yeast. This will facilitate detection of mutants that are hypersensitive to UV light.	4. Colonies of yeast would be replica plated onto multiple petri plates, which would then be exposed to different levels of UV irradiation by controlling the time of UV exposure. The time of exposure that causes a slight decrease in the number of wild-type yeast colonies would be chosen as the UV dose for the genetic screen.

(continued)

Solution Strategies	Solution Steps
Solve	
5. Answer the problem by describing a genetic screen using the procedures described above.	5. **Answer:** Mutagenize yeast cells with a chemical mutagen (for example, EMS) and allow the survivors of mutagenesis to form colonies on agar medium. Replica plate the colonies to two plates and expose one of them to UV irradiation using the dose that does not kill wild type. Colonies that form on the non-irradiated plate but not on the irradiated plate are candidates for mutants with defects in repair of UV-induced DNA damage.

16.03 Solutions to End-of-Chapter Problems

1.   *Evaluate:* This is type 1. This problem concerns the function of elements in the pUC18 plasmid cloning vector. *Deduce > Solve:* The product of the *β-lactamase* gene (β-lactamase) provides resistance to ampicillin, which allows for selection of bacterial carrying pUC18 on medium containing ampicillin. The product of the *lacZ* gene (β-galactosidase) catalyzes the reaction that converts colorless X-gal into a blue pigment. Bacteria containing recombinant pUC18 plasmids are visually identified as white colonies on X-gal–containing medium, because they do not express β-galactosidase.

2a.   *Evaluate:* This is type 2. This problem requires calculation of restriction enzyme site frequencies. **TIP:** When not specified by the problem, assume that the sequence content of DNA is 50% GC and 50% AT. *Deduce:* The number of fragments expected from digestion of the DNA with a restriction enzyme is given by the equation genome size (bp) × restriction site frequency. The frequency of specific sequences in DNA that is 50% GC is given by the equation $\left(\frac{1}{4}\right)^N$, where N is the number of bases in the site. *Solve:* For *Sau*3A, there would be $(3 \times 10^9) \left(\frac{1}{4}\right)^4 = 1.17 \times 10^7$ DNA fragments; for *Bam*HI and *Eco*RI, there would be $(3 \times 10^9) \left(\frac{1}{4}\right)^6 = 7.32 \times 10^5$ fragments; and for *Not*I, there would be $(3 \times 10^9) \left(\frac{1}{8}\right)^4 = 4.58 \times 10^4$ fragments.

2b.   *Evaluate:* This is type 2. This problem requires calculation of restriction enzyme site frequencies. **TIP:** When the DNA sequence content is 40% GC, the frequency of a restriction site is given by the equation $\left(\frac{2}{10}\right)^{GC} \times \left(\frac{3}{10}\right)^{AT}$, where GC is the number of GC base pairs and AT is the number of AT base pairs. *Deduce:* The number of fragments expected from digestion of the DNA with a restriction enzyme is given by the equation genome size (bp) × restriction site frequency. *Solve:* For *Sau*3A, there would be $(3 \times 10^9) \left(\frac{2}{10}\right)^2 \times \left(\frac{3}{10}\right)^2 = 1.17 \times 10^7$ DNA fragments; for *Bam*HI, there would be $(3 \times 10^9) \left(\frac{2}{10}\right)^4 \times \left(\frac{3}{10}\right)^2 = 4.32 \times 10^5$ DNA fragments; for *Eco*RI, there would be $(3 \times 10^9) \left(\frac{2}{10}\right)^2 \times \left(\frac{3}{10}\right)^4 = 9.72 \times 10^5$ DNA fragments; and for *Not*I, there would be $(3 \times 10^9) \left(\frac{2}{10}\right)^8 = 4.58 \times 10^4$ DNA fragments.

3.   *Evaluate:* This is type 1. This problem concerns the strategy of cloning DNA using plasmids. *Deduce > Solve:* Treating the plasmid vector with a phosphatase after restriction digestion and before ligation would prevent the ends of the plasmids from being ligated because the 5′ phosphates are required for the ligation reaction. This strategy is used to prevent the formation of nonrecombinant plasmids when attempting to insert DNA into a plasmid vector digested with a single restriction enzyme.

4a.   *Evaluate:* This is type 1. This problem concerns the sequence content of genomic and cDNA libraries. Recall that cDNA libraries contain DNA copies of mRNAs expressed in the sample tissue, whereas genomic libraries contain DNA representing the entire genome regardless of the sample tissue type. *Deduce > Solve:* Since cDNA represents only sequences expressed as mRNA and genomic DNA contains the entire genome, the two genomic libraries would have greater sequence diversity than either cDNA library.

4b. *Evaluate:* This is type 1. This problem concerns the sequence content of genomic and cDNA libraries. Recall that cDNA libraries contain DNA copies of mRNAs expressed in the sample tissue, whereas genomic libraries contain DNA representing the entire genome regardless of the sample tissue type. *Deduce > Solve:* Since genomic DNA contains the entire genome, including all genes expressed in all tissue, the two genomic libraries would be expected to have the same DNA sequences and therefore would completely overlap, and each genomic library would contain all sequences present in the cDNA libraries. Since cDNA libraries contain only those sequences expressed as mRNA in the sample tissue, and brain tissue and muscle tissue are expected to have only partially overlapping sets of mRNAs, the two cDNA libraries would be expected to only partially overlap.

5a–b. *Evaluate:* This is type 1. This problem concerns the expected frequency that a particular DNA sequence will be present in a given genomic library. **TIP:** Genomic libraries created using DNA from different tissues typically do not differ. *Deduce:* The frequency of a given sequence in a genomic library can be estimated by the equation $\frac{\text{average insert size}}{\text{genome size}}$. *Solve:* The frequency of the myostatin gene (or any other typical gene) in a genomic library from either brain or muscle will depend on the size of the library (i.e., how many clones it contains divided by 30,000). Therefore, for both **(a)** muscle and **(b)** brain, on average one in every 30,000 clones will represent myostatin.

5c–d. *Evaluate:* This is type 1. This problem concerns the expected frequency that a particular DNA sequence will be present in a given cDNA library. **TIP:** The frequency of a particular cDNA clone depends on the relative abundance of the corresponding mRNA in the tissue used for cDNA cloning. *Deduce > Solve:* **(c)** The frequency of the myostatin cDNA clone will depend on the relative abundance of the myostatin mRNA in the muscle tissue used to prepare the cDNA library. **(d)** Since myostatin is expressed only in muscle cells, the frequency of myostatin cDNA in a brain cDNA library is expected to be zero. The myostatin cDNA is not expected to be present.

6. *Evaluate:* This is type 2. This problem concerns how frequently a particular sequence is expected to occur within a genome of specific size. **TIP:** For PCR primers to be specific for amplification of a single band from genomic DNA, the expected frequency of occurrence of each primer in random DNA should be ≤1. *Deduce:* The frequency of occurrence of a PCR primer in genomic DNA is given by the equation genomic DNA length × sequence frequency. The sequence frequency is given by the equation $\left(\frac{1}{4}\right)^N$, where N corresponds to the number of nucleotides in the site. *Solve:* The length of a primer sequence, N, that occurs once in the human genome, is given by the equation $3 \times 10^9 \left(\frac{1}{4}\right)^N = 1$. To solve for N: $\left(\frac{1}{4}\right)^N = \frac{1}{3 \times 10^9}$; $\log\left(\frac{1}{4}\right)^N = \left(\frac{\log 1}{3 \times 10^9}\right)$; $N\left(\log\frac{1}{4}\right) = -9.477$; $N(-0.602) = -9.477$; $N = \frac{-9.477}{-0.602} = 15.7$. The primer should be at least 16 nucleotides long.

7. *Evaluate:* This is type 1. This problem tests your understanding of cloning genes using information from protein sequence alignments. *Deduce > Solve:* One approach would be to amplify the β-globin cDNA by PCR using a degenerate PCR primer based on a newt β-globin amino acid sequence. That primer, in combination with an oligo dT primer, would then be used to amplify the 3′ end of the newt cDNA using a cDNA library as the template. Since the newt β-globin amino acid sequence is not known, it could be inferred from the amino acid sequence alignment shown in Figure 16.20 in the textbook. One candidate sequence is LIVYPWTQ, which would be encoded by the DNA sequence 5′-CTN/TTN ATT/A/C GTN TAT/C CCN TGG ACN CAA/G-3′. A 21-nucleotide-long degenerate primer sequence that has an unambiguous 3′ end would be 5′-TN ATT/A/C GTN TAT/C CCN TGG AC-3′.

8a.   *Evaluate:* This is type 1. This problem concerns genetic and physical genome maps and the methods used to create them. *Deduce > Solve:* Genetic maps are constructed by measuring genetic distance, which is determined by measuring meiotic recombination frequencies between genetic loci along a chromosome. Since the unit of distance in a genetic map is recombination frequency, the distance between genes will depend on the relative frequency of recombination between loci, which is not necessarily directly proportional to the physical distance. Physical maps are constructed by cloning and characterizing chromosomes and determining the distance between genetic loci in base pairs. The physical distance between loci will not necessarily correlate directly with genetic distance.

8b.   *Evaluate:* This is type 1. This problem concerns genetic and physical genome maps and the methods used to create them. *Deduce > Solve:* Since genetic maps are constructed by measuring meiotic recombination frequencies between genetic loci along a chromosome, there will likely be slight differences in genetic maps constructed by different laboratories due to stochastic differences in recombination frequencies in different experiments. These differences could be exacerbated if the laboratories perform the experiments using different strains or experimental conditions. Since physical maps are constructed by cloning and characterizing chromosomes and by determining the distance between genetic loci in base pairs, it is likely that physical maps constructed by different laboratories would be close to identical. This assumes that the source of genomic DNA used by the laboratories does not differ in structural aberrations such as deletions, inversions, or translocation.

8c.   *Evaluate:* This is type 1. This problem concerns genetic and physical genome maps and the methods used to create them. *Deduce > Solve:* Genetic and physical maps can be combined by determining the location of genetically mapped loci on the physical map. This is typically done by determining the sequence of the genetically mapped locus and by finding the sequence on the physical map. The combined map is actually two maps superimposed on each other, both showing the location of the same genetic loci, with one showing distance between loci in recombination frequency and the other showing distance in base pairs.

9.   *Evaluate:* This is type 1. This problem concerns the directional, or orientation-specific, cloning of a cDNA. *Deduce > Solve:* cDNAs (or any DNA fragments) can be cloned into a plasmid in a specific orientation by using two different restriction enzyme sites at the ends of the DNA and within the polylinker of pUC18 (pUC19 has an inverted MCS relative to pUC18 to allow for either orientation to be selected). To add two different restriction sites at the ends of the cDNA, one site can be included as the 5' end of the oligo dT primer used to prime first-strand cDNA synthesis (e.g., *Bam*HI GGATCC), and the second can be ligated onto the ends of the cDNA after synthesis (e.g., *Eco*RI GAATTC). Digestion of the cDNA with both enzymes will generate a cDNA with an *Eco*RI sticky end at the 5' end and a *Bam*HI sticky end at the 3' end. Digestion of pUC18 with both *Eco*RI and *Bam*HI will generate a linear plasmid with one *Eco*RI sticky end and one *Bam*HI sticky end. Ligation of the cDNA with the pUC18 will result in cDNA clones, all of which have their 5' ends at the *Eco*RI side of the pUC18 and their 3' ends at the *Bam*HI side.

10.   *Evaluate:* This is type 3. This problem concerns the type of DNA sequences represented in genomes sequenced using Sanger sequencing and massively parallel sequencing technologies. **TIP:** Recall that sequencing using the Sanger method requires cloning target sequences, whereas the newer sequencing technologies do not. *Deduce > Solve:* Because Sanger sequencing requires cloning DNA in order to be sequenced, genomic sequences that are difficult to clone (e.g., repetitive DNA and DNA sequences that are toxic to bacteria) are underrepresented or absent from Sanger-derived genome sequences. Since newer sequencing technology does not require cloning, these sequences will be represented in the data generated by massively parallel sequencing technologies.

11.   *Evaluate:* This is type 4. This problem concerns the effect of genome size on the strategy employed to clone genes based on position. *Deduce > Solve:* Positional cloning relies on accurately mapping the location of the gene of interest to a reasonably small genetic interval by combining genetic and physical map information. As the amount of DNA per cM increases, the need for a more precise genetic map increases. For example, to limit the location of a gene to a 100-kb region of a chromosome, the gene would have to be located to a 0.3-cM region in *C. elegans* (3 cM/1 Mb); a 0.15-cM region in *D. melanogaster* and *D. rerio* (1.5 cM/1 Mb); a 0.05-cM region in *M. musculus* (0.5 cM/1 Mb); and a 0.09-cM region in *H. sapiens* (0.9 cM/1 Mb). More precise genetic maps require greater numbers of mating events (either deduced from pedigrees or conducted by controlled matings); therefore, organisms with large kb/cM values will require more genetic mapping data than those with smaller kb/cM values.

12. *Evaluate:* This is type 1. This problem concerns the impact of advances in DNA synthesis on strategies for cloning relatively small DNA fragments. *Deduce > Solve:* The ability to synthesize DNA fragments of 10 kb allows the custom design of any DNA molecule without need for isolating that molecule from cells. The only limitation to this technique, besides size, is that the sequence must be known or predicted in order to be synthesized. If a wild-type gene or mutant allele of unknown sequence is desired, cells containing that DNA will still be necessary, and the gene will have to be cloned using standard techniques.

13. *Evaluate:* This is type 3. This problem concerns the ethics of having personal genome sequence information available. *Deduce > Solve:* A complete genome sequence is expected to benefit mankind by identifying the genetic components of health and disease in each of us, which would allow a genetically based, custom-designed approach to health care. There is also concern about the need for privacy and the abuse of genetic information if privacy is not ensured. Ethical concerns about personal genome sequencing include the use of genetic information in a discriminatory way by healthcare providers and employers. There is also concern for the use of genetics to determine race or gender against the individual's will. These concerns underscore the need for the development of public policy that protects the rights of individuals against genetically based discrimination and privacy invasion.

14a. *Evaluate:* This is type 2. This problem concerns restriction analysis of linear and circular forms of the bacteriophage λ chromosome. *Deduce:* The number of fragments that are generated by cutting a linear molecule is given by the equation number of sites + 1 = number of fragment. The number of fragments generated by cutting a circular molecule is the same as the number of sites. There is one *Xho*I site and one *Xba*I site in the λ chromosome. *Solve:* For *Xho*I, there will be 1 + 1 = 2 fragments when λ is linear and 1 fragment when it is circular. The same is true for *Xba*I. For *Xho*I plus *Xba*I, there will be 2 + 1 = 3 fragments when λ is linear and 2 fragments when it is circular.

14b. *Evaluate > Deduce > Solve:* This is type 2. This problem concerns restriction analysis of linear and circular forms of the bacteriophage λ chromosome. See figure.

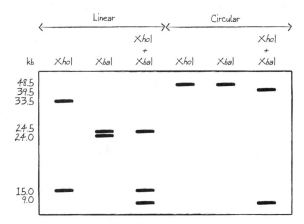

15. *Evaluate:* This is type 2. This problem concerns the sequence of restriction enzyme recognition sites. *Deduce:* Sticky ends generated by restriction digests can be ligated if they are complementary. The overhangs generated by *Xho*I and *Sal*I are complementary; therefore, they can be ligated. The result of ligation is the sequence 5'-CTCGAC-3' or 5'-GTCGAG-3'; these are neither *Xho*I sites nor *Sal*I recognition sites. *Solve:* Yes, the sticky ends of *Xho*I and *Sal*I can be ligated; but the resulting site will not be cut by *Xho*I or *Sal*I.

16. *Evaluate:* This is type 2. This problem concerns restriction mapping. *Deduce:* The chromosome is circular; therefore, there is one *Pst*I site, one *Psi*I site, and two *Dra*I sites. The distances between *Pst*I and *Psi*I are given by that double digest, and they are 2308 bp in one direction and 3078 bp in the other direction. The distances between the *Dra*I sites are given by the results of *Dra*I digestion, and they are 4307 bp in one direction and 1079 bp in the other direction. The *Pst*I *Dra*I double digest indicates that the *Pst*I site is within the 4307-bp *Dra*I fragment. The *Psi*I *Dra*I double digest indicates that the *Psi*I site is also within the 1079-bp *Dra*I fragment. *Solve:* See figure.

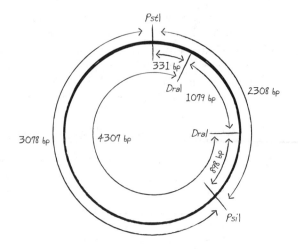

17. *Evaluate:* This is type 2. This problem concerns restriction mapping. *Deduce:* The cDNA was cloned as an *Eco*RI fragment into an *Eco*RI site of the 2.961-kb plasmid. Since there is only 1 *Hind*III site in the MCS of the plasmid vector and digestion of the clone with *Hind*III generates two fragments, there must be 1 *Hind*III site in the cDNA. The orientation of the clone is such that the cDNA *Hind*III site is 0.3 kb from the polylinker *Hind*III site. *Solve:* See figure.

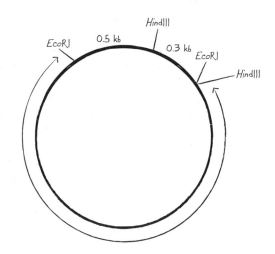

18a. *Evaluate:* This is type 2. This problem concerns restriction mapping. *Deduce > Solve:* A genomic clone contains all of the DNA that is present in the corresponding cDNA and typically all of the restriction sites present in the cDNA. Therefore, the *Hind*III site in the cDNA probably corresponds to one of the *Hind*III sites in the genomic clone.

18b. *Evaluate:* This is type 1. This problem concerns the relationship between genomic and cDNA clones. *Deduce > Solve:* cDNAs contain sites that are formed by fusion of exons, which creates a DNA sequence in the cDNA that is not present in the genomic DNA. If this happens to create a restriction site, the site will be present in the cDNA but not in the genomic DNA. Genomic restriction sites located in introns will not be present in the cDNA clone. Therefore, it is possible that restriction sites in the cDNA may not be in the genomic clone, and vice versa.

19. *Evaluate:* This is type 2. This problem concerns restriction mapping. *Deduce:* Each of the sites being mapped is present once in the MCS of the plasmid vector. *Xba*I, *Xho*I, and *Sal*I cut the clone twice, so there is one site for each enzyme within the genomic clone. *Hind*III cuts the clone three times, so the genomic clone contains two *Hind*III sites. The *Xba*I alone cuts the clone into 4.5-kb and 9.5-kb fragments, and the *Xba*I plus *Eco*RI double digest cuts the clone into 4.5-kb, 6.5-kb, and 3.0-kb fragments. This indicates that the 9.5-kb *Xba*I fragment is the 3.0-kb plasmid vector plus 6.5-kb of the insert, which places the genomic *Xba*I site 4.5 kb in from the *Xba*I site in the MCS. *Xho*I alone cuts the clone into 13.2-kb and 0.8-kb fragments, and the *Xho*I plus *Eco*RI double digest cuts the clone into 10.2-kb, 3.0-kb, and 0.8-kb fragments. This indicates that the 13.2-kb *Xba*I fragment contains the 3.0-kb plasmid vector plus 10.2 kb of the insert, which places the genomic *Xho*I site 0.8 kb in from the *Xho*I site in the MCS. *Sal*I alone cuts the clone into 6.0-kb and 8.0-kb fragments,

and the *Sal*I plus *Eco*RI double digest cuts the clone into 6.0-kb, 5.0-kb, and 3.0-kb fragments. This indicates that the 8.0-kb *Sal*I fragment contains the 3.0-kb plasmid vector plus 5.0 kb of insert, which places the genomic *Sal*I site 6.0 kb in from the *Sal*I site in the MCS. *Hin*dIII cuts the clone into 12.0-kb, 1.5-kb, and 0.5-kb fragments. The 12-kb *Hin*dIII fragment must contain a 3.0-kb plasmid vector plus 9 kb of insert, which places one genomic *Hin*dIII site 12.0 kb away from the *Hin*dIII site in the MCS. The remaining genomic *Hin*dIII site is either 0.5 kb or 1.5 kb in from the MCS *Hin*dIII site: both locations are consistent with the results. An *Xho*I plus *Hin*dIII double digest would distinguish between the two possible *Hin*dIII maps, yielding a 12.0-kb fragment plus a 1.0-kb and two 0.5-kb fragments, or one 12.0-kb fragment plus three 0.5-kb fragments. *Solve:* See figure.

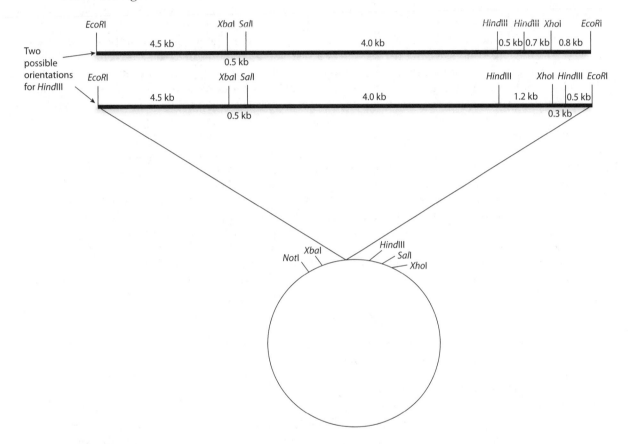

20a. *Evaluate:* This is type 2. This problem concerns the DNA sequence analysis of a cDNA clone. *Deduce > Solve:* **(a)** The sequence from the T3 primers contains a run of T residues, which corresponds to the 3′ poly-A tail of the mRNA; therefore, the T3-primer-derived sequence represents the 3′ end of the cDNA and the T7-primer-derived sequence represents the 5′ end of the cDNA. **(b)** Since the stretch of T's in the sequence corresponds to the mRNA poly-A tail, and the poly-A tail is not encoded by the gene, the stretch of T's will not be present in the genomic clone. **(c)** The T7-primer-derived sequence includes the *Eco*RI site used in cloning; therefore, the GAATTC *Eco*RI sequence and everything before it is derived from the plasmid vector. The T3-primer-derived sequence includes the *Xho*I site used in cloning; therefore, the CTCGAG *Xho*I sequence and everything before it is derived from the plasmid vector. **(d)** The start codon should be the first ATG after the *Eco*RI site in the T7-primer-derived sequence, which corresponds with the ATG just after position 140. The sequence before the start codon and after the *Eco*RI site represents the 5′ UTR of the mRNA.

21a. *Evaluate:* This is type 4. This problem concerns the assignment of mutants to genes using complementation analysis. *Deduce > Solve:* Each mutant would be mated to wild type and to every other mutant to create diploid strains. The diploids would be assayed for growth at permissive and restrictive temperature. Diploids formed by mating a mutant to a wild type that can grow at restrictive temperatures identify the mutation as recessive. Dominant mutations cannot be studied using complementation analysis. Diploids formed by mating two recessive mutants identify mutations in the same gene if the diploid cannot grow at restrictive temperature (non-complementation), and they identify mutations in different genes if the diploids can grow at restrictive temperature (complementation). All possible pairwise diploid combinations will identify how many complementation groups (genes) were identified by the temperature-sensitive secretion mutants.

21b. *Evaluate:* This is type 4. This problem concerns strategies to clone genes identified by genetic analysis in yeast. *Deduce > Solve:* The yeast mutations are temperature-sensitive lethal; therefore, these genes can be cloned by complementation of the temperature-sensitive lethal phenotype. This would entail transforming each mutant yeast by using a genomic DNA library made from wild-type DNA and selecting for growth at a restrictive temperature. This approach could also be attempted for cloning the human homologs if a cDNA library made from human mRNA were cloned into a yeast expression vector. This approach will work if the human cDNA complements the yeast mutations. A second approach to clone human homologs of the yeast genes would be to use the yeast genes as a probe to screen a human cDNA library. This approach will work for human genes that are sufficiently homologous at the DNA sequence level.

22. *Evaluate:* This is type 4. This problem concerns strategies for a genetic screen to identify genes involved in meiosis. *Deduce:* Yeast would be an appropriate model organism for identifying meiotic mutants. The screen should be based on the expected phenotype for a mutant unable to perform meiosis. In yeast, this would mean screening for mutants that, when diploid, would be unable to complete meiosis and make viable ascospores. The simplest screen would be for dominant mutations, since the phenotype would need to be expressed in diploids. *Solve:* First, mutagenize yeast and collect a large number of survivors of mutagenesis. Each survivor is assumed to contain a mutation in a different gene in the genome and is therefore a potential yeast meiotic mutant. Each mutant would then be mated to a wild type, and each resulting diploid would be placed on sporulation medium and screened for the ability to produce viable ascospores.

23a. *Evaluate:* This is type 4. This problem tests your understanding of strategies using a genetic screen to identify genes involved in *Drosophila* eye development. **TIP:** Review genetic screens using *Drosophila,* including Figure 16.3 in the textbook. *Deduce: Drosophila* eyes are not essential for viability; therefore, mutants that perturb or prevent eye development could be isolated. Such screens are typically done one chromosome at a time using balancer chromosomes that allow a single, mutagenized chromosome to be followed through multiple crosses. An F_3 screen for recessive mutations on chromosome 3 that affect eye development is described here. *Solve:* Male flies homozygous for dark body (*db*) are fed EMS to induce mutations and then mated to females heterozygous for a balancer chromosome 3 containing the *db* allele and the dominant marker *CyO*. Individual F_1 males that have curly wings and dark bodies are selected and mated back to a female from the chromosome 3 balancer strain to create many independent F_2 lines. F_2 progeny with curly wings and dark bodies will be heterozygous for the new mutations. Interbreeding the F_2 will create many independent F_3 populations. The straight-winged flies within each population will be homozygous recessive for new mutations on chromosome 3, and the rest will be heterozygous for the mutation. F_3 that have straight-winged flies with eye abnormalities identify fly lines with mutations on chromosome 3 affecting eye development. **(b)** Genes that are important for development of adult structures may also be important at earlier stages of development. Mutations in these genes may be embryonic lethals or may lead to developmental abnormalities that prevent observation of their roles in adult eye development.

24. *Evaluate:* This is type 4. This problem concerns strategies for cloning the *Drosophila* genes *dunce* and *rutabaga*. *Deduce:* Positional cloning or transposon-tagging strategies would be appropriate for cloning the *dunce* and *rutabaga* genes since cloning by complementation is typically not done with *Drosophila,* and researchers have no information on the sequence of the genes to use in homology-based techniques. Since transposon tagging would require making new mutants and mapping the mutations to the *dunce* and *rutabaga* locus, positional cloning using the existing mapping data would be the most straightforward approach. *Solve:* The genetic map locating *dunce* and *rutabaga* would be used to identify the candidate genes on the physical map. Cloned DNA containing those genes would be obtained from the *Drosophila* genome center and then used in P-element-mediated transformation experiments to determine which clone complements the mutant phenotype. Sequencing of the clone will identify the wild-type gene sequence, which can be used to design a PCR approach to amplify the same locus from *dunce* and *rutabaga* mutants. Sequence analysis indicating that the DNA from the mutant carries a non-wild-type allele of the candidate gene would confirm the identity of the *dunce* and *rutabaga* genes.

25a.   *Evaluate:* This is type 4. This problem concerns strategies for selecting the best candidate genes within the genetic interval identified as containing the gene mutated in cystic fibrosis. *Deduce > Solve:* Since cystic fibrosis is a disease that affects specific tissues, you would consider only candidate genes that are expressed in those tissues. Also, since cellular secretions are affected, you would pay special attention to candidate genes that are involved in secretion (for example, membrane proteins or secretory pathway genes).

25b.   *Evaluate:* This is type 4. This problem concerns what constitutes proof that a particular candidate gene is the gene mutated in a human disease. *Deduce > Solve:* If the candidate gene is the cystic fibrosis gene, then every person with cystic fibrosis should be homozygous for a mutant allele of the gene, and their parents should be heterozygous for that allele. If multiple families with cystic fibrosis all contain what appear to be mutant variants likely to disrupt formation of the same functional protein, then this would constitute strong evidence supporting the identity of the gene as the cystic fibrosis gene. Formal proof would require evidence about the mechanism by which the mutations cause the disease.

26a.   *Evaluate:* This is type 4. This problem concerns the strategies used to sequence a 4.5-kb cDNA clone and a 250-kb genomic BAC clone. *Deduce > Solve:* The 4.5-kb cDNA clone can be sequenced by first sequencing in from the ends of the clone and then using that sequence to design primers to sequence further in toward the middle of the clone. This "primer walking" technique allows for sequencing relatively small clones without the need for generating additional smaller clones (sub-clones). The 250-kb BAC clone would be sequenced by shotgun sequencing, which involves fragmentation of the clone DNA, sub-cloning, and random sequencing of large numbers of clones. The individual sequence reads from each clone would then be assembled by computer programs into one contiguous sequence representing the intact BAC clone.

26b.   *Evaluate:* This is type 4. This problem concerns the strategy determining the DNA sequence of mutant *CFTR* alleles from cystic fibrosis patients. *Deduce > Solve:* Since most of the mutations affecting *CFTR* function will be in exons, sequencing cDNA clones from patients would be the most efficient way to search for *CFTR* mutations. For patients that show no mutations in cDNA sequence, isolation of the genomic sequence would then become necessary.

27.   *Evaluate:* This is type 4. This problem concerns the design of a genetic screen to identify recessive lethal mutations in *Drosophila*. **TIP:** Review genetic screens using *Drosophila*, including Figure 16.3 in the textbook. *Deduce:* Screens for lethal mutations are typically done one chromosome at a time using balancer chromosomes that allow a single, mutagenized chromosome to be followed through multiple crosses. An F_3 screen for recessive lethal mutations on chromosome 3 is described here. *Solve:* Male flies homozygous dark body (*db*) for are fed EMS to induce mutations and then mated to females heterozygous for a balancer chromosome 3 containing the *db* allele and the dominant marker, *CyO*. Individual F_1 males that have curly wings and dark bodies are selected and mated back to a female from the chromosome 3 balancer strain to create many independent F_2 lines. F_2 progeny with curly wings and dark bodies will be heterozygous for the new mutations. Interbreeding the F_2 will create many independent F_3 populations. If the F_2 were heterozygous for a recessive embryonic lethal mutation, there would be no straight-winged flies among the F_3. The curly-winged F_3 will be heterozygous for the mutations and can be used to generate fly stocks carrying the new mutant allele.

28a–b. *Evaluate:* This is type 4. This problem concerns the design of a genetic screen to identify genes required for female gametophyte development in *Arabidopsis*. *Deduce: Arabidopsis* gametophytes, which contain the egg, sperm, and embryo sac, develop from haploid spores (female megaspores and male microspores) that are the products of meiosis. Plants heterozygous for mutations affecting gametophyte development or functions will produce half normal and half mutant gametophytes because the mutant allele will be uncovered during gametophyte development. Such plants will have reduced fertility, which can be used as the phenotype in the screen. Seed from wild-type plants would be treated with a mutagen, allowed to develop and self, and then the progeny would be screened for reduced fertility. *Solve:* **(a)** Plants showing reduced fertility could be studied to determine if the reduced fertility is due to defects in ovule development. **(b)** Plants showing reduced fertility could be studied for defects in pollen development, or pollen could be collected and tested for viability in a cross to a wild-type plant.

29a. *Evaluate:* This is type 4. This problem tests your understanding of how to design a genetic screen to identify *Drosophila* mutants that have an altered circadian rhythm. *Deduce:* Using eclosion at times other than dawn as the phenotype, you could conduct an F_3 screen for recessive mutations that affects the time pupae eclose. An F_3 screen for circadian rhythm mutations on chromosome 3 is described here. *Solve:* Male flies homozygous for dark body (*db*) are fed EMS to induce mutations and then mated to females heterozygous for a balancer chromosome 3 containing the *db* allele and the dominant marker, *CyO*. Individual F_1 males that have curly wings and dark bodies are selected and mated back to a female from the chromosome 3 balancer strain to create many independent F_2 lines. F_2 progeny with curly wings and dark bodies will be heterozygous for the new mutations. Interbreeding the F_2 will create many independent F_3 populations. If the F_2 were heterozygous for a circadian rhythm mutation, about one-third of the progeny would eclose at times other than dawn. Rather than looking for pupae to eclose, you could sleep past dawn and then screen flies for the presence of recently emerged virgin females, which have a greenish meconium plug in their abdomen.

29b. *Evaluate:* This is type 4. This problem concerns the design of a genetic screen to identify *Arabidopsis* mutants that have an altered circadian rhythm. *Deduce:* Since some genes involved in photosynthesis are expressed in a circadian manner, you could screen for mutants with altered patterns of expression of these genes or for visible traits that reflect circadian defects (for example, defective starch utilization at night). *Solve:* Seed from wild-type plants would be treated with a mutagen, allowed to develop and self, and the progeny collected and screened for gene expression profile defects or defects in visible traits related to circadian control (e.g., stomatal opening).

29c. *Evaluate:* This is type 4. This problem concerns the strategies needed to clone the *Drosophila* and *Arabidopsis* genes identified by the genetic screen in parts (a) and (b). *Deduce:* Positional cloning could be used to clone the genes in both organisms. *Solve:* First you would need to genetically map the mutations, and you would then use the map locations to identify the candidate genes on the physical map. Cloned DNA containing those genes would be obtained and used in transformation experiments to determine which clone complements the mutant phenotype. Sequencing of the clone identifies the wild-type gene sequence, which you could then use to design a PCR approach to amplify the same locus from the mutants. Sequence analysis indicating that the DNA from the mutant carries a non-wild-type allele of the candidate gene would confirm the identity of the genes.

16.04 Test Yourself

Problems

1. Describe a genetic screen for yeast genes that are required for anaphase chromosome segregation during mitosis.

2. Describe how you would clone the genes corresponding to the anaphase mutants identified in the previous problem.

3. The figure shows a fragment of a chromosome that contains genes *X, Y,* and *Z.* The DNA sequences of the top strand of the chromosome in a small portion of the region that contains the *Hind*III restriction sites (5′-AAGCTT-3′) are shown. **(a)** Use this figure to draw a gel showing the results of the following restriction digests of this fragment: *Eco*RI; *Bam*HI; *Hind*III; *Eco*RI plus *Bam*HI; *Eco*RI plus *Hind*III; and *Bam*HI plus *Hind*III. **(b)** Use this figure to design 20-nucleotide-long PCR primers that could be used to amplify a fragment containing Gene *Y.*

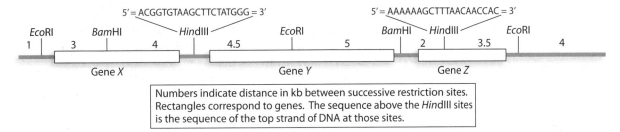

Numbers indicate distance in kb between successive restriction sites. Rectangles correspond to genes. The sequence above the *Hind*III sites is the sequence of the top strand of DNA at those sites.

4. You wish to clone Genes *X, Y,* and *Z* (shown in the map for Problem 3) into the plasmid vector, pTY. pTY contains three dominant genes that confer resistance to antibiotics in *E. coli.* AMP-R confers resistance to ampicillin, TET-R confers resistance to tetracycline, and CM-R confers resistance to chloramphenicol. The figure also shows the results of agarose gel electrophoresis of molecular weight markers (outside lanes) and of six clones; each clone number is indicated above the gel lanes. **(a)** You cut the chromosome fragment from Problem 3 and pTY with *Eco*RI, mix the digested DNAs together, perform a ligation reaction, transform *E. coli* with a sample of the ligation reaction, and plate the cells on LB + AMP. You isolate plasmid DNA from four clones (clones 1, 2, 3, and 4), digest the plasmids with *Bam*HI, and run each digest in a separate lane as indicated in the figure. Based on these results, which of these clones contains a complete copy of gene *X,* gene *Y,* or gene *Z*? **(b)** Draw a map of the recombinant plasmid corresponding to clone 1. **(c)** You perform a similar cloning experiment as in part (a), except that you use *Bam*HI to cut the chromosome and pTY. You isolate plasmid DNA from two clones (clones 5 and 6), digest them with *Eco*RI, and run each digest in a separate lane as indicated in the figure. Your analysis of these results indicates that both clones contain a complete copy of gene *Y.* Why do the results of the restriction digests of clones 5 and 6 differ? **(d)** If you mixed up the *E. coli* cells that contain clones 1 and 5, how could you tell them apart without isolating DNA from them? Briefly explain what you would do, and describe the expected result for each clone.

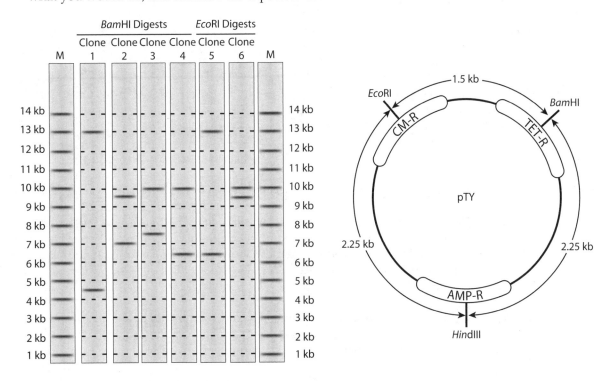

Solutions

1. *Evaluate:* This is type 4. This problem concerns the design of a genetic screen for yeast mutants defective in anaphase of mitosis. *Deduce:* Anaphase is the stage of mitosis when sister chromatids separate and migrate toward the spindle poles. Mutants defective in anaphase will reach metaphase and then arrest cell division with condensed chromosomes attached to a metaphase spindle. This is a lethal phenotype; therefore, the screen will search for conditional mutations. *Solve:* Mutagenize yeast with EMS, plate the survivors of mutagenesis on solid medium such that they form about 100 colonies per petri dish, and incubate the plates at permissive temperature. Screen for conditional lethal mutations by replica plating to two fresh plates, incubating one at permissive and one at restrictive temperature. Colonies that form at permissive but not restrictive temperature carry temperature-sensitive (ts) lethal mutations. Culture each ts mutant at permissive temperature and then shift the culture to restrictive temperature until all cells have traversed at least one cell cycle (about 100 minutes), and then fix and stain the cells to observe DNA and microtubules. Those mutants that accumulate cells with a single mass of DNA and a metaphase-sized mitotic spindle are candidates for ts anaphase mutants.

2. *Evaluate:* This is type 1. This problem concerns cloning of genes by complementation of a mutant phenotype. *Deduce:* The yeast mutants defective in anaphase of mitosis carry temperature-sensitive lethal mutations. *Solve:* The genes corresponding to each mutation can be cloned by transformation of the mutants with a yeast genomic DNA library and selection of transformants that can form colonies at restrictive temperature.

3a. *Evaluate:* This is type 2. This problem concerns the analysis of DNA by restriction digestion and gel electrophoresis. *Deduce:* Digestion with *Eco*RI creates restriction fragments that are 1, 4, 10.5, and 11.5 kb in length. Digestion with *Bam*HI creates restriction fragments that are 4, 9.5, and 13.5 kb in length. Digestion with *Hin*dIII creates restriction fragments that are 7.5, 8, and 11.5 kb in length. Digestion with *Eco*RI and *Bam*HI creates restriction fragments that are 1, 3, 4, 5, 5.5, and 8.5 kb in length. Digestion with *Eco*RI and *Hin*dIII creates restriction fragments that are 3.5, 4, 4.5, and 7 kb in length, and two fragments of 7 kb are created. Digestion with *Bam*HI and *Hin*dIII crates restriction fragments of 2, 4, 7.5, and 9.5 kb in length, and two fragments of 4 kb are created. *Solve:* See figure.

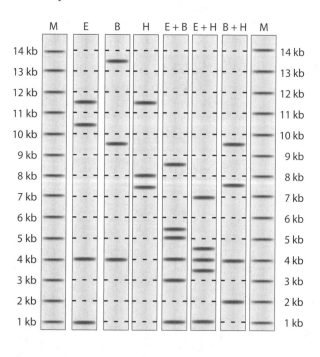

3b. *Evaluate:* This is type 2. This problem concerns the design of PCR primers. *Deduce:* The forward (left to right) primer should hybridize to the bottom strand at the leftmost *Hin*dIII site. The sequence of that primer is the same as the sequence shown in the figure. The reverse (right to left) primer should hybridize to the top strand at the rightmost *Hin*dIII site. The sequence of that primer should be the reverse complement of the sequence shown in the figure. *Solve:* The forward primer sequence is 5'-ACGGTGTAAGCTTCTATGGG-3'. The reverse primer sequence is 5'-GTGGTTGTTAAAGCTTTTTT-3'.

4a. *Evaluate:* This is type 2. This problem concerns the construction and analysis of recombinant DNA clones. *Deduce:* Two different genomic *Eco*RI fragments could have been cloned. One is the 11.5-kb fragment that contains gene *X* and part of gene *Y*, and the other is the 10.5-kb fragment that contains part of gene Y and all of gene Z. The total DNA in a pTY of gene X is 17.5 kb, whereas the total DNA in the pTY clone of gene Z is 16.5 kb. No *Eco*RI clone can contain all of gene Y. *Solve:* The fragments created by *Bam*HI digestion of clones 1 and 3 total 17.5 kb; therefore, clones 1 and 3 contain gene X. The fragments generated by *Bam*HI digestion of clones 2 and 4 total 16.5 kb; therefore, clones 2 and 4 contain gene Z.

4b.   *Evaluate:* This is type 2. This problem concerns restriction mapping. *Deduce:* Clone 1 contains the 11.5-kb genomic *Eco*RI fragment inserted into the *Eco*RI site of pTY. This fragment could be inserted in one of two orientations. If the gene X clone is oriented such that the *Bam*HI site within gene X is closest to the *Bam*HI site in pTY, then the *Bam*HI digestion pattern would be a 4.5-kb fragment and a 13.5-kb fragment. If the gene X clone is in the opposite orientation, then the *Bam*HI digestion pattern would be a 7.5-kb fragment and a 10-kb fragment. The *Bam*HI digestion pattern of clone 1 is 4.5 kb and 13.5 kb. *Solve:* See figure.

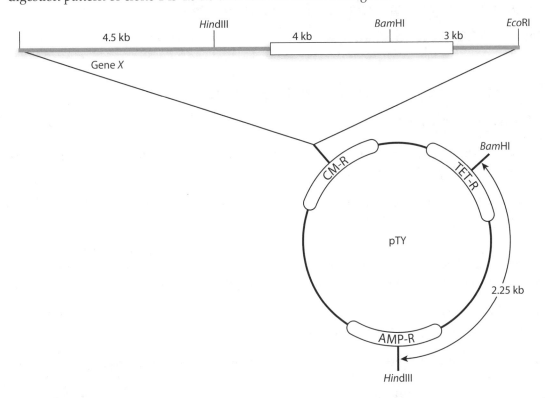

4c.   *Evaluate:* This is type 2. This problem concerns the analysis of a recombinant DNA clone. *Deduce > Solve:* Clones 5 and 6 contain the same insert but in opposite orientations. In clone 5, the gene Y clone is oriented with the gene Y *Eco*RI site 6.5 kb from the pTY *Eco*RI site, whereas in clone 6, the gene Y *Eco*RI site is 8.5 kb from the pTY *Eco*RI site.

4d.   *Evaluate:* This is type 2. This problem concerns the phenotypic analysis of *E. coli* transformants. *Deduce:* Clone 1 is a recombinant pTY plasmid with an insert in its *CM-R* gene; therefore, *E. coli* carrying that plasmid would be resistant to tetracycline and ampicillin but sensitive to chloramphenicol. Clone 5 is a recombinant pTY plasmid with an insert in its *TET-R* gene; therefore, *E. coli* carrying that plasmid would be resistant to chloramphenicol and ampicillin but sensitive to tetracycline. *Solve:* Plate samples of each culture on ampicillin-, tetracycline-, and chloramphenicol-containing plates. The cells carrying clone 1 will grow on tetracycline- and ampicillin-containing medium but not chloramphenicol-containing medium. Cells carrying clone 5 will grow on ampicillin- and chloramphenicol-containing medium but not on tetracycline-containing medium.

17

Applications of Recombinant DNA Technology and Reverse Genetics

17.01 Genetics Problem-Solving Toolkit

Key Terms and Concepts

Allele knock-in (gene knock-in): The endogenous copy of a gene in an organism's genome is replaced with an allele generated using recombinant DNA technology in vitro.

Animal models for human diseases: Selecting or creating mutations in the animal orthologs of genes known to be involved in heritable human diseases.

Enhancer trap: Experiment that uses a reporter gene lacking an enhancer to identify genomic locations that contain tissue-specific enhancers by randomly integrating the reporter gene throughout the genome and identifying transgenic animals that show tissue-specific reporter gene expression.

Gene knockout: The endogenous copy of a gene in an organism's genome is replaced with a positive selectable marker in order to create a loss-of-function allele.

Homologous recombination: Crossing over involving homologous DNA sequences.

Illegitimate recombination: Crossing over involving non-homologous DNA sequences.

Reporter genes: Genes that encode proteins easily detectable in cell, tissue, or whole organisms, for example, *lacZ* (β-galactosidase) and *GFP* (green fluorescent protein).

RNAi: RNA that knocks down the expression of a gene by posttranscriptional mechanisms through the introduction of dsRNAs into cells.

Site-specific recombination: Crossing over between specific sequences that are recognized by a site-specific recombinase, which catalyzes the recombination event.

Transgene: A gene from one organism transferred into another, creating a *transgenic organism*.

Transcriptional fusions: Chimeric genes that are generated using recombinant DNA technology in vitro and that contain the promoter and regulatory sequences from one gene fused to the coding sequence of another.

Translational fusions: Chimeric genes that are generated using recombinant DNA technology in vitro and that contain the transcriptional regulatory sequences and at least part of the coding sequence from one gene fused to the coding sequence of another.

Key Analytical Tools

Types of transformation vectors and their use in model genetic organisms

Transformation Vector	Applicable Model Organisms
Circular plasmids (pUC19, Ti plasmid, *E. coli*–yeast shuttle plasmids)	Bacteria (*E. coli*, *Agrobacterium*), yeast, plants
Transposons (*P* elements)	Fruit flies (*Drosophila*)
Linear molecules (knockout and knock-in constructs)	Bacteria, yeast, plants, mice

Reverse Genetic Approaches in Model Genetic Organisms (See Table 17.2 in textbook.)

17.02 Types of Genetics Problems

1. **Apply recombinant DNA technology to create transgenic organisms.**
2. **Apply recombinant DNA technology to study gene expression.**
3. **Apply recombinant DNA technology to create mutants.**

1. Apply recombinant DNA technology to create transgenic organisms.

This type of problem tests your understanding of the use of recombinant DNA technologies to create transgenic organisms. You may be asked to design strategies to introduce DNA into bacteria, yeast, fruit flies, plants, or mice. Success in answering these problems requires an understanding of DNA vectors, which vary between different types of organisms, and the genetic markers used to select for or identify transformants for each organism. The problem may require integration of the transforming DNA at specific chromosomal sites by homologous recombination or to random sites by transposition or illegitimate recombination. The problem may require expression of a recombinant protein in specific tissue of the transgenic organism. See Problem 8 for an example.

Example Problem: Transgenes are typically constructed in plasmids using *E. coli* and then transferred to specialized vectors for introduction into other types of host cells. In some cases, one vector can be useful for both purposes. Design a plasmid vector that could be maintained in *E. coli* and would also be useful for expressing human genes in yeast.

Solution Strategies	Solution Steps
Evaluate	
1. Identify the topic this problem addresses, and explain the nature of the requested answer.	1. This problem tests your understanding of bacterial and yeast transformation vectors. The answer requires a description of the critical components of such a DNA molecule.
2. Identify the critical information given in the problem.	2. The vector must be capable of propagating in *E. coli* and facilitating human gene expression in yeast.

(continued)

Solution Strategies	Solution Steps
Deduce	
3. List the plasmid components that are required for their maintenance and replication in cells.	3. The components (a) have an origin of DNA replication recognized by the host cell, and (b) have a dominant selectable marker appropriate for selecting in those host cells.
4. Describe plasmid elements that are important for inserting genes and promoting gene expression in yeast.	4. The components are (a) multiple cloning sites, (b) a promoter, and (c) regulatory sequences for controlling transcription.
Solve	
5. Answer the problem by drawing a map of the plasmid that shows all the required sequence elements.	5. **Answer:** See figure.

2. Apply recombinant DNA technology to study gene expression.

This type of problem tests your ability to apply recombinant DNA technology to study regulation of gene expression. You may be asked to design experimental strategies to study gene expression using gene fusions and reporter genes or to interpret the results of experiments analyzing reporter gene expression; to discuss the strategies for choosing between transcriptional and translation gene fusions or to evaluate the relative merit of results obtained using both methods; and to consider the potential caveats of studying gene expression using gene fusions, including how conclusions drawn from reporter gene analysis may differ from the expression pattern of the endogenous gene. See Problem 16a for an example.

Example Problem: You have constructed a transcriptional fusion of the promoter and regulatory region of a muscle-specific mouse gene to *GFP* and then created transgenic mice that carry this construct. All transgenic mice with this version of the fusion gene show bright green fluorescence in muscle tissue. Transgenic mice that contain a version of the gene fusion that lacks a 100-bp sequence at the 5′ end of the regulatory region show faint green fluorescence in muscle tissue. Transgenic mice that contain a version of the gene fusion that lacks a 100-bp sequence at the 3′ end of the regulatory region show no fluorescence in any tissue. Transgenic mice that contain a version of the gene fusion that contains only the 5′ 100 bp and 3′ 100 bp of the regulatory regions show bright green fluorescence in muscle tissue. Based on these results, which elements of the regulatory region were identified as important, and what is their function?

Solution Strategies	Solution Steps
Evaluate	
1. Identify the topic this problem addresses, and explain the nature of the requested answer.	1. This problem tests your ability to interpret the results of gene expression studies using transcriptional fusions to *GFP*. The answer requires relating the cell type and amount of fluorescence in transgenic animals to the structure of the promoter and regulatory sequences of the transgene.
2. Identify the critical information given in the problem.	2. The fluorescence in four types of transgenic animals was analyzed. Bright, muscle-specific fluorescence was seen in the animals carrying the full-length transcriptional fusion or the fusion deleted for all but the terminal 100-bp sequences. Dim, muscle-specific fluorescence was seen in the animals carrying a deletion of the 5' 100 bp of the regulatory region. No fluorescence was detected in animals carrying a deletion of the 3' 100 bp of the regulatory region.
Deduce	
3. Describe the relationship between fluorescence and the function of the regulatory sequences controlling fusion gene expression.	3. The cell types that show fluorescence reveal the cell-type specificity of the regulatory sequences driving GFP expression. The fluorescence intensity indicates the level of transcription (transcription rate) directed by the regulatory sequences.
4. Describe the relationship between the expression pattern of an endogenous gene and the full-length transgene, and discuss the implications of this relationship on the regulatory sequences in the transgene.	4. The full-length transgene faithfully reproduces the muscle-specific expression pattern of the endogenous gene. This suggests that the full-length transgene contains the regulatory sequences necessary for normal regulation of endogenous gene.
5. Relate the expression pattern of the deleted versions of the transgene to the function of the deleted sequences. Correlate regulatory sequence function to known types of cis-acting regulatory sequences (e.g., enhancers, silencers, or promoters).	5. Of the three deletions, only two altered the transgene's expression pattern. The 5' 100-bp region is required for high-level but not basal transcription, similar to an enhancer. The 3' 100-bp region is required for any transcription, similar to the promoter. The rest of the regulatory region is not necessary for transcription.
Solve	
6. Answer the problem by presenting a complete description of the function of regulatory sequences identified in step 5.	6. **Answer:** The transgene contains an enhancer-like element at the 5' end and a promoter-like element at the 3' end. These sequences are separated by DNA whose function is not revealed in these studies.

3. Apply recombinant DNA technology to create mutants.

This type of problem tests your understanding of an experimental approach called reverse genetics, whereby the function of a gene is investigated by using recombinant DNA and transgenic technologies to create organisms carrying mutations in the gene. You may be asked to describe how to create transgenic strains of bacteria, yeast, or mice that carry loss-of-function (gene knockout) mutations or gain-of-function (allele knock-in) mutations. For both knockout and knock-in experiments, you may be asked to describe the structure of the transgene, explain how to use existing transgenes to generate the transgenic organisms, outline how to perform gene knockout experiments using RNAi in fruit flies or plants, predict the results of a reverse genetic analysis, or interpret the results from a reverse genetic experiment. See Problem 16b for an example.

Example Problem: How does the frequency of illegitimate recombination in an organism affect the strategy used to perform gene knockout experiments? Give examples of organisms that differ in illegitimate recombination frequency.

Solution Strategies	Solution Steps
Evaluate	
1. Identify the topic this problem addresses, and explain the nature of the requested answer.	1. This problem tests your understanding of illegitimate and homologous recombination. The answer will relate recombination mechanisms to the choice of strategies for performing gene knockout experiments and provide examples of organisms with low and high illegitimate recombination frequencies.
2. Identify the critical information given in the problem.	2. The problem implies that the frequency of illegitimate recombination in an organism affects the strategy employed to conduct gene knockout experiments.
Deduce	
3. Define illegitimate recombination, and describe the alternative recombination.	3. Illegitimate recombination refers to crossing over (genetic exchange) involving non-homologous DNA sequences. The alternative to illegitimate recombination is called homologous recombination, where crossing over involves only homologous DNA sequences.
4. Describe the role of homologous recombination in a gene knockout experiment, and give an example of an organism that primarily uses homologous recombination to insert transgenes into its genome. **TIP:** See Figure 17.5b in the textbook.	4. In a gene knockout, a selectable marker contained in the transgene replaces (knocks out) a gene on a chromosome. The replacement occurs through a double crossover between the transgene and the chromosome. The crossovers involve homologous DNA sequences that flank the gene on the chromosome and the selectable marker on the transgene. Yeast use homologous recombination as the primary mechanism for incorporating transgenes into their genome.

(continued)

Solution Strategies	Solution Steps
5. Describe the consequences of illegitimate recombination on the success of the gene knockout experiments, and give an example of an organism that primarily uses illegitimate recombination to insert transgenes into its genome. **TIP:** See Figure 17.12a in the textbook.	**5.** Illegitimate recombination involving the transgene and chromosomal DNA leads to insertion of the selectable marker at locations other than the target gene. This leads to transformants that still contain (i.e., wild-type for) the target gene. Mice use illegitimate recombination as the primary mechanism for incorporating transgenes into their genome.
6. Describe the approach used to select against illegitimate recombination for gene knockouts. **TIP:** See Figure 17.12a in the textbook.	**6.** Addition of a negative selectable marker to one end or the other (or both) of the transgene is used to select against transformants that result from illegitimate recombination of the transgene into the genome. Transformants are cultured in medium that selects for the positive selectable marker (e.g., resistance to a drug such as neomycin) and against the negative selectable marker (e.g., sensitivity to a nucleotide analog such as ganciclovir). Only transformants that arise via recombination involving the homologous sequences that flank the target gene will incorporate the positive selectable marker but not the negative selectable marker into their genome; therefore, only these transformants will survive.
Solve **7.** Answer the problem by summarizing the difference in gene knockout strategies and by giving examples of organisms for which each strategy is commonly used.	**7.** **Answer:** When illegitimate recombination frequency is low, as in yeast, the knockout can be performed using a positive selectable marker flanked by DNA sequences homologous to the sequences that flank the target gene. When illegitimate recombination frequency is high, as in mice, the transgene should contain an additional, negative selectable marker outside of the sequences homologous to target-gene flanking sequences.

17.03 Solutions to End-of-Chapter Problems

1. *Evaluate:* This is type 2. This problem tests your understanding of the reporter genes *GFP* and *lacZ*. Recall that GFP is green fluorescent protein, and that *lacZ* encodes β-galactosidase. *Deduce > Solve:* GFP is observed using fluorescence microscopy, and its expression can be detected in living cells, tissues, whole organisms, or in fixed samples, directly, without further manipulation of the sample. Detection of *lacZ* requires application of substrates for β-galactosidase or antibodies to the enzyme. Thus, *GFP* is often advantageous over *lacZ* when wishing to analyze whole live animals. This is particularly true for *C. elegans*, which are transparent, and for larval and some adult surface structures in *Drosophila* and mice. If GFP is expressed at a low level and is difficult to detect above background fluorescence, or if the tissue of expression is not observable in the whole animal, which is true for many mouse tissues, then detection of *lacZ* expression by β-galactosidase activity assays or by using anti–β-galactosidase antibodies to stain tissue sections could be advantageous.

2. *Evaluate:* This is type 2. This problem requires comparison of gene expression analysis using transcriptional gene fusions to those using translation gene fusions. *Deduce > Solve:* Transcriptional fusions determine the cell types in which the gene is transcribed, whereas translational fusions determine the cell types in which transcription and translation occur and can also identify regulation due to elements in part of the mRNA and protein itself. The problem states that the transcriptional fusion detects expression in all cells, while the translational fusion detects protein accumulation in only one cell type. This information indicates that cell-type-specific expression of the gene is controlled posttranscriptionally, either by regulating initiation of translation, mRNA stability, or protein stability.

3a. *Evaluate:* This is type 3. This problem tests your understanding of reverse genetic approaches to study mouse gene function. *Deduce > Solve:* The first step is to identify the mouse ortholog of human DMD using bioinformatics. If mutant alleles of the mouse DMD (M-DMD) already exist, then these mutants can be used as the model for DMD in mice. If no M-DMD mutants are already available, then mutants can be created by gene knockout or allele knock-in strategies. For example, the gene knockout would be performed by creating a transgene that contains DNA flanking an essential DMD exon with a positive selectable marker in the place of the DMD exon. A negative selectable marker would be included outside the flanking DNA (e.g., the order of DNA elements would be flanking DNA on the left, positive selectable marker, flanking DNA on the right, negative selectable marker). Injection of this construct into mouse ES cells and selection for the positive selectable marker and against the negative selectable marker would yield ES clones containing the knockout of at least one M-DMD allele. These ES cells would then be injected into blastocysts of a mouse whose coat-color genotype is different from that of the mouse used as the ES cell source, and the blastocysts would then be transplanted into a pregnant female. Chimeric mouse pups would be identified by coat color. These mice would then be bred to determine which ones carry the knockout construct in their germ line. These mice, which would most likely be heterozygous for the knockout, would be bred to produce litters that include homozygous mutant knockout pups, whose development and adult phenotype could be studied to characterize the phenotype of loss of M-DMD function.

3b. *Evaluate:* This is type 3. This problem tests your understanding of reverse genetic approaches to study *Drosophila* gene function. *Deduce > Solve:* The first step is to identify the *Drosophila* ortholog of human DMD using bioinformatics. If mutant alleles of *Drosophila* DMD (D-DMD) are already available, then these mutants can be used as the model for DMD in *Drosophila*. If no D-DMD mutants are known, then RNAi technology can be employed to knock out D-DMD expression and create a model for at least partial loss of D-DMD function. For example, stem-loop-containing, double-stranded RNA corresponding to a D-DMD exon could be injected into embryos and the phenotype of these embryos examined. Alternatively, transgenes that drive expression of the double-stranded RNA under control of a heat-shock-inducible promoter could be created and introduced into flies using *P* element–mediated transformation. Flies carrying the transgene in their germ line could be identified by breeding, and flies from that line could be heat shocked at various times of development to determine the phenotype of partial loss of D-DMD function.

4. *Evaluate:* This is type 1. This problem tests your understanding of homologous recombination and its use in creating transgenic organisms. *Deduce > Solve:* Creation of transgenic yeast or mice via homologous recombination requires the transforming DNA to contain regions of homology with the genome that flank a positive selectable marker. This organization of the transforming DNA (order: homologous segment, positive selectable marker, homologous segment) is sufficient in yeast because homologous recombination is common, and illegitimate (non-homologous) recombination is rare. Because mice cells favor illegitimate recombination over homologous recombination, a negative selectable marker must be included on one side or the other of the homologous regions. This transformation construct (order: homologous segment, positive selectable marker, homologous segment, negative selectable marker) will yield viable transformants only when two recombination events occur, one involving each homologous segment.

5.   *Evaluate:* This is type 4. This problem tests your understanding of the use of recombinant DNA technology to produce recombinant protein. *Deduce > Solve:* To produce human insulin in the milk of sheep, the human insulin gene would first have to be cloned and used to create a transgene that allows for expression of insulin in milk; then that transgene would be used to create transgenic sheep. Transgene construction would require the identification of sheep regulatory sequences that direct gene expression in cells that contribute to milk and not in any other cell type (inappropriate expression of human insulin elsewhere in sheep may be harmful). The fusion construct would include sequences that direct proper cell-type-specific expression and perhaps sequences that direct secretion of insulin (although the secretory signals present in the human gene may suffice). Transgenic sheep could be created by techniques analogous to those used with mice.

6.   *Evaluate:* This is type 1. This problem tests your understanding of the use of recombinant DNA technology to generate transgenic animals. *Deduce > Solve:* Gene therapy requires the introduction of transgenes into human cells, which requires that those cells are accessible to manipulation. Bone marrow cells that give rise to blood cells can be isolated, cultured and manipulated in vitro, and reintroduced into the same individual to avoid immune responses. Furthermore, the transgenic blood cells produced could be used to deliver recombinant proteins to all regions of the body infiltrated by the vascular system. This is in contrast to cells that are in tissues not directly accessible to manipulations (e.g., heart, liver, other internal organs) or cell types that permanently reside in many different parts of the body (e.g., skeletal muscle and bone).

7.   *Evaluate:* This is type 3. This problem tests your understanding of RNAi technology. *Deduce > Solve:* RNAi technology results in knockout in expression of the targeted gene. For diseases caused by recessive alleles, which typically involve loss of gene function or reduction in gene function, further reduction in gene expression using RNAi would be of little value. For diseases that are dominant, which are typically due to gain of gene function or increase in gene function, reduction in gene expression using RNAi could be effective. In situations where RNAi therapy could be effective, the major obstacle to using RNAi in therapy would be the ability to accurately deliver the double-stranded RNA molecule to all of the correct cells and no other cell type. In many cases, cells affected by human genetic disorders are located throughout the body, are in close proximity to unaffected cells, and are not always amenable to experimental manipulation.

8.   *Evaluate:* This is type 1. This problem tests your understanding of transgenic technology for *Drosophila melanogaster* and plants. *Deduce > Solve:* Transformation uses naturally occurring genetic elements in both cases. For plants, the element is from the *Agrobacterium* Ti plasmid and is called T-DNA. T-DNA mediates the transfer of DNA from *A. tumefaciens* to plant cells and the random integration of the transforming DNA into the plant genome. Any DNA cloned into the T-DNA region in *A. tumefaciens* cells is transferred into the plant by those cells. Transfer occurs naturally in dicotyledonous plants and can be chemically induced in monocots. For many species, the transgenes can be selected in callous tissue culture cells with the aid of a selectable marker and whole, regenerated plants that carry the transgene. For *D. melanogaster*, the element is from the *P* element transposon and includes the direct repeats at the ends of the *P* element along with the *P* element transposase gene. Any DNA cloned between *P* element repeats is randomly integrated into the *Drosophila* genome if injected into *Drosophila* embryos along with the *P* element transposase gene.

9.   *Evaluate:* This is type 3. This problem tests your understanding of reverse genetic approaches involving insertions and TILLING. *Deduce > Solve:* The pros and cons of insertion mutations versus point mutations from TILLING screens depend on the goal of the investigation. If researchers wish to determine the loss-of-function phenotype, then insertion alleles are useful. If the insertion is a transposon or T-DNA, it can also be advantageous in using the mutant as a source of DNA for cloning the mutated gene. For example, the cloning of a gene disrupted by a *P* element insertion can be accomplished by using an inverse PCR approach. If multiple

alleles are desired, including partial loss-of-function alleles as well as null alleles, then TILLING is advantageous. TILLING requires that the sequence of the organism's genome is known, however. Since neither method is targeted to specific genes, both share the possibility that mutations induced in other genes may confound the results.

10. *Evaluate:* This is type 2. This problem tests your understanding of the application of recombinant DNA technology to study gene expression. *Deduce > Solve:* The cloned mouse ortholog to human HD (M-HD) could be used to study M-HD expression in a couple of ways. One would be to use the gene to generate a probe for in situ hybridization analysis of mRNA accumulation. Another would be to create a fusion of *GFP* or *lacZ* to the control region of the M-HD and then to introduce the transgene into the mouse genome by homologous recombination. This would place the M-HD–*GFP* or –*lacZ* fusion under control of the M-HD regulatory sequences at the M-HD locus. Analysis of *GFP* or *lacZ* expression would then be used to study M-HD expression.

11a–b. *Evaluate:* This is type 2. This problem requires an understanding of the application of recombinant DNA technology to study gene expression. *Deduce > Solve:* **(a)** Genes expressed in response to exposure to low temperature could initially be identified using DNA microarray experiments. The temporal and spatial pattern of the low-temperature-induced genes could be accomplished by creating transcriptional or translation fusions of selected, temperature-regulated genes to a reporter gene such as *GFP* or *lacZ*. The gene fusions would then be introduced into plants using T-DNA–mediated transformation to create transgenic plants. **(b)** Farmers could observe gene expression in the field by using *GFP* reporter-gene fusions, although they would require a UV light source. Alternatively, the reporter gene could encode an enzyme that produced a visible pigment.

12a. *Evaluate:* This is type 3. This problem tests your understanding of reverse genetic analysis in yeast. *Deduce > Solve:* Loss-of-function mutations in each of the five yeast genes identified as salt-induced would be studied individually at first. Each gene would be deleted by gene knockout using homologous recombination to replace the gene with a positive selectable marker, creating five mutant strains. The mutants would then be grown under low-salt conditions, shifted to high-salt conditions, and observed for a mutant phenotype. If individual mutations had no effect, then crosses could be performed to create double, triple, quadruple, and quintuple deletion mutant strains for analysis. PCR with appropriate primers can be used to verify the presence of disrupted genes. Gain-of-function mutations in each gene could be created by increasing the copy number of each gene to induce overexpression, followed by examining each overexpresser for a mutant phenotype. Gain-of-function mutations could also be created by placing the coding sequence of each gene under control of yeast *GAL* gene regulatory elements, generating transgenic yeast strains, and observing the phenotype of the strains after shifting from glucose medium (Gal promoter off) to galactose medium (Gal promoter induced to high levels).

12b. *Evaluate:* This is type 3. This problem tests your understanding of reverse genetic analysis in plants. *Deduce > Solve:* RNAi would have to be used to test the effects of loss-of-function of tomato genes because TILLING and knockout library approaches are, to date, impractical in that organism. The salt-induced genes would have to be cloned and their sequence determined to provide the necessary information for designing stem-loop-containing dsRNA constructs. The effect of RNAi-mediated knockout of each gene would be studied individually by creating transgenic tomato plants that contain transgenes expressing the double-stranded RNA of each gene under control of a constitutive plant promoter. Five transgenic plant lines would be developed, each carrying the RNAi construct of one gene, and the plants would be studied under low- and high-salt conditions for potential mutant phenotypes. For gain-of-function mutations, each gene would be placed under control of a highly active promoter and then used to generate transgenic plants. Each plant would overexpress one gene and could be studied under low- and high-salt conditions for potential mutant phenotypes.

13. *Evaluate:* This is type 1. This problem tests your ability to analyze transgenic plants. *Deduce:* Each transformant contains one or more copies of a dominant selectable herbicide resistance marker in its genome. Each locus of insertion can be given the genotype Rr, where R stands for the integrated dominant herbicide resistance marker and r stands for the absence of that marker on the homologous chromosome. The $T_1 \times$ wild-type crosses will be a monohybrid test cross for transformants with a single copy of the marker ($Rr \times rr$, which yields a progeny ratio of 1:1 resistance: sensitivity) or a dihybrid test cross for transformants containing two unlinked copies of the marker ($R_1r_1 \, R_2r_2 \times r_1r_1 \, r_2r_2$, which yields a ratio of 3:1 resistance: sensitivity). The T_2 self-crosses will be a self-cross of a monohybrid strain in the case of single-copy T_1 transformants ($Rr \times Rr$, which yields a ratio of 3:1 resistance: sensitivity). For double-copy T_1 transformants, the T_2 self-cross results depend on the T_2 plant chosen. For a T_2 plant with two copies of the marker, the cross will be a dihybrid self-cross ($R_1r_1 \, R_2r_2 \times R_1r_1 \, R_2r$, which yields a ratio of 15:1 resistance: sensitivity). For a T_2 plant with only one copy of the marker, the cross will be a monohybrid self-cross ($R_1r_1 \, r_2r_2 \times R_1r_1 \, r_2r_2$ or $r_1r_1 \, R_2r_2 \times r_1r_1 \, R_2r_2$, both of which yield a ratio of 3:1 resistance: sensitivity). *Solve:* For transformant Line 1, the results of the $T_1 \times$ wild type and T_2 self-cross are consistent with a single insertion of the herbicide selectable marker. For the transformant Line 2, the results of the $T_1 \times$ wild type and T_2 self-cross are consistent with two unlinked insertions of the selectable marker. For transformant Line 3, the results of the crosses fit neither model. Line 3 may contain more than two insertions of the selectable marker or two selectable markers integrated at linked sites on opposite homologs of the same chromosome $\left(\frac{R_1r_2}{r_1R_2}\right)$. Chimeric regions in the reproductive tissues could also lead to unusual ratios—as could transgene-induced gene silencing, which often occurs when multiple copies of a gene are inserted into a plant genome.

14. *Evaluate:* This is type 3. This problem tests your understanding of the application of recombinant DNA technology to study bacterial gene function. *Deduce > Solve:* Two approaches are possible, one determining which genes are required for crude oil metabolism in the strain mentioned in the problem, and another determining which genes are sufficient to impart oil metabolism to a strain that does not normally metabolize oil. The former approach can be executed by randomly mutagenizing the oil-metabolizing strain and isolating mutants that cannot metabolize oil. These genes can be mapped and cloned to identify the oil metabolism genes. The latter approach can be executed by generating a recombinant DNA library using the oil-metabolism strain as a source of DNA and then transforming the library DNA into a strain that cannot metabolize oil. Transformants could be plated on medium that requires metabolism of oil for growth and, if any transformant received all the genes necessary to metabolize oil, it would grow on the plate. Since bacterial genes that are related in metabolic function are often clustered together in operons, there is a chance that a clone could contain the entire operon and therefore be identified by this strategy.

15. *Evaluate:* This is type 1. This problem tests your understanding of transgenic technology in plants. *Deduce > Solve:* To create transgenic plants that express Bt toxin only in response to the plant being fed upon by insects, the constitutive promoter presently used for the Bt toxin gene should be replaced by one that is induced only in response to insect attack. One could identify endogenous plant genes that are induced specifically in response to insect attack and then create a transcriptional fusion of the promoter from one of those genes to the Bt toxin coding sequence. To create transgenic plants that lack the selectable marker, the selectable marker would have to be removed from the genome of the transgenic plant. This could be accomplished by using the *Cre–lox* system for deleting DNA. In this case, the selectable marker would be flanked by *loxP* sites and then the gene for Cre-recombinase would be transiently introduced to promote excision of the marker. Transient introduction of Cre could be accomplished by crossing the transgenic line containing the "floxed" marker and the inducible Bt toxin gene with a line containing the *Cre* gene at an unlinked locus. Progeny that contain *Cre* and the Bt toxin gene but have lost the selectable marker could be identified and then crossed to wild type to create a transgenic line that contains only the inducible Bt toxin gene.

16a.   *Evaluate:* This is type 2. This problem requires an understanding of the application of recombinant DNA technology to study gene expression. ***Deduce > Solve:*** Expression of *Hoxa7* and *Hoxb7* could be studied in wild-type (non-transgenic) animals by in situ hybridization to identify tissues containing the mRNAs or by using antibodies to detect the proteins. Transcriptional or translational gene fusions of *GFP* or *lacZ* to *Hoxa7* and *Hoxb7* could also be performed; however, the gene fusions would have to be knocked in to their normal loci in order to most accurately study their expression. The knock-in alleles would therefore have to be functional, since they replace the endogenous alleles. Alternatively, if loss of *Hoxa7* and *Hoxb7* function were recessive, you could create heterozygotes containing one wild-type and one knock-in fusion allele. Regardless of whether wild-type or transgenic lines are used, multiple samples using animals at different stages of development would be needed to assess the temporal pattern of gene expression.

16b.   *Evaluate:* This is type 3. This problem tests your understanding of reverse genetic techniques for mouse. ***Deduce > Solve:*** Loss-of-function alleles of *Hoxa7* and *Hoxb7* could be created by performing gene knockouts for each gene. The gene knockout would be performed by creating a transgene that contains DNA flanking an essential *Hox7* exon with a positive selectable marker in the place of the DMD exon. A negative selectable marker would be included outside the flanking DNA (for example, the order of DNA elements would be flanking DNA on the left, positive selectable marker, flanking DNA on the right, negative selectable marker). Injection of this construct into mouse ES cells and selection for the positive selectable marker and against the negative selectable marker would yield ES clones containing the knockout of at least one *Hox7* gene. These ES cells would then be used to generate mice that carry the knockout construct in their germ line.

16c.   *Evaluate:* This is type 3. This problem requires an understanding of reverse genetic approaches to investigate redundant gene function. ***Deduce > Solve:*** If *Hoxa7* and *Hoxb7* have one or more redundant functions, then a strain that is homozygous for loss of function mutations in both genes $\left(\frac{hoxa7^-}{hoxa7^-}\,\frac{hoxb7^-}{hoxb7^-}\text{ double mutants}\right)$ will have a more severe or different phenotype than the combinations of phenotypes of strains homozygous for loss-of-function mutations in each gene $\left(\frac{hoxa7^-}{hoxa7^-}\text{ and }\frac{hoxb7^-}{hoxb7^-}\text{ single mutants}\right)$. Each single mutant would be created by performing gene knockouts and then the single mutant would be crossed to isolate double mutants.

17a.   *Evaluate:* This is type 2. This problem tests your understanding of enhancer trap technology and recombinant DNA cloning. ***Deduce > Solve:*** The enhancer trap line contains a *P* element inserted near an enhancer that directs gene expression in wing imaginal disks. Genes near this location can be cloned by using two alternative approaches using the genomic DNA from the enhancer trap line. In one approach, a genomic DNA library can be created from the line, which will contain recombinant clones that include the *P* element and flanking DNA sequences. These clones can be identified using *P* element DNA as a probe in a colony hybridization technique. Alternatively, PCR can be used to amplify sequences flanking the inserted *P* element in an approach called inverse PCR. In this technique, genomic DNA from the enhancer trap line is (1) digested with an enzyme that does not cut the *P* element, (2) ligated under condition where the ends of each fragment are ligated to form circles, (3) digested with a restriction enzyme (RE) that cuts in the middle of the *P* element, and (4) amplified by PCR using primers that hybridize to the ends of the *P* element. The product from this PCR reaction will be a linear fragment with *P* element DNA at its ends and genomic DNA on each side of the element from the enhancer trap line between them. See figure for illustration of inverse PCR.

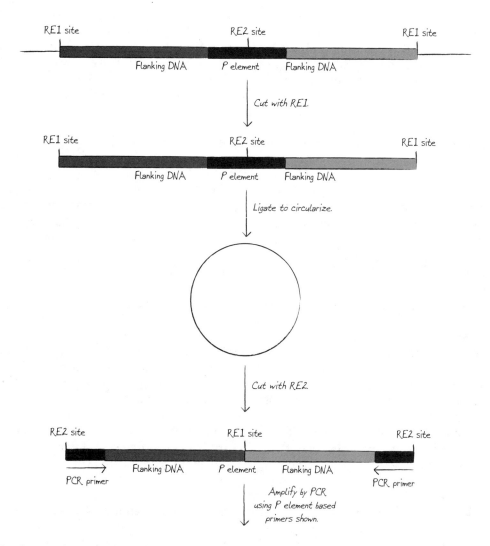

17b. *Evaluate:* This is type 2. This problem requires an understanding of the application of recombinant DNA technology to the study of gene expression. *Deduce > Solve:* Expression of the endogenous gene identified in the enhancer trap line could be studied by in situ hybridization. Alternatively, the gene fusions could be constructed using *GFP* or *lacZ*, the transgenes could be transformed into *Drosophila* using *P* element transformation, and expression of the gene fusions could be studied.

18a. *Evaluate:* This is type 3. This problem requires an understanding of reverse genetics technology in yeast. *Deduce > Solve:* To create loss-of-function *fun* gene mutants in yeast, transgenes for knocking out the *fun* gene would first be constructed and then used to replace the endogenous *fun* gene by transformation and homologous recombination. The gene knockout transgene would contain DNA flanking the *fun* gene on either side of a selectable marker (order of DNA elements would be flanking DNA, selectable marker, flanking DNA).

18b. *Evaluate:* This is type 2. This problem requires understanding of the use of reporter genes to study protein localization. *Deduce > Solve:* To determine the subcellular localization of *fun* gene proteins, *fun* gene–*GFP* translational fusions would be created. The translational fusion should include the entire *fun* gene coding sequence with the *GFP* sequence added either at the carboxyl terminus or amino terminus of the fun protein. The gene fusion could be introduced into yeast on a plasmid, integrated randomly into the genome, or used to replace the endogenous *fun* gene. GFP fluorescence would then indicate the subcellular localization of the fun protein.

19. *Evaluate:* This is type 2. This problem requires understanding of translation gene fusions and genetic complementation. *Deduce > Solve:* The translational gene fusion should contain most if not all of the coding sequence of the gene in question. This gene fusion could be introduced into a strain that is homozygous for a loss-of-function mutation of this gene to create a transgenic organism in which the only copy of the gene in question is the gene fusion. If the transgenic strain has the wild-type phenotype, then the gene fusion is functional. If the transgenic strain has the mutant phenotype, then the gene fusion is nonfunctional. If the transgenic strain has an intermediate phenotype, then the gene fusion has partial function.

20.   *Evaluate:* This is type 1. This problem tests your understanding of creating recombinant DNA clones for expression of protein in *E. coli*. *Deduce > Solve:* Both sequences encode the β-chain of human insulin. The top, recombinant sequence differs from the bottom, human sequence in that an ATG start codon has been added to its 5′ end, and it contains different nucleotides at the third position of many of the codons. The ATG was added for translation initiation and expression in *E. coli*: the human sequence is part of the preproinsulin sequence whose ATG start codon is further upstream. The third position of many codons is different from the native gene sequence because it was selected without knowledge of the human codon used and perhaps with an eye toward codons preferred by *E. coli*.

21.   *Evaluate:* This is type 3. This problem tests your understanding of transposons as transformation vectors and of enhancer trap techniques. *Deduce:* The transposon vector described would be inserted randomly in the genome, placing nearby genes under control of the specific promoter and enhancer included in the vector. In some instances, the transposon vector would prevent nearby genes from being expressed at their normal time and in their normal tissue, causing expression at inappropriate times and in inappropriate tissues. Inappropriate expression could result in a loss-of-function mutation if the loss-of-normal expression causes a mutant phenotype while gain-of-abnormal expression does not. The gain-of-abnormal expression could also cause a mutant phenotype even if the loss-of-normal expression does not. And if the loss-of-normal expression and the gain-of-abnormal expression both result in mutant phenotypes, the inserted transposon vector could conceivably result in both loss- and gain-of-function mutations.

22.   *Evaluate:* This is types 3 and 4. This problem tests your understanding of site-directed mutagenesis, recombinant protein expression in *E. coli*, and allele knock-in technology in mice. *Deduce:* Two versions of the mouse *RAS* gene are needed to accomplish the goals provided by this problem. A genomic clone would be needed for site-directed mutagenesis to create a mutant allele, which would then be used to knock in to the mouse genome to create an oncogenic RAS mouse mutant. A cDNA clone would be needed for site-directed mutagenesis to create a mutant allele, which would then be cloned into a bacterial expression vector for production of recombinant protein for biochemical analysis of oncogenic RAS protein. *Solve:* Mutant versions of the *RAS* gene and cDNA would be constructed using the same strategy, which involves PCR amplification of the *RAS* sequence using one PCR primer that hybridizes to the site of the wild-type Gly codon but that contains the mutant Val codon in its place (the middle G of the wild-type Gly codon would be replaced with a T of the mutant Val codon). This PCR reaction would create mutant genomic or cDNA clones. The mutant genomic clone would then be used to construct a knock-in transgene, which would also include genomic DNA flanking the mutated exon, a positive selectable marker, and a negative selectable marker (the order of DNA elements would be flanking DNA, mutant exon, positive selectable marker, flanking DNA, negative selectable marker). This construct would be injected into mouse ES cells, and clones in which the mutant allele has replaced one of the wild-type alleles would be identified by selection for the positive selectable marker and selection against the negative selectable marker. These ES cells would be used to generate transgenic mice that contain the mutant knock-in allele in their germ line. The mutant cDNA would be cloned into a bacterial expression vector such that the coding sequence is under control of a bacterial promoter. This clone would then be introduced into bacterial cells to create a transgenic strain that would be used to produce recombinant mutant RAS protein for use in in vitro studies.

23.   *Evaluate:* This is type 1. This problem tests your understanding of transgenic technology in *Arabidopsis*. *Deduce > Solve:* VTE4 encodes an enzyme that converts α-tocopherol to α-tocopherol, and *Arabidopsis* plants accumulate γ-tocopherol in seeds. This information suggests that plants with high levels of α-tocopherol in seeds could be created by expression of the *VTE4* gene in seeds. A transcriptional fusion between a promoter from a seed-specific gene and the *VTE4* coding region could be created and then introduced to *Arabidopsis* by T-DNA–mediated transformation. A constitutive promoter such as the CaMV 35s promoter should also work.

24.   *Evaluate:* This is type 4. This problem tests your ability to analyze gene expression data obtained from studying transgenic *Drosophila* carrying gene fusions. *Deduce > Solve:* All transgenic lines contain a minimal promoter that includes the region from the transcription start site to –0.04 kb. Since lines that contain only this region fail to express *lacZ*, this region is not sufficient for expression. No lines lack this region; therefore, it cannot be determined from this study whether the region is required for expression, although that is likely to be true given its proximity to the transcription start site. All transgenic lines that express *lacZ* at any level contain at least the region from about –1.0 kb to –1.1 kb; therefore this region, together with the +1 to –0.04 region, is required for low-level expression. Lines that contain from –1.0 to –1.55 in addition to +1 to –0.04 give wild-type-like expression; therefore, these regions are identified by this study as containing sequences that determine the expression of *even-skipped* in segment 2.

25a.   *Evaluate:* This is type 1. This problem tests your understanding of recombinant protein production using transgenic *E. coli*. *Deduce > Solve:* The coding sequence of the lipid-degrading enzyme would be cloned into an *E. coli* expression plasmid to create a transcriptional and translational fusion. Transcription of the chimeric protein would be under control of an inducible promoter, and translation would include an *E. coli* secretion signal sequence at the amino terminus of the enzyme. Bacteria carrying the plasmid would be cultured under conditions that prevent expression of the transgene and then transferred to conditions that induce the transgene production. The enzyme should be easily isolated due to its section into the culture medium.

25b.   *Evaluate:* This is type 1. This problem tests your understanding of optimizing recombinant protein production in transgenic *E. coli*. *Deduce > Solve:* When expression of the recombinant protein is successful but the level of expression is low, poor transcriptional or translational efficiency are commonly underlying reasons. Since the promoters used in these experiments are standard and known to work well, translational efficiency is the more likely culprit. Translation efficiency can be improved by changing codon usage from those in the original gene to those that are not are preferred in *E. coli*. For example, the Gly codons GGA and GGG are rarely used in *E. coli*, whereas GGC and GGT are predominantly used. Altering the lipid-degrading enzyme's Gly codons from GGA or GGG to GGC or GGT may increase the efficiency of translation and improve the level of enzyme production in *E. coli*.

17.04 Test Yourself

Problems

1.   Draw a table that that lists the organisms described in this chapter (bacteria, yeast, fruit fly, plants, and mice) in one column, the corresponding transformation vector(s) in another, and the selectable markers used for transformation in a third.

2.   Describe the general process by which transgenic fruit flies are generated.

3.   Describe the general process by which transgenic plants are generated using *Agrobacterium*.

4.   Describe the general process by which transgenic mice are generated by homologous recombination.

5. In experiments using gene fusions to study temporal (developmental time) and spatial (the tissue specificity) gene expression in mice, it is important to integrate the gene fusion at the same locus as the endogenous gene as opposed to insertion at random sites. Why?

6. Many animal genes are essential for the development and function of both embryonic and adult tissue. Combine your understanding of transcriptional gene fusions and the function of the *Cre* recombinase gene to create a line of mice that undergoes adult tissue–specific deletion of a transcription factor gene as an adult. Assume that you already have a germ-line-transgenic mouse that is homozygous for an allele of the transcription factor gene flanked by *loxP* sites (the gene is floxed).

Solutions

1. *Evaluate:* This is type 1. This problem tests your general knowledge of transgenic technology as presented in the textbook. *Deduce > Solve:* See table.

Organism	Vector	Selectable Marker
Bacteria	Circular plasmids	Antibiotic-resistant genes
Yeast	Plasmids, linear DNA molecules	Drug-resistant genes
Fruit flies	*P* element	w^+ gene and ry^+ gene
Plants	*Agrobacterium* T-DNA–containing plasmids	Antibiotic-resistant gene for bacteria; herbicide-resistant gene for plants
Mice	Linear DNA molecules	Drug-resistant genes; ganciclovir sensitivity gene (thymidine kinase)

2. *Evaluate:* This is type 1. This problem tests your understanding of the procedure for generating transgenic fruit flies (*Drosophila*). *Deduce > Solve: Drosophila* embryos are injected with two DNA molecules: in one molecule, the ends of the *P* element transposon flank the transgenes, and it also includes a selectable marker (e.g., ry^+); the second molecule is a plasmid carrying the *P* element transposase gene. Expression of transposase catalyzes transposition of the transgene into random sites in the genome. Flies that express the selectable marker are bred to identify germ-line transformants.

3. *Evaluate:* This is type 1. This problem tests your understanding of transgenic plant formation. *Deduce > Solve: Agrobacteria* are transformed with two plasmids: one contains the genes necessary for virulence and conjugative transfer of T-DNA to plants cells, and the other contains the transgenes, including a plant selectable marker (herbicide-resistant gene), flanked by T-DNA border sequences. These bacteria are then placed in contact with plants cells, where they mobilize the T-DNA segment into the plant cells. If plant-cell culture is used, transformed plant cells are allowed to regenerate whole plants, all of which will be germ-line transformants. For *Arabidopsis*, flowers can be exposed to *Agrobacteria*, and seeds that contain the transgenes can be selected to identify transgenic plants, all of which will be germ-line transformants.

4. *Evaluate:* This is type 1. This problem tests your understanding of the procedure for generating transgenic mice. *Deduce > Solve:* The transforming DNA is linear and contains the transgene disrupted by a positive selectable marker (e.g., neomycin-resistant gene) if creating knockouts or along with the marker if creating knock-ins. The DNA to be inserted is flanked by sequences homologous to the chromosomal position chosen as the site for insertion of the transgenes. The DNA molecule also contains a negative selectable marker (e.g., thymidine kinase for sensitivity to ganciclovir) at one end, outside of the homologous flanking DNA. This DNA is injected into mouse ES cells from a brown-furred (*BB*) mouse, and these cells are then cultured in medium that selects for expression of the positive selectable marker and against expression of the negative selectable marker. The chromosomal DNA of transformed cells is analyzed using PCR or a Southern blot to identify those that contain a single copy of transgenes integrated at the correct chromosomal locus. These ES cells are implanted into blastocysts of a black-furred (*bb*) mouse embryo, and transformed mice pups are identified as chimera, containing patches of brown fur. These are bred with black-furred mice. Progeny that are completely brown-furred are heterozygous, germ-line transformants. Heterozygotic germ-line transformants are bred to produce homozygous germ-line transformants.

5.   *Evaluate:* This is type 2. This problem tests your understanding of the use of transgenic animals to study gene expression. ***Deduce > Solve:*** Gene expression in eukaryotes and specifically in animals is affected by chromatin structure, which can differ from one chromosomal location to the next. Random integration of a gene fusion will generate a population of transgenic animals that differ with respect to the site of integration, and this result often leads to variation in the pattern or level of fusion gene expression among animals. Directed integration of the gene fusion at a specific locus removes this variability from the analysis, and placement of the gene fusion at the same site as the endogenous gene places the gene fusion in a chromatin environment most similar to the endogenous gene.

6.   *Evaluate:* This is type 3. This problem tests your understanding of reverse genetics and site-specific recombination using the *Cre–lox* system. ***Deduce > Solve:*** The floxed transcription factor gene can be deleted in specific adult tissue if the Cre recombinase is expressed in that tissue. To direct Cre expression in only the desired adult tissue, a promoter and enhancer specific to that tissue in adults would be used to create a transcriptional fusion to the *Cre* recombinase coding sequence. A transgene containing this gene fusion would be used to transform ES cells from an otherwise wild-type mouse, and transgenic animals from that mouse could be screened for those that contain the transgene in their germ line. Those mice would be screened for transgenic animals that express Cre only in the correct adult tissue. These mice would be mated to mice that contain the floxed transcription factor gene to generate mice that carry both the *Cre* gene and the floxed transcription factor gene. These mice would be bred to generate a line that is homozygous for the floxed transcription factor gene and contains at least one copy of the *Cre* gene fusion. These mice can be examined via PCR to confirm loss of the transcription factor genes in the appropriate adult tissue, and the phenotype of such mice would provide information about the function of the transcription factor in that adult tissue.

18

Genomics: Genetics from a Whole-Genome Perspective

18.01 Genetics Problem-Solving Toolkit

Key Terms and Concepts

Basic local alignment search tool (BLAST): The computer program used to search for DNA, RNA, and protein sequences within databases that are similar to a sequence of interest called a query.

Contig: DNA sequence assembled by combining overlapping, individual sequence reads.

Expression array: DNA microarray that contains oligonucleotide sequences corresponding to all or most known genes in a genome and is used to characterize an organism's transcriptome.

Genome annotation: Information explaining the structure and function of genetic elements in a genome sequence.

Paired-end sequencing: The sequencing of both ends of large DNA clones, which is used to identify clones that span gaps in a genome sequence.

Scaffold: Multiple-sequence contigs that can span a portion or all of a chromosome.

Whole-genome tiling array: DNA microarray that contains oligonucleotides corresponding to an entire genome and is used to characterize an organism's transcriptome and identify DNA polymorphisms.

Key Analytical Tools

Genomic Analysis	Goals and Objectives of the Analyses	Experimental Approaches or Tools Employed
Structural genomics	Sequence genomes Annotate genomes • Identify genes • Identify regulatory sequences • Identify chromosomes • Identify chromosome replication and segregation sequences	Shotgun sequencing Clone-by-clone approach Paired-end reads Metagenomic approach High-throughput cDNA sequencing Bioinformatic analysis • Gene-finding programs • BLAST searches

(continued)

Genomic Analysis	Goals and Objectives of the Analyses	Experimental Approaches or Tools Employed
Evolutionary genomics	Identify DNA polymorphisms Understand species relationships Understand gene relationships Annotate genomes	BLAST searches Phylogenetic analyses • Species trees • Gene trees
Functional genomics	Characterize the transcriptome Identify genetic interactions Identify protein–protein interactions Indentify essential genes Determine the function of genes Identify transcription-factor binding sites	Expression microarray analysis High-throughput cDNA sequencing Genetic screens for synthetic–lethal interactions Two-hybrid screens for protein–protein interactions Gene knockout studies Chromatin immunoprecipitation analysis (ChiP analysis)

18.02 Types of Genetics Problems

1. Describe how to determine, assemble, and annotate genomes or analyze genome sequences.
2. Describe how genes and genomes evolve, or analyze the results of gene and genome comparisons.
3. Describe experimental approaches for analyzing gene function, or interpret the results of functional genome analyses.

1. Describe how to determine, assemble, and annotate genomes or analyze genome sequences.

This type of problem tests your understanding of how genome sequences are determined, assembled, and annotated. You may be asked to argue for or against specific strategies for sequencing a particular genome. You may also be asked to describe genome sequencing approaches, including clone-by-clone and metagenomic, or approaches for filling in gaps within genomic sequences created by repetitive DNA. You may be asked to assemble contiguous genomic sequences from individual sequence reads. See Problem 2 for an example.

Example Problem: The sequences below include three short contigs that do not overlap (sequences A, B, and C) and two sequences that represent paired-end reads from a larger, scaffolding clone from the same genome (end X and end Y). Which two of the three contigs do the paired-end reads indicate are part of the same scaffold? Draw a figure showing the scaffold including both contigs and the larger, scaffolding sequence.

Sequence A

    5'-GCTGCGCATCAGTATTTCCCGCTGGCAGCTGGATCTCAGTGCGCTGCTGG-3'

Sequence B

    5'-GATTACGCCCGTGCCTTATCCGGAGAGGATGAATGACGCGACAGGAAGAA-3'

Sequence C

    5'-TGGACCCTATGGTCAAGCAGCGCCGAAAAGGCCTGTGTCAAGGCCTACCA-3'

Paired-end X

    5'GCTGGCAGCTGGATCTCAGT 3'

Paired-end Y

    5'-CAGGCCTTTTCGGCGCTGCT-3'

Solution Strategies	Solution Steps
Evaluate	
1. Identify the topic this problem addresses, and explain the nature of the requested answer.	1. This problem tests your understanding of genome assembly. The answer requires the identification of contigs linked together in a scaffold by paired-end reads of a scaffolding clone.
2. Identify the critical information given in the problem.	2. There are three contig sequences and one pair of paired-end reads from a scaffolding clone in the same genome sequencing project. The sequences are listed.
Deduce	
3. Write the double-stranded sequence for each contig.	3. Sequence A top strand 5'-GCTGCGCATCAGTATTTCCCGCTGGCAGCTG GATCTCAGTGCGCTGCTGG-3' bottom strand 3'-CGACGCGTAGTCATAAAGGGCGACCGTCGA CCTAGAGTCACGCGACGACC-5' Sequence B top strand 5'-GATTACGCCCGTGCCTTATCCGGAGAGGATG AATGACGCGACAGGAAGAA-3' bottom strand 3'-CTAATGCGGGCACGGAATAGGCCTCTCCTAC TTACTGCGCTGTCCTTCTT-5' Sequence C top strand 5'-TGGACCCTATGGTCAAGCAGCGCCGAAAA GGCCTGTGTCAAGGCCTACCA-3' bottom strand 3'-ACCTGGGATACCAGTTCGTCGCGGCTTTTC CGGACACAGTTCCGGATGGA-5'
4. Determine which two contigs contain each paired-end sequence and which strand is identical to the paired-end read.	4. Sequence A contains the paired-end read X in the top strand: 5'-GCTGGCAGCTGGATCTCAGT-3' Sequence C contains the paired-end read Y in the bottom strand: 5'-CAGGCCTTTTCGGCGCTGCT-3'
5. Use the paired-end reads to determine how the two contigs identified in step 4 and the scaffolding clone are oriented relative to each other in the scaffold. TIP: The paired-end reads are oriented 5' to 3' from each end of the scaffolding clone.	5. Contig sequence A is to the left of contig sequence C as they are given in the problem; the paired-end read X is at the left end of the scaffolding clone, and read Y is at the right.

(*continued*)

Solution Strategies	Solution Steps
Solve	
6. Answer the problem by illustrating the contigs and scaffold in a figure.	6. **Answer:** See figure. 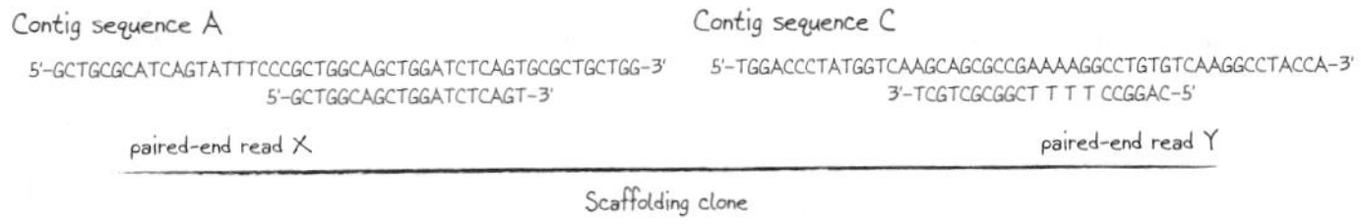

Contig sequence A

5'-GCTGCGCATCAGTATTTCCCGCTGGCAGCTGGATCTCAGTGCGCTGCTGG-3'
5'-GCTGGCAGCTGGATCTCAGT-3'

paired-end read X

Contig sequence C

5'-TGGACCCTATGGTCAAGCAGCGCCGAAAAGGCCTGTGTCAAGGCCTACCA-3'
3'-TCGTCGCGGCT T T T CCGGAC-5'

paired-end read Y

Scaffolding clone

2. Describe how genes and genomes evolve, or analyze the results of gene and genome comparisons.

This type of problem tests your understanding of how genes and genomes evolve. You may be asked to interpret phylogenetic trees constructed from genes or species, and identify orthologous and paralogous genes or predict the relative similarity of genes within a gene family that has evolved from a single ancestral gene. You may be asked to propose a hypothesis for the evolution of a gene by duplication and subfunctionalization or neofunctionalization, to describe how evolutionary genomics can be used to facilitate genome annotation, or to interpret the results of gene or genome comparisons in order to annotate a genome sequence. See Problem 16 for an example.

Example Problem: The figure shows the result of a blastp search using the predicted protein sequence of a segment of a newly sequenced fungal genome. Label the predicted cellular and molecular function of the sequence (see Figure 18.7), and provide a suggested name for the protein.

Sequences producing significant alignments:

Accession	Description	Max score	Total score	Query coverage	E value	Max ident
XP_002372304.1	telomerase reverse transcriptase, putative [Aspergillus flavus NRRL3357] >dbj|BAE55253.1| unnamed protein product [Aspergillus oryzae RIB40] >gb|EED56692.1| telomerase reverse transcriptase, putative [Aspergillus flavus NRRL3357]	431	431	100%	9e-145	64%
XP_001817254.2	hypothetical protein AOR_1_266174 [Aspergillus oryzae RIB40]	429	429	100%	2e-138	64%
XP_001273297.1	telomerase reverse transcriptase [Aspergillus clavatus NRRL 1] >gb|EAW11871.1| telomerase reverse transcriptase [Aspergillus clavatus NRRL 1]	426	426	99%	3e-137	63%
XP_001261379.1	telomerase reverse transcriptase [Neosartorya fischeri NRRL 181] >gb|EAW19482.1| telomerase reverse transcriptase [Neosartorya fischeri NRRL 181]	410	410	100%	2e-131	61%
XP_749051.2	telomerase reverse transcriptase [Aspergillus fumigatus Af293] >gb|EAL87013.2| telomerase reverse transcriptase, putative [Aspergillus fumigatus Af293]	406	406	100%	9e-130	60%
EDP48298.1	telomerase reverse transcriptase, putative [Aspergillus fumigatus A1163]	405	405	100%	3e-129	60%

Solution Strategies	Solution Steps
Evaluate	
1. Identify the topic this problem addresses, and explain the nature of the requested answer.	1. This problem tests your ability to use the results of a BLAST search to annotate an unknown gene sequence. The answer requires identifying the appropriate cellular and biochemical functions and proposing a name for the gene.
2. Identify the critical information given in the problem.	2. The figure shows the top hits from a blastp search using the predicted protein sequence encoded by a newly sequenced fungal gene.
Deduce	
3. State the names of the proteins identified by the blastp search.	3. The list shows six fungal genes: five of them are called telomerase reverse transcriptase, and one of them is a hypothetical protein with no name.
4. Determine whether the similarity of the sequences identified to the query sequence is statistically significant. TIP: The "E value" is the probability of the similarity resulting from random chance for the size of the database queried.	4. The *largest* E value is $3e^{-129}$; therefore, the similarity of the query sequence and those identified are unlikely to result from random chance.
Solve	
5. Answer the problem by using the statistically significant search results to select a cellular function, biochemical function, and name for the newly sequenced gene.	5. **Answer:** The newly sequenced gene is proposed to have the cellular function "cell growth, cell division, and DNA synthesis" and the biochemical function "nucleic acid enzyme." It should be called telomerase reverse transcriptase.

3. Describe experimental approaches for analyzing gene function, or interpret the results of functional genome analyses.

This type of problem tests your understanding of the experimental analysis of gene and genome function. You may be asked to describe experimental approaches appropriate for analyzing gene function or to interpret the results of such analyses. These analyses include expression microarrays, tiling microarrays, two-hybrid interactions, and gene knockouts. See Problem 22 for an example.

Example Problem: A two-hybrid analysis was performed to determine whether protein A and protein B physically interact. In the first experiment, protein A was fused to Gal4-DB and protein B was fused to Gal4-AD. Both hybrid proteins were expressed in a his^- yeast strain that contains the corresponding HIS^+ gene under control of the UAS_{GAL4}. The result was that the yeast strain had a his^- phenotype. In the second experiment, protein B was fused to Gal4-DB and protein A was fused to Gal4-AD. Both hybrid proteins were expressed in the same his^- strain as in experiment 1, and the result was that the yeast strain had a his^+ phenotype. Assuming that all the hybrid proteins were properly expressed in yeast, what is the proper interpretation of these results? If these results cannot be unambiguously interpreted, suggest an additional experiment involving yeast that would provide more conclusive information.

Solution Strategies	Solution Steps
Evaluate	
1. Identify the topic this problem addresses, and explain the nature of the requested answer.	1. This problem tests your understanding of the two-hybrid assay. The answer will be an interpretation of the results and, if possible, a statement of whether they support the hypothesis that proteins A and B physically interact. If a conclusive interpretation of the results is not possible, the answer should suggest an additional experiment to clarify the situation.
2. Identify the critical information given in the problem.	2. The problem states that yeast expressing one combination of two-hybrid proteins are *his*$^-$, whereas yeast expressing the alternate combination of two-hybrid proteins are *his*$^+$.
Deduce	
3. Describe how the *his*$^-$ and *his*$^+$ phenotype of the yeast strain relates to the expression of the reporter *HIS*$^+$ gene in the two-hybrid system.	3. In the absence of *HIS*$^+$ reporter gene expression, the yeast strain has a *his*$^-$ phenotype. If the *HIS*$^+$ reporter gene is expressed, the yeast strain will have a *his*$^+$ phenotype.
4. Describe the mechanism by which *HIS*$^+$ reporter gene expression is activated in the two-hybrid system.	4. *HIS*$^+$ reporter gene transcription is transcribed only when a transcriptional activator binds to UAS$_{GAL4}$. In the two-hybrid system, a transcriptional activator binds to UAS$_{GAL4}$ only when the two-hybrid proteins bind together or when the protein fused to the Gal4-DB contains a transcriptional activation domain.
5. Interpret the result obtained when the two hybrid proteins were protein A + Gal4-DB and protein B + Gal4-AD.	5. Yeast expressing protein A + Gal4-DB and protein B + Gal4-AD were *his*$^-$; therefore, this result suggests that the two-hybrid proteins do not interact.
6. Interpret the result obtained when the two-hybrid proteins were protein A + Gal4-AD and protein B + Gal4-DB.	6. Yeast expressing protein B + Gal4-DB and protein A + Gal4-AD were *his*$^+$; therefore, this result suggests that the two-hybrid proteins do interact or, alternatively, that protein B contains a transactivation domain.
7. Suggest an experiment that would rule out the alternative interpretation provided in step 6 and the two possible results.	7. Express the protein B + Gal4-DB alone in the *his*$^-$ yeast strain. If protein B contains a transactivation domain, then the yeast strain will be *his*$^+$. If protein B does not contain a transactivation domain, the yeast strain will be *his*$^-$.
Solve	
8. Answer the problem by summarizing your interpretation of the results provided by the problem, including the alternatives proposed in step 7.	8. **Answer:** If expression of protein B + Gal4-DB alone activates the *HIS*$^+$ reporter, then the two-hybrid result protein B + Gal4-DB with protein A + Gal4-AD is misleading, and the interpretation is that proteins A and B do not interact. If expression of the protein B + Gal4-DB alone does not activate the *HIS*$^+$ reporter, then the two-hybrid result protein A + Gal4-DB with protein B + Gal4-AD is probably misleading, and the interpretation is that proteins A and B likely interact.

18.03 Solutions to End-of-Chapter Problems

1a-b. *Evaluate:* This is type 1. This problem tests your understanding of the genomic sequencing strategies. *Deduce > Solve:* **(a)** A whole-genome shotgun sequencing approach would be appropriate for sequencing the prokaryote's genome because (1) its genomic DNA free of contaminating DNA can be prepared in the laboratory from pure cultures of the bacterium and (2) the genome is unlikely to have significant amounts of repetitive DNA sequences. **(b)** A metagenomic approach would be appropriate for obtaining genome sequence of the prokaryote if it could not be cultured in the laboratory, although a method to identify the genome of the prokaryote in question from other genomic sequences in the sample would be needed.

2a–c. *Evaluate:* This is type 1. This problem tests your understanding of the consequences of repetitive DNA on genome sequencing and assembly. **TIP:** Sequences from individual DNA clones are assembled into a contiguous genomic DNA sequence by identifying common (overlapping), unique sequences in different clones. *Deduce > Solve:* **(a)** Repetitive DNA sequences exist at many different locations in a genome; therefore, clones containing repetitive DNA will appear to match cloned DNA sequences from many different genomic locations. This situation makes it difficult or impossible to unambiguously assign a repetitive DNA-containing sequence to its correct genomic location. **(b)** Dispersed repetitive DNA sequences that are too long to cross by a single sequencing reaction are the most problematic because they will not have unique (single copy) DNA sequences on either side to aid in contig assembly. **(c)** Sequencing both ends of large clones (paired-end sequencing) can help overcome genome assembly problems caused by repetitive DNA sequences because it can provide unique DNA sequence information that flanks the repetitive DNA sequence on either side.

3a–d. *Evaluate:* This is type 1. This problem tests your understanding of genome sequence assembly. *Deduce > Solve:* **(a)** A contig is a contiguous DNA sequence that is made up of many, short, overlapping individual sequence runs. Contigs end when no further sequences derived from small clones overlap and extend beyond the ends of the contig sequence. A scaffold is a collection of contigs that are grouped together by the paired-end sequences from large clones (BACs or YACs) that span the gaps between contigs. Since a scaffold must have at least two contigs, but typically contains many, the number of scaffolds will always be less than the number of contigs. **(b)** A physical gap is a gap between scaffolds and represents genomic DNA sequence that is not known to be present in a clone. A sequence gap is a gap between contigs that is contained within larger clones that define the scaffold in which the contig resides. **(c)** Since a physical gap exists because no clone containing the DNA in the gap has been identified, the gap can be closed only by identifying a novel DNA clone that spans the gap. The gap-spanning DNA may be identified by the paired-end reads of a new clone generated using a different approach from that used previously. If the physical gap is small, the gap-spanning DNA could be produced by PCR using sequences on either side of the gap to design PCR primers. **(d)** Since a sequence gap is a gap between contigs, which are contained within larger clones, the sequence of the gap can be determined by sequencing the corresponding region of the larger clone. If the sequence gap is small, the gap-spanning DNA could be produced by PCR using sequences on either side of the gap to design PCR primers.

4. *Evaluate:* This is type 1. This problem tests your understanding of genome annotation. *Deduce > Solve:* cDNA clones provide the sequence of regions of the genome that are transcribed and exist as exons in mRNA. This information identifies regions of the genome that contain genes and, for each gene, it identifies the transcribed region, the precise boundaries between exons and introns, the start site of transcription, and the poly-A addition site. Different full-length cDNA, which includes some but not all exons with identical nucleotide sequences and in the same order, provide information for alternatively spliced mRNAs.

5.   *Evaluate:* This is type 1 and 4. This problem tests your understanding of genome annotation. *Deduce > Solve:* The genomes of related species are most likely to be similar (evolutionarily conserved) in regions of functional importance, which include gene sequences and sequences that regulate the transcription of genes. Therefore, comparison of a new genomic sequence to that of an annotated sequence of a closely related species is likely to provide information on the identity and locations of all genes and regulatory sequences conserved between the two species.

6.   *Evaluate:* This is type 1. This problem tests your ability to use bioinformatics to identify genes during genome annotation and your understanding of differences in the genomes of eukaryotes and prokaryotes. *Deduce > Solve:* The key differences in gene structure between prokaryotes and eukaryotes are (1) most eukaryotic genes contain introns, whereas prokaryotic genes do not; and (2) many prokaryotic genes are operons, whereas eukaryotic genes are not. A computer algorithm for prokaryotes would, therefore, identify genes primarily by searching for open reading frames that begin with an ATG, include a nearby potential ribosome binding site and end with a stop codon sequence (TGA, TAA, or TAG). It would identify operons by finding multiple open reading frames located close together without intervening promoter sequences. The algorithm for eukaryotic genes would identify open reading frames that begin with ATG as well as those that begin with other codons and then would search for sequences that define the intervening regions as introns (e.g., the 3′ and 5′ splice junctions).

7.   *Evaluate:* This is type 1. This problem tests your understanding of the differences between prokaryotic and eukaryotic genome structure. *Deduce:* Bacterial genes are short and do not contain introns, some bacterial genes are operons, the gene regulatory sequences are short, and the genes are closely packed closely together with little to no intergenic DNA. Bacterial genomes do not contain much repetitive DNA. Eukaryotic genes are large and consist primarily of short exons separated by large introns; there are no operons and the genes contain larger, complex regulatory regions. Eukaryotic genomes contain many interspersed, repetitive DNA elements. *Solve:* The 100-kb *Bacillus* genome sequence will contain many more annotated genes than the 100-kb *Gorilla* sequence because prokaryotic genes are more compact, contain short regulatory sequences, do not have introns, and are packed tightly together; in contrast, eukaryotic genes are larger due to the presence of large introns, contain larger regulatory regions, and are separated by large intergenic DNA sequences. The *Bacillus* genome segment will likely contain some operons and little repetitive DNA, whereas the *Gorilla* segment will lack operons and contain interspersed repetitive DNA.

8.   *Evaluate:* This is type 1, 2, and 3. This problem tests your understanding of structural, evolutionary, and functional genomics. *Deduce > Solve:* One method would be to use bioinformatics and employ gene-finding algorithms to identify potential genes based on DNA sequence alone. This method requires no experimentation and can be performed on any genome sequence using existing computer programs; however, it is also the least authoritative because it is not based on any experimental evidence. Direct experimental validation of the genome annotation would still be required. A second method would be to perform a whole-genome comparison with a related genome that is fully annotated. This method also requires no further experimentation and can be performed on any genome using existing computer programs. This approach has an advantage over the gene-finding programs because gene identity and gene function assignments are based on evolutionary relationships to known genes; however, the reliability of those assignments decreases with increasing evolutionary distance between the species. Direct experimental validation of the genome annotation would still be required. A third method is to conduct functional analyses of the genome (e.g., cDNA cloning and sequencing, microarray analysis, and gene-knockout studies). This approach is the most informative of the three but is also the most time consuming and costly, and mutational studies may not be feasible for the organism being studied.

9.   *Evaluate:* This is type 2. This problem tests your understanding of gene and pseudogene evolution. *Deduce > Solve:* Pseudogenes will be homologous to known genes but will contain (typically many) mutations that destroy their coding capacity—for example, nonsense and

frameshift mutations. Some (processed) pseudogenes also lack introns and promoters present in their functional homologs. Pseudogenes arise by multiple mechanisms—including gene and genome duplications and reverse transcription of RNA to DNA—but all diverge from their functional homologs by the accumulation of mutations.

10. *Evaluate:* This is type 2. This problem tests your understanding of species and gene evolution. *Deduce > Solve:* Humans and fungi diverged from a more recent common ancestor than plants; therefore, human proteins will be more similar to fungal proteins than to plant proteins, and plant proteins will be equally related (or unrelated) to human and fungal proteins.

11. *Evaluate:* This is type 2. This problem tests your understanding of species and gene evolution. *Deduce > Solve:* Orthologous genes are genes in different species that are derived from a single ancestral gene in the two species' last common ancestor. This condition can be determined only by constructing the evolutionary history of the gene family using phylogenetic methods, although it is commonly assumed that orthologous members of a gene family are also the closest related in structure and function. Since orthologs are established based on sequence comparisons, additional members of the gene family not included in the phylogenetic analysis may exist if the genomes of the species in question are not completely sequenced. If some members of the gene family are missing, paralogs can be misidentified as orthologs.

12. *Evaluate:* This is type 3. This problem tests your understanding of transcriptome analysis. *Deduce > Solve:* Transcriptome analysis, which seeks to analyze gene expression by identifying and quantifying the RNA present in cells, can be accomplished by hybridization-based approaches using DNA microarrays or by high-throughput cDNA sequencing. DNA microarrays, which contain synthetic oligonucleotide sequences corresponding to segments of the genome on a solid support called a DNA chip, can be either expression arrays or tiling arrays. Expression arrays contain only known gene sequences and are used to quantify the relative transcript levels for these genes. Genome-tiling arrays contain the entire genome sequence and are used to identify all portions of a genome that are transcribed. They are also used to identify novel genes and RNA splice variants and to quantify relative gene expression levels. High-throughput sequencing of cDNA can provide information on relative expression levels as well as identify novel genes and RNA splice variants; thus, this process has the potential to supplant microarray-based methods for transcriptome analysis. cDNA-sequencing-based approaches also have the potential to identify polymorphisms in or between populations, which is difficult to do with an array-based approach.

13. *Evaluate:* This is type 1. This problem tests your understanding of yeast two-hybrid analysis. *Deduce > Solve:* The yeast two-hybrid technique determines whether two proteins can physically bind to each other if expressed in the same subcellular compartment of the same cell. The two-hybrid assay works by coupling protein–protein binding to the activation of transcription of a reporter gene, whose expression can be detected by a simple colorimetric enzyme assay or by colony growth. The two-hybrid assay involves the engineering of two hybrid genes using recombinant DNA techniques and expressing both genes in yeast containing one or more reporter genes. Each hybrid gene is a translational fusion of one of the proteins being studied to one-half of the yeast *GAL4* transcriptional activator. One protein is fused to the GAL4 DNA binding domain (*GAL4-BD*), and the other is fused to the *GAL4* trans-activating domain (*GAL4-AD*). Expression of the reporter gene is controlled by the UAS$_{GAL4}$ regulatory sequence; therefore, reporter gene expression depends on the binding of the GAL4-BD fusion protein to UAS$_{GAL4}$, and activation requires that the protein fused to the *GAL4-AD* domain interact with the protein fused to the *GAL4-BD*. Activation and reporter gene expression occurs only if the two hybrid proteins bind together, which depends on the binding of the two proteins being studied. Proteins that cannot bind to each other do not activate reporter gene transcription, proteins that bind to each other weakly activate low-level reporter gene transcription, and proteins that bind tightly to each other activate high-level reporter gene transcription. A false-positive result would be one in which two proteins interact in the two-hybrid assay but not in their normal cellular environments. This could occur if two proteins are never expressed in the same cell type, are

sequestered from each other in different subcellular compartments, or if another bound protein is required to act as a bridge. A false-negative result would occur if two proteins that normally bind to each other in their normal cellular environments do not interact in the two-hybrid assay. This most likely could occur if the gene fusions prevent normal binding interactions, for example, by disrupting the structure of a domain on one or the other protein required for binding.

14.   *Evaluate:* This is type 1. This problem tests your understanding of and ability to use the bioinformatics program, BLAST. *Deduce > Solve:* A BLAST search using the sequence against the nr/nt database finds several entries corresponding to synthetic genes for human insulin B chain. The sequence differs from the natural human insulin gene sequence because it employs *E. coli* codons for efficient expression of the gene and insulin protein production in *E. coli*.

15.   *Evaluate:* This is type 1. This problem tests your understanding genome sequence assembly and annotation. *Deduce > Solve:* The double-stranded sequences corresponding to each of the sequences given in the problem are shown in the figure. The third and fifth sequences have been inverted and flipped in order to orient all overlapping sequences in the same 5'-to-3' direction. The regions of the overlap are shown by the alignment of the sequences, and overlapping regions within each sequence are underlined for emphasis. The contig is shown below the alignments as a single sequence with the regions of overlap still underlined. From inspection of the sequence alone, it is not clear whether the contig represents transcribed or translated sequence, although it does contain open reading frames. A blastn search of the nr/nt database using the contig as the query finds multiple *Drosophila melanogaster Ultrabithorax* sequences including mRNA sequences, which suggests that the contig represents transcribed DNA. A tblastx search of the nr/nt database using the contig as a query also finds multiple *D. melanogaster Ultrabithorax* sequences, indicating that multiple reading frames within the contig are present in *D. melanogaster Ultrabithorax* transcripts. A tblastx search of the nr/nt database limited to *Homo sapiens* finds multiple sequences corresponding to homeobox (*Hox*) genes, which are evolutionarily related to *Ultrabithorax*. A proposed annotation of an open reading frame in the contig is shown in the figure. This open reading frame corresponds to an ORF in the *D. melanogaster Ultrabithorax* mRNA.

```
atctgacccgcagacggagaatcgagatgg   Inverted orientation of
tagactgggcgtctgcctcttagctctacc   sequence five

      ggagaatcgagatggcgcacgcgctatgcc   Sequence four
      cctcttagctctaccgcgtgcgcgatacgg

            acgcgctatgcctgacggagcggcagatca   Inverted orientation of
            tgcgcgatacggactgcctcgccgtctagt   sequence three

  Sequence two   gagcggcagatcaagatctggttccagaac
                 ctcgccgtctagttctagaccaaggtcttg

                    Sequence one   ttccagaaccggcgaatgaagctgaagaag
                                   aaggtcttggccgcttacttcgacttcttc
```

Assembled contig

```
atctgacccgcagacggagaatcgagatggcgcacgcgctatgcctgacggagcggcagatcaagatctggttccagaaccggcgaatgaagctgaagaag
```

Annotated contig

```
atctgacccgcagacggagaatcgagatggcgcacgcgctatgcctgacggagcggcagatcaagatctggttccagaaccggcgaatgaagctgaagaag
 L  T  R  R  R  R  I  E  M  A  H  A  L  C  L  T  E  R  Q  I  K  I  W  F  Q  N  R  R  M  K  L  K  K
```

16a–c.   *Evaluate:* This is type 2. This problem tests your understanding of evolutionary genomics. *Deduce > Solve:* (a) *AX, BX,* and *CX* are all derived from a single common ancestor by speciation. This means they are homologous genes and also orthologs. (b) Genes *AY1* and *AY2* were created by a duplication event within the lineage leading to species A. Genes *AY1, BY,* and *CY* are derived from a single common ancestor by speciation events. Therefore, genes *AY1* and *AY2* are paralogs, genes *AY1* and *BY* are orthologs, and genes *BY* and *CY* are orthologs. **(c)** Genes *AZ1* and *AZ2* were created by a duplication event within the lineage leading to species A; therefore, *AZ1* and *AZ2* are paralogs. Genes *BZ1*, *BZ2*, and *BZ3* were created by duplication events within the lineage leading to species B; therefore, *BZ1*, *BZ2*, and *BZ3* are paralogs. Genes *CZ1* and *CZ2* were created by a duplication event within the lineage leading to species C; therefore, genes *CZ1* and *CZ2* are paralogs. Genes *AZ1* and *AZ2* and genes *BZ1*, *BZ2*, and *BZ3* are derived from a single common ancestor by speciation events; therefore, *AZ1* and *AZ2* are orthologs of *BZ1*, *BZ2*, and *BZ3*. Genes *AZ1* and *AZ2* and genes *CZ1* and *CZ2* are derived from a single common ancestor by speciation events; therefore, *AZ1* and *AZ2* are orthologs of *CZ1* and *CZ2*. Genes *BZ1* and *BZ3* and gene *CZ1* are derived from a common ancestor by a speciation event; therefore, *BZ1* and *BZ3* are orthologs of *CZ1*. Gene *BZ2* and gene *CZ2* are derived from a common ancestor by a speciation event; therefore, *BZ2* and *CZ2* are orthologs.

17a–b. *Evaluate:* This is type 2. This problem tests your understanding of evolutionary and functional genomics. *Deduce > Solve:* **(a)** To identify regulatory elements that regulate the isolated gene in mammary glands, compare the chromosome regions containing the gene in the human, mouse, dog, chicken, and pufferfish genomes. DNA sequences immediately 5′ and 3′ to the coding sequence of the gene that are conserved in human, mouse, and dog but not conserved in chicken and pufferfish would be candidate regulatory sequences controlling mammary-gland-specific expression. These sequences could then be characterized functionally by deletion analysis or gene fusion experiments to determine whether any are necessary or sufficient for mammary-gland-specific transcription. **(b)** The presence of orthologs of the gene in chicken and pufferfish, which do not produce milk, indicates that the gene's function is not limited to milk production and suggests that the ancestral form of this gene was likely involved in some other biological function. For example, the gene could encode a transcriptional regulatory protein that has evolved a function in milk production within the lineage common to mammals after it diverged from those leading to birds and fish.

18. *Evaluate:* This is type 2. This problem tests your understanding of evolutionary genomics. *Deduce > Solve:* Large-scale rearrangements of chromosomes have been rare during primate evolution, leading to high conservation of overall chromosome structure. Smaller-scale changes (indels and point mutations), on the other hand, have been more common since the lineages leading to humans and chimpanzees diverged as well as within the lineage leading to present-day humans.

19. *Evaluate:* This is type 1. This problem tests your understanding of structural genomics and knowledge of bacterial lifestyles. *Deduce > Solve: Rickettsia prowazekii* is an obligate parasite, not a free-living eubacterium like the other eubacteria listed in Table 18.1. *R. prowazekii* does not need genes for the biosynthesis of all required nutrients, nor does it need to respond to environmental changes that are required for survival of free-living bacteria like *E. coli* and *A. tumefaciens*. Therefore, *R. prowazekii* has far fewer genes than either of those bacteria.

20. *Evaluate:* This is type 1. This problem tests your understanding of functional genomics. *Deduce > Solve:* Segmental duplications of chromosomes lead to gene duplications, where each copy of a particular gene is identical. Such genes are said to be functionally redundant. Individuals who are homozygous for loss-of-function alleles of one copy of the gene may not display a mutant phenotype due to the compensatory action of the second copy. When this is the case, all copies of the gene must be identified and transgenic organisms that are homozygous for loss-of-function alleles of all gene copies must be created. Another approach would be to create transgenic organisms carrying gain-of-function alleles of one gene copy and look for a dominant phenotype that may be informative as to gene function.

21. *Evaluate:* This is type 3. This problem tests your understanding of functional genomics.
Deduce > Solve: See figure.

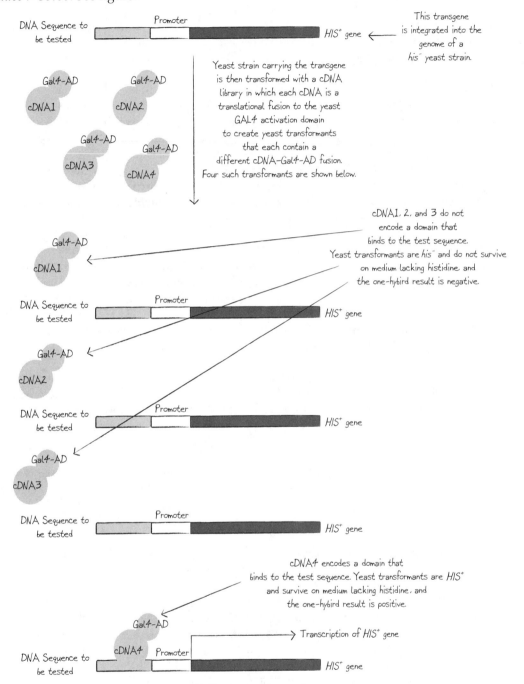

22a–b. *Evaluate:* This is type 3. This problem tests your understanding of functional genomics.
Deduce > Solve: **(a)** Multiple approaches can be used to test the function of yeast genes: (1) create a loss-of-function allele by gene replacement and examine the mutant phenotype; (2) create a gain-of-function allele by elevated expression of the gene and examine the mutant phenotype; (3) create a reporter gene translational fusion allele to determine the subcellular location of the gene product; (4) perform a two-hybrid screen to identify interacting proteins; (5) perform a transcriptome analysis to examine the expression pattern of the gene under various environmental conditions. **(b)** There are fewer approaches to characterize the function of a human gene: (1) perform a two-hybrid screen to identify interacting proteins or (2) perform a transcriptome analysis to determine the tissue-specific pattern of the gene's expression using tissue samples available at research hospitals.

23.   *Evaluate:* This is type 2. This problem tests your understanding of evolutionary genomics. *Deduce > Solve:* The gene duplication event that led to the creation of the human β-globin and the human δ-globin genes predates the divergences of human and chimpanzee lineages: both species contain the entire globin gene cluster. Therefore, the human β-globin and chimpanzee β-globin genes would exhibit a higher level of sequence similarity than the human β-globin and the human δ-globin genes.

24.   *Evaluate:* This is type 3. This problem tests your understanding of functional genomics. *Deduce > Solve:* The first step in the analysis would be to group together the genes that respond to both conditions in the same manner. Genes whose expression increases under both conditions are *b, p,* and *s*. Genes whose expression decreases under both conditions are *a, c, d, e, f, g, i, j, k, n, q,* and *r*. Genes whose expression increases only in response to high salt but decreases in response to high temperature are *l* and *m*. Genes whose expression increases in response to high temperature but decreases in response to high salt are *h* and *o*. This sorting or clustering of genes provides insight into whether the genes are responding to stressful environments in general (heat and salt) or are specific to one kind of stress (heat or salt).

25a.   *Evaluate:* This is type 2. This problem tests your understanding of functional genomics. *Deduce > Solve:* If the genes whose expression decreases in response to high salt are regulated by a common transcription factor, they should have common regulatory DNA sequences. You could inspect the regions 5′ to the coding sequence of the gene to identify those genes that share a regulatory DNA element. This condition could then be tested by coupling the element to a reporter gene and showing that its expression is also repressed in high-salt medium as well as standard growth medium.

25b.   *Evaluate:* This is type 2. This problem tests your understanding of evolutionary genomics. *Deduce > Solve:* You could perform whole-genome comparisons between *S. cerevisiae* and the other yeast species to search for conserved noncoding sequences located near the genes whose expression decreases in response to high-salt conditions. Conserved noncoding sequences would be candidates for regulatory sequences for these genes.

26.   *Evaluate:* This is type 3. This problem tests your understanding of the two-hybrid assay. *Deduce > Solve:* The two-hybrid assay detects physical binding of proteins. The problem indicates that protein A binds to proteins B and C; protein B binds to proteins A, D, and E; protein C binds only to protein A; protein D and E both bind only to protein B; protein G and F bind only to each other.

27.   *Evaluate:* This is type 3. This problem tests your understanding of functional genomics. *Deduce:* From Problem 26, proteins A, B, C, D, and E appear to be components of a single complex (ABCDE); F and G are members of a different complex (FG). The similar phenotype (80% growth rate) due to null mutants lacking any individual component of the ABCDE complex or lacking both A and any other component suggests that each component is essential for ABCDE function and that ABCDE function is required for maximum growth rate but not survival. The lack of a phenotype due to loss of F or G function indicates that loss of FG complex function has no obvious phenotypic consequence under standard growth conditions. The fact that loss of both A and F or A and G is lethal suggests that F and G are both essential components of the FG complex and that having at least one of the complexes, ABCDE or FG, is required for viability. *Solve:* The ABCDE complex is required for maximum growth rate. The FG complex is not important in otherwise normal cells; however, when the ABCDE complex is inactive, FG complex function is required for viability.

28.   *Evaluate:* This is type 2. This problem tests your understanding of evolutionary genomics. *Deduce > Solve:* Since the predicted PEG10 protein sequence is related to sequences present in the retrotransposon, PEG10 is likely derived from an ancient retrotransposon that was inserted into this chromosomal locus within the lineage leading to placental and marsupial mammals after the divergence of the monotreme lineage. Retrotransposons encode a reverse transcriptase, which converts RNA to DNA. It is possible that PEG10 retains this biochemical function or a function related to RNA metabolism that has evolved to suit a purpose during placental development.

18.04 Test Yourself

Problems

1. The BLAST site at NCBI (http://blast.ncbi.nlm.nih.gov/Blast.cgi) offers several BLAST options including blastn, blastp, blastx, tblastn, and tblastx. Describe each blast program in terms of the nature of the query, the database, and what, if any, manipulation is done to the query or database sequences in performing the search.

2. Which one or two of the BLAST search options described in Test Yourself Problem 1 would be appropriate for the purpose of identifying potential protein-coding genes in a newly sequenced, unannotated genome?

3. Which one or two of the BLAST search options described in Test Yourself Problem 1 would be most appropriate for the purpose of identifying potential orthologs to a protein sequence?

4. Organize the following list of terms in order of size, from largest to smallest: contig, sequence read, chromosome, scaffold, and genome.

5. A 150-bp stretch of DNA in a fungal genome is identified as containing an open reading frame that begins with an ATG. Describe evidence that could be provided by bioinformatics and experimental analyses that would support the hypothesis that this open reading frame is part of an evolutionarily conserved gene required for maximum growth rate.

6. Deletion of the yeast gene *A* is not a lethal mutation; however, the *A* deletion mutant has an increased rate of certain types of mutation. Deletion mutant of the yeast gene *B* is not lethal, but gene *B* deletion mutants have an increased rate of the same types of mutations as gene *A* mutants. Combining the *A* and *B* mutations is lethal. How would you interpret these results if **(a)** the gene *A* and gene *B* proteins were shown to physically interact, and if **(b)** the gene *A* and gene *B* proteins were shown not to physically interact?

Solutions

1. *Evaluate:* This is type 1, 2, and 3. This problem tests your understanding of BLAST searches. *Deduce > Solve:* In a blastn search, the query is a DNA sequence and the database is a set of DNA sequences. In a blastp search, the query is a protein sequence and the database is a set of protein sequences. In a blastx search, the query is a DNA sequence translated into six protein sequences using all possible reading frames, and the database is a set of protein sequences. In a tblastn search, the query is a protein sequence and the database is a set of DNA sequences translated in all possible reading frames. In a tblastx search, the query is a DNA sequence translated into six protein sequences using all possible reading frames, and the database is a set of DNA sequences translated in all possible reading frames.

2. *Evaluate:* This is type 1. This problem tests your understanding of the use of BLAST searches to annotate a genome sequence. *Deduce > Solve:* Since the genome sequence is DNA and the goal is to identify protein-coding genes, the genome sequence query needs to be translated as part of the BLAST search, and the database must be either protein sequences or predicted protein sequences within a DNA sequence database. The appropriate searches are blastx and tblastx to identify potential protein coding genes in a genome sequence.

3. *Evaluate:* This is type 2. This problem tests your understanding of the use of BLAST searches to find orthologs of a protein sequence. *Deduce > Solve:* Since the query sequence is protein and the databases need to be protein or translated DNA sequences, a blastp or tblastn search should be performed to identify potential orthologs to a protein sequence.

4. *Evaluate:* This is type 2. This problem tests your understanding of structural genomics. *Deduce > Solve:* From largest to smallest, the list should be genome, chromosome, scaffold, contig, and sequence read.

5.   *Evaluate:* This is type 2. This problem tests your ability to apply evolutionary and functional genomics to determine the function of a genomic sequence. *Deduce > Solve:* If the 150-nucleotide-long open reading frame is part of an evolutionarily conserved gene required for maximum growth rate, then the encoded protein sequence will be similar to sequences found in other genomes, it will be part of a transcribed sequence, and its expression will be required for a rapid growth rate in that fungus. Evidence for evolutionary conservation of the sequence would include results from a blastp or tblastx search that identifies proteins or protein-coding sequences in other organisms using the protein sequences as a query. Evidence for expression of the sequence would include the identification of a cDNA sequence from the fungus (or a relative) that includes those 150 base pairs or results from an expression microarray analysis showing that mRNA corresponding to the open reading frame accumulates in cells of the fungus. Evidence for a function of the sequence in growth would include showing that mutants with a defect in the sequence have a reduced growth rate or the generation of a knockout mutation for that sequence resulting in reduced growth rate.

6.   *Evaluate:* This is type 3. This problem tests your understanding of genetic interactions. *Deduce > Solve:* Since *A* and *B* single mutants are viable but an *A, B* double mutant is dead, genes *A* and *B* are said to show a synthetic–lethal genetic interaction. If the proteins encoded by these genes physically interact, then one reasonable interpretation would be that the proteins are part of the same complex involved in DNA synthesis or in a repair pathway that is essential for viability. Both proteins are required for optimal function of the complex; however, the complex still functions when one or the other is absent or nonfunctional. In the absence of both proteins, the complex cannot function; thus, the DNA synthesis or repair pathway ceases to operate, leading to lethality. If the proteins encoded by *A* and *B* do not physically interact, then *A* and *B* would be proposed to function as part of distinct pathways required for optimal DNA synthesis or repair. The pathways are partially redundant such that loss of one or the other increases mutation rates. Lethality results only when both pathways are defective.

19

Cytoplasmic Inheritance and the Evolution of Organelle Genomes

Section 19.01 Genetics Problem-Solving Toolkit

Section 19.02 Types of Genetics Problems

Section 19.03 Solutions to End-of-Chapter Problems

Section 19.04 Test Yourself

19.01 Genetics Problem-Solving Toolkit

Key Terms and Concepts

Chloroplast DNA: A prokaryotic-like genome that contains some but not all the genes necessary for chloroplast function and reproduction.

Endosymbiont theory: Theory that explains the evolutionary similarity of mitochondria and chloroplasts to extant prokaryotes as being due to the origin of these organelles as prokaryotic endosymbionts of an ancient, eukaryotic ancestor.

Heteroplasmy: Presence of more than one mitochondrial or chloroplast DNA sequence in a cell or organism.

Homoplasmy: Presence of a single mitochondrial or chloroplast DNA sequence for all copies of the organelle genome in a cell or organism.

Horizontal gene transfer: The transfer of genes between organisms by means other than reproduction.

Maternal inheritance: Inheritance of a trait from the female parent; typically due to inheritance of a cytoplasmic DNA element.

Mitochondrial DNA: Prokaryotic-like genome that contains some but not all the genes necessary for mitochondrial function and reproduction.

Reduced penetrance of a maternally inherited trait: Can be due to genetic interactions, environmental factors, or heteroplasmy.

Uniparental inheritance: Inheritance of a trait from one parent; typically due to inheritance of organellar DNA.

Key Analytical Tools

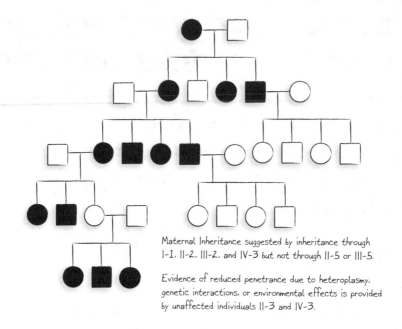

Maternal Inheritance suggested by inheritance through I-1, II-2, III-2, and IV-3 but not through II-5 or III-5.

Evidence of reduced penetrance due to heteroplasmy, genetic interactions, or environmental effects is provided by unaffected individuals II-3 and IV-3.

19.02 Types of Genetics Problems

1. Interpret or predict the patterns of transmission for traits that exhibit uniparental inheritance.
2. Describe or explain mitochondrial and chloroplast genetics, function, and evolution.

1. Interpret or predict the patterns of transmission for traits that exhibit uniparental inheritance.

This type of problem tests your understanding of uniparental inheritance. You may be asked to interpret pedigrees for patterns of maternal inheritance or differentiate between maternal and sex-linked inheritance. Where maternal inheritance is evident but discrepancies exist, you will be asked to consider genetic interactions, environmental effects, or heteroplasmy as possible explanations. You may be asked to predict the proportion of progeny genotypes and phenotypes for crosses involving cytoplasmic and nuclear genetic elements. See Problem 12 for an example.

Example Problem: A fruit-fly line is pure-breeding for a mutant trait called mutant. A mutant female was mated to a male from a pure-breeding wild-type line. A female F_1 was selected and backcrossed to her wild-type father. Describe the phenotypic distribution of the F_2 progeny under the assumption that the trait shows **(a)** maternal inheritance, and the line is homoplasmic for the mutant mitochondrion; and **(b)** X-linked dominant inheritance.

Solution Strategies	Solution Steps
Evaluate	
1. Identify the topic this problem addresses, and explain the nature of the requested answer.	1. This problem tests your ability to discriminate between maternal and X-linked dominant inheritance. The answer will be the phenotypic distribution of progeny expected for each pattern of inheritance.

(continued)

Solution Strategies	Solution Steps
2. Identify the critical information given in the problem.	2. The mutant and wild-type lines are pure-breeding, and for maternal inheritance, the line is homoplasmic. The crosses are (1) mutant female × wild-type male and (2) F_1 female × wild-type male.
Deduce	
3. Write the genotypes of the parents under each assumed pattern of inheritance.	3. For maternal inheritance, the mutant female carries a mutant mitochondrion ($mito^-$), and the wild-type male carries a wild-type mitochondrion ($MITO^+$). For X-linked recessive inheritance, the mutant female is $X^D X^D$ (where D is the dominant allele), and the male is $X^d Y$.
4. Write the genotypes and phenotypes of female F_1 under each assumed pattern of inheritance.	4. For maternal inheritance, the F_1 female is $mito^-$. For X-linked dominant inheritance, the female is $X^D X^d$.
5. Write the genotypes for the cross, female F_1 × male wild-type parent, under each assumed pattern of inheritance.	5. For maternal inheritance, the cross is female $mito^-$ × male $MITO^+$. For X-linked dominant inheritance, the cross is female $X^D X^d$ × male $X^d Y$.
Solve	
6. Answer the problem by determining the progeny types and distributions for each cross described in step 5.	6. **Answer:** For maternal inheritance, all F_2 will be mutant. For X-linked dominant inheritance, one-half of the F_2 will be mutant and one-half will be wild type.

2. Describe or explain mitochondrial and chloroplast genetics, function, and evolution.

This type of problem tests your understanding of the endosymbiont theory as it explains the genetics, function, and evolution of mitochondria and chloroplasts. You may be asked to describe how organelle DNA sequence analysis can be used to study genetic relationships between species or between individuals. You may be asked to describe the genetic control of organelle function or reproduction, considering the relative roles of the organelle and nuclear genomes. See Problem 8 for an example.

Example Problem: Human mitochondrial DNA encodes the 12S and 16S mitochondrial rRNAs but none of the mitochondrial ribosomal proteins. Describe how mitochondrial ribosomal subunits are assembled. Start your description at a time after all necessary genes have been transcribed.

Solution Strategies	Solution Steps
Evaluate	
1. Identify the topic this problem addresses, and explain the nature of the requested answer.	1. This problem tests your understanding of human mitochondrial genetics. The answer will be a description of all posttranscriptional steps in the process of mitochondrial ribosome biogenesis.
2. Identify the critical information given in the problem.	2. The human mitochondrial rRNA genes are in the mitochondrion, whereas the mitochondrial ribosomal protein genes are in the nucleus.

(continued)

Solution Strategies	Solution Steps
Deduce	
3. Describe the process resulting in mitochondrial rRNA production.	3. The 12S and 16S mitochondrial rRNAs are present in two transcripts and are cleaved from those transcripts by a ribonuclease to release the separate rRNAs.
4. Describe the process by which ribosomal proteins are synthesized.	4. mRNA for mitochondrial ribosomal proteins are transcribed in the nucleus and exported to the cytosol, where they are translated to produce mitochondrial ribosomal proteins.
5. Describe the process by which the mitochondrial ribosomal proteins enter the mitochondrion.	5. Mitochondrial proteins include a signal peptide that causes them to be transported into the mitochondrion through pores in the outer and inner mitochondrial membranes.
Solve	
6. Answer the problem by summarizing the steps in mitochondrial ribosomal subunit assembly.	6. **Answer:** rRNA is transcribed and processed in the mitochondrion. Ribosomal proteins are synthesized in the cytosol and then transported into the mitochondrion. The ribosomal subunits assemble in the mitochondrion.

19.03 Solutions to End-of-Chapter Problems

1. *Evaluate:* This is type 1. This problem tests your ability to distinguish between sex-linked inheritance and maternal inheritance patterns. *Deduce > Solve:* The most likely explanations for reciprocal crosses that yield different results are sex-linked or uniparental (cytoplasmic) inheritance. Parts (a) and (b) of the figure show pedigrees for a maternally inherited mammalian trait. All progeny of the affected female show the trait in part (a), whereas no progeny of the affected male do in part (b). This pattern is easily distinguished from the pedigrees in parts (c) and (d) of the figure, which show inheritance of a recessive, X-linked trait. All male but no female progeny of the affected female show the trait in part (c), whereas no progeny of the affected male show the trait in part (d).

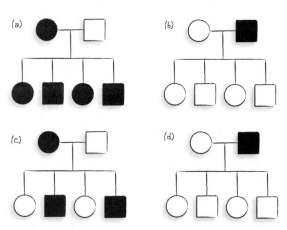

2. *Evaluate:* This is type 2. This problem tests your understanding of the endosymbiont theory for evolution of mitochondria and chloroplasts. *Deduce > Solve:* The structure of the organelles and the structure and mechanism of expression of the genetic material of mitochondria and chloroplast are explained by their evolutionary origin as bacterial endosymbionts. For example, their genomes are most often circular DNA molecules whose structure and packaging are similar to bacterial nucleoids. Organelle genes can code for polycistronic mRNAs that lack introns, and the DNA, RNA, and protein-synthesis enzymes are more closely related to bacteria than to those of their eukaryotic host.

3. *Evaluate:* This is type 2. This problem tests your understanding of mitochondrial translation. *Deduce > Solve:* The human mitochondrial genome encodes only 22 tRNAs and the *Plasmodium* mitochondrial genome encodes none, yet both mitochondria carry out protein synthesis in their

matrix space. Human mitochondria accomplish this feat by using only mitochondria-encoded tRNAs but allowing more flexibility for base pairing in the third, wobble position of codons. *Plasmodium* mitochondria use nuclear-encoded tRNAs, which are imported from the cytosol.

4.   *Evaluate:* This is type 2. This problem tests your understanding of lateral gene transfer. *Deduce:* Nuclear genes derived from mitochondria are detected by their similarity to bacterial and mitochondrial genes. Those genes transferred to the nuclear genome long ago have diverged more from their prokaryotic ancestor than those transferred more recently. *Solve:* Genes recently transferred from mitochondrial to nuclear chromosomes are more similar to bacterial genes than those transferred long ago. Also, some NUMPTS and NUPTS found in sequenced genomes have a higher level of similarity to the organelle genome sequences than would be expected for ancient transfers.

5.   *Evaluate:* This is type 2. This problem tests your understanding of mitochondrial DNA replication. *Deduce:* Nuclear DNA replicates only during S phase of the cell cycle, doubling the level of DNA content in G_1 by the time it reaches G_2. Mitochondrial DNA replicates throughout the cell cycle, roughly doubling in amount before the cell divides. Mitochondrial DNA replication is coordinated with an increase in mitochondrial number to ensure approximately equal distribution of mitochondria to daughter cells. *Solve:* See figure.

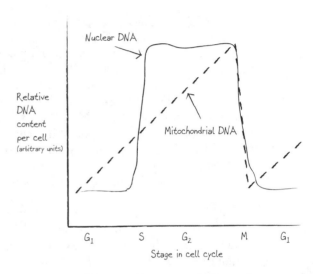

6.   *Evaluate:* This is type 2. This problem tests your understanding of mitochondrial genome evolution. *Deduce > Solve:* The UGA universal stop codon codes for Trp in many but not all mitochondria. The AUA Ile codon codes for Met in vertebrates and yeast. The CUN Leu codon codes for Thr in yeast. The AGG and AGA Arg codons code for Ser in yeast. The changes of AUA to Met, UGA to Trp, and AGG and AGA to Ser all make third-position mutations in those codons less likely to cause missense effects without the need for additional tRNAs. Specifically for UGA → Trp, the change could also mean fewer release factors would be required.

7.   *Evaluate:* This is type 2. This problem tests your understanding of the endosymbiont theory for organelle evolution. *Deduce > Solve:* The sequence of mitochondrial genes is more closely related to α-proteobacteria than to eukaryotic genes. The sequence of chloroplast genes is more closely related to cyanobacteria than to eukaryotic genes.

8.   *Evaluate:* This is type 2. This problem tests your understanding of the endosymbiont theory for organelle evolution. *Deduce > Solve:* First, the coding sequence is transferred to the nucleus and inserted into a chromosome. Then, it must obtain a promoter and transcriptional regulatory sequences in order to be transcribed. This could occur by accumulation of mutations in flanking sequences, chromosome rearrangements, or transposition events. If the mitochondrial open reading frame contains UGA Trp codons, mutation to UGG or another sense codon would be necessary. Finally, the amino terminal segment of the coding sequence would need to evolve into a mitochondrial targeting sequence so that the protein, after synthesis in the cytosol, would be imported into the mitochondrion.

9.   *Evaluate:* This is type 2. This problem tests your understanding of the endosymbiont theory of organelle evolution. *Deduce > Solve:* Secondary endosymbiosis refers to the engulfment of a eukaryotic cell by another eukaryotic cell; the organelle in the engulfed eukaryotic cell becomes the secondary endosymbiont. In the case of brown algae, this condition is manifest

in their photosynthetic genes being more closely related to those of eukaryotic red algae than to those of green algae. This is shown in Figure 19.18 by the vertical line leading from the brown algal lineage (rhodophytes) to the common ancestor of alveolates, stramenopiles, and chromalveolates, which include brown algae.

10.   *Evaluate:* This is type 2. This problem tests your understanding of mitochondrial protein complex assembly and function. *Deduce > Solve:* Organelles, like all cellular components, are produced and maintained by the coordinated synthesis, modification, and assembly of many individual gene products to form a functional complex. Because mitochondrial genes reside in the nuclear genome as well as the mitochondrial genome, transcription, translation, transport, and assembly of mitochondrial components are coordinated by regulatory pathways that act in the nucleus, cytosol, and mitochondrion.

11.   *Evaluate:* This is type 2. This problem tests your understanding of mitochondria-based phylogenetic analysis of human evolution. *Deduce > Solve:* Since mitochondrial DNA is inherited intact from a mother, mitochondrial DNA sequence comparisons provide unambiguous data for the analysis of human lineages. While mutations occur that can consequently be used to track subpopulations, all extant mitochondrial DNA sequences can be traced to one mitochondrial genome of a woman (referred to as Mitochondrial Eve) from which *Homo sapiens* descended. Analysis of human mitochondrial DNA supports the "recent African origin" hypothesis for human evolution, which states that humans first evolved in Africa and only recently migrated from Africa to replace all other hominid species in locations around the world. This hypothesis has strong support because more diversity of mtDNA is found in Africa, as would be expected for a location where mutant types have had the greatest amount of time to accumulate.

12.   *Evaluate:* This is type 1. This problem tests your understanding of maternal inheritance. *Deduce > Solve:* The pedigree describes the inheritance of a mitochondrial determined trait that is highly, although not completely, penetrant. III-1, III-2, and III-3 are affected, having inherited mutant mitochondrial DNA from their affected mother. Since III-1 is a male, you would advise him that there is no chance that his children will have the trait if he mates with an unaffected woman. Since III-2 and III-3 are both affected women, you would advise them that all of their children will have the trait. You could mention that there is a small chance that one or more of III-2's or III-3's children may not show the trait because it is not 100% penetrant; however, the frequency of such an event is not predictable based on the information provided.

13.   *Evaluate:* This is type 1. This problem tests your understanding of the inheritance of defects associated with mitochondrial dysfunction. *Deduce > Solve:* The wild-type F_1 and 3:1 (wild type: mutant) F_2 ratio indicate segregation of a recessive allele of a nuclear gene. Since the necrotic tissue phenotype is associated with instability of mitochondrial DNA, the defective gene must be a nuclear-encoded mitochondrial protein required for mitochondrial DNA synthesis or maintenance.

14.   *Evaluate:* This is type 1. This problem tests your understanding of inheritance patterns. *Deduce > Solve:* The pedigree shows transmission from females only, although the sample size is limited. The pattern is entirely consistent with maternal inheritance and autosomal dominant inheritance. Autosomal recessive inheritance is also possible; however, I-2 and II-5 would have to be carriers, which is unlikely if the trait is rare. III-1 rules out X-linked dominant inheritance unless the trait is considered to be incompletely penetrant. I-2 and II-5 rule out X-linked recessive inheritance unless the trait is thought to show incomplete penetrance.

15.   *Evaluate:* This is type 1. This problem tests your understanding of organelle DNA inheritance in *Chlamydomonas*. **TIP:** *Chlamydomonas* progeny inherit their chloroplast DNA from their mt^+ parent and their mitochondrial DNA from their mt^- parent, where mt^- mt^- are alleles of a nuclear gene controlling mating type. *Deduce > Solve:* mt^+ str^R hyg^S × mt^- str^S hyg^R yields $\frac{1}{2}$ mt^+ str^R hyg^R and $\frac{1}{2}$ mt^- str^R hyg^R progeny. mt^+ str^S hyg^R × mt^- str^R hyg^S yields $\frac{1}{2}$ mt^+ str^S hyg^S and $\frac{1}{2}$ mt^- str^S hyg^S progeny.

16. *Evaluate:* This is type 1. This problem tests your understanding of organelle DNA inheritance in yeast. *Deduce > Solve:* Yeast *petite* mutants are caused by defects in mitochondrial function, which can be due to nuclear or mitochondrial genes. The *pet1* × wild-type cross yields 2:2 (*petite*: wild type), indicating that *pet1* is a mutation in a nuclear gene and that *pet1* is a *segregational petite*. The *pet2* × wild-type cross yields all wild-type progeny, indicating that *pet2* is a mutation in a mitochondrial gene and that *pet2* is a *neutral petite*. A *pet1* × *pet2* cross is expected to yield 2:2 (*petite* : wild type) progeny because the nuclear *pet1* and *PET1*⁺ alleles will segregate 2:2, and the mutant mitochondrial DNA of the *pet2* strain will not be passed on to any progeny.

17. *Evaluate:* This is type 1. This problem tests your understanding of inheritance patterns. *Deduce > Solve:* Maternal inheritance. There are several discrepancies: I-female must be "carrier"; II-only 2 is affected, others should be carriers; III-3, 4, 12, and 14 should be carriers. The affected individuals tend to be males, suggesting involvement of alleles of a second gene, on the X chromosome, that influences the penetrance of the phenotype of the mitochondrial allele.

18. *Evaluate:* This is type 1. This problem tests your understanding of mitochondrial DNA inheritance. *Deduce > Solve:* MELAS is a disease associated with mutations in mitochondrial DNA; therefore, MELAS will be maternally inherited. In the scenario where the man has MELAS and his wife is unaffected, there is no need to be concerned that their children will develop MELAS. In the scenario where the woman has MELAS and her husband is unaffected, there is certainly a chance the children will develop the disorder, depending on the extent of heteroplasmy in the woman.

19. *Evaluate:* This is type 1. This problem tests your understanding of mitochondrial DNA inheritance. *Deduce > Solve:* Since LHON is maternally inherited, one of two possibilities exist. First, I-1 may not have the LHON mutation and II-3 could be due to spontaneous germ-line mutation in the mitochondrial DNA she inherited. The second possibility is that I-1 carries the LHON mutation but is heteroplasmic and has a sufficient percentage of wild-type mitochondria that it does not show the trait. Therefore, II-3 unfortunately would have developed from an egg that contained a sufficient percentage of mutant mitochondria for her to express the trait and pass it on to all her children.

20. *Evaluate:* This is type 1. This problem tests your understanding of mitochondrial DNA inheritance. *Deduce > Solve:* MERFF is due to mutations in mitochondrial DNA and, therefore, is maternally inherited. Since II-2 is an affected woman, all of her children are likely to inherit MERFF. Since II-5 is a man mating with an unaffected woman, none of his children will inherit MERFF.

21. *Evaluate:* This is type 1. This problem tests your understanding of mitochondrial inheritance and mitochondrial DNA-based human population genetics. *Deduce > Solve:* One hypothetical scenario is that New Zealand and Polynesia were settled by individuals from Taiwan, most of whom carried the mitochondrial DNA polymorphism. The different frequencies of the polymorphism in the populations could be explained by slightly different polymorphism frequencies among the Taiwanese settlers, or it could be explained by a change in the frequencies of the polymorphism within each population after they were established.

22. *Evaluate:* This is type 1. This problem tests your understanding of inheritance patterns. *Deduce:* All affected individuals are males with unaffected parents whose mothers are related. The simplest explanation is that the trait is a sex-linked recessive with chance segregation, giving all five sons of carrier females being affected a 1 in 32 probability. An alternative could be maternal inheritance with the absence of affected females, suggesting that the trait is sex limited, penetrant only in males, similar to cytoplasmic male sterility in plants. Sex-influenced inheritance with the trait dominant in males and recessive in females such as pattern baldness would also be possible. Autosomal recessive inheritance is unlikely given that affected individuals are in generations II and III (II-1 and II-2 would have to be unrelated carriers). Autosomal dominant is possible, again with the sex-limited phenotype; however, this type

of inheritance would require that II-5 and II-7 did not pass the allele to their sons. *Solve:* The pedigree is consistent with sex-linked recessive, sex-influenced inheritance, or with maternal inheritance of a male-limited trait.

23. *Evaluate:* This is type 1. This problem tests your understanding of mitochondrial DNA inheritance. *Deduce > Solve:* The presumptive Anastasia's mitochondrial DNA should be identical to that of her mother, so differences in the sequence could eliminate the claim; but identity would not prove the claim, since mitochondrial haplotypes are not unique to one family.

24. *Evaluate:* This is type 2. This problem tests your understanding of the use of mitochondrial DNA analysis to study evolutionary relationships. *Deduce > Solve:* Two characteristics of mitochondrial DNA make it a good choice for analyzing evolutionary relationships. First, it is inherited maternally without recombination and therefore can be used to directly trace evolutionary lineages. Second, it is abundant in essentially every cell type; therefore, some trace of mitochondrial DNA is typically detectable even in fossilized remains.

25. *Evaluate:* This is type 1. This problem tests your understanding of maternal inheritance and cytoplasmic male sterility. *Deduce > Solve:* The CMS genes are maternally inherited mitochondrial alleles, whereas the *Rf* loci are nuclear. The F_1 will be CMS1 $\frac{Rf1}{Rf1} \frac{rf2}{rf2}$ and will be male sterile. All F2 will be CMS1, with the nuclear *Rf* Loci segregating independently:

9 CMS1 *RF1_Rf2_*   male fertile
3 CMS1 *Rf1_rf2 rf2*   male fertile
3 CMS1 *rf1 rf1 Rf1_*   male sterile
1 CMS1 *rf1 rf1 rf2 rf2*   male sterile

26. *Evaluate:* This is type 2. This problem tests your understanding of how to use mitochondrial DNA to determine phylogenetic relationships. *Deduce:* If interbreeding did not occur, then all coyote mitochondrial DNA sequences would be more closely related to each other than to wolf, and vice versa. This was not the case: the mitochondrial DNA of wolf 1 and wolf 2 are more closely related to coyote sequences than to that of other wolves. *Solve:* The mitochondrial DNA analysis demonstrates that interspecific hybridization occurred with wolf as the female in the crosses detected here.

27. *Evaluate:* This is type 2. This problem tests your understanding of organelle evolution. *Deduce > Solve:* The malarial parasite, *Plasmodium falciparum,* contains chloroplast DNA, which encodes ribosomes that are sensitive to aminoglycoside antibiotics. If the *P. falciparum* chloroplasts are functional and important for reproduction, then aminoglycoside antibiotics may be effective for treating malaria.

28. *Evaluate:* This is type 2. This problem tests your understanding of the endosymbiont theory of organelle evolution. *Deduce > Solve:* If the sea slug has acquired chloroplast genes by horizontal gene transfer from the algal genome, then the sea slug genome would contain nuclear chloroplast genes, and these genes would be more closely related to the nuclear chloroplast genes of algae than they would be to the nuclear-encoded genes of any other species.

19.04 Test Yourself

Problems

1. A fruit-fly line is pure-breeding for a mutation. A female mutant was crossed to a pure-breeding wild-type male. All the progeny were mutant. Would the reciprocal cross determine whether the trait is mitochondrial or X-linked dominant? If so, describe the result. If not, describe a cross that could be done to determine which is correct.

2. Mitochondria are essential components of all eukaryotic cells. How is it possible for a person to have mitochondrial DNA that contains a null mutation in an essential mitochondrial gene?

3. Most human cells contain and require mitochondria, yet mitochondrial defects are primarily associated with defects in muscle and neuronal tissue. Provide a reasonable explanation of this observation.

4. How does replicative segregation explain the passage of a maternal disease from a mother to only half her children?

5. You have two strains of *Chlamydomonas*. One is mt^+ mating type, which is sensitive to drug A but resistant to drugs B and C. The other is mt^- mating type, which is resistant to drug A but sensitive to drugs B and C. You crossed the two strains and analyzed 100 progeny. Predict the results of the cross, assuming that sensitivity or resistance to drug B is due to a nuclear gene that is unlinked to the mating type gene (*mt*), sensitivity or resistance to drug A is due to a chloroplast gene, and sensitivity or resistance to drug C is due to a mitochondrial gene.

Solutions

1. *Evaluate:* This is type 1. This problem tests your understanding of maternal versus sex-linked inheritance. *Deduce > Solve:* The reciprocal cross would produce all normal progeny if the trait is maternal (pure-breeding normal female by pure-breeding mutant male), whereas the females will be mutant and the males normal if the trait is X-linked dominant.

2. *Evaluate:* This is type 1. This problem tests your understanding of heteroplasmy. *Deduce > Solve:* Individuals with mitochondrial disorders are often heteroplasmic, carrying both wild-type and mutant mitochondrial DNA. The wild-type mitochondria in these individuals are present in sufficient quantities to make the individual viable.

3. *Evaluate:* This is type 2. This problem tests your understanding of mitochondrial function and human disease. *Deduce > Solve:* Individuals with mutant mitochondria either contain a mixture of functional and nonfunctional mitochondria and are heteroplasmic or contain mutant mitochondria that are partially functional. Either way, all cells have at least partial mitochondrial function. Those tissues that exhibit the greatest defect will be those that have the largest demand for mitochondrial function and ATP production, and they are often nervous and muscle tissue.

4. *Evaluate:* This is type 2. This problem tests your understanding of replicative segregation. **TIP:** Refer to Figure 19.8 in the textbook. *Deduce > Solve:* Replicative segregation refers to random segregation of cytoplasmic genetic elements during cell reproduction. Because of replicative segregation, a woman who is heteroplasmic for a mitochondrial mutation can produce eggs that differ in the percentage of wild-type and mutant mitochondria. Children that result from fertilization of eggs that contain predominantly wild-type mitochondria may develop into unaffected individuals, whereas children that result from fertilization of eggs that contain predominantly mutant mitochondria may develop into affected individuals.

5. *Evaluate:* This is type 1. This problem tests your understanding of uniparental inheritance of mitochondria and chloroplasts in *Chlamydomonas*. *Deduce > Solve:* Mating type and sensitivity or resistance to drug B will assort independently to give $\frac{1}{4}$ mt^+ $drugB^R$, $\frac{1}{4}$ mt^- $drugB^R$, $\frac{1}{4}$ mt^+ $drugB^S$, and $\frac{1}{4}$ mt^- $drugB^S$. All the progeny will inherit the mitochondria from the mt^- parent and the chloroplasts from the mt^+ parent; therefore, they will all be sensitive to drugs A and C.

20

Developmental Genetics

20.01 Genetics Problem-Solving Toolkit

Key Terms and Concepts

Coordinate genes: Maternally expressed regulatory genes that function in the egg to establish the embryonic axes in *Drosophila* (often called maternal effect genes).

Gap genes: *Drosophila* developmental regulatory genes, which are regulated by coordinate genes and define broad regions of the embryo.

Pair-rule genes: *Drosophila* developmental regulatory genes that define segments of the embryo.

Segment-polarity genes: *Drosophila* developmental regulatory genes that define the anterior and posterior regions of individual segments.

Homeotic genes: Evolutionarily conserved developmental regulatory genes that determine the fate or identity of adult structures.

Homeobox genes: Homeotic genes of animals that code for containing a highly conserved DNA-interacting protein domain (the homeobox).

MADS-box genes: Homeotic genes of plants with conserved domains for DNA-interacting proteins.

Induction: Promotion of differentiation of a cell by a neighboring cell producing a positive-acting signaling molecule.

Lateral inhibition: Inhibition of differentiation of a cell by a neighboring cell producing a negative-acting signaling molecule.

Key Analytical Tools

Please study and reference Figures 20.3, 20.5, 20.11, 20.14, 20.16, and 20.20 in the textbook.

20.02 Types of Genetics Problems

1. Describe the principles governing cellular differentiation, pattern formation, and the evolutionary history of genes controlling development.
2. Describe the developmental genetics of *Drosophila* or interpret results from genetic analysis of *Drosophila* development.

1. Describe the principles governing cellular differentiation, pattern formation, and the evolutionary history of genes controlling development.

This type of problem tests your understanding of the function and evolution of genes that control development in a variety of organisms. You may be asked to describe the principles underlying cellular differentiation and pattern formation or use your understanding of these principles to interpret results from genetic analyses of development. You may also be asked to explain the evolution of diversity in organismal form and function in terms of variation in patterns of gene expression. See Problem 22 for an example.

Example Problem: A worm that was homozygous for a loss-of-function mutation in *let-60* was crossed to a worm that was heterozygous for a dominant, *lin-23* allele. The progeny were one-half wild type and half multi-vulvate. What are the expected phenotypic proportions of the F_2 if one of the multi-vulvate F_1 were backcrossed to the *let-60* mutant parent?

Solution Strategies	Solution Steps
Evaluate	
1. Identify the topic this problem addresses, and explain the nature of the requested answer.	1. This problem tests your understanding of cell fate determination during vulval development in *C. elegans*. The answer will describe the types and proportions of phenotypes among the progeny of the cross described.
2. Identify the critical information given in the problem.	2. The F_1 were derived from a cross of a homozygous recessive, *let-60* vulva-less mutant and a multi-vulval mutant heterozygous for a gain-of-function *lin-23* allele. The F_1 were one-half wild type and half multi-vulvate.
Deduce	
3. Deduce the genotype of the multi-vulvate F_1.	3. The vulva-less parent was *let-60⁻/let-60⁻* and all the progeny had vulvas; therefore, the F_1 were *let-60⁻/let-60⁺*. The F_1 were one-half multi-vulvate and half wild type; therefore, the vulva-less parent was *lin-23⁺/lin-23⁺* and the multi-vulvate F_1 was *lin-23^D/lin-23⁺*.
4. Determine the genotypic distribution among the F_2.	4. The cross was *lin-23^D/lin-23⁺, let-60⁺/let-60⁻ × lin-23⁺/lin-23⁺, let-60⁻/let-60⁻*. The progeny will be $\frac{1}{4}$*lin-23^D/lin-23⁺, let-60⁺/let-60⁻*; $\frac{1}{4}$*lin-23^D/lin-23⁺, let-60⁻/let-60⁻*; $\frac{1}{4}$*lin-23⁺/lin-23⁺, let-60⁺/let-60⁻*; $\frac{1}{4}$*lin-23⁺/lin-23⁺, let-60⁻/let-60⁻*.
5. Recall the epistatic relationship between *lin-23^D* and *let-60⁻*.	5. *let-60* functions downstream of *lin-23*; therefore, loss of *let-60* function will be epistatic to gain of *lin-23* function.
Solve	
6. Answer the problem by assigning phenotypes to each genotype from step 4.	6. **Answer:** The F_2 will be $\frac{1}{4}$ multi-vulvate, $\frac{1}{2}$ vulva-less, and $\frac{1}{4}$ wild type.

2. Describe the developmental genetics of *Drosophila* or interpret results from genetic analysis of *Drosophila* development.

This type of problem tests your understanding of the developmental genetics of *Drosophila*. You may be asked to describe the effects of mutations in maternal genes, gap genes, pair-rule genes, segment-polarity genes, or homeobox genes on development; predict the results of crosses involving flies carrying one or more of those mutations; apply your knowledge of *Drosophila* developmental genetics to generate a hypothesis concerning the developmental regulation in other organisms. See Problem 14 for an example.

Example Problem: A recessive *Drosophila* mutation causes developmental abnormalities characterized by defects of a portion of every segment. Which class of developmental regulatory genes is likely to be affected by the mutation, and what are two specific genes that could be mutant?

Solution Strategies	Solution Steps
Evaluate	
1. Identify the topic this problem addresses, and explain the nature of the requested answer.	1. This problem tests your understanding of genetic control of *Drosophila* development. Answers will identify a specific class of *Drosophila* developmental mutants and list two specific examples of genes in that class.
2. Identify the critical information given in the problem.	2. The mutation is recessive and causes a defect in a portion of each segment.
Deduce	
3. List the different classes of genes that regulate *Drosophila* development, and state the developmental defect associated with each class. TIP: See Figures 20.5 and 20.11.	3. Mutations in coordinate genes cause defects in the embryonic axis of the embryo. Mutations in gap genes cause the loss of contiguous segments. Mutations in pair-rule genes cause loss of parts of adjacent segment pairs in alternating segments. Mutations in segment-polarity genes cause defects in the anterior or posterior portions of each segment. Mutations in homeotic genes cause defects in the identity of one or more segments.
4. Identify two specific examples of genes within each category.	4. *bicoid* and *nanos* are examples of coordinate genes. *Krüppel* and *hunchback* are examples of gap genes. *even-skipped* and *odd-skipped* are examples of pair-rule genes. *gooseberry* and *hedgehog* are examples of segment-polarity genes. *Antennapedia* and *Ultrabithorax* are examples of homeobox genes.
Solve	
5. Answer the problem, stating the class of gene and the two specific gene examples from steps 3 and 4 that match the mutant described by the problem.	5. **Answer:** The mutant is most likely a defect in a segment-polarity gene, examples of which include *gooseberry* and *hunchback*.

20.03 Solutions to End-of-Chapter Problems

1. *Evaluate:* This is type 1. This problem tests your understanding of the relationship between development and gene regulation. *Deduce > Solve:* Cell differentiation is a fundamental aspect of development and results from changes in gene expression. Thus, genes that control development do so by controlling gene expression through the regulation of transcription. Transcription factors and signaling molecules play key roles in the regulation of transcription.

2. *Evaluate:* This is type 2. This problem tests your understanding of organizers and induction. *Deduce > Solve:* The transplanted neural crest cells developed into structures resembling the source, rather than the host, indicating that neural crest cell development is autonomous. Some cells of the host tissue also developed characteristics resembling the source of the neural crest cells, indicating that these cells develop non-autonomously. The neural crest cells acted as organizers and controlled the development of host tissue by induction.

3. *Evaluate:* This is type 2. This problem tests your understanding of *Drosophila* early embryonic development. *Deduce > Solve: Drosophila* eggs are produced with the anterior–posterior and dorsal–ventral axis predetermined by maternal effect genes. *bicoid* and *nanos* are maternal effect genes that control the anterior–posterior axis. *bicoid* encodes a transcription factor that induces expression of genes such as *hunchback*, which in turn induce the development of anterior structures. *nanos* is a translation inhibitor that prevents expression of mRNAs, such as hunchback, that determine anterior fate. High-to-low gradients of *bicoid* and *nanos* determine anterior–posterior body axis. *bicoid* protein is present in high levels at the anterior; *nanos*, at the posterior.

4a. *Evaluate:* This is type 2. This problem tests your understanding of early embryonic development in *Drosophila*. *Deduce > Solve:* The syncytial blastoderm is a single, large cell with nuclei distributed throughout, and patterning is done within regions of this one cell rather than among different cells. Patterning cannot, therefore, be determined by the expression of specific extracellular signaling molecules that react with corresponding cell membrane-bound receptors. Instead, patterning is determined by mechanisms that affect the location and expression of regulatory molecules within the syncytium, by establishing intracellular gradients of transcriptional and translational regulators.

4b. *Evaluate:* This is type 1. This problem tests your ability to apply the principles derived from studying *Drosophila* patterning to mechanisms of pattern formation in unrelated organisms. *Deduce > Solve:* Analysis of *Drosophila* development indicates that gradients of intracellular molecules determine the anterior–posterior and dorsal–ventral axes of the syncytium. This knowledge can be applied to cellular blastoderm patterning by proposing that gradients of extracellular signaling molecules and their corresponding membrane-bound receptors establish cell identity along the embryonic axes.

5a–c. *Evaluate:* This is type 2. This problem tests your understanding of pair-rule gene expression in *Drosophila*. *Deduce > Solve:* **(a)** The regulatory region controlling *eve* expression in stripe 2 is modular, with binding sites for the positive activators *Hunchback* and *bicoid* and the negative regulators *Giant* and *Krüppel*. *Hunchback* and *bicoid* are present and *Giant* and *Krüppel* absent only in the region corresponding to stripe 2; therefore, *eve* expression is limited to this region. **(b)** *Giant, Hunchback,* and *bicoid* are bound to their sites in parasegement 2; only *Hunchback* and *bicoid* are bound to their sites in parasegment 3; and *Krüppel, Hunchback,* and *bicoid* are all bound to their sites in parasegment 4. The binding of either *Giant* or *Krüppel* is sufficient to repress transcription; therefore, *eve* is transcribed only in parasegment 3. **(c)** In a *Krüppel* mutant, *Hunchback* and *bicoid* are bound in parasegment 4 but *Krüppel* is not, resulting in expression of *eve* in both parasegments 3 and 4. In a *Giant* mutant background, *Hunchback* and *bicoid* are bound to their sites in parasegments 2 but *Giant* is not, resulting in *eve* expression in both parasegments 2 and 3. In *Hunchback* and *bicoid* mutants, expression in parasegment 3 is either lost (*bicoid*) or diminished (*Hunchback*).

6. *Evaluate:* This is type 2. This problem tests your understanding of *Drosophila* development. *Deduce > Solve:* Segments correspond to the anatomical divisions seen in larvae and adults, whereas parasegments span regions containing the posterior part of one segment and the anterior part of its neighboring segment. Parasegments are subdivisions of the embryo that correspond to regions or domains of gene expression.

7. *Evaluate:* This is type 2. This problem tests your understanding of the role of *Hox* genes in development and the role of *Ultrabithorax* in *Drosophila* development. *Deduce > Solve:* Loss of *Hox* gene expression leads to loss of body parts; this result is more likely to be lethal than gain of extra parts, as is seen with gain-of-function mutations. The *Ultrabithorax* (*Ubx*) gene is a *Hox* gene that is required for embryonic development; therefore, loss of *Ubx* function is lethal. *Ubx* expression is regulated by a combination of enhancers: *Ultrabithorax*^{bithorax} and *Ultrabithorax*^{postbithorax}. Mutations in these enhancer elements alter the pattern of *Ubx* expression but do not prevent expression; therefore, they are not lethal.

8. *Evaluate:* This is type 1. This problem tests your understanding of the genetic control of development. *Deduce > Solve:* Floral organ development involves repeated structures that are analogous to segments in *Drosophila*, and floral organ structures also develop under the control of a family of transcription factors that act in a combinatorial manner to specify the fate of each structure. On the other hand, the transcriptional regulator *Hox* genes in *Drosophila* are clustered, whereas *MADS-box* genes are scattered throughout the genome in *Arabidopsis*. Also, *Hox* genes are expressed along a linear head-to-tail axis, giving rise to differential regions of overlapping expression while the *MADS-box* genes are expressed in whorls from cells in the shoot meristem.

9. *Evaluate:* This is type 1. This problem tests your understanding of differences between maternal effect genes and zygotic genes in the control of development. *Deduce > Solve:* Maternal effect genes are expressed during egg development and their products, either RNA or protein or both, are deposited in the egg and present at the time of fertilization. Presence of an inhibitor of transcription in the fertilized egg does not affect the function of these genes. Since Actinomycin D does not inhibit embryonic development in frogs until just before gastrulation, development from fertilization until that point must be largely under control of maternal effect genes. Since Actinomycin D does block development past that point, zygotic gene expression is required for gastrulation and beyond.

10a–b. *Evaluate:* This is type 2. This problem tests your understanding of vulval development in *C. elegans*. *Deduce > Solve:* **(a)** *let-23* encodes the receptor that receives the *let-3* signal from the anchor cells. In the absence of the anchor cell, loss of *let-23* function will have no further consequence and the individual will be vulva-less. **(b)** *let-23* gain-of-function causes the signal for vulval development to be sent regardless of the presence of the *let-3* signal sent by the anchor cell. The absence of the anchor cell will not affect the *let-23* gain-of-function mutation, and this individual will have a multi-vulvate phenotype.

11. *Evaluate:* This is type 2. This problem tests your understanding of lateral inhibition during vulval development in *C. elegans*. *Deduce > Solve:* Activation of *let-23* or *let-60* results in differentiation of a vulval precursor cell (VPC) into a 1° cell, which then expresses *lag-2* to inhibit adjacent cells from becoming 1° cells themselves. Thus, *let-23* and *let-60* gain-of-function mutants will induce some VPCs to become 1° cells, but these 1° cells will induce neighboring cells to become 2° cells. Adjacent cells, therefore, will be 1° and 2°, not 1° and 1° or 2° and 2°.

12a–b. *Evaluate:* This is type 1. This problem tests your understanding of cell differentiation during development. *Deduce > Solve:* **(a)** For an early embryo to split into two embryos both with the potential to complete normal development, the split must occur while embryonic cells are still totipotent. **(b)** Fusion of two different embryos results in formation of a single individual containing cells of related but distinct genotypes. Such an individual will be a genetic mosaic, and many examples of such individuals have been documented and have even led to accusations that a mother is not the biological parent of her child based on DNA tests.

13a–c. *Evaluate:* This is type 3. This problem tests your understanding of the maternal effect *bicoid* gene. *Deduce > Solve:* **(a)** Since *bicoid* is a maternal effect gene, females heterozygous for *bicoid* produce normal eggs that develop normally. Because the cross is +/*bicoid* × +/*bicoid*, one-fourth of the progeny will be homozygous *bicoid* mutants and the females with this genotype will be sterile. **(b)** The homozygous *bicoid* female will produce all defective eggs, which develop into headless embryos that do not complete development. **(c)** Homozygous *bicoid* mutant females (and males) can be produced by crossing male and female +/*bicoid* heterozygotes. Homozgyous *bicoid* mutant males that were produced by a female heterozygous for *bicoid* will be phenotypically normal.

14. *Evaluate:* This is type 2. This problem tests your understanding of the role of *bicoid* in *Drosophila* development. *Deduce > Solve:* Increased dosage of the *bicoid* gene will result in an increase in the amount of *bicoid* mRNA and therefore *bicoid* protein. This increase would alter the gradient of *bicoid* protein in the egg and, in turn, alter the expression of genes downstream of *bicoid* function. For example, *Hunchback* is activated by *bicoid*, so *Hunchback* expression also increases and the zone of expression is pushed further posterior in a fly with a higher *bicoid* gene dosage. The entire body plan of the mutant flies would likely be altered, resulting in more anterior-like segment fates being adopted by what would normally be more posterior-like segments.

15. *Evaluate:* This is type 2. This problem tests your understanding of *Drosophila* development. *Deduce > Solve: Krüppel* is a gap gene, and mutations in gap genes cause the loss of a number of contiguous body segments: in this case, a loss of head, thorax, and anterior abdominal segments. *odd-skipped* is a pair-rule gene, and mutations in pair-rule genes cause a loss of alternate segments: in this case, a loss of the odd-numbered segments. *hedgehog* is a segment-polarity gene, and mutations in segment-polarity genes cause defects in patterning within segments leading to polarity defects in embryonic and adult structures. *Ultrabithorax* is a homeotic gene, and mutations in homeotic genes cause defects in the specification of segmental identity: in this case, a shift in the identity of the third thoracic segment and the first abdominal segment to an identity resembling the second thoracic segment.

16a–b. *Evaluate:* This is type 2. This problem tests your understanding of the genetic analysis of *Drosophila* development. *Deduce > Solve:* **(a)** *fushi tarazu* is a pair-rule gene that is expressed in the seven, even-numbered segments. *engrailed* is a segment-polarity gene, which is expressed in the anterior part of each parasegment. Pair-rule genes might be expected to influence the expression of the segment-polarity genes, which act at a later time in development, predicting that *engrailed* expression would be altered in a *fushi tarazu* mutant but not necessarily vice versa. **(b)** The *fushi tarazu* single mutant likely has a loss of the even-numbered parasegments, and the *engrailed* single mutant has defects in the anterior part of each parasegment. You might predict that the double mutant would be a combination of these two single mutant phenotypes.

17. *Evaluate:* This is type 2. This problem tests your understanding of homeotic gene function in *Drosophila*. *Deduce > Solve:* The genes of the *Antennapedia* complex drive formation of segments that contain legs, whereas genes of the *bithorax* complex promote development of abdominal segments that lack legs. Since centipedes have legs on all segments posterior to the head, *bithorax* gene expression may be expected to be repressed and *Antennapedia* expression to be induced in all the leg-containing segments.

18. *Evaluate:* This is type 2. This problem tests your understanding of lateral inhibition. *Deduce > Solve:* A generalized description of the scenario posed by the problem is that a specific differentiated cell type (bristle cell) is never adjacent to a cell just like itself. This fits the pattern expected if the bristle cell produces signaling molecules that prevent adjacent cells from differentiating into bristle cells. This regulatory process is called lateral inhibition.

19. *Evaluate:* This is type 2. This problem tests your understanding of regulation of segment identity in plants. *Deduce > Solve:* The problem describes a plant homeotic mutant in which one whorl that normally develops into stamens instead develops into carpels, similar to the adjacent segment.

Segment identity in plants is controlled by transcription factors of the *MADS-box* variety; therefore, you might expect the homeotic mutation to be in a *MADS-box* gene equivalent to *apetela3* or *pistillata* in *Arabidopsis*.

20a–b. *Evaluate:* This is type 2. This problem tests your understanding of the genetic analysis of vulval development in *C. elegans*. *Deduce > Solve:* **(a)** *lin-3* is a signal molecule produced by the anchor cell; therefore, loss-of-function *lin-3* mutants do not signal the VPE. Gain-of-function mutations in downstream genes, such as the *lin-3* receptor (*lin-23*) or the intracellular signal molecule activated by *lin-23* (*let-60*), would suppress the *lin-3* null mutation because they induce vulval development (resulting in multi-vulval development) independently of *lin-3*. **(b)** Loss-of-function mutations in downstream genes, such as the *let-60* gene, will suppress the gain-of-function *lin-23* mutant because it prevents vulval development (resulting in vulva-less development); this happens because *lin-23* acts through activation of *let-60*.

21. *Evaluate:* This is type 1. This problem tests your understanding of the genetic control of development. *Deduce:* Development is controlled by multiple genes. The number of genes controlling a trait can be determined in a series of crosses where parents with extreme phenotypes are crossed to produce F_1 hybrids that are then selfed to produce an F_2 population. The number of genes segregating in the F_2 can be estimated by the equation: fraction of progeny with an extreme phenotype $= \frac{1}{4^N}$, where N is the number of genes that are heterozygous in the F_1. The problem states that $\frac{1}{1000}$ F_2 have the extreme phenotype. *Solve:* N is approximately five because $\frac{1}{4^5} = 1024$, indicating that relatively few genes are affecting corn development, but each gene has a relatively large effect. Thus, allelic changes at only five loci could result in a $\frac{1}{1000}$ segregation, suggesting that the different architectures in maize and teosinte are determined by changes with large effect in only a few genes.

22a–b. *Evaluate:* This is type 1. This problem tests your understanding of *Hox* gene control of digit development. *Deduce > Solve:* **(a)** Ectopic expression of *Hoxd10* throughout the developing limb bud would add *Hoxd10* expression to *Hoxd9* in what should be a thumb, converting the thumb into an index finger. Ectopic expression of *Hoxd11* would add *Hoxd11* to *Hoxd9* and *Hoxd10* in what should be the index finger, converting the index finger into a middle finger, and it could affect thumb development as well. Ectopic expression of *Hoxd10* and *Hoxd11* would add both genes to *Hoxd9* in what would be the thumb, converting the thumb into a middle finger. It would also add *Hoxd11* to *Hoxd9* and *Hoxd10* in what would be the index finger, converting the index finger into a middle finger. **(b)** To prevent *Hoxd9–13* expression only in the limb bud, you would need to construct a conditional mutant using the *cre-lox* system described in Chapter 19. In this approach, the *Hoxd* gene cluster would be flanked by *loxP* sites and expression of *cre*-recombinase would be driven by a limb-bud-specific promoter.

23a–b. *Evaluate:* This is type 1. This problem tests your understanding of the concept that "evolution behaves like a tinkerer." *Deduce > Solve:* **(a)** Species with different developmental patterns often accomplish this result with the same developmental regulatory genes by altering their pattern of expression. In this case, the difference between the two species of sticklebacks would be regulatory sequences controlling *Eda* expression. **(b)** The phenotype of loss of *Eda* function in humans indicates that *Eda* has been co-opted by evolution to control different aspects of development, as it does in sticklebacks. It also indicates that *Eda* controls multiple aspects of human development, either directly by controlling each tissue type independently or by controlling a tissue that is a developmental precursor to the affected tissue types.

24. *Evaluate:* This is type 1. This problem tests your understanding of developmental regulatory pathways. *Deduce:* Loss-of-function *tra-1* or *tra-2* alleles cause male development in XX individuals, suggesting that *tra* genes are negative regulators of male development and/or positive regulators of hermaphrodite development. Loss-of-function of *her-1* causes hermaphroditic development in XO individuals, suggesting that *her-1* is a negative regulator of hermaphrodite development or a positive regulator of male development. Dominant, gain-of-function *tra-1* mutations induce hermaphrodite development, which is also consistent with *tra* genes as repressors of male development and inducers of hermaphrodite development. Furthermore, the dominant *tra-1* allele is epistatic to loss of *tra-2* function, suggesting that *tra-1* is downstream of *tra-2*. Two models consistent with these results are shown in the figure. *Solve:* Loss of *tra-1* or *tra-2* function is epistatic to loss of *her-2* function (the double

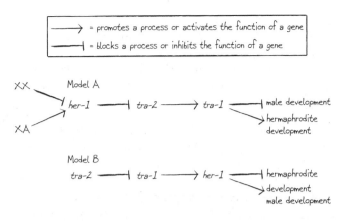

mutants are male), which places *tra* genes downstream of *her-2*. This rules out Model B and is consistent with Model A. The X : A ratio could regulate *her-2*, and XX (1 : 1) would inhibit *her-2*, which would allow the *tra* genes to prevent male development and/or promote hermaphrodite development. An X : A of 0.5 : 1 would activate *her-2*, which in turn blocks the *tra* genes from preventing male development and/or promoting hermaphrodite development.

25. *Evaluate:* This is type 1. This problem tests your understanding of the concept that "evolution behaves like a tinkerer." *Deduce > Solve:* The difference between the two related plants is one developmental structure, the central whorl of stamens. If the central stamen whorl of one species is replaced by a whorl with a different fate in the other, then a homeotic transformation occurs, which could be due to mutation in a *MADS-box* regulatory gene. If the central whorl that is the stamens in one species is altogether absent in the other species, then the mutation could be in a regulator, analogous to a *Drosophila* gap gene (loss of segments). The male-only flowers in the dioecious species suggest lack of expression of a C-class MADS gene required for carpel development.

26. *Evaluate:* This is type 1. This problem tests your understanding of developmental regulatory pathways. *Deduce > Solve:* **(a)** In an otherwise wild-type background, the effects of *agamous* mutations are confined to the third and fourth whorls. Likewise, the effects of *apetala2* mutations are limited to the first and second whorls. In *agamous apetala2* double mutants, all four whorls are affected as expected but are not simply additive: the *agamous* mutation affects the first and second whorls, and the *apetala2* mutation affects the third and fourth. This result suggests that wild-type *agamous* prevents the effect of *apetala2* mutations in the third and fourth whorls (*agamous* is epistatic to *apetala2*) and that wild-type *apetala2* prevents the effect of *agamous* mutations in the first and second whorls (*apetala2* is epistatic to *agamous*). The mutually epistatic interactions are consistent with a whorl-specific antagonistic (inhibitory) relationship between the two developmental regulators. **(b)** Comparing homeotic control of floral organ identity with animal development along the anterior–posterior axis uncovers similarities, such as the interaction between *apetala2* and *agamous* previously described and the mutual regulation in animal *Hox* genes where the more posteriorly expressed *Hox* genes repress expression of the more anteriorly expressed *Hox* genes. Both systems use combinatorial overlapping expression patterns to define locations where specific organ development will occur. Differences between homeotic control of plant and animal development are observed in the location of genes: plants use *MADS-box* genes that are dispersed throughout the genome, whereas animals have clusters of *Hox* genes.

27. *Evaluate:* This is type 1. This problem tests your understanding of homeotic regulation of *Drosophila* development. *Deduce > Solve: Drosophila* is homozygous for two recessive mutations affecting *Ultrabithorax* expression, *Ubx*[bx pdx], that prevent expression of *Ubx* in the third thoracic segment causing the halteres on that segment to be converted into wings. Dipterans could have evolved from four-winged insects by mutations that prevent a homeobox gene like *Ubx* from being expressed in the third thoracic segment.

28. *Evaluate:* This is type 1. This problem tests your understanding of the genetic analysis of development. *Deduce > Solve:* To identify the genes required for patterning, screen for mutants with altered development; for example, mutants that are viable but did not make mushrooms (sexual reproductive structures) or that made aberrant mushrooms. The genes identified in such a screen are required for mushroom formation and are either regulators or targets of regulators. Ectopic or over-expressed genes could identify regulatory genes, which would likely cause genes normally limited to mushroom structures to be expressed, and may cause the formation of mushrooms under conditions typically preventing mushroom formation. Genes shown to be both necessary and sufficient for development are likely candidates for developmental regulatory genes.

29a. *Evaluate:* This is type 1. This problem tests your understanding of the genetic analysis of *Drosophila* development. *Deduce:* The figure shows a Punnett square for the cross. *Solve:* The expected phenotypic distribution in the F_2 is (1) legs on head and normal mating, (2) legs on head and abnormal mating, (3) no legs on head and normal mating, (4) no legs on head and abnormal mating. Calculations for F_2 phenotypes 1–3 are shown below:

	$Ant^d\,fru^+$ 0.425	$Ant^+\,fru^-$ 0.425	$Ant^d\,fru^-$ 0.075	$Ant^+\,fru^+$ 0.075
$Ant^d\,fru^+$ 0.425	$Ant^d\,fru^+/$ $Ant^d\,fru^+$ 0.181	$Ant^d\,fru^+/$ $Ant^d\,fru^-$ 0.181	$Ant^d\,fru^+/$ $Ant^d\,fru^-$ 0.0319	$Ant^d\,fru^+/$ $Ant^+\,fru^+$ 0.0319
$Ant^+\,fru^-$ 0.425	$Ant^+\,fru^-/$ $Ant^d\,fru^+$ 0.181	$Ant^+\,fru^-/$ $Ant^+\,fru^-$ 0.181	$Ant^+\,fru^-/$ $Ant^d\,fru^-$ 0.0319	$Ant^+\,fru^-/$ $Ant^+\,fru^+$ 0.0319
$Ant^d\,fru^-$ 0.075	$Ant^+\,fru^-/$ $Ant^d\,fru^+$ 0.0319	$Ant^d\,fru^-/$ $Ant^+\,fru^-$ 0.0319	$Ant^d\,fru^-/$ $Ant^d\,fru^-$ 0.0056	$Ant^d\,fru^-/$ $Ant^+\,fru^+$ 0.0056
$Ant^+\,fru^+$ 0.075	$Ant^+\,fru^+/$ $Ant^d\,fru^+$ 0.0319	$Ant^+\,fru^+/$ $Ant^+\,fru^-$ 0.0319	$Ant^+\,fru^+/$ $Ant^d\,fru^-$ 0.0056	$Ant^+\,fru^+/$ $Ant^+\,fru^+$ 0.0056

(1) $= 0.181 + 0.181 + 0.0319 + 0.0319 + 0.181 + 0.319 + 0.0056 + 0.0319 + 0.0056 = 0.6818$;

(2) $= 0.0319 + 0.0319 + 0.0056 = 0.0694$;

(3) $= 0.0319 + 0.0319 + 0.0056 = 0.0694$;

(4) $= 0.181$.

29b. *Evaluate:* This is type 1. This problem tests your understanding of the genetic analysis of *Drosophila* development. *Deduce > Solve:* The only types of progeny recovered are non-recombinant progeny. Since the cross involves hybrids, the lack of non-recombinants could be caused by a hybrid heterozygous for an inversion of a chromosomal region including *fru* and *Ant*. Crossover within this region would not be recovered among the progeny, since the resulting duplications and deletions of the large segment of chromosome would be lethal. You could test this hypothesis by checking whether salivary gland polytene chromosomes of the F_1 hybrids show an inversion loop.

29c. *Evaluate:* This is type 1. This problem tests your understanding of regulation of *Drosophila* development. *Deduce > Solve:* A mutation consistent with the chromosomal rearrangement proposed in part (b) would be an inversion that placed a regulatory sequence driving *Ant* expression in the cells normally destined to become antennae. Such a rearrangement could be caused by X-rays, which were used to induce the dominant Ant^d mutation.

20.04 Test Yourself

Problems

1. A pair of wild-type fruit flies of unknown origin are mated and produce wild-type F_1. The F_1 are allowed to interbreed and produce all wild-type F_2. The F_2 are allowed to interbreed and produce an F_3 population that is $\frac{3}{4}$ wild type and $\frac{1}{4}$ mutant. If the mutant phenotype is due to a recessive mutation in an autosomal maternal effect gene, what are the genotypes of the original parents?

2.   A pure-breeding vulva-less line of *C. elegans* is crossed to a pure-breeding wild-type line, and all the progeny are multi-vulvate. A multi-vulvate hermaphrodite is allowed to self-fertilize and produces $\frac{9}{16}$ multi-vulvate, $\frac{4}{16}$ vulva-less, and $\frac{3}{16}$ wild type. A cross of any vulva-less F_2 to a tester strain homozygous for loss of *let-60* function produced all vulva-less progeny. What was the genotype of the original pure-breeding vulva-less parent?

3.   Expression of a newly cloned gene is examined using a *lacZ* reporter fused to the gene's promoter. *lacZ* expression is seen in seven stripes within the developing embryo. What class of *Drosophila* developmental regulatory gene most likely corresponds to the newly cloned gene?

4.   Expression of a newly cloned gene is examined using a *lacZ* reporter fused to the gene's promoter. *lacZ* expression is seen in a broad band located at the approximate center of the anterior–posterior axis. What class of *Drosophila* developmental regulatory gene most likely corresponds to the newly cloned gene, and how is it regulated?

Solutions

1.   *Evaluate:* This is type 1. This problem tests your understanding of maternal effect genes and autosomal recessive inheritance. ***Deduce > Solve:*** The mothers of individuals that show a mutant phenotype due to mutations in maternal effect genes must be homozygous recessive for the mutant allele. Since 1/4 of the F_3 show the mutant phenotype, then 1/4 of the F_2 females were homozygous recessive (m/m), and 3/4 carried at least one wild-type allele ($M/\_$). The F_2 genotypic distribution—1/4 m/m and 3/4 $M/\_$—is consistent with the F_1 being heterozygous (M/m). Heterozygotic F_1 could result only by mating parents that are M/M and m/m. Since the F_1 are all wild type, the female parent could not have been m/m—all F_1 would have shown the mutant phenotype—indicating that the male parent was m/m and the female was M/M.

2.   *Evaluate:* This is type 1. This problem tests your understanding of the signaling pathway controlling cell fate during *C. elegans* vulval development. ***Deduce > Solve:*** The test cross of the vulva-less F_2 indicates that the original vulva-less parent was homozygous for loss of *let-60* function. The appearance of multi-vulvate F_1 and the 9 : 4 : 3 phenotypic ratio among the F_2 indicate that the original vulva-less parent was also homozygous for a gain-of-function mutation in a gene upstream of *let-60*. *lin-23* is upstream of *let-60* and dominant, gain-of-function mutations in *lin-23* are known to cause a multi-vulvate phenotype in *let-60*+ *C. elegans*. The original vulva-less parent was homozygous for *lin-23*^D and *let-60*^− mutations.

3.   *Evaluate:* This is type 2. This problem tests your understanding of *Drosophila* developmental genetics. ***Deduce > Solve:*** The seven stripes of gene expression most likely represent expression in odd or even parasegments of the embryo. This pattern is consistent with a pair-rule gene, such as *even-skipped* or *odd-skipped*, depending on which set of parasegments shows *lacZ* expression.

4.   *Evaluate:* This is type 2. This problem tests your understanding of *Drosophila* developmental genetics. ***Deduce > Solve:*** Expression in the center of the anterior–posterior axis most likely corresponds to a gap gene, similar to *Krüppel*. The gene would be activated in regions that contain moderate levels of *bicoid* but low levels of *hunchback, knirps,* and *giant.*

21

Genetic Analysis of Quantitative Traits

Section 21.01 Genetics Problem-Solving Toolkit

Section 21.02 Types of Genetics Problems

Section 21.03 Solutions to End-of-Chapter Problems

Section 21.04 Test Yourself

21.01 Genetics Problem-Solving Toolkit

Key Terms and Concepts

Artificial selection: Selective breeding of a subset of individuals from a population for the purpose of creating a new population whose mean value for a quantitative trait is higher or lower than the original population.

Broad sense heritability: Proportion of population variance that is due to all types of genetic factors.

Concordance rate: Percentage of pairs of individuals that have the same phenotype.

Multifactorial traits: Phenotypes that are controlled by multiple genes and by environmental factors.

Narrow sense heritability: Proportion of population variance that is due to simple additive genetics.

Quantitative traits: Multifactorial traits that are characterized by a number and display a continuous variation from one extreme to the other, for example, body weight or height.

Quantitative trait locus (QTL): Genetic locus that is associated with one or more genes that contribute to a quantitative trait.

Simple additive genetics: Genetic control of a polygenic trait in which each gene contributes equally and additively to the phenotype.

Threshold traits: Multifactorial traits that are characterized by a few discrete, qualitative characters, for example, diabetes or schizophrenia.

Key Analytical Tools

$M' = M + h^2(M_s - M)$, where M is the mean of the original population, M_s is the mean of the selected population, M' is the mean of the resulting population, and h^2 is the narrow sense heritability.

$(a + b)^N$ is the general equation for a binomial distribution, which can be used to calculate the genotypic frequency distribution in a population derived from a hybrid cross where N additive genes are involved and the frequencies of the alternative additive alleles are a and b.

$$V_P = V_G + V_E; \qquad V_G = V_A + V_D + V_I; \qquad H^2 = \frac{V_G}{V_P}; \qquad h^2 = \frac{V_A}{V_P}$$

21.02 Types of Genetics Problems

1. Define or explain the fundamental concepts and principles governing inheritance of multifactorial traits.
2. Calculate the mean, variance, standard deviation, and heritability of quantitative traits.
3. Predict the results of crosses involving polygenic traits.
4. Analyze the results of artificial selection experiments.

1. Define or explain the fundamental concepts and principles governing inheritance of multifactorial traits.

This type of problem tests your understanding of multifactorial inheritance. You may be asked to define or explain concepts such as polygenic inheritance, quantitative traits, threshold traits, multifactorial inheritance in humans, or quantitative trait loci. You may also be asked to analyze or predict the analysis of human twin studies. See Problem 16 for an example.

Example Problem: What is the relationship between quantitative trait loci (QTLs) and the genes that control quantitative traits?

Solution Strategies	Solution Steps
Evaluate	
1. Identify the topic this problem addresses, and explain the nature of the requested answer.	1. This problem tests your understanding of quantitative trait loci (QTLs). The answer requires an understanding of QTLs, genes, and the relationship between the two.
2. Identify the critical information given in the problem.	2. The problem asks you to describe the relationship between QTLs and genes.
Deduce	
3. Describe the nature of genes that control quantitative traits.	3. Genes that control quantitative traits do so by encoding gene products that contribute to complex phenotypes by interacting with the products of other genes.
4. Define the term *quantitative trait locus*.	4. A QTL is a genetic locus, typically a DNA polymorphism that shows statistically significant association with the trait being studied. The association study demonstrates linkage of the QTL to the trait.
Solve	
5. Answer the problem by relating QTLs to genes that control a quantitative trait.	5. **Answer:** QTLs are genetic loci that are linked to genes that control a quantitative trait but are not necessarily located in or part of those genes.

2. Calculate the mean, variance, standard deviation, and heritability of quantitative traits.

This type of problem tests your ability to analyze quantitative traits. You may be asked to calculate the mean, variance, and standard deviation using experimental data on a quantitative trait. You may be asked to dissect the population variance, V_p, in the trait into its component parts, V_G and V_E, and genetic variance, V_G, into its component parts V_A, V_D, and V_I. You may be asked to calculate the broad sense (H^2) or narrow sense (h^2) heritability of a trait. See Problem 6 for an example.

Example Problem: The variance in body weight for two inbred insect populations is 2.4 mg^2 and 2.2 mg^2. An F_1 hybrid population shows a variance of 2.3 mg^2. The F_1 are interbred to produce an F_2 population that has a variance of 6 mg^2. Calculate V_G, V_E, and H^2.

Solution Strategies	Solution Steps
Evaluate	
1. Identify the topic this problem addresses, and explain the nature of the requested answer.	1. This problem tests your understanding of quantitative inheritance. The answers will state the values for genetic variance, environmental variance, and heritability of body weight in the F_2 insect population.
2. Identify the critical information given in the problem.	2. The variance in the F_1 population is 2.3 mg^2. The variance in body weight in the F_2 population is 6 mg^2.
Deduce	
3. State the relationship between V_p, V_G, V_E, and H^2.	3. $V_p = V_G + V_E$ and $H^2 = \dfrac{V_G}{V_p}$
4. Identify V_p and V_E from the information given in the problem.	4. The F_1 population is genetically homogeneous; therefore, the variance of the F_1 is entirely due to environment. $V_E = 2.3$ mg^2. The F_2 population is genetically heterogeneous, therefore, the variance of the F_2 is due to both genetics and environment. $V_p = 6$ mg^2.
Solve	
5. Calculate V_G and H^2.	5. $V_G = V_p - V_E = 6 - 2.3 = 3.7$ mg^2. $H^2 = \frac{3.7}{6} = 0.62$.
6. Answer the problem by stating V_p, V_E, V_G, and H^2.	6. **Answer:** $V_p = 6$ mg^2, $V_E = 2.3$ mg^2, $V_G = 3.7$ mg^2, and $H^2 = 0.62$.

3. Predict the results of crosses involving polygenic traits.

This type of problem tests your understanding of polygenic inheritance. You may be asked to use a model for genetic control of a quantitative trait to predict the phenotype associated with specific genotypes. You may be asked to predict the distribution of phenotypes among progeny of a cross or calculate the probability of a specific phenotypic class. See Problem 8 for an example.

Example Problem: Four genes contribute to body weight in an insect. There are two alleles of each gene: the "1" allele contributes 1 mg to body weight and the "2" allele contributes nothing to body weight. The weight of an insect with the genotype $A_2A_2B_2B_2C_2C_2D_2D_2$ is 10 mg. What is the distribution of body weight among an insect bred by mating two insects with the genotype $A_1A_2B_1B_2C_1C_2D_1D_2$?

Solution Strategies	Solution Steps
Evaluate	
1. Identify the topic this problem addresses, and explain the nature of the requested answer.	1. This problem tests your understanding of polygenic control of a quantitative trait. The answers will describe the distribution of phenotypes in a population created by the mating specified.
2. Identify the critical information given in the problem.	2. Each "1" allele of each gene contributes 1 mg to body weight whereas each "2" allele contributes nothing. An insect with all 2 alleles of the genes weighs 10 mg. The cross is a tetrahybrid cross.
Deduce	
3. Describe how the distribution of progeny from the tetrahybrid cross can be calculated in terms of fraction of progeny with 1, 2, 3, 4, 5, 6, 7, or 8 copies of 1 alleles. TIP: Since each gene contributes equally, the distribution of 1 and 2 alleles can be calculated as a binomial distribution with the probability of 1 and 2 alleles each being $\frac{1}{2}$.	3. The binomial distribution is given by expanding the equation $(a + b)^8$, where a and b are both $\frac{1}{2}$. This equation has 9 terms, each given by the equation $\frac{N!}{S!T!}\left(\frac{1}{2}\right)^S\left(\frac{1}{2}\right)^T$, where S and T are the number of 1 and 2 alleles in each category. $\left(\frac{1}{2}\right)^S\left(\frac{1}{2}\right)^T$ will always be $\left(\frac{1}{2}\right)^8$, and the coefficient can be taken from Pascal's triangle.
4. Write out the binomial expansion that describes the genotypic distribution.	4. $1\left(\frac{1}{2}\right)^8 + 8\left(\frac{1}{2}\right)^8 + 28\left(\frac{1}{2}\right)^8 + 56\left(\frac{1}{2}\right)^8 +$ $70\left(\frac{1}{2}\right)^8 + 56\left(\frac{1}{2}\right)^8 + 28\left(\frac{1}{2}\right)^8 + 8\left(\frac{1}{2}\right)^8 + 1\left(\frac{1}{2}\right)^8$
5. Assign genotypes and phenotypes to each term in the expanded binomial. TIP: Each genotype can be listed in terms of the number of 1 and 2 alleles. The phenotype will be 10 mg + the weight added by the genotype.	5. $1\left(\frac{1}{2}\right)^8$ corresponds to eight 1 and no 2 alleles, which gives $10 + 8 = 18$ mg and also corresponds to no 1 and eight 2 alleles, which gives $10 + 0 = 10$ mg. $8\left(\frac{1}{2}\right)^8$ corresponds to seven 1 and one 2 allele, which gives $10 + 7 = 17$ mg and also corresponds to one 1 and seven 2 alleles, which gives $10 + 1 = 11$ mg. $28\left(\frac{1}{2}\right)^8$ corresponds to six 1 and two 2 alleles, which gives $10 + 6 = 16$ mg, and also corresponds to two 1 and six 2 alleles, which gives $10 + 2 = 12$ mg. $56\left(\frac{1}{2}\right)^8$ corresponds to five 1 and three 2 alleles, which gives $10 + 5 = 15$ mg and also corresponds to three 1 and five 2 alleles, which gives $10 + 3 = 13$ mg. $\frac{1}{256}$ corresponds to four 1 and four 2 alleles, which gives $10 + 4 = 14$ mg.
Solve	
6. Calculate the values for each term in the expanded binomial in step 4.	6. $\frac{1}{256} + \frac{8}{256} + \frac{28}{256} + \frac{56}{256} + \frac{70}{256} + \frac{56}{256} + \frac{28}{256} + \frac{8}{256} + \frac{1}{256}$
7. Answer the problem by stating the frequency of each phenotype.	7. **Answer:** The phenotypic distribution will be $\frac{1}{256}$ weighing 18 mg, $\frac{8}{256}$ weighing 17 mg, $\frac{28}{256}$ weighing 16 mg, $\frac{56}{256}$ weighing 15 mg, $M' = M + h^2(M_s - M)$ weighing 14 mg, $\frac{56}{256}$ weighing 13 mg, $\frac{28}{256}$ weighing 12 mg, $\frac{8}{256}$ weighing 11 mg, and $\frac{1}{256}$ weighing 10 mg.

4. Analyze the results of artificial selection experiments.

This type of problem tests your understanding of artificial selection. You may be asked to use the relationships between narrow sense heritability (h^2), selection differential (S), and response to selection (R) to calculate one of these terms given the other two. Alternatively, you may also be asked to examine experimental data to determine h^2, S, or R values. You may also be asked to comment on the relative potential for successful use of artificial selection given h^2 values or the ability to calculate h^2. See Problem 14 for an example.

Example Problem: A population of insects has a mean body weight of 24 mg. A subset of this population with a mean weight of 34 mg was selected to breed a new population. The mean weight of insects in the new population was 27 mg. What is the narrow sense heritability for body weight in this original population of insects?

Solution Strategies	Solution Steps
Evaluate	
1. Identify the topic this problem addresses, and explain the nature of the requested answer.	1. This problem tests your understanding of artificial selection. The answers will be a value between 1 and 0.
2. Identify the critical information given in the problem.	2. The mean weight of the original population (M) was 24 mg. The mean weight of the selected population (M_s) was 34 mg. The mean weight of the resulting population (M') was 27 mg.
Deduce	
3. State the relationship between the original population mean, the selected population mean, the resulting population mean, and the narrow sense heritability.	3. $M' = M + h^2(M_s - M)$
4. Rearrange the equation to solve for h^2.	4. $h^2 = \frac{(M' - M)}{M_s - M}$
Solve	
5. Answer the problem by solving for h^2.	5. **Answer:** $h^2 = \frac{(M' - M)}{(M_s - M)} = \frac{(27 - 24)}{(34 - 24)} = 0.3$

21.03 Solutions to End-of-Chapter Problems

1a–e. *Evaluate:* This is type 1. This problem tests your understanding of quantitative traits.
Deduce > Solve: Quantitative traits include **(a)** body weight in chickens, **(b)** growth rate in sheep, **(c)** milk production in cattle, and **(d)** fruit weight in tomatoes. **(e)** Coat color in dogs is a qualitative trait, but it is controlled by multiple genes that interact.

2. *Evaluate:* This is type 1. This problem tests your understanding of multifactorial traits.
Deduce > Solve: Body weight in chickens, growth rate in sheep, milk production in cattle, and fruit weight in tomatoes are all multifactorial—that is, affected by multiple genes and environmental factors. Two likely environmental factors that may play a role in phenotypic variation for all four traits include nutrition and nutrient sources.

3. *Evaluate:* This is type 1. This problem tests your understanding of heritability as it applies to complex traits. *Deduce > Solve:* Complex traits are determined by multiple genes, environmental factors, and interactions between genetic and environmental factors. Quantitative traits are complex traits that are described by a distribution of values in a population and represented by statistical measurements such as mean, variance, and standard deviation. Heritability measures the proportion of variance in a quantitative trait caused by genetic factors. Broad sense heritability, H^2, is the proportion of variance that results from all types of genetic factors (V_A, V_D, V_I); narrow sense heritability h^2 is that which results only from genetics that is described as simple, additive genetics (V_A). For example, if the variance for weight in an insect population is 100 mg^2, 50% of the variance is due to all types of genetics, and 25% is due to simple, additive genetics, then H^2 is $0.5 \times 100 = 50$ mg^2 and narrow sense heritability is $0.25 \times 100 = 25$ mg^2.

4. *Evaluate:* This is type 2. This problem tests your ability to calculate V_G and V_E given information from a series of crosses. *Deduce:* The F_1 are genetically homogenous because their parents were pure-breeding; therefore, the variance in the F_1, 2.25 g, is V_E. The genes controlling tomato weight segregate in the F_2; therefore, the variance in the F_2, 5.40 g, results from the formula $V_E + V_G = V_P$. *Solve:* To calculate V_G, subtract V_E from V_P: 5.40g – 2.25 g = 3.15 g, and to calculate broad sense heritability, use the formula $H^2 = \frac{V_G}{V_E}$, which gives $\frac{3.15}{5.40} = 0.583$.

5. *Evaluate:* This is type 1. This problem tests your understanding of continuous and discontinuous polygenic traits. *Deduce > Solve:* Discontinuous phenotypic variation is characterized by discrete, easily distinguished phenotypic categories, for example, unaffected versus affected or white versus red versus pink. Continuous phenotypic variation is characterized by a continuous range of phenotypes, for example, fruit weight or milk volume. Threshold traits are discontinuous (schizophrenia versus normal or diabetic versus normal) but are explained by an underlying quantitative distribution of risk factors that are genetic and environmental. Polygenic inheritance can explain both continuous and discontinuous traits. For example, if the variation in weight in a population is explained by the segregation of 10 genes that control weight in a simple additive genetic fashion, the distribution will be continuous from the lowest value (homozygous recessive at all 10 genes) to the highest value (homozygous dominant at all 10 genes). Alternatively, if the variation in a threshold trait, such as in a population, is explained by the segregation of 10 genes and a threshold of 7 dominant alleles causes the trait to be expressed, then only those members of the population with at least 7 dominant alleles will show the trait, and the rest will be wild type.

6. *Evaluate:* This is type 2. This problem tests your ability to apply statistical analyses to a quantitative trait data set. *Deduce > Solve:* The mean is given by the equation $\bar{X} = \frac{\sum_{i=1}^{n} X_i}{n}$, which gives $\bar{X} = \frac{161 + 172 + 155 + 173 + 149 + 177 + 156 + 174 + 158 + 162 + 171 + 181}{12} = 165.75$ ounces.

The variance is given by the equation $\sigma_X^2 = \frac{\sum_{i=1}^{n}(X_i - \bar{X})^2}{n-1}$, which gives $\frac{1137.22}{11} = 103.38$. The standard deviation is $\sigma_X = \sqrt{\sigma_X^2}$, = which gives $\sqrt{103.38} = 10.17$.

7a–f. *Evaluate > Deduce > Solve:* This is type 1. This problem tests your understanding of key principles of complex traits. **(a)** Additive genes are genes that contribute equally to the value of a quantitative trait. Additive genes have two types of alleles; additive alleles, each of which contributes equally to the value of a quantitative trait, and nonadditive alleles, which do not contribute anything to the quantitative trait. **(b)** The concordance values for twins are the percentage of twin pairs in which both have the same phenotype for a trait. **(c)** Multifactorial inheritance refers to traits that are determined by many genes and environmental factors. **(d)** Polygenic inheritance refers to traits that are determined by multiple genes. **(e)** A quantitative trait locus is a location on a chromosome that shows statistically significant association with a quantitative trait. **(f)** A threshold trait is a discontinuous trait for which the underlying inheritance is multifactorial and is expressed only when a threshold level of genetic and/or environmental factors exists.

8a–f. *Evaluate:* This is type 3. This problem tests your understanding of polygenic control of a quantitative trait. *Deduce > Solve:* **(a)** A plant with the genotype $A_1A_1B_1B_1C_1C_1$ will be 36 cm tall. A plant with the genotype $A_2A_2B_2B_2C_2C_2$ will be 18 cm tall. **(b)** The progeny of the cross will have the genotype $A_1A_2B_1B_2C_1C_2$ and will be 27 cm tall. **(c)** A plant with the genotype $A_1A_2B_2B_2C_1C_2$ will be 24 cm tall. **(d)** A plant that is 33 cm tall will have 5 alleles that contribute 6 cm to height and one that contributes 3 cm to height. The possible genotypes are $A_1A_1B_1B_1C_1C_2$, $A_1A_1B_1B_2C_1C_1$, or $A_1A_2B_1B_1C_1C_1$. **(e)** There are $(3)^3 = 27$ possible genotypes. **(f)** There are six different phenotypes: 36 cm, 33 cm, 30 cm, 27 cm, 24 cm, 21 cm, and 18 cm.

9a–d. *Evaluate:* This is type 3. This problem tests your understanding of polygenic control of a quantitative trait. *Deduce > Solve:* **(a)** A plant with the genotype $A_1A_1B_1B_1C_1C_1$ will be 30 cm tall. **(b)** A plant with the genotype $A_2A_2B_2B_2C_2C_2$ is 12 cm. **(c)** The F_1 will have the genotype $A_1A_2B_1B_2C_1C_2$ and will be 30 cm tall. **(d)** The F_2 will be $\frac{27}{64}$ $A_1\_B_1\_C_1\_$, which are 30 cm tall; $\frac{9}{64}$ $A_1\_B_1\_C_2C_2$, which are 24 cm tall; $\frac{9}{64}$ $A_1\_B_2B_2C_1\_$, which are 24 cm tall; $\frac{9}{64}$ $A_2A_2B_1\_C_1\_$, which are 24 cm tall; $\frac{3}{64}$ $A_1\_B_2B_2C_2C_2$, which are 18 cm tall; $\frac{3}{64}$ $A_2A_2B_1\_C_2C_2$, which are 18 cm tall; $\frac{3}{64}$ $A_2A_2B_2B_2C_1\_$, which are 18 cm tall, $\frac{1}{64}$ $A_2A_2B_2B_2C_2C_2$, which are 12 cm tall.

10a–b. *Evaluate:* This is type 2. This problem tests your ability to calculate variance and heredity. *Deduce > Solve:* **(a)** The F_1 are genetically homogeneous; therefore, F_1 $V_P = V_E = 3.5$ g. The genes controlling seed weight segregate in the F_2, therefore, F_2 $V_P = V_E + V_G = 7.4$ g. Since $V_G = V_P - V_E$, $V_G = 7.4 - 3.5 = 3.9$ g. **(b)** $H^2 = \frac{V_G}{V_P} = \frac{3.9}{7.4} = 0.53$.

11a–c. *Evaluate:* This is type 2. This problem tests your ability to apply statistical analyses to a quantitative trait data set. *Deduce > Solve:* **(a)** See figure. **(b)** The mean height in men is

$$\overline{X} = \frac{\sum_{i=1}^{n} X_i}{n} = \frac{1392}{20} = 69.6 \text{ in.}$$ The mean height for women is $\overline{X} = \frac{1296}{20} = 64.8$ in. The mean weight in men is $\overline{X} = \frac{3587}{20} = 179.4$ lb. The mean weight in women is $\overline{X} = \frac{2740}{20} = 137$ lb. The variance in height for men is $\sigma_X^2 = \frac{\sum_{i=1}^{n}(X_i - \overline{X})^2}{n-1} = 5.14$ in. The variance in height for women is 7.36 in. The variance for weight in men is 1028 lb. The variance in weight for women is 447 lb. The standard deviation for height in men is $\sigma_X = \sqrt{\sigma_X^2} = \sqrt{5.14} = 2.27$ in. The standard deviation for height in women is 2.71 in. The standard deviation for weight in men is 32.07 lb. The standard deviation for weight in women is 21.15 lb. **(c)** Yes. The graphs show higher average height and weight in men than women, corresponding to the means for these traits. The apparent distribution of weight in men is broader than that in women, corresponding to the larger variance and standard deviation.

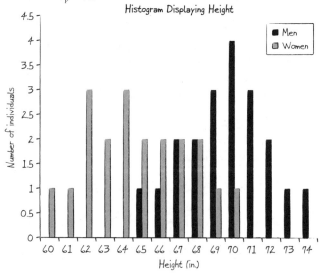

Histogram Displaying Height

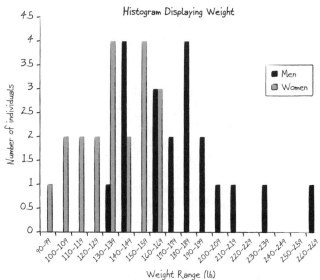

Histogram Displaying Weight

The apparent distribution of height in women is broader than that in men, corresponding to the larger variance and standard deviation values.

12. *Evaluate:* This is type 4. This problem tests your ability to calculate H^2 given information from a series of crosses. *Deduce > Solve:* The F_1 are genetically homogeneous; therefore, the F_1 $V_p = V_E = 3.62$ mm. The genes controlling corolla length segregate in the F_2; therefore, F_2 $V_p = V_E + V_G = 38.10$ mm. Since $V_G = V_p - V_E$, $V_G = 38.10 - 3.62 = 34.48$ mm. $H^2 = \frac{V_G}{V_p} = \frac{34.48}{38.10} = 0.905$

13. *Evaluate:* This is type 3. This problem tests your understanding of artificial selection. *Deduce:* Recall from the problem that $M = 28$ cm, $M_s = 34$ cm, and $h^2 = 0.7$. *Solve:* $S = M_s - M$, which gives 6 cm. $R = M' - M$. M' is found by the equation $M_s + h^2(M_s - M)$, which gives $28 + 0.7(6) = 32.2$ cm. By plugging in the numbers, we get $R = 32.2 - 28$, which is 4.2.

14a–b. *Evaluate:* This is type 3. This problem tests your understanding of artificial selection. *Deduce > Solve:* (a) Recall from the problem that $M = 16$ g, M_s for first cross = 12 g, M_s for the second cross = 24 g. Therefore, S for the first cross is $M_s(12) - M(16) = -4$ g. S for the second cross = $M_s(24) - M(12) = 12$ g. (b) $h^2 = 0.8$ and $R = M' - M$. For the first cross, M' is found by the equation $M + h^2(M_s - M)$. $M' = 16 + 0.8 (-4) = 12.8$ g; therefore, $R = 12.8 - 16 = -3.2$. For the second cross, $M' = 16 + 0.8 (12) = 25.6$ g; therefore, $R = 25.6 - 16 = 9.6$.

15a–d. *Evaluate:* This is type 2. This problem tests your understanding of polygenic control of a quantitative trait. *Deduce > Solve:* (a) The number of genes segregating in the F_2 can be estimated by the equation fraction of progeny with an extreme phenotype $= \frac{1}{4^N}$, where N is the number of genes. The extreme phenotypes are dark red (9) and white (12), and the total number of F_2 was 160. Plugging in the numbers, you get $\frac{9}{160}$, which is approximately $\frac{1}{17}$, and $\frac{12}{160}$, which is approximately $\frac{1}{13}$; therefore, $4^N = 13$ to 17, or about 2. (b) For two genes, there will be 4 alleles. (c) The white parent will be $A_2A_2B_2B_2$, where each allele contributes nothing to the trait. The dark red parent will be $A_1A_1B_1B_1$, where each allele contributes equally to red pigment production. The F_1 will be $A_1A_2B_1B_2$, which is pink. The F_2 will be as follows: $\frac{1}{16}$ dark red with the genotype $A_1A_1B_1B_1$; $\frac{4}{16}$ red with the genotypes $A_1A_2B_1B_1$ and $A_1A_1B_1B_2$; $\frac{6}{16}$ pink with the genotypes $A_1A_1B_2B_2$, $A_1A_2B_1B_2$ and $A_2A_2B_1B_1$; $\frac{4}{16}$ light pink with the genotype $A_1A_2B_2B_2$ and $A_2A_2B_1B_2$; $\frac{1}{16}$ white with the genotype $A_2A_2B_2B_2$. (d) $F_1 \times$ dark red $= A_1A_2B_1B_2 \times A_1A_1B_1B_1$ yields the following: $\frac{1}{4}$ white with the genotype $A_1A_1B_1B_1$; $\frac{1}{2}$ light pink with the genotypes $A_1A_2B_1B_1$ and $A_1A_1B_2B_1$; $\frac{1}{4}$ pink with the genotype $A_1A_2B_1B_2$.

16. *Evaluate:* This is type 4. This problem tests your understanding of heritability of complex traits in humans. *Deduce:* Traits that have a large heritable component will show higher concordance values in monozygotic twins than in dizygotic twins. *Solve:* Blood type, manic depression, schizophrenia, diabetes, cleft lip, and club foot all meet this criterion and therefore are traits in which genetics plays a relatively large role. There is no difference in concordance for chicken pox between mono- and dizygotic twins; therefore, genetics has minimal influence on this trait.

17. *Evaluate:* This is type 1. This problem tests your understanding of genetic and environmental influences on complex traits. *Deduce > Solve:* You should emphasize that heritability is the proportion of the variation within a population that is explained by genetic variation. Unless H^2 is 1.0, environmental influences are also present on the trait within the population. You could point out that H^2 describes a pattern at the population level, rather than for any one person. Thus, even for a highly heritable trait such as height, the environment (nutrition, for example) of an individual can still play a role in determining a person's height.

18. *Evaluate:* This is type 4. This problem tests your understanding of quantitative trait loci. *Deduce > Solve:* The horse geneticists would have to identify DNA polymorphisms that span the entire horse genome and then genotype a large number of horses shown to run at different speeds. DNA polymorphisms that show a statistically significant association with fast horses identify genomic regions that are linked to one or more genes associated with fast running speed.

19a–b. *Evaluate:* This is type 3. This problem tests your understanding of artificial selection.
Deduce > Solve: **(a)** The phenotypic change in oil content during selection indicated that genetic variation existed in the original population and that this variation decreased in each successive generation. Populations with increased oil or protein content contained a higher frequency of alleles favoring those traits while populations with decreased oil or protein content contained a higher frequency of alleles favoring those traits. **(b)** Since oil and protein content in the artificially selected populations was reversible, we can deduce that these populations still contained significant variation in the genes controlling those traits. The populations with high oil or protein content still contained alleles that favor low oil or protein content and vice versa, such that artificial selection in the reverse direction was possible. Furthermore, it is theoretically possible to carry out the initial selection until variation in oil or protein content due to genetics is zero; reversal of direction by artificial selection in that population would be impossible.

20a–d. *Evaluate:* This is type 2. This problem tests your understanding of polygenic inheritance.
Deduce > Solve: **(a)** The genotype $G_1G_2M_1M_2T_1T_2$ would be 9 units of color. **(b)** In this genetic model, the only genotype that determines 9 units of color is $G_1G_2M_1M_2T_1T_2$. In a trihybrid cross, $\frac{1}{8}$ of the progeny will be trihybrids; therefore, $\frac{1}{8}$ of the progeny will have 9 units of color. **(c)** In this threshold model for kernel color, 8 units of color or less gives white kernels, and 9 or more units of color gives colored kernels. Any combination of genotypes that contains at least three 1 alleles will have 9 or more units of color. For example, $G_1G_1M_1M_2T_2T_2$ gives 6 + 3 + 1 = 10 units, $G_1G_2M_1M_2T_1T_2$ gives 3 + 3 + 3 = 9 units. Any genotype that contains two or fewer 1 alleles will have 8 units of color or less. For example, $G_1G_1M_2M_2T_2T_2$ gives 6 + 1 +1 = 8 units and $G_1G_2M_1M_2T_2T_2$ gives 3 + 3 + 1 = 7 units. The histogram in Figure 21.2 shows that $\frac{1}{64} + \frac{6}{64} + \frac{15}{64} = \frac{22}{64}$ will have two or fewer 1 alleles and will be white. **(d)** The cross $G_1G_2M_1M_2T_2T_2 \times G_1G_2M_1M_2T_1T_2$ can be considered as a dihybrid cross ($G_1G_2M_1M_2 \times G_1G_2M_1M_2$) and then as a monohybrid test cross ($T_2T_2 \times T_1T_2$). For the dihybrid cross, the distribution of 1 alleles is $\frac{1}{16}$ with 4, $\frac{4}{16}$ with 3, $\frac{6}{16}$ with 2, $\frac{4}{16}$ with 1, and $\frac{1}{16}$ with none. Among the $\frac{6}{16}$ with two 1 alleles, $\frac{1}{2}$ will be T_1T_2 and $\frac{1}{2}$ will be T_2T_2. This leaves $\left(\frac{1}{2} \times \frac{6}{16}\right) + \frac{4}{16} + \frac{1}{16} = \frac{16}{32}$ of the progeny with two 1 alleles or less; therefore, $\frac{16}{32}$ or $\frac{1}{2}$ of the progeny will be white and $\frac{1}{2}$ will be colored.

21a. *Evaluate:* This is type 1. This problem tests your understanding of the heritability of quantitative traits. *Deduce:* To calculate broad sense heritability, use the equation $H^2 = \frac{V_G}{V_P}$; for narrow sense heritability, use the equation $h^2 = \frac{V_A}{V_P}$. *Solve:* For body mass, $H^2 = 0.48$ and $h^2 = 0.17$. For body fat, $H^2 = 0.42$ and $h^2 = 0.15$. For body length, $H^2 = 0.51$ and $h^2 = 0.16$.

21b. *Evaluate:* This is type 3. This problem tests your understanding of artificial selection.
Deduce > Solve: The response to selection is $R = S(h^2)$; therefore, as the value for h^2 increases, so does the response to selection. All three traits have a similar h^2 value; therefore, all three traits will exhibit a similar response to selection. Body mass will be slightly higher than body length, which will be slightly higher than body fat.

22a–b. *Evaluate:* This is type 3. This problem tests your understanding of artificial selection.
Deduce > Solve: **(a)** The selection differential is calculated using the equation $S = M_1 - M$, which is 22.7 – 20.2 = 2.5% for protein content and 7.4 – 6.5 = 0.9% for butterfat content. **(b)** The selection response is calculated using the equation $R = S(h^2)$, which is (2.5) × 0.6 = 1.5% for protein content and (0.9) × 0.8 = 0.72% for butterfat content. The total change is greater for protein content than for butterfat content, whereas the percentage increase for butterfat content, $\frac{0.72}{6.5} = 0.11$, is higher than the percentage increase for protein content, $\frac{1.5}{20.2} = 0.07$.

23a–b. *Evaluate:* This is type 1. This problem tests your understanding of threshold traits.
Deduce > Solve: **(a)** Midface clefting disorders are due to a particular combination of defective genes and environmental factors. The number of defective genes and the severity of the gene defects contribute to the genetic component of the risk in having a child with these defects. The recurrence risk of having a child with this disorder can be thought of as the chance that someone will pass on to their child a sufficient number of genes that are sufficiently defective. The defective genes are present throughout the population, however; individuals from families with a history

of midface clefting disorders are more likely to carry enough of the defective genes that are sufficiently defective than are individuals from families without a history of the disorder. **(b)** Cleft lip with cleft palate is a more severe form of the same disorder as cleft lip alone. Families with a history of cleft lip with cleft palate therefore carry a greater number of defective genes, or more defective forms of these genes, than families with a history of cleft lip alone. Thus individuals from families with a history of cleft lip and cleft palate are more likely to have children with cleft lip with or without cleft palate than are individuals from a family with cleft lip disorder alone.

24a–b. *Evaluate:* This is type 4. This problem tests your understanding of concordance rates. *Deduce > Solve:* **(a)** Since blood type is determined by genetics alone, the concordance rate in monozygotic twins, which are genetically identical, will be 100%. **(b)** The concordance rate for a genetically determined trait is the same as the probability that two independent fertilization events produce the same genotype. For this mating, the rate is $\frac{1}{2}$ or 50%, since there are two equally probable blood types.

25a–b. *Evaluate:* This is type 4. This problem tests your understanding of concordance rates. *Deduce > Solve:* **(a)** The concordance rate for a trait that is substantially determined by genes will be higher for monozygotic twins than for dizygotic twins because the monozygotic twins are genetically identical, whereas dizygotic twins are, on average, only 50% identical. **(b)** The concordance rate for a trait that is primarily determined by environment will be about the same for mono- and dizygotic twins because both types of twins share essentially the same environment. Possible reasons that could lead to a greater concordance in monozygotic twins than dizygotic twins would be a more similar intrauterine environment shared by monozygotic twins, the tendency of parents and others to treat monozygotic twins as more similar than dizygotic twins, and the fact that half of dizygotic twins are different sexes.

21.04 Test Yourself

Problems

1.   Plant height is controlled by three genes, *A*, *B*, and *C*. There are two alleles for each gene in the population being studied, where the 1 allele contributes 2 cm to plant height and the 2 allele contributes 0 cm to height. Plants that are $A_2A_2B_2B_2C_2C_2$ are 20 cm tall. The frequency of each allele is $\frac{1}{2}$. What is the distribution of height in a population generated by self-fertilizing a trihybrid?

2.   Using the model in Problem 1, what is the probability that a progeny of the cross $A_1A_2B_1B_2C_2C_2 \times A_2A_2B_2B_2C_2C_2$ will be 24 cm tall?

3.   A bag of seeds is derived from the self-fertilization of a plant from a pure-breeding line of plant. Half the seeds were planted in a plot on the north side of town; the other half were planted in a plot on the south side of town. The mean height of plants in both populations was 24 cm. The variance in the north plot was 18 cm^2, and the variance in the south plot was 8 cm^2. Why was the mean the same in both plots, and why was the variance different?

4.   The narrow sense heritability of height in a plant population is 0.5. The mean height of that population was 25 cm. Seeds derived from self-fertilization of an unknown subset of plants from that population were planted in the same environment as the original population and produced plants with a mean height of 30 cm. What was the mean height of the plants used to produce those seeds?

Solutions

1.   *Evaluate:* This is type 3. This problem tests your understanding of polygenic control of a quantitative trait. *Deduce:* Three genes contribute equally to height; therefore, the distribution

of genotypes can be calculated by expanding the binomial $(a + b)^6$, where a and b are $\frac{1}{2}$. **Solve:** This predicts $\frac{1}{64}$ with six 1 alleles and no 2 alleles, which are 32 cm tall; $\frac{6}{64}$ with five 1 alleles and one 2 allele, which are 30 cm tall; $\frac{15}{64}$ with four 1 alleles and two 2 alleles, which are 28 cm tall; $\frac{20}{64}$ with three 1 alleles and three 2 alleles, which are 26 cm tall; $\frac{15}{64}$ with two 1 alleles and four 2 alleles, which are 24 cm tall; $\frac{6}{64}$ with one 1 allele and five 2 alleles, which are 22 cm tall; and $\frac{1}{64}$ with no 1 alleles and six 2 alleles, which are 20 cm tall.

2.   *Evaluate:* This is type 3. This problem tests your understanding of polygenic control of a quantitative trait. *Deduce:* Since the *C* locus in both strains is homozygous for *C2*, the cross is a dihybrid cross, and the probability of a progeny plant of 24 cm in height is the probability of progeny with two 1 alleles and two 2 alleles. *Solve:* This probability is given by the corresponding term from the expanded binomial $(a + b)^N$, which is $6\left(\frac{1}{2}\right)^4 = \frac{6}{16}$.

3.   *Evaluate:* This is type 3. This problem tests your understanding of the statistical analysis of quantitative traits. *Deduce > Solve:* The plants being studied are genetically homogenous; therefore, the variance in each population is due to environmental factors. The difference in variance in the two populations is due to differences between the two environments. The means of the two populations are the same because environmental variations do not necessarily affect the mean of quantitative traits.

4.   *Evaluate:* This is type 4. This problem tests your understanding of artificial selection. *Deduce:* Since $M' = M + h^2(M_s - M)$, this formula can be rearranged to $h^2 = \frac{V_A}{V_P}$, and you can calculate M_s (the mean of the selected population) using the information given. *Solve:* $\frac{30 - 25}{0.5} + 25 = 35$. The mean height of the selected population was 35 cm.

22

Population Genetics and Evolution

Section 22.01 Genetics Problem-Solving Toolkit

Section 22.02 Types of Genetics Problems

Section 22.03 Solutions to End-of-Chapter Problems

Section 22.04 Test Yourself

22.01 Genetics Problem-Solving Toolkit

Key Terms and Concepts

Balanced polymorphism: A polymorphism that is maintained by selection against each homozygous genotype.

Coefficient of inbreeding (*F*): (1) The chance that an inbred individual is identical by descent of one genetic locus. (2) The fraction of genetic loci in an inbred individual that are identical by descent.

Directional natural selection: An evolutionary force that selects against one polymorphic allele.

Genetic bottleneck: A dramatic reduction in the number of individuals in a population that randomly alters the frequency of the alleles in the population.

Hardy-Weinberg equilibrium: Principle that the frequency of alleles in a population characterized by random mating does not change over time.

Inbreeding depression: A consequence of the increased percentage of homozygous loci caused by inbreeding that leads to a reduction in the relative fitness of individuals in the population.

Population: A group of interbreeding individuals.

Relative fitness: The reproduction rate of a genotype relative to the most successfully reproducing genotype. The value of relative fitness (ω) is between 1 and 0.

Selection coefficient: The selection coefficient (*s* or *t*) is the difference between the 1 and the relative fitness ($s = 1 - \omega$).

Key Analytical Tools

Identification of Inbred Individuals and Inbreeding Loops in Pedigrees

Calculating allele frequencies from genotype frequencies (fq):

fq(AA) = (fq A)^2

fq(Aa) = 2 × (fq A)(fq a)

fq(aa) = (fq a)^2

Equilibrium frequency due to a balanced polymorphism:

$$fq(A) = \frac{s}{(s+t)} \quad fq(a) = \frac{t}{(s+t)}$$

where s = selection coefficient against aa, and t = selection coefficient against AA.

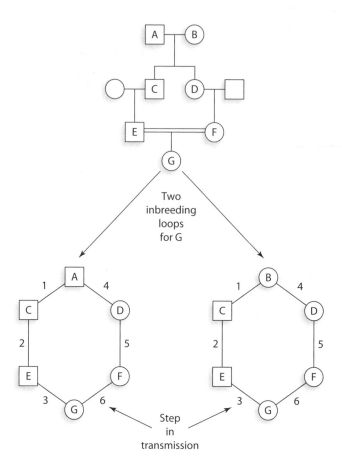

22.02 Types of Genetics Problems

1. Define, explain, or apply the principles of population genetics and evolution.
2. Calculate allele, genotype, or phenotype frequencies from population data under the assumptions of Hardy-Weinberg equilibrium.
3. Calculate allele frequencies under conditions of directional selection or selection favoring heterozygotes.
4. Analyze inbreeding using pedigrees, and calculate inbreeding coefficients.

1. Define, explain, or apply the principles of population genetics and evolution.

This type of problem tests your understanding of the fundamental principles, concepts, and terminology of population and evolutionary genetics. The problems may ask you to explain or define the terms or to apply the principles to understand qualitative questions concerning allele frequencies and how evolutionary forces drive changes in those frequencies. They may also ask you to describe evolutionary forces that drive speciation and to identify the mechanism and patterns of speciation based on information in the problem. See Problem 6 for an example.

Example Problem: A randomly mating population has a genotype frequency of 0.25 AA, 0.50 Aa, and 0.25 aa. If the pattern of mating changed from random mating to inbreeding, what would be the expected short-term effect on the genotypic frequencies and the long-term effect on the relative fitness of individuals in the population?

Solution Strategies	Solution Steps
Evaluate	
1. Identify the topic this problem addresses, and explain the nature of the requested answer.	1. This problem tests your understanding of inbreeding. The answer should address the effects of inbreeding on genotypic frequencies and relative fitness.
2. Identify the critical information given in the problem.	2. The genotype frequencies were 0.25 *AA*, 0.50 *Aa*, and 0.25 *aa* during random mating.
Deduce	
3. Define inbreeding, and describe the general effect inbreeding has on genotypic frequency.	3. Inbreeding is the mating of individuals with like genotypes, and it reduces the frequency of heterozygotic loci at the expense of homozygotic loci.
4. Relate the effect of inbreeding on the frequency of homozygotic genes to the relative fitness of individuals in the population.	4. The increase in loci that are homozygous will include homozygous recessive genotypes that decrease the health or reproductive potential of individuals. As more loci become homozygous, the number of homozygous recessive genotypes increases and the overall relative fitness of the population decreases.
Solve	
5. Answer the problem by stating the short-term effects of inbreeding on the frequency of genotypes at the *A* locus and its long-term effect on relative fitness.	5. **Answer:** Over the short term, the frequency of *Aa* genotypes will decrease, and the frequency of *AA* and *aa* genotypes will increase. This occurs for all genetic loci and leads to an increase in the number of homozygous recessive genotypes that decrease relative fitness.

2. Calculate allele, genotype, or phenotype frequencies from population data under the assumptions of Hardy-Weinberg equilibrium.

This type of problem tests your understanding of population genetics data and your ability to manipulate the data to calculate allele, genotype, or phenotype frequencies under the assumption of random mating. The problems may provide you with the number of individuals with specific genotypes in a population and ask you to calculate allele frequencies. You may have to determine whether calculated or given allele frequencies are consistent with Hardy-Weinberg equilibrium allele frequencies by performing chi-square tests. You may also be asked to use allele frequency data to calculate genotype and phenotype frequencies or the frequency of matings between specific genotypes. See Problem 10 for an example.

Example Problem: The frequency of an autosomal recessive disorder is 1 in every 1000 individuals. Determine the frequency of carriers of this disorder, and comment on the fraction of the recessive alleles that are phenotypically detectable in this population.

Solution Strategies	Solution Steps
Evaluate	
1. Identify the topic this problem addresses, and explain the nature of the requested answer.	1. This problem tests your ability to analyze population genetics data. The answer requires calculation of the frequency of heterozygous genotypes and the fraction of recessive alleles present in homozygous recessive individuals versus heterozygotes.
2. Identify the critical information given in the problem.	2. The disorder shows autosomal recessive inheritance and is present in 1 out of every 1000 individuals.
Deduce	
3. Determine the frequency of the recessive (mutant) and dominant (wild type) alleles. TIP: Assume Hardy-Weinberg equilibrium conditions exist.	3. The frequency of individuals with homozygous recessive genotypes is 0.001. The frequency of the recessive allele, q, is the square root of the genotype frequency, or 0.01. The frequency of the dominant allele, p, is 1 – the frequency of the recessive allele, or 0.99.
Solve	
4. Calculate the frequency of heterozygotes.	4. The frequency of heterozygotes is $2pq$, which is $2(0.99)(0.01) = 0.0198$.
5. Calculate the fraction of recessive alleles in persons who are affected and those who are carriers.	5. The fraction of the recessive allele in affected individuals is $\frac{2 \times \text{frequency of affected individuals}}{(2 \times \text{frequency of affected individuals}) + (\text{frequency of carriers})}$. By plugging in the numbers given in the problem, we get $\frac{2(0.001)}{2(0.001) + (0.0198)} = \frac{0.002}{0.0218} = 0.092$. The fraction of the recessive allele that is in carriers is, then, $1 - 0.092 = 0.908$.
6. Answer the problem by stating the carrier frequency and commenting on the fraction of the recessive alleles that are phenotypically detectable.	6. **Answer:** The frequency of carriers is 0.0198, which is about 20 times the frequency of affected individuals. This result indicates that most recessive alleles are present in carriers and therefore are not phenotypically detectable.

3. Calculate allele frequencies under conditions of directional selection or selection favoring heterozygotes.

This type of problem tests your understanding of natural selection and the use of selection coefficients to calculate allele frequencies. You may be asked to calculate the effect of directional selection on allele frequencies after successive generations of selection or to calculate equilibrium allele frequencies due to selection that favors heterozygotes. See Problem 20 for an example.

Example Problem: The genotypic frequencies in a population that are not under selection are 0.25 *AA*, 0.50 *Aa*, and 0.25 *aa*. The environment changes and reduces the relative fitness of *Aa* individuals to 0.9 and *aa* individuals to 0.5. After one generation of selection against adults, the survivors mate at random to produce a new generation. What will the genotypic frequencies be in this new population?

Solution Strategies	Solution Steps
Evaluate	
1. Identify the topic this problem addresses, and explain the nature of the requested answer.	1. This problem tests your ability to calculate the effect of directional selection on genotypic frequencies. The answer will consider the effect of one generation of selection to arrive at a new genotypic frequency in the resulting population.
2. Identify the critical information given in the problem.	2. The genotypic frequencies are $AA = 0.25$, $Aa = 0.50$, and $aa = 0.25$. The relative fitness of the genotypes are $AA = 1.0$, $Aa = 0.9$, and $aa = 0.5$.
Deduce	
3. Calculate the frequency at which each genotype will contribute to the next generation after selection has occurred. **TIP:** Do this in three steps: (1) adjust for the effect of the relative fitness on each genotype frequency by calculating (genotype frequency) × (genotype relative fitness); (2) sum those frequencies; and (3) divide each adjusted genotype frequency by that sum. (The sum of these frequencies will equal 1.)	3. The adjusted frequency of AA is $(0.25)(1.0) = 0.25$. The adjusted frequency of Aa is $(0.50)(0.90) = 0.45$. The adjusted frequency of aa is $(0.25)(0.5) = 0.125$. The sum of these adjusted frequencies is $0.25 + 0.45 + 0.125 = 0.825$. The frequency at which AA will contribute to the next generation is $\frac{0.25}{0.825} = 0.303$. For Aa, this is $\frac{0.45}{0.825} = 0.545$; and for aa, this is $\frac{0.125}{0.825} = 0.152$.
Solve	
4. Calculate the frequency of the A and a alleles in the new generation.	4. The frequency of A is $(0.303) + (0.5) \times (0.545) = 0.05755$. The frequency of a is $1 - fq(A) = 1 - 0.5755 = 0.4245$.
5. Answer the problem by using the allele frequencies and assuming random mating to determine the genotype frequencies in the next generation.	5. **Answer:** The frequency of AA will be $(0.5755)^2 = 0.3312$, the frequency of Aa will be $2(0.5755)(0.4245) = 0.4886$, and the frequency of aa will be $(0.4245)^2 = 0.1802$.

4. Analyze inbreeding using pedigrees, and calculate inbreeding coefficients.

This type of problem tests your understanding of pedigree analysis with an emphasis on identifying inbred individuals. You may be asked to identify an inbreeding loop in a pedigree and determine which individual is inbred as well as the person's common ancestors. You may be asked to calculate the inbreeding coefficient, or to interpret inbreeding coefficients to infer the structure of the corresponding pedigree. See Problem 36 for an example.

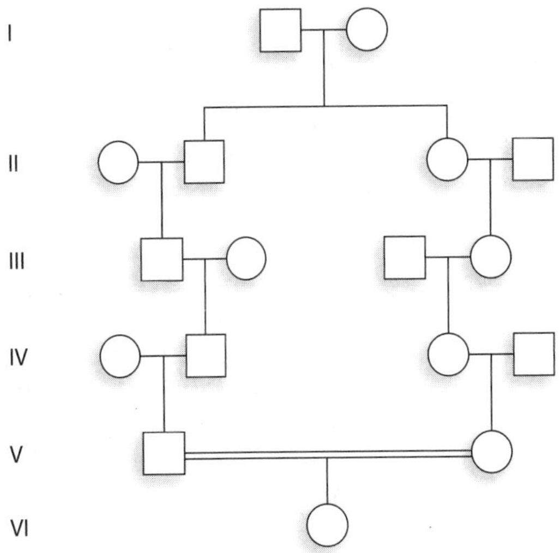

Example Problem: Examine the pedigree shown in the figure, and identify all inbred individuals. For each inbred individual, identify all of his or her common ancestors and calculate their inbreeding coefficient.

Solution Strategies	Solution Steps
Evaluate	
1. Identify the topic this problem addresses, and explain the nature of the requested answer.	1. This problem tests your ability to analyze pedigrees for evidence of inbreeding. The answer will identify the inbred individual and his or her common ancestors as well as provide his or her inbreeding coefficient.
2. Identify the critical information given in the problem.	2. The pedigree shows six generations of a family.
Deduce	
3. Identify all inbred individuals. TIP: Inbred individuals have parents who are related.	3. VI-1 is inbred because her parents are related.
4. Identify all common ancestors of the inbred individual, and relate that to the number of inbreeding loops they identify.	4. I-1 and I-2 are both common ancestors of VI-1, each identifying one inbreeding loop.
5. Determine the number of steps of transmission in each inbreeding loop. TIP: The steps of transmission can be determined by counting the number of vertical lines connecting a common ancestor with the inbred individual.	5. The inbreeding loops are the same, each with 10 steps of transmission.
Solve	
6. Calculate the inbreeding coefficient. TIP: The inbreeding coefficient (F) for each inbreeding loop is $2\left(\frac{1}{2}\right)^{n}$, where n is the number of steps of transmission in the loop. The inbreeding coefficients (F) for multiple loops are added to determine the F value for the inbred individual.	6. The inbreeding coefficient F for VI-1 is $$2\left(\frac{1}{2}\right)^{10} + 2\left(\frac{1}{2}\right)^{10} = 0.0039.$$
7. Answer the problem by stating the identity of the inbred individual, her common ancestors, and her inbreeding coefficient.	7. **Answer:** The only inbred individual is VI-1, her common ancestors are I-1 and I-2, and her inbreeding coefficient is 0.0039.

22.03 Solutions to End-of-Chapter Problems

1a–f. *Evaluate:* This is type 1. This problem tests your understanding of the principles of population genetics. *Deduce > Solve:* (a) A population is a group of interbreeding individuals, whereas a gene pool is the sum of all genes in a population. (b) Random mating involves individuals chosen by chance alone regardless of their phenotype or genotype, whereas inbreeding refers to mating between related individuals. (c) Natural selection and random genetic drift are both evolutionary forces, or phenomena that alter allele frequencies. Natural selection alters allele frequency by favoring reproduction of individuals with a specific heritable phenotype, thereby increasing the frequency of the associated alleles. Genetic drift alters allele frequency in a random manner, increasing the frequency of alleles based on chance. **(d)** A polymorphic trait is a phenotype that exists in more than one form in a population. A polymorphic gene is a gene for which there are two or more alleles present in a population. **(e)** A founder effect explains the allele frequencies in a population as a result of the allele frequency of its founding members. A genetic bottleneck explains the allele frequency of a population as a result of the allele frequency present in the few members that survive some cataclysmic event. Both events can cause the allele frequency of the resulting population to differ from that of the source population, and in both cases the change is typically due to chance.

2.   *Evaluate:* This is type 1. This problem tests your understanding of inbreeding. *Deduce > Solve:* Inbreeding is the mating between like individuals; therefore, inbreeding increases the frequency of homozygotes and decreases the frequency of heterozygotes. Although inbreeding does not alter allele frequencies, the frequency of individuals homozygous for rare recessive alleles will increase, causing an increase in the members of the population expressing the recessive trait.

3.   *Evaluate:* This is type 1. This problem tests your understanding of the evolutionary forces that drive changes in allele frequencies. *Deduce > Solve:* The four evolutionary forces that can cause changes in allele frequency are mutation, natural selection, migration, and random genetic drift. Mutation is the ultimate source of genetic variation, changing one allele into another and creating new alleles. Natural selection alters allele frequency by favoring reproduction of individuals with a specific heritable phenotype, thereby increasing the frequency of the associated alleles. Random genetic drift alters allele frequency in a random manner, increasing the frequency of alleles based on chance.

4.   *Evaluate:* This is type 3. This problem tests your understanding of balanced polymorphisms. *Deduce > Solve:* Natural selection that favors heterozygotes results in a balanced polymorphism of allele frequencies by selecting against both homozygous genotypes. For example, the sickle cell allele, β^S, is in balance with the wild-type allele, β^A, in areas where malaria is endemic. In this environment, both $\beta^S\beta^S$ homozygotes and $\beta^A\beta^A$ homozygotes are less fit than $\beta^A\beta^S$ heterozygotes. The equilibrium allele frequency of β^S is given by the equation $\frac{s}{(s+t)}$; the equilibrium frequency of β^A is $\frac{t}{(s+t)}$, where s is the selection coefficient against $\beta^S\beta^S$ and t is the selection coefficient against $\beta^A\beta^A$ homozygotes.

5.   *Evaluate:* This is type 1. This problem tests your understanding of polymorphisms. *Deduce > Solve:* A locus is polymorphic if there are at least two alleles in the population, which requires mechanisms for introducing new alleles into the population and for maintaining them. One possibility is the continuous migration of new alleles into the population at a rate that exceeds loss by genetic drift or purifying selection. Another would be introduction of new alleles into the population by a rate that exceeds loss of the alleles by genetic drift or purifying selection.

6.   *Evaluate:* This is type 1. This problem tests your understanding of the relationship between random genetic drift and population size. *Deduce > Solve:* In small populations, nonrandom mating among a relatively small number of individuals will have a significant impact on the allele frequencies in the next generation because each individual contributes significantly to a population's gene pool. In large populations, nonrandom mating among a relatively small number of individuals will have little effect on the population's allele frequency because each individual contributes little to the population's gene pool.

7.   *Evaluate:* This is type 1. This problem tests your understanding of the relationship between random genetic drift and population size. *Deduce > Solve:* In a small population, genetic drift will cause allele frequencies to fluctuate significantly, eventually leading to fixation of loss of alleles at each locus.

8.   *Evaluate:* This is type 1. This problem tests your understanding of the effect of genetic bottlenecks on allele frequencies. *Deduce > Solve:* A genetic bottleneck substantially reduces the number of organisms in a population, randomly altering allele frequencies and typically leading to loss of and fixation of alleles. This results in a reduction in genetic diversity, which is typically still displayed in the population after it has increased in size. The postbottleneck population will be genetically less diverse than the prebottleneck population.

9.   *Evaluate:* This is type 2. This problem tests your understanding of Hardy-Weinberg equilibrium for allele frequencies. *Deduce > Solve:* Yule did not consider random matings among all genotypes. If he had, he would have come to the conclusion that, under the random mating assumption and in the absence of selection, allele frequencies and thus phenotypic frequencies do not change from one generation to the next.

10a–c. *Evaluate:* This is type 2. This problem tests your ability to calculate allele frequencies. **TIP:** Calculate allele frequencies assuming Hardy-Weinberg equilibrium. *Deduce > Solve:* (a). Let q be the frequency of the recessive allele and p be the frequency of the dominant allele. Since the inability to taste PTC is recessive, and $(0.08)^2 = 0.006$, then $\sqrt{\text{(nontaster frequency)}} = q = \sqrt{\frac{140}{500}} = 0.53$. (b) $p = 1 - q = 1 - 0.53 = 0.47$. (c) Homozygous dominant is calculated as $p^2 = (0.47)^2 = 0.221$; heterozygotes are calculated as $2(0.92)(0.08) = 0.147$; and homozygous recessive is calculated as $(q)^2 = (0.47)^2 = 0.221$.

11a–d. *Evaluate:* This is type 2. This problem tests your ability to calculate allele frequencies. **TIP:** The frequency of males with an X-linked recessive trait is the same as the frequency of a recessive allele. *Deduce > Solve:* (a) The frequency of males with the recessive trait is calculated as $\frac{40}{500} = 0.08$. (b) The frequency of males with the dominant trait is calculated as $1.0 - 0.08 = 0.92$. (c) The frequency of homozygous dominant females is calculated as $(0.92)^2 = 0.846$. The frequency of female heterozygotes is calculated as $2(0.92)(0.08) = 0.147$. And, the frequency of homozygous recessive females is calculated as $(0.08)^2 = 0.006$. (d) See the following table.

Mating for Color Blindness (given $p = 0.92$ and $q = 0.08$)

Mating	Mating Frequency	Progeny Genotype Frequencies				
		Female Progeny			Male Progeny	
		CC	Cc	cc	CY	cY
CC X CY	(0.846)(0.92) = 0.778	0.778	0	0	0.778	0
Cc X CY	(0.147)(0.92) = 0.135	0.0675	0.0675	0	0.0675	0.0675
cc X CY	(0.006)(0.92) = 0.005	0	0.005	0	0	0.005
CC X cY	(0.846)(0.08) = 0.068	0	0.068	0	0.068	0
Cc X cY	(0.147)(0.08) = 0.012	0	0.006	0.006	0.006	0.006
cc X cY	(0.006)(0.08) = 0.0005	0	0	0.0005	0	0.0005
		0.8455	0.1465	0.0065	0.9195	0.079

12.   *Evaluate:* This is type 1. This problem tests your understanding of evolution by natural selection. *Deduce > Solve:* The widespread use of antibiotics creates environments where rare, antibiotic-resistant mutants have a selective advantage over wild-type, antibiotic-sensitive bacteria. These rare mutants will increase in population frequency under these conditions, eventually replacing the wild-type variants. Thus, infections are more likely to result from antibiotic-resistant strains.

13a–c. *Evaluate:* This is type 1. This problem tests your understanding of the effect of migration and random genetic drift on allele frequency. *Deduce > Solve:* **(a)** Since the two populations interbreed, the allele frequencies of both populations likely will be similar. **(b)** When migration is prevented, the populations become isolated and their allele frequencies are free to drift independently. Eventually, the allele frequencies in the two populations will be expected to differ. **(c)** The smaller, island population would be expected to change the most dramatically since the force driving change is random genetic drift, which has a larger impact on smaller populations.

14. *Evaluate:* This is type 1. This problem tests your understanding of the relationship between genetic variation and evolution. *Deduce > Solve:* Once an allele is fixed, there is no longer genetic variation at that locus (assuming mutations do not subsequently occur). In the absence of genetic variation, no evolution can occur, including evolution by directional selection.

15. *Evaluate:* This is type 1. This problem tests your understanding of inbreeding depression. *Deduce > Solve:* Inbreeding depression refers to the reduced fitness or viability of a species caused by inbreeding, which drives genetic loci to homozygosity. Homozygosity for deleterious alleles is thought to be the basis of the reduced fitness or viability. This is a concern for species-conservation breeding programs because they typically begin with only a few individuals, and breeding continues among the small population; this situation leads to inbreeding depression, which is counterproductive for restoring the species.

16. *Evaluate:* This is type 1. This problem tests your understanding of inbreeding depression. *Deduce > Solve:* For both populations, care should be taken to monitor or control breeding such that breeding occurs between the most distantly related individuals in each population.

17. *Evaluate:* This is type 2. This problem tests your understanding of Hardy-Weinberg equilibrium allele frequencies. *Deduce > Solve:* The number of M alleles is calculated as $\sqrt{\frac{336}{1000}} = 0.580$; the number of N alleles is calculated as $(187 \times 2) + 500 = 874$. The M allele frequency is calculated as $\frac{1184}{1184 + 874} = 0.57$, and the N allele frequency is calculated as $1 - 0.57 = 0.43$. The expected MM frequency is calculated as $(0.57)^2 = 0.325$, the expected MN frequency is calculated as $2(0.43)(0.57) = 0.490$, and the expected NN frequency is calculated as $(0.43)^2 = 0.185$. The expected number of MM is calculated as $0.325 \times 1029 = 334.43$, the expected number of MN is calculated as $0.49 \times 1029 = 504.21$, and the expected number of NN is calculated as $0.185 \times 1029 = 190.37$. The chi-square calculation is $\frac{(342 - 334.43)^2}{334.43} + \frac{(500 - 504.21)^2}{504.21} + \frac{(187 - 190.37)^2}{190.37} = 0.266$; for 2 degrees of freedom, the p value is > 0.95; therefore, the hypothesis that the population is at Hardy-Weinberg equilibrium cannot be rejected.

18. *Evaluate:* This is type 2. This problem tests your understanding of Hardy-Weinberg equilibrium allele frequencies. *Deduce > Solve:* The frequency of C_1C_1 black rabbits will be $(0.7)^2 = 0.49$; the frequency of C_1C_2 tan-colored rabbits will be $2(0.7)(0.3) = 0.42$; and the frequency of C_2C_2 white rabbits will be $(0.3)^2 = 0.09$.

19a–b. *Evaluate:* This is type 2. This problem tests your ability to calculate allele frequencies and apply Hardy-Weinberg equilibrium to determine genotype frequencies. *Deduce > Solve:* **(a)** The frequency of the β^S allele is $\frac{8}{100} = 0.08$, and the frequency of the β^A allele is $1 - 0.08 = 0.92$. **(b)** The frequency of carriers is $2(0.08)(0.92) = 0.147$.

20. *Evaluate:* This is type 3. This problem tests your ability to calculate allele frequencies under heterozygote advantage. *Deduce > Solve:* The selection coefficient against homozygous wild type is $s = 1 - 0.82 = 0.18$. The selection coefficient against those with SCD is $t = 1 - 0.32 = 0.68$. The equilibrium frequency of the β^A allele will be $\frac{t}{s + t} = \frac{0.68}{0.18 + 0.68} = 0.791$, and the equilibrium frequency of the β^S allele will be $\frac{s}{s + t} = \frac{0.18}{0.18 + 0.68} = 0.209$.

21a–b. *Evaluate:* This is type 2. This problem tests your ability to calculate allele frequencies and apply Hardy-Weinberg equilibrium to determine genotype frequencies. *Deduce > Solve:*
(a) In population A, the *t* allele frequency is $\sqrt{(0.36)} = 0.6$, and the *T* allele frequency is $1 - 0.6 = 0.4$. In population B, the *t* allele frequency is $\sqrt{(0.25)} = 0.5$, and the frequency of *T* is $1 - 0.5 = 0.5$. In population C, the *t* allele frequency is $\sqrt{(0.09)} = 0.3$, and the *T* allele frequency is $1 - 0.3 = 0.7$.
(b) In population A, the frequency of *TT* is $(0.4)^2 = 0.16$, the *Tt* frequency is $2(0.6)(0.4) = 0.48$, and the *tt* frequency is $(0.6)^2 = 0.36$. In population B, the frequency of *TT* is $(0.5)^2 = 0.25$, the *Tt* frequency is $2(0.5)(0.5) = 0.5$, and the *tt* frequency is $(0.5)^2 = 0.25$. In population C, the frequency of *TT* is $(0.7)^2 = 0.49$, the *Tt* frequency is $2(0.7)(0.3) = 0.42$, and the *tt* frequency is $(0.3)^2 = 0.09$.

22a–d. *Evaluate:* This is type 2. This problem tests your ability to calculate allele frequencies and apply Hardy-Weinberg equilibrium to determine genotype frequencies. *Deduce > Solve:*
(a) A genetic bottleneck is a reduction in population size that decreases genetic variation and alters allele frequencies. (b) The pogroms reduced the Ashkenazi Jewish population randomly with respect to genotype. The Tay-Sachs allele frequency in the reduced population was greater than that in the original population and remained high as that population re-expanded.
(c) The frequency of the recessive Tay-Sachs allele is $\sqrt{\frac{1}{750}} = 0.0365$. (d) The frequency of a carrier is $2(0.9635)(0.0365) = 0.0703$; therefore, the probability that both individuals in the couple will be carriers is $(0.0703)^2 = 0.0049$.

23. *Evaluate:* This is type 2. This problem tests your ability to calculate allele frequencies and apply Hardy-Weinberg equilibrium to determine genotype frequencies. *Deduce > Solve:* The frequency of the recessive *cf* allele is $\sqrt{\frac{1}{2000}} = 0.022$, and the frequency of carriers is $2(0.022)(0.978) = 0.043$.

24a–d. *Evaluate:* This is type 2. This problem tests your ability to calculate allele frequencies and apply Hardy-Weinberg equilibrium to determine genotype frequencies. *Deduce > Solve:* (a) The frequency of the recessive factor VIII allele is the same as the frequency of males with the disorder, which is $\frac{1}{2000} = 0.0005$, and the wild-type allele is $1 - 0.0005 = 0.9995$. (b) The frequency of female hemophiliacs is $(0.0005)^2 = 0.00000025$, and the frequency of carriers is $2(0.9995) \times (0.0005) = 0.001$. (c) Only a female can be a hemophilia carrier, because only females can be heterozygous at this X-linked locus. (d) For 1 million people, there are 500 affected males, 0.25 affected females, and 1000 carrier females.

25a–d. *Evaluate:* This is type 3. This problem tests your ability to calculate allele frequencies under conditions of heterozygote advantage. *Deduce > Solve:* (a) The original population distribution is $C_1C_1 = (0.8)^2 = 0.64$, $C_1C_2 = 2(0.2)(0.8) = 0.32$, and $C_2C_2 = (0.2)^2 = 0.04$. After one generation of selection against the surviving population fraction, the allele frequency is $(0.64)(0.3) + (0.32)(0.6) + (0.04)(1.0) = 0.424$. The fraction of survivors that are C_1C_1 will be $\frac{(0.64)(0.3)}{0.424} = 0.453$, the survivors that are C_1C_2 will be $\frac{(0.32)(0.6)}{0.424} = 0.453$, and the survivors that are C_2C_2 will be $\frac{0.04}{0.424} = 0.094$. The frequency of the C_1 allele is the frequency of $C_1C_1 + (0.5)$(frequency of C_1C_2) $= 0.453 + (0.5)(0.453) = 0.6795$. The frequency of C_2 is $1 - 0.6795 = 0.3205$. (b) Random mating of the second generation is expected to produce a generation containing $C_1C_1 = (0.6795)^2 = 0.4617$; $C_1C_2 = 2(0.6795)(0.3205) = 0.4356$; and $C_2C_2 = (0.3205)^2 = 0.1027$. (c) After a second generation of selection, the surviving population will be $(0.6795)(0.3) + (0.4356)(0.6) + (0.1027)(1.0) = 0.5679$. The fraction of survivors that are C_1C_1 will be $\frac{(0.6795)(0.3)}{0.5679} = 0.3590$, the survivors that are C_1C_2 will be $\frac{(0.4563)(0.6)}{0.5679} = 0.4602$, and the survivors that are C_2C_2 will be $\frac{0.1027}{0.567} = 0.1808$. The frequency for C_1 will be the frequency of $C_1C_1 + (0.5)$(frequency of C_1C_2) $= 0.3590 + (0.5)(0.4602) = 0.5891$. The frequency of C_2 is $1 - 0.5891 = 0.4109$. (d) Continued selection will eliminate C_1 and fix C_2 at a frequency of 1.0.

26a–c. *Evaluate:* This is type 3. This problem tests your ability to calculate allele frequencies under conditions of heterozygote advantage. *Deduce > Solve:* (a) The original population distribution is $C_1C_1 = (0.8)^2 = 0.64$, $C_1C_2 = 2(0.2)(0.8) = 0.32$, and $C_2C_2 = (0.2)^2 = 0.04$. After one generation of selection against the surviving population fraction, the allele frequency is $(0.64)(0.4) + (0.32)(1.0) + (0.04)(0.8) = 0.608$. The fraction of survivors that are C_1C_1 will be $\frac{(0.64)(0.4)}{0.608} = 0.421$, the survivors that are C_1C_2 will be $\frac{(0.32)(1.0)}{0.608} = 0.526$, and the survivors that are C_2C_2 will be $\frac{(0.04)(0.8)}{0.608} = 0.053$. The frequency of the C_1 allele is the frequency of C_1C_1 + (0.5)(frequency of C_1C_2) = 0.421 + (0.5)(0.526) = 0.684. The frequency of C_2 is 1 – 0.684 = 0.316. **(b)** Random mating of the second generation is expected to produce a generation containing $C_1C_1 = (0.684)^2 = 0.4679$; $C_1C_2 = 2(0.684)(0.316) = 0.4323$; $C_2C_2 = (0.316)^2 = 0.0999$. **(c)** The selection coefficient for C_1 is s = 1 – 0.4 = 0.6. The selection coefficient for C_2 is t = 1 – 0.8 = 0.2. The equilibrium C_1 allele frequency is $\frac{t}{s+t} = \frac{0.2}{0.6+0.2} = 0.25$, and the equilibrium C_2 frequency is $\frac{s}{s+t} = \frac{0.6}{0.6+0.2} = 0.75$.

27. *Evaluate:* This is type 2. This problem tests your ability to calculate genotype and phenotype frequencies under Hardy-Weinberg assumptions. *Deduce > Solve:* The frequency of I^AI^A will be $(0.3)^2 = 0.09$, the frequency of I^BI^B will be $(0.15)^2 = 0.0225$, the frequency of I^AI^B will be 2(0.3)(0.15) = 0.09, the frequency of I^OI^O will be $(0.55)^2 = 0.3025$, the frequency of I^AI^O will be 2(0.3)(0.55) = 0.33, and the frequency of I^BI^O will be 2(0.15)(0.55) = 0.165. The frequency of type A blood will be 0.09 + 0.33 = 0.42. The frequency of type B blood will be 0.0225 + 0.165 = 0.1875. The frequency of type AB blood will be 0.09. The frequency of type O blood will be 0.3025.

28. *Evaluate:* This is type 2. This problem tests your ability to determine allele frequencies given phenotypic frequencies. *Deduce > Solve:* The O blood type must have the genotype I^OI^O and their frequency is $(r)^2 = 0.366$, where r is the frequency of I^O. This gives $r = \sqrt{0.366} = 0.580$. Since type A includes I^AI^A (given by p^2) plus I^AI^O (given by $2pr$), then $0.421 = p^2 + 2pr$. Adding r^2 to each side of this equation gives $0.421 + 0.366 = p^2 + 2pr + r^2$, which factors down to $0.421 + 0.366 = (p + r)^2$ and $\sqrt{0.421 + 0.366} = (p + r)$. This gives $0.870 = (p + r)$; therefore, $p = 0.870 – 0.58 = 0.29$. Since $p + q + r = 1$, then $q = 1 – 0.870 = 0.13$.

29a–b. *Evaluate:* This is type 2. This problem tests your ability to calculate allele frequencies given genotype frequencies and to perform a chi-square calculation to test for fit to Hardy-Weinberg equilibrium. *Deduce > Solve:* **(a)** There are (2 × 225) + 175 = 625D_1 alleles and (2 × 100) + 175 = 375 D_2 alleles. The D_1 allele frequency is 0.625, and the D_2 allele frequency is 0.375. **(b)** The expected D_1D_1, D_1D_2, and D_2D_2 genotype frequencies are $(0.625)^2 = 0.391$, 2(0.625)(0.375) = 0.469, and $(0.375)^2 = 0.141$, respectively. The expected numbers of each genotype would be 234.5 D_1D_1, 195.5 D_1D_2, and 70.5 D_2D_2. The chi-square calculation is $\frac{(225-234.5)^2}{234.5} + \frac{(175-195.5)^2}{195.5} + \frac{(100-70.5)^2}{70.5} = 14.9$. For 2 degrees of freedom, the P value is < 0.01; therefore, the hypothesis that this population is at Hardy-Weinberg equilibrium must be rejected.

29c. *Evaluate:* This is type 1. This problem tests your understanding of inbreeding. *Deduce > Solve:* Inbreeding leads to a reduction in the frequency of heterozygotes relative to the frequency of homozygotes. This population does not reflect this pattern; therefore, inbreeding is not a likely explanation for the observed deviation from Hardy-Weinberg equilibrium.

30a–b. *Evaluate:* This is type 2. This problem tests your ability to calculate genotype frequencies from allele frequencies. *Deduce > Solve:* **(a)** The frequency of $DD = (0.62)^2 = 0.3844$, Dd = 2(0.62)(0.38) = 0.4712, $dd = (0.38)^2 = 0.1444$, $TT = (0.76)^2 = 0.5776$, Tt = 2(0.76)(0.24) = 0.3648, and $tt = (0.24)^2 = 0.0576$. **(b)** The frequency of dimpled tasters is $D\_T\_$ = (0.3844 + 0.4712)(0.5776 + 0.3648) = 0.8063. The frequency of dimpled nontasters is D_tt = (0.3844 + 0.4712)(0.0576) = 0.0493. The frequency of non-dimpled tasters is $ddT\_$ = (0.1444)(0.5776 + 0.3648) = 0.1361. The frequency of non-dimpled nontasters is $ddtt$ = (0.1444)(0.0576) = 0.0083.

31a–d. *Evaluate:* This is type 2. This problem tests your ability to calculate allele frequencies from genotypic frequencies. *Deduce > Solve:* **(a)** The recessive albinism allele frequency is $\sqrt{\frac{1}{4000}} = 0.0158$. The frequency of the recessive, wild-type brachydactyly allele is 0.9999. **(b)** The frequency of the dominant albinism allele is 1 – 0.0158 = 0.9842. The frequency of the dominant, brachydactyly allele is 0.0001. **(c)** The frequency of individuals heterozygous for albinism is 2(0.0158)(0.9842) = 0.0311. The frequency of individuals heterozygous for brachydactyly is 2(0.0001)(0.9999) = 0.0002. **(d)** The frequency of matings between individuals heterozygous for albinism is $(0.0311)^2 = 0.00097$.

32a–b. *Evaluate:* This is type 2. This problem tests your ability to calculate genotype frequencies given allele frequencies. *Deduce > Solve:* **(a)** The frequency of 16/18 heterozygotes at *D3S1358* is 2(0.229)(0.162) = 0.0742. The frequency of 14/18 heterozygotes at *VWA* is 2(0.131)(0.189) = 0.0495. The frequency of 23/26 heterozygotes at *FGA* is 2(0.131)(0.018) = 0.0047. **(b)** Each STRP-locus is assumed to assort independently, and the frequency of each genotype is calculated using Hardy-Weinberg principles. Thus, the frequency of any combination of STRP-loci genotypes can be combined using the product rule to illustrate the likelihood (typically minute) that an individual could have a given genotype by random chance.

33a–b. *Evaluate:* This is type 4. This problem tests your ability to determine inbreeding coefficients from analysis of a pedigree. *Deduce > Solve:* **(a)** IV-1 is inbred, and her common ancestor is I-2 because IV-1 could have two copies of an allele derived from I-2 (making her alleles identical by descent). **(b)** The number of transmission events from I-2 to IV-1 is 6, and IV-1 could be identical for either of the two alleles at each locus; therefore, $F = 2\left(\frac{1}{2}\right)^6 = \frac{1}{32} = 0.0313$.

34a–c. *Evaluate:* This is type 4. This problem tests your ability to determine inbreeding coefficients from analysis of a pedigree. *Deduce > Solve:* **(a–b)** V-1, V-2, and V-3 are inbred, and their common ancestors are I-1 and I-2 because all three could have two copies of an allele derived from either parent. **(c)** The inbreeding coefficients are the same for each inbred individual: $F = 4\left(\frac{1}{2}\right)^8 = \frac{4}{256} = 0.0156$.

35. *Evaluate:* This is type 4. This problem tests your ability to determine inbreeding coefficients from analysis of a pedigree. *Deduce > Solve:* Alice is inbred, and her common ancestors are I-1 and I-2; therefore she is part of two inbreeding loops. There are six transmission steps in each inbreeding loop. George VI is inbred, and his common ancestors are I-4 and I-5; therefore, he is part of two inbreeding loops. There are nine transmission steps in each inbreeding loop. Charles is inbred, and his common ancestors are Queen Victoria (III-2) and Prince Albert, I-4, and I-5; therefore, Charles is part of four inbreeding loops. There are 10 transmission steps in the two loops involving Queen Victoria and her spouse, and there are 13 transmission steps in the two loops involving I-4 and I-5.

36a–d. *Evaluate:* This is type 4. This problem tests your ability to deduce the structure of a pedigree displaying inbreeding from the equation for the inbreeding coefficient of an individual. *Deduce > Solve:* See figure.

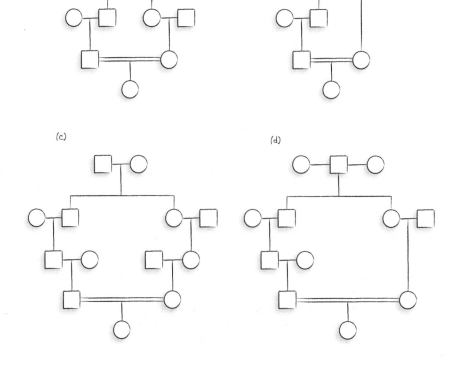

37.   *Evaluate:* This is type 1. This problem tests your understanding of natural selection.
Deduce > Solve: The higher frequency of alleles leading to darker skin and the lower frequency of alleles leading to lighter skin in African populations is consistent with increased fitness of dark-skinned individuals in that environment. The selection is against the alleles leading to lighter skin pigmentation. There is no such selective pressure (or far less selection) in European populations; therefore, the alleles leading to lighter skin color are more frequent.

38.   *Evaluate:* This is type 1. This problem tests your understanding of genetic bottlenecks.
Deduce > Solve: The relatively high frequency of the mutant achromatopsia allele in the island population is due to a genetic bottleneck created by the typhoon. After the typhoon, the allele frequency was at least 1/40, corresponding to one male heterozygote among the 20 individuals surviving the storm. This male contributed to about 5% of the population in the next generation; therefore his alleles, along with those of the other reproducing survivors, become predominant, frequent alleles in the expanding population.

39a–c.   *Evaluate:* This is type 1. This problem tests your understanding of speciation. *Deduce > Solve:*
(a) Assuming that males and females of the two species can physically breed and gametes can join, the resulting zygotes will be aneuploid and nonviable. **(b)** The mechanism of speciation described most likely corresponds to postzygotic speciation due to hybrid inviability. **(c)** The pattern of speciation described corresponds to sympatric speciation because the two new species evolve in the absence of a physical barrier separating them.

22.04 Test Yourself

Problems

1.   A population of *Drosophila* is grown on food that has relatively high ethanol content. ADH^F is an allele of the alcohol dehydrogenase gene that promotes the reproduction of *Drosophila* in ethanol-containing environments. Before exposure to the high-ethanol environment, the frequency of the ADH^F allele was 0.5. In this new environment, the relative fitness of ADH^F ADH^F homozygotes is 1.0, ADH^F *adh* heterozygotes is 0.9, and *adh adh* homozygotes is 0.6. What are the genotypic and allelic frequencies after one generation in this new environment?

2.   In a population of flowers, there are three alleles at the *C* locus, which control flower color. C_1 and C_2 are codominant, and both are dominant to C_3. C_1C_1 plants have red-and-white-striped flowers, C_1C_2 plants have purple (a mixture of red and blue)-and-white-striped flowers, C_2C_2 plants have blue-and-white-striped flowers, and C_3C_3 plants are white without stripes. In a random sample of 1000 plants, there are 320 plants with red-and-white stripes, 320 plants with blue-and-white stripes, 320 plants with purple-and-white stripes, and 40 with no stripes. Use this information to determine the frequency of the C_1, C_2, and C_3 alleles under the assumptions of Hardy-Weinberg equilibrium.

3.   A new insect invaded the environment inhabited by the plants in the previous problem. This insect eats flowers that have stripes such that it reduces the relative fitness of these plants from 1.0 to 0.8. Determine the C_1, C_2, and C_3 allele frequencies after one generation of predation by this insect.

Solutions

1.   *Evaluate:* This is type 3. This problem tests your ability to calculate the effect of directional selection on genotype and allele frequencies. *Deduce > Solve:* The genotype frequencies before exposure to the new environment are ADH^F ADH^F = 0.25, ADH^F *adh* = 0.50, and *adh adh* = 0.25. After one generation, the relative proportion of ADH^F ADH^F homozygotes will be $\frac{(0.25)(1.0)}{0.85} = 0.2941$, the relative proportion of ADH^F adh will be $\frac{(0.50)(0.9)}{0.85} = 0.5294$, and the relative proportion of *adh adh* will be $\frac{(0.25)(0.6)}{0.85} = 0.1765$. Assigning p to the frequency of ADH^F and q to the frequency of *adh*, $p = $ fq $ADH^F\ ADH^F + \frac{1}{2}$ (fq $ADH^F\ adh$) , which gives $0.2941 + \frac{1}{2}(0.5294)$ $= 0.5588$ and $q = 1 - p = 1 = 0.5588 = 0.4412$.

2.   *Evaluate:* This is type 2. This problem tests your ability to apply Hardy-Weinberg equilibrium principles to determine the frequency of alleles at a locus with three alleles segregating in the population. *Deduce > Solve:* Let the frequency of C_1, C_2, and C_3 alleles be designated by p, q, and r, respectively. The frequency of plants with no stripes is r^2; therefore, $r^2 = 0.04$, and $r = 0.2$. The frequency of red-and-white-striped plants is the sum of plants that are C_1C_1 (given by p^2) and those that are C_1C_3 (given by $2pr$). This value can be written as $0.32 = p^2 + 2pr$. Adding r^2 to the left of that equation and 0.04 to the right gives $0.32 + 0.04 = p^2 + 2pr + r^2$, which factors to $0.32 + 0.04 = (p + r)^2$. Therefore, $\sqrt{0.32 + 0.04} = p + r$. To solve for p, substitute 0.2 for r, which gives $p = 0.6 - 0.2 = 0.4$. Since $p + q + r = 1$, then $q = 1 - 0.6 = 0.4$.

3.   *Evaluate:* This is type 3. This problem tests your understanding of directional selection and your ability to determine the effect of selection acting at a locus with three alleles. *Deduce > Solve:* The genotype frequencies can be determined from the allele frequencies calculated in the previous problem. $C_1C_1 = (0.4)^2 = 0.16$, $C_1C_2 = 2(0.4)(0.4) = 0.32$, $C_1C_3 = 2(0.4)(0.2) = 0.16$, $C_2C_2 = (0.4)^2$ $= 0.16$, $C_2C_3 = 2(0.4)(0.2) = 0.16$, and $C_3C_3 = (0.2)^2 = 0.04$. Since predation acts on plants whose flowers have stripes of any kind, the relative fitness of C_3C_3 is 1.0 and that of all the others is 0.8. The proportion of this population that reproduces will be $(0.8 \times 0.96) + 0.04 = 0.808$ because 0.96 of the population have a relative fitness of 0.8, and 0.04 of the population have a relative fitness of 1.0. The proportion of each genotype in the reproducing population is $C_1C_1 = \frac{(0.16)(0.8)}{0.808}$ $= 0.1584$. Since 0.16 is the starting genotype frequency for C_2C_2, C_1C_3, and C_2C_3, they will each contribute to the next generation at a frequency of 0.1584. The proportion of the reproducing population that is C_1C_2 is $\frac{(0.32)(0.8)}{0.808} = 0.3168$. The proportion of the reproducing population that is C_3C_2 is $\frac{(0.04)(1.0)}{0.808} = 0.0495$. The allele frequencies can be determined from the phenotypic frequencies, where $p = $ fqC_1, $q = $ fqC_2, and $r = $ fqC_3. The frequency of no stripes is fq$C_3C_3 = r^2 = 0.0495$; therefore, $r = 0.2225$, and $p + q = 1 - 0.2225 = 0.7775$. Since selection exerted by the insects was against striped flowers of either color, the frequency of the C_1 and C_2 alleles will be equally affected. Also, since C_1 and C_2 were present at equal frequencies before predation, they will be present at equal frequencies after predation. Thus, $p = q = \frac{0.7775}{2} = 0.38875$.